Life Atomic

A SERIES IN THE HISTORY OF CHEMISTRY, BROADLY CONSTRUED, EDITED BY ANGELA N. H. CREAGER, ANN JOHNSON, JOHN E. LESCH, LAWRENCE M. PRINCIPE, ALAN ROCKE, E. C. SPARY, AND AUDRA J. WOLFE, IN PARTNERSHIP WITH THE CHEMICAL HERITAGE FOUNDATION

Life Atomic

A History of Radioisotopes in Science and Medicine

ANGELA N. H. CREAGER

THE UNIVERSITY OF CHICAGO PRESS   CHICAGO AND LONDON

ANGELA N. H. CREAGER is the Philip and Beulah Rollins Professor of History at Princeton University.

The University of Chicago Press, Chicago 60637
The University of Chicago Press, Ltd., London

Printed in the United States of America

22 21 20 19 18 17 16 15 14 13   1 2 3 4 5

ISBN-13: 978-0-226-01780-8 (cloth)
ISBN-13: 978-0-226-01794-5 (e-book)
DOI: 10.7208/CHICAGO/9780226017945.001.0001

Library of Congress Cataloging-in-Publication Data

Creager, Angela N. H.
Life atomic: a history of radioisotopes in science and medicine/Angela N. H. Creager.
pages; cm. — (Synthesis: a series in the history of chemistry, broadly construed)
Includes bibliographical references and index.
ISBN-13: 978-0-226-01780-8 (cloth : alkaline paper)—ISBN-10: 978-0-226-01794-5 (e-book)
1. Radioisotopes in research—History. 2. Radioisotopes in medical diagnosis—History.
3. Nuclear medicine—History. 4. Radioisotopes—Industrial applications—History.
I. Title. II. Series: Synthesis (University of Chicago. Press).
QC798.A1C743 2013
660'.2988409—DC23

2013006680

♾ This paper meets the requirements of ANSI/NISO Z39.48-1992 (Permanence of Paper).

FOR MY PARENTS

Contents

Preface

I first encountered radioisotopes in the biochemistry and molecular biology laboratories I worked in during college and graduate school. They were a routine part of many experimental procedures, from enzyme assays and protein labeling to nucleic acid sequencing and Southern blots. Most radioisotopes in the Schachman lab in Berkeley arrived from New England Nuclear, almost always as radiolabeled compounds, especially sulfur-35-labeled dideoxy nucleotides for DNA sequencing and carbon-14-labeled substrates for measuring enzyme activity. I never gave much thought to the provenance of radioisotopes, or why so many methods depended on them, until the mid-1990s. Then, at the prompting of an astute potential editor, I wrote an abstract on the development of radiolabeling for a proposed volume on biophysics and instrumentation (which, due to lack of interest, never materialized). I was intrigued, and surprised, to discover that in the US radioisotopes were initially sold to scientists by the Manhattan Project. As it turned out, the timing of my interest was fortuitous; in 1994 President Bill Clinton appointed a scholarly panel to examine the US government's role in experiments (both secret and open) that exposed humans to radiation, and Department of Energy Secretary Hazel O'Leary oversaw the declassification of thousands of relevant government documents. The story of radioisotopes was ripe for the telling, though it took me years to complete the harvest.

Several US government grants funded my research on the book and the leave time necessary to write it: a National Science Foundation CAREER award, SBE 98-75012, from 1999 to 2006; a National Endowment for the Humanities Fellowship Award in 2006–7; and, from the National Institutes of Health, a National Library of Medicine Grant for Scholarly Works in Biomedicine and Health, 5G13LM9100, from 2007 to 2011. Any opinions, findings, and conclusions or recommendations expressed in this material are my own (or those of authors quoted) and do not necessarily reflect the views of the National Science Foundation or other agencies. Princeton University provided additional research support through the University Committee on Research in the Humanities and Social Sciences and the Department of History, as well as priceless sabbatical time. I am grateful to the research assistants who collected sources, created databases, scanned images, and edited files: Sultana Banulescu, Edna Bonhomme, Dan Bouk, Stephen Feldman, Brooke Fitzgerald, Dan Gerstle, Evan Hepler-Smith, Greg Kennedy, Jennifer Weber, and Doogab Yi.

Pieces of this book have appeared in print elsewhere, and I acknowledge their publishers for allowing them to reappear here. An early version of material now in chapters 2, 3, and 4 appeared in "Tracing the Politics of Changing Postwar Research Practices: The Export of 'American' Radioisotopes to European Biologists," *Studies in History and Philosophy of the Biological and Biomedical Sciences* 33C (2002): 367–88. Copyright 2002, with permission from Elsevier. A brief synopsis of chapter 5 was published as "Atomic Transfiguration," *Lancet* 372 (2008): 1726–27. Chapters 3 and 7 include material from "Nuclear Energy in the Service of Biomedicine: The U.S. Atomic Energy Commission's Radioisotope Program, 1946–1950," *Journal of the History of Biology* 39 (2006): 649–84, © Springer 2006, reprinted with kind permission from Springer Science+Business Media B.V. Chapter 7 includes material from "Phosphorus-32 in the Phage Group: Radioisotopes as Historical Tracers of Molecular Biology," *Studies in History and Philosophy of the Biological and Biomedical Sciences* 40 (2009): 29–42. Copyright 2009, with permission from Elsevier. An early version of chapter 6 appeared in *The Science–Industry Nexus: History, Policy, Implications*, edited by Karl Grandin, Nina Wormbs, and Sven Widmalm, and is reprinted here by permission of Science History Publications/USA & The Nobel Foundation. Most of chapter 4 was published as "Radioisotopes as Political Instruments, 1946–1953," *Dynamis* 29 (2009): 219–39. Passages from chapters 7 and 9 may be found in "Timescapes of Radioactive Tracers in Biochemistry and Ecology," *History and Philosophy of the Life Sciences* 35 (2013): 83–90, and appear here by courtesy of the journal. Lastly, part of chapter 8 is reprinted by permission of the publishers from "Molecular Surveillance: A History of Radioimmunoassays," in *Crafting Immunity: Working Histories of Clinical Immunology*, ed. Kenton Kroker, Jennifer Keelan, and Pauline M. H. Mazumdar (Farnham, UK: Ashgate, 2008), pp. 201–30. Copyright © 2008.

For permission to quote from documents in their collections, I thank the Bancroft Library, the Cushing Memorial Library and Archives of Texas

A&M University, the Herbert Hoover Presidential Library, the Archives and Records Office of Lawrence Berkeley National Laboratory, the Institute Archives and Special Collections of MIT Libraries, and Special Collections at the J. Willard Marriott Library of the University of Utah. On a more personal note, I wish to thank Julia Beach, Elizabeth Bennett, Marjorie Ciarlante, David Farrell, Lee Hiltzig, David Hollander, Jennifer Isham, Stan Larson, Tab Lewis, Nora Murphy, Pamela Patterson, Charles Reeves, Tom Rosenbaum, Susan Snyder, and John Stoner for helping me chase down myriad documents and photographs.

The other intellectual debts I have accumulated are too numerous to adequately recount. Michael Gordin was my closest interlocutor about all things atomic during the years I have been working on this project; I thank him for his insights and encouragement. Early on, Peter Westwick offered important tips and generously shared research notes. Heinrich von Staden was my host during a wonderful formative year (2002–3) when I was a visitor at the School of Historical Studies at the Institute for Advanced Study. John Krige, who was in Princeton as a Davis Center fellow in 2006–7, kept me company during another leave and shaped my perspective on science in American foreign policy, particularly Atoms for Peace. Conversations with Lynn Nyhart, especially during a visit to Madison in 2008, were inspiring and her friendship inestimable. Many other colleagues responded to talks or offered comments on my papers. For their questions and suggestions, I thank Matthew Adamson, Ken Alder, Gar Allen, Carl Anderson, Nancy Anderson, Itty Abraham, Crispin Barker, John Beatty, Paola Bertucci, Bill Bialek, Tom Broman, Andrew Brown, Peter Brown, Soraya Boudia, Sydney Brenner, Bryn Bridges, Richard Burian, Luis Campos, Nathaniel Comfort, Ruth Schwartz Cowan, Robert Crease, Soraya de Chadarevian, Freeman Dyson, Laura Engelstein, Raphael Falk, Alex Gann, Dan Garber, Jean-Paul Gaudillière, Joel Hagen, Néstor Herran, Jeff Hughes, Bill Jordan, Dave Kaiser, Joshua Katz, Barbara Kimmelman, Robert Kohler, Dan Kevles, Alison Kraft, Kenton Kroker, Jerry Kutcher, Edward Landa, Hannah Landecker, Susan Lederer, Richard Lewontin, Ilana Löwy, Liz Lunbeck, Jay Malone, Erika Milam, Tania Munz, Cyrus Mody, Staffan Müller-Wille, Naomi Oreskes, Robert Proctor, Jeff Peng, Karen Rader, Bill Rankin, Nick Rasmussen, Carsten Reinhardt, Jessica Riskin, Dan Rodgers, Naomi Rogers, Hans-Jörg Rheinberger, Xavier Roqué, María Jesús Santesmases, Eric Schatzberg, Sonja Schmid, Alexander Schwerin, Suman Seth, Steve Shapin, Matthew Shindell, Asif Siddiqi, Alistair Sponsel, Richard Staley, Bruno Strasser, Edna Suarez, William Summers, Helen Tilley, Emily Thompson, Simone Turchetti, Judith Walzer,

John Harley Warner, James Watson, Gilbert Whittemore, Norton Wise, Matt Wisnioski, and Jan Witkowski.

As the book was coming into final form, a number of colleagues, some of whom had never met me, offered invaluable feedback. For their comments on chapters, I thank Peder Anker, Étienne Benson, Robert Blankenship, Stephen Bocking, Randy Brill, Graham Burnett, Gene Cittadino, Alan Covich, Henry Cowles, Will Deringer, Bridget Gurtler, Jacob Hamblin, John Heilbron, Evan Hepler-Smith, Karl Hubner, David Jones, Kärin Nickelsen, Rachel Rothschild, Judy Johns Schloegel, Laura Stark, Keith Wailoo, Sam Walker, Michael Welch, Ward Whicker, and the participants in Princeton's History of Science Program Seminar. Several intrepid souls read the entire manuscript and offered excellent suggestions, namely James Adelstein, Yaacob Dweck, Michael Gordin, Susan Lindee, Karen-Beth Scholthof, Audra Wolfe, Katherine Zwicker, and the anonymous referees for the press. Karen Merikangas Darling provided astute guidance and encouragement, along with Audra Wolfe and the board of *Synthesis*. I thank Mary Corrado for her meticulous job in copyediting the manuscript. In the end, of course, I alone am responsible for the interpretation taken and any remaining errors.

Lastly there are more personal debts to acknowledge. Jenny Weber and Michael Keevak provided both diversion and reinforcement at critical junctures. Cynthia and Edward Peterson hosted my illuminating trip to Oak Ridge, where Fred Strohl provided an unforgettable tour of the X-10 graphite reactor building. Beth and Ted Streeter graciously put me up during my numerous visits to Berkeley, where I also had stimulating conversations with my graduate advisor, Howard Schachman. My children, Elliot, Jameson, and Georgia, must regard me as thoroughly contaminated with the subject of radioactivity, but they never complained about my obsession, and magnanimously watched atomic age movies and documentaries with me. The support of my wonderful husband, Bill, has been crucial from the outset, when I started my marathon trips to the National Archives a year after Georgia was born. Not only did he hold our household together with love and good cheer, but also certain figures exist only because of his expertise and help. Lastly, my parents, Bill and Jan Hooper, have always been enthusiastic in their support of my work, reading grant proposals and articles along the way and listening to me talk endlessly about finishing the book. It is finally done, and I dedicate it to them.

Princeton
September 2012

Abbreviations

Acc	Accession number
ABCC	Atomic Bomb Casualty Commission
ACBM	Advisory Committee for Biology and Medicine
ACHRE	Advisory Committee on Human Radiation Experiments
Aebersold papers	Paul C. Aebersold papers, Cushing Memorial Library and Archives, Texas A&M University
AEC	US Atomic Energy Commission
BEAR	Biological Effects of Atomic Radiation
BNL	Brookhaven National Laboratory
CEW	Clinton Engineer Works
CF	Central Files
Ci, mCi, μCi	curie, millicurie (10^{-3} curie), microcurie (10^{-6} curie)
DC-LBL files	Donald Cooksey Files Administrative (Director's Office), Accession Number 434-90-20, ARO-1537, Lawrence Berkeley National Laboratory Archives and Records Office, 1 Cyclotron Rd. MS: 69R0102, Berkeley, California 94720
DoD	US Department of Defense
DOE Info Oak Ridge	Department of Energy Information Center, 475 Oak Ridge Turnpike, Oak Ridge, TN, 37830

OpenNet	Department of Energy OpenNet database of declassified documents, accessible at https://www.osti.gov/opennet/; these can be identified by either Accession number or Document number
ERDA	US Energy Research and Development Administration
EOL papers	Ernest Orlando Lawrence papers, 72/117c, Film 2248, Bancroft Library, University of California, Berkeley
Evans papers	Robley Duglinson Evans Papers, MC 80, Massachusetts Institute of Technology, Institute Archives and Special Collections, Cambridge, Massachusetts
FBI	Federal Bureau of Investigation
GAC	General Advisory Committee of US AEC
Gen Corr	General Correspondence
Hutchinson papers	G. Evelyn Hutchinson papers, MS 649, Manuscripts and Archives, Sterling Memorial Library, Yale University
Hickenlooper papers	Bourke B. Hickenlooper papers, Herbert Hoover Presidential Library, West Branch, Iowa
JCAE	US Congress Joint Committee on Atomic Energy
JHL papers	John Hundale Lawrence papers, 87/86c, Film 2005, Bancroft Library, University of California, Berkeley
JHL-LBL Administrative	Radioisotope Studies R&D Administrative Files of Dr. John H. Lawrence, FRC Accession No. 434-90-168A, ARO-1537, Lawrence Berkeley National Laboratory Archives and Records Office, 1 Cyclotron Rd. MS: 69R0102, Berkeley, California 94720
JHL-LBL Technical	Nuclear Medicine R&D Technical Documents of Dr. John H Lawrence, 434-92-66, ARO-2225, Lawrence Berkeley National

	Laboratory Archives and Records Office, 1 Cyclotron Rd. MS: 69R0102, Berkeley, California 94720
Jones papers	Hardin B. Jones Papers, 79/112c, Bancroft Library, University of California, Berkeley
MED	Manhattan Engineer District
MeV	10^6 electron volts
μ-units	10^{-6} units (of hormone activity)
MIT President's Papers	Massachusetts Institute of Technology, Office of the President, records of Karl Taylor Compton and James Rhyne Killian, AC 4, Massachusetts Institute of Technology, Institute Archives and Special Collections, Cambridge, Massachusetts
MMES/X-10	Martin Marietta Energy Systems, Incorporated, contractor for X-10 plant, Oak Ridge National Laboratory, 1984–95
NARA College Park	US National Archives and Records Administration II, College Park, Maryland*
NARA Atlanta	US National Archives and Records Administration, Southeast Region, Atlanta, Georgia*
NAS	National Academy of Sciences
NATO	North Atlantic Treaty Organization
NCRP	National Committee on Radiological Protection
NRC	National Research Council
Oppenheimer papers	J. Robert Oppenheimer Papers, Manuscript Division, Library of Congress
ORINS	Oak Ridge Institute of Nuclear Studies

*At both the College Park and Southeast Region branches of the National Archives, I focused on record group 326, Atomic Energy Commission. At College Park, the papers within each record group are organized by entry number. I refer extensively to textual records in E67A, AEC Secretariat Files, 1947–1951, and E67B, AEC Secretariat Files, 1951–1958. At the Southeast Region facility, papers in RG 326 are organized by accession number. In the notes for these documents I have provided both the accession number and the abbreviated title for each collection.

ORNL	Oak Ridge National Laboratory
OROO	Oak Ridge Operations Office
OSRD	Office of Scientific Research and Development
RAC	Rockefeller Archive Center, Pocantico Hills, New York
Rad Lab	Radiation Laboratory, Berkeley
RF	Rockefeller Foundation records
RG	Record Group
RIA	radioimmunoassay
RSB	Radiological Safety Branch
Stannard papers	J. Newell Stannard Papers, MS-2020, Special Collections University of Tennessee, Knoxville–Libraries
Stent papers	Gunther S. Stent papers, 99/149z, Bancroft Library, University of California, Berkeley
Strauss papers	Lewis L. Strauss papers, Herbert Hoover Presidential Library, West Branch, Iowa
TBI	Total Body Irradiation
TVA	Tennessee Valley Authority
Wintrobe papers	Maxwell M. Wintrobe papers, Accession number 954, Special Collections, J. Willard Marriott Library, University of Utah

CHAPTER ONE

Tracers

Quite unlike atomic power, radioactive isotopes, tracers, and radiations can be available to science, medicine and technology on a scale adequate to all anticipated needs, and in most cases can be so today.—US Atomic Energy Commission, 1947[1]

At the close of World War II, the nuclear detonations over Hiroshima and Nagasaki demonstrated the devastating power of the atom. As Americans became aware of their country's secret development of nuclear weapons, the US government swiftly turned attention to the peaceful benefits of nuclear knowledge. Foremost was harnessing the energy of atomic fission for electrical power and transportation, but these applications would require time and technology to realize. Another byproduct of atomic energy, however, was ready immediately. Nuclear reactors could be used to generate radioactive isotopes—unstable variants of chemical elements that give off detectable radiation.[2] Scientists began using radioisotopes in biomedical experiments two decades before the atomic age, but their availability remained small-scale until nuclear reactors were

1. Draft Note on Atomic Power from the General Advisory Committee, 29 Jul 1947, enclosed in memorandum from Carroll L. Wilson to the Commissioners, "Note on Atomic Power," 23 Sep 1947, NARA College Park, RG 326, E67A, box 1, folder 5 Press Releases and Public Statements.

2. A note on terminology: In 1947, Truman Kohman ("Proposed New Word: Nuclide" [1947]) observed that isotope by definition "refers to a species of a particular and designated element, and emphasizes its relationship to other isotopes of that element." For this reason he introduced "nuclide," or radionuclide, to refer to a species of atom, arguing that it was preferable to "radioisotope." However, for historical accuracy I generally follow original sources in using the term radioisotope; the term radionuclide was not widely used until the 1960s and is still not used uniformly. In referring to specific radioisotopes, I will spell out the element, such as carbon-14, except when it is part of a prefix, e.g., ^{14}C-labeled-CO_2.

developed for the bomb project. In planning for postwar atomic energy, leaders of the Manhattan Project proposed converting a large reactor at Oak Ridge, part of the infrastructure for the bomb project, into a production site for radioisotopes for civilian scientists. The US Atomic Energy Commission (AEC) inherited this plan and oversaw an expansive program making isotopes available for research, therapy, and industry. This book is an account of the uses of radioisotopes as a way to shed new light on the consequences of the "physicists' war" for postwar biology and medicine.[3]

By the end of the war, the Manhattan Engineer District had constructed laboratories and plants at more than a dozen sites throughout the country. On January 1, 1947, these were transferred to the AEC, a civilian agency charged with continuing US atomic weapons production while developing the peaceful uses of atomic energy. Five presidentially appointed Commissioners directed the AEC. The appointees tended to be directors of other agencies, lawyers, businessmen, and physical scientists.[4] The chair of the Commission served as a spokesperson for this group, but decisions were made by majority vote. A general manager oversaw day-to-day operations. There were limits, both practical and legal, on the civilian orientation of the AEC. Since it produced the growing arsenal of atomic weapons, the agency remained closely tied to the Armed Forces. A Military Liaison Committee provided the organizational interface between the agency and the military, though more extensive interactions developed. The rapid escalation in atomic weapons production and testing through the 1950s, and the shift to thermonuclear bombs, kept the AEC focused heavily on military applications.[5] Even so, the Commission, particularly its first chair David J. Lilienthal, viewed the development of civilian applications of atomic energy as central to the agency's mandate, as well as politically expedient. The Commission's radioisotope distribution program was vital to this mission. In fact, by the late 1940s, the AEC had little else to hold out as evidence of the atom's peaceful benefits; hopes for rapid devel-

3. Kevles, *Physicists* (1978), ch. 20. For an insightful history of biophysics along this line, see Rasmussen, "Mid-Century Biophysics Bubble" (1997).

4. There was usually one physicist or chemist on the Commission at any time. Physical scientists also had input to the Commission through the General Advisory Committee. Sylves, *Nuclear Oracles* (1987).

5. Between 1946 and 1961, military applications made up two-thirds of all research, development, and construction undertaken by the AEC. Clarke, "Origins of Nuclear Power" (1985), p. 477.

opment of an atomic power industry faded and the nuclear arms race took off with the hardening of the Cold War.[6]

The political value of radioisotopes derived from their scientific and medical utility. Isotopes differ from ordinary atoms by having an alternate (often greater) number of neutrons. Stable isotopes persist indefinitely with this extra nuclear baggage, and can be identified on account of their increased atomic mass. Radioisotopes, by contrast, can be detected when they decay to another—usually stable—form, by emitting at least one of three kinds of radiation. Alpha particles (each made up of two protons and two neutrons) do not go far, and cannot pass through paper. Beta particles (high-energy electrons or positrons) are more penetrating, but can be stopped by wood. Gamma rays have high energy and travel the longest distance in air; they are stopped only by dense materials such as lead or concrete. The radiation hazard associated with each radioisotope depends on how frequently it decays (its half-life), and on the kind and energy of the particles emitted when it does.

Already in the 1930s, scientists and physicians distinguished two ways of using radioisotopes. Like radium or x-ray machines, they could be employed as a source of radiation, such as in cancer therapy. This generally required a significant amount of radioactivity (measured in curies).[7] More innovatively, and generally requiring less radioactivity, radioisotopes could be used as molecular tracers. (So could stable isotopes.) Isotopes gave researchers a way to tag compounds by replacing an ordinary atom with its radioactive sibling. One could then follow the labeled molecule through chemical reactions or biological systems by detecting the radiation emitted as radioisotopic atoms decayed. Consequently, previously imperceptible molecular processes could be traced, leading observers to compare radioisotopes with microscopes. The AEC's 1948 semiannual report emphasized the revolutionary character of radioisotopes: "As tracers, they are proving themselves the most useful new research tool since the invention of the microscope in the 17th Century; in fact, they represent that rarest of all scientific advances, a new mode of perception."[8] A 1952 report reiterated: "Because of the special ability to chart their course through

6. On the history of the AEC, see Hewlett and Anderson, *New World* (1962); Hewlett and Duncan, *Atomic Shield* (1969); Hewlett and Holl, *Atoms for Peace and War* (1989); Mazuzan and Walker, *Controlling the Atom* (1985); and Walker, *Containing the Atom* (1992).

7. A curie is a unit of radioactivity corresponding to 3.7×10^{10} disintegrations per second.

8. AEC, *Fourth Semiannual Report* (1948), p. 5.

living organisms and intact objects, isotopes have been called the most important scientific tool developed since the microscope."[9]

In contrast to the microscope, however, the aim of using isotopic tracers was not to bring into view anatomical structures so much as dynamic transformations. Biochemists used radioisotopes to reveal the sequence of chemical reactions in metabolism. Physiologists followed the assimilation and turnover of key nutrients and tagged molecules such as insulin to track the movement and activity of hormones. Molecular biologists labeled nucleic acids with radioisotopes to follow the replication and expression of genes. Physicians utilized radioisotopes such as radioiodine and radiophosphorus to diagnose thyroid function and detect tumors. Ecologists profited as well, using phosphorus-32 to trace nutrient cycling through the living and nonliving parts of aquatic and terrestrial landscapes, giving concrete meaning to the notion of an ecosystem.

Underlying these diverse applications of isotopic tracers was a common tactic. Biologists of all stripes employed radioisotopes to trace out the movement and chemical transformation of key molecules, charting the circulation of materials and energy through cells, organisms, and communities. As labels, radioisotopes could be used to follow compounds through separation techniques (centrifugation, electrophoresis, chromatography) or through biological processes, such as the synthesis of proteins or the movement of phosphorus from phytoplankton into inorganic debris. These tools and their representations—as maps, pathways, and cycles—invited new questions about the economy and regulation of life, informed by the concepts of cybernetics.[10] Radioisotopes were key ingredients of a postwar episteme of understanding life in molecular terms.[11]

By analogy, *Life Atomic* uses radioisotopes as historical tracers, analyzing how they were introduced into systems of scientific research, how

9. AEC, *Some Applications of Atomic Energy* (1952), p. 100. This was not a new trope. In the 1930s, British physiologist A. V. Hill asserted that just as the microscope had revealed cells, isotopes would "permit the biologist in effect to see the atoms." "Atom Smashing and the Life Sciences," RAC, RF 1.1, Trustees Bulletin 1937, p. 12. For other uses of the metaphor, Hamilton, "Use of Radioactive Tracers" (1942); Woodbury, *Atoms for Peace* (1955), p. 168.

10. Keller, *Refiguring Life* (1995); Creager and Gaudillière, "Meanings in Search of Experiments" (1996), pp. 6–15; Kay, *Who Wrote the Book of Life?* (2000); Landecker, "Hormones and Metabolic Regulation" (2011).

11. On the importance of tracking and tracing practices to twentieth-century biology, see Rheinberger, *Toward a History* (1997); Griesemer, "Tracking Organic Processes" (2007).

they circulated, and what new developments they enabled.[12] I analyze the movement of radioisotopes through government facilities, laboratories, and clinics, both in the United States and around the world, as a way to make visible key transformations in the politics and epistemology of postwar biology and medicine.[13] The launching of a US government distribution system led to a remarkable penetration of radioisotopes into American laboratories. From 1946 to 1955, the AEC's Oak Ridge contractor sent out nearly 64,000 shipments of radioactive materials to more than 2,400 laboratories, companies, and hospitals.[14] This number underestimates by several-fold the number of ultimate recipients, because many bulk shipments went to companies that sold radiolabeled compounds and radiopharmaceuticals. These radioisotopes were used in more than 10,000 scientific publications during that first decade of the AEC's program.[15] The vast majority of these radioisotopes originated in the Oak Ridge reactor that had been part of the Manhattan Project. (See figure 1.1.)

The availability of this new research tool was tied up with the politics of atomic energy. Most importantly, this is why radioisotope usage took off so quickly—the AEC did everything it could to encourage scientists and physicians to use these "by-products" of the bomb project. By making laboratories, clinics, and companies the beneficiaries of the government's nuclear largesse, the AEC hoped to build public support. In 1950, Commissioner Henry DeWolf Smyth assessed the impact of the isotope distribution program in the following terms:

12. For a related approach, see Herran and Roqué, "Tracers of Modern Technoscience" (2009). This is the introduction to a special collection in *Dynamis* on "Isotopes: Science, Technology and Medicine in the Twentieth Century." Several contributions are cited below but the entire set deserves mention.

13. My work builds on an extensive literature that addresses the role of materials, instruments, and objects in science. Influential examples include Clarke and Fujimura, *Right Tools* (1992); Kohler, *Lords of the Fly* (1994); Pickering, *Mangle of Practice* (1995); Galison, *Image and Logic* (1997); Rheinberger, *Toward a History* (1997); Daston, *Biographies of Scientific Objects* (2000); Joerges and Shinn, *Instrumentation* (2001). This approach has been especially fruitful in studies of biology and medicine: e.g., Creager, *Life of a Virus* (2002); Rader, *Making Mice* (2004); Landecker, *Culturing Life* (2007); Friese and Clarke, "Transposing Bodies" (2011). I have also taken inspiration from the studies of "drug trajectories" in following radioisotopes: Gaudillière, "Introduction" (2005).

14. AEC, *Eight-Year Isotope Summary* (1955), p. 2. By 30 Nov 1966 the total number of shipments had increased to 156,236. Eisenbud, *Environmental Radioactivity* (1963), p. 234.

15. AEC, *Eight-Year Isotope Summary* (1955), p. 1.

FIGURE 1.1. A remotely controlled device lifting a shipping bottle of a radioisotope at the shipping and storage room at Oak Ridge. Credit: Oak Ridge Operations Office. National Archives, RG 326-G, box 4, folder 7, AEC-54-5054.

When [the AEC] is asked, "What are the peacetime uses of atomic energy?" it can reply "Isotopes." Not that they will be useful *sometime* but that they are already useful. The isotope distribution program is enormously valuable because it reveals the Atomic Energy Commission as more than just a weapons organization. This is true not only in this country, but abroad where the

> foreign distribution of isotopes has had a very good effect on our foreign relations.[16]

Smyth is remembered chiefly not as an AEC Commissioner, but as the physicist who wrote the official history of the Manhattan Project.[17] The Smyth Report is the starting point for a vast historiography of the atomic bomb oriented around its origins in and consequences for the physical sciences.[18] *Life Atomic* builds on a more recent scholarship examining how the Manhattan Project and the atomic age shaped postwar biology and medicine.[19] For example, the AEC's interest in understanding radiation damage led to significant funding for genetics research, as well as for studies of Japanese survivors by the Atomic Bomb Casualty Commission.[20] In the 1950s, concerns about the environmental effects of radioactive waste led the AEC's Oak Ridge National Laboratory to organize a large ecology research group, which was instrumental to the development of radioecology.[21] Research into a wide range of biological, medical, and environmental problems took place at the AEC's national laboratories, and at many universities and hospitals with grants from the Commission.[22] Compared with these other AEC initiatives, the radioisotope program is notable for its early origin—the program was established by the Manhattan Project in advance of the Atomic Energy Act of 1946—and the sheer number of research sites that it affected.[23] Though rarely acknowledged, the AEC

16. Summary of the Proceedings of the Advisory Committee on Isotope Distribution, 23–24 Mar 1950, NARA College Park, RG 326, E67A, box 32, folder 12 Advisory Committee on Isotope Distribution, p. 1. Emphasis in original.

17. Smyth, *Atomic Energy for Military Purposes* (1945).

18. For a few representatives: Hewlett and Anderson, *New World* (1962); Rhodes, *Making of the Atomic Bomb* (1986); Hoddeson et al., *Critical Assembly* (1993); Norris, *Racing for the Bomb* (2002); Rotter, *Hiroshima* (2008).

19. On nuclear medicine in particular, see Kutcher, *Contested Medicine* (2009) and Leopold, *Under the Radar* (2009).

20. Beatty, "Genetics in the Atomic Age" (1991); Lindee, *Suffering Made Real* (1994); Barker, *From Atom Bomb* (2008).

21. Bocking, "Ecosystems, Ecologists, and the Atom" (1995); idem, *Ecologists and Environmental Politics* (1997).

22. Rasmussen, "Mid-Century Biophysics Bubble" (1997); Westwick, *National Labs* (2003); Schloegel and Rader, *Ecology, Environment, and "Big Science"* (2005); Rader, "Hollaender's Postwar Vision" (2006).

23. Others that have highlighted the importance of the radioisotope distribution program are ACHRE, *Final Report* (1996), ch. 6; Lenoir and Hays, "Manhattan Project for Biomedicine" (2000); and Rheinberger, "Putting Isotopes to Work" (2001).

shaped life science and medicine as profoundly as it did physics and engineering. Hans-Jörg Rheinberger has aptly described the dissemination of radioisotopes as "big science coming in small pieces."[24]

Smyth focused on the important role of the AEC's radioisotope supply in demonstrating that atoms could be helpful as well as harmful. As he indicated, it was not only domestic politics that the agency targeted; the US government saw radioisotope shipments abroad as a key means of aiding diplomacy. Exports were initially justified, over the objections of Congressional critics, as part of the Marshall Plan. By the mid-1950s, the international reach of the isotope program received special attention from President Eisenhower, whose "Atoms for Peace" initiative focused on the foreign development of atomic energy.[25] The United States was competing with other nuclear powers, most notably the Soviet Union, in providing radioisotopes and reactors as a means to wield geopolitical influence. Other Western nations building atomic energy infrastructures launched national companies to supply radioisotopes or develop nuclear power.[26] The AEC, in contrast, was charged with fostering "free enterprise," despite the fact that the 1946 Atomic Energy Act forbade private ownership of fissionable material and most patents on nuclear technologies. This led to a convoluted attempt by the AEC to involve companies in its operations despite the absence of anything like a free market. Moreover, the AEC's national security–related requirements for radioisotope exports put the United States at a disadvantage in comparison with the British and Canadian governments, which sold radioisotopes to foreigners with fewer restrictions.

Although the 1946 Atomic Energy Act made provision for distributing the "by-products" of nuclear reactors, most of the isotopes scientists and physicians desired were not typical fission by-products. Rather, the

24. Rheinberger, "Physics and Chemistry of Life" (2004), p. 224. The term "big science" is attributed to the long-time director of Oak Ridge National Laboratory: Weinberg, "Impact of Large-Scale Science" (1961); Galison and Hevly, *Big Science* (1992).

25. Here I follow John Krige's emphasis on the use of science and technology in American foreign policy during the Cold War. Krige, "Atoms for Peace" (2006); *American Hegemony* (2006); "Techno-Utopian Dreams" (2010). See also Osgood, *Total Cold War* (2006).

26. Hecht, *Radiance of France* (1998); Kraft, "Between Medicine and Industry" (2006); Gaudillière, "Normal Pathways" (2006); Adamson, "Cores of Production" (2009); Herran, "Isotope Networks" (2009); Santesmases, "Peace Propaganda" (2006); idem, "From Prophylaxis to Atomic Cocktail" (2009); Schwerin, "Prekäre Stoffe" (2009); idem, "Österreichs im Atomzeitalter" (2011).

neutron flux of reactors was employed to irradiate target materials. The specific radioisotopes generated through irradiation were usually chemically purified for sale. Thus the production and distribution of radioisotopes put the US government in a peculiar role, as the manufacturer of a perishable laboratory good.[27] Even as Congressional debates over the appropriate relationship of the federal government to university science delayed the establishment of the National Science Foundation until 1950, AEC-produced radioisotopes represented the support of the US government for scientific research in strikingly tangible terms.[28] The importance of the Cold War in shaping developments in biology and medicine should be understood not only in terms of ideology, but also in terms of infrastructure.[29] The significance of the politics of atomic energy for postwar science, in other words, can be traced using the radioisotopes that left the AEC's nuclear reactors and entered laboratories, clinics, and companies.

Sources and Story

While radioisotopes were important to research and medicine through most of the twentieth century (particularly if natural radioisotopes such as radium-226 are included), the ensuing chapters emphasize developments from 1945 to 1965, when artificial radioisotopes first achieved widespread utilization. This early postwar period was a crucial juncture for the diffusion of radioisotopes as a research technology, for reasons beyond technical utility.[30] After the first atomic weapons were detonated over Japan, the US government's attempts to both exploit and justify atomic energy resulted in the vast uptake of radioisotopes into laboratories, clinics, and the environment. In this sense, by tracing the pathways along which

27. Given that the first Chair of the Commission was David Lilienthal, former head of the Tennessee Valley Authority (TVA), one might contend that the role of the government as a supplier of radioisotopes—and even more so nuclear power—continued the controversial involvement of the government in energy utilities during the New Deal. On a related note, Thomas Hughes has compared the construction of the Manhattan Project facilities to the construction of the TVA. Hughes, "Tennessee Valley" (1989).

28. Kevles, "National Science Foundation" (1977); Greenberg, *Politics of Pure Science* (1967); Appel, *Shaping Biology* (2000).

29. Creager and Landecker, "Technical Matters" (2009).

30. Joerges and Shinn, *Instrumentation* (2001).

radioisotopes moved, one sees how intertwined were the political, military, economic, and scientific aspects of atomic energy.

That said, there is an important caveat to this book's attempt to narrate postwar science, medicine, and politics through radioisotopes. The AEC generated a massive volume of published and unpublished documents; these provide the main source-base for *Life Atomic*, undeniably shaping its perspective.[31] The promise and perils of atomic energy presented here are those voiced by government officials and scientists allied with the AEC. To be sure, one finds sharp points of disagreement within this elite group, their vantage points reflecting not only their varying backgrounds, political convictions, and disciplinary affiliations, but also their locations, whether based in Oak Ridge or in Washington, DC, in the AEC's headquarters or on the floors of Congress. Yet they shared, nearly without exception, a commitment to the civilian development of atomic energy and a confidence in the ability of scientists and engineers to safely manage nuclear materials, by-products, and wastes. This underlying consensus is especially important for understanding why the biological hazards and environmental problems associated with radiation, which were not completely unknown, did not sway the government's determination to disseminate radioisotopes and develop nuclear power.[32]

The AEC itself funded extensive research into the biological effects of radiation, which eventually yielded evidence that no dose of ionizing radiation is low enough to be innocuous, yet such investigations occurred alongside, rather than prior to, the development of atomic energy for military, civilian, and industrial use. These studies included human experiments with radioisotopes and radiation sources, many of which remained classified until the 1990s, when new investigative journalism led the Clinton Administration to appoint a panel to evaluate the government's role in these experiments, and to declassify and publicize the relevant government documents they discovered. The Advisory Committee on Human Radiation Experiments (ACHRE) detailed a wide variety of medical experiments using radiation sources, both military and civilian, public and private. Their report contextualizes these diverse activities within the less

31. The unpublished documents concerning the AEC's radioisotope distribution, however, are strikingly dispersed; the Advisory Committee on Human Radiation Experiments described these papers as "the most fugitive records [it] desired." ACHRE, *Final Report, Supp. Vol. 2a* (1995), p. 58.

32. Hamblin, *Poison in the Well* (2008).

stringent human subjects guidelines (and lack of federal regulation) that characterized postwar medical research, even as ACHRE criticized researchers who did not follow existing guidelines and government agencies for not providing more oversight. In response, President Clinton offered a formal apology to individuals who were harmed by radiation tests that the US government conducted or supported.[33]

Human uses of radioisotopes form a subset of the "human radiation experiments" that journalists, government-appointed scholars, and the media considered and critiqued in the 1990s, and this book engages those discussions and debates.[34] But focusing specifically on the representation and dissemination of radioisotopes also recasts the history of these experiments in a new light. Radioisotopes had already proven valuable for biomedical research and therapy before the Manhattan Project, and scientists pointed to radioisotopes as evidence that the atom could cure as well as kill. More generally, the AEC and its advisors believed the benefits of atomic energy outweighed its costs in terms of health risks or environmental contamination, which would exist anyway due to the continued production and testing of nuclear weapons. To be sure, by the 1950s there emerged scientific and public dissent from the government's representation of low-level radiation exposure as negligible and manageable, though this was and remains an area of technical uncertainty and dispute.[35] As critics observed, the US government tended to present new information about the hazards of radiation in ways that did not undermine its policy of continuing to develop atomic energy. In fact, the conflict of interest between the AEC's promotional and regulatory responsibilities is a major theme of its historiography, and the agency's distribution program for radioisotopes manifested this problem even earlier than the domestic nuclear industry.[36] The story of how radioisotopes spread and what they signified relied upon—and reveals—this postwar mindset that valorized atomic energy development, even as the resulting scientific knowledge corroded this optimistic view.

The narrative of *Life Atomic* begins with how radioisotopes were produced and used before nuclear reactors existed. In the late 1930s,

33. ACHRE, *Final Report* (1996); Kutcher, *Contested Medicine* (2009), ch. 8.

34. See chapter 8.

35. Semendeferi, "Legitimating a Nuclear Critic" (2008); Vaiserman, "Radiation Hormesis" (2010).

36. Mazuzan, "Conflict of Interest" (1981); Walker, *Containing the Atom* (1992).

physicists could produce many artificial radioisotopes, if in limited amounts, in cyclotrons. Chapter 2 follows developments around E. O. Lawrence's cyclotrons, which supplied radioisotopes to researchers and physicians, through personal contacts and requests, both within and beyond the Radiation Laboratory. In Berkeley, biological research with radioisotopes as tracers proceeded alongside therapeutic experimentation using radioisotopes as radiation sources. The wartime mobilization of Lawrence's laboratory interfered with its ability to provide radioisotopes to physicians and scientists outside the Manhattan Project. In addition, new priorities shaped the ongoing human experiments conducted by Lawrence's colleagues (such as John Lawrence and Joseph Hamilton), as Berkeley scientists began investigating the toxicity and metabolism of fission products for the military.

During the final year of World War II, leaders in the Manhattan Project laid the groundwork for the government's mass-production of radioisotopes in peacetime. After Enrico Fermi's demonstration that one could achieve criticality in nuclear fission with his improvised reactor in Chicago, the US Army built a larger, permanent reactor in Oak Ridge, Tennessee, the site of several isotope separation plants used in the Manhattan Project. This reactor was a pilot plant for the plutonium-producing reactors built in Hanford, Washington, and its postwar fate was uncertain. Scientists proposed dedicating it as a production reactor for radioisotopes for civilians, with the dual aims of benefiting postwar science and justifying a long-term national laboratory in Tennessee. Chapter 3 covers the establishment of the civilian AEC, the launching of radioisotope distribution—still under the auspices of the Manhattan Project until January 1, 1947—and the US government's public relations efforts that were staged around the early shipments. Through the 1950s, the AEC sought to establish a one-stop isotope shop (with stable and radioactive isotopes, as well as irradiation services) for scientists and clinicians. As it turned out, the same Oak Ridge reactor that was producing radioisotopes for civilian purchasers was simultaneously producing materials for radiological warfare experiments and other classified research projects.

The fourth chapter explores the ways in which radioisotopes were used as political instruments—both by the federal government in world affairs, and by critics of the civilian management of atomic energy. Congress established an agency for atomic energy outside of the military, with support from scientists, with the expectation that peacetime benefits would materialize. But the controversies the AEC faced in the early years of the

Cold War, particularly whether to ship radioisotopes to foreign scientists, demonstrate the agency's political vulnerabilities. The core of this chapter analyzes this debate, which prevented shipments from being sent abroad during the first year of the program. Even after exports commenced, radioisotope shipments to foreign physicists and engineers raised worries that the United States was aiding weapons programs abroad. Moreover, the AEC's critics frequently equated the export of radioisotopes with the dissemination of nuclear information, which was explicitly prohibited by the 1946 Atomic Energy Act. However, the demise of the American nuclear monopoly meant that foreign requests could be met outside of the AEC's supply. In the mid-1950s Eisenhower's Atoms for Peace program, seeking to reclaim the image of American beneficence, gave new visibility to the AEC's exports of radioisotopes as emblems of humanitarianism, against the backdrop of an escalating nuclear arms race.

Following World War II, the publication of accounts such as John Hersey's *Hiroshima* documented the devastating effects of nuclear weaponry on inhabitants of the two Japanese cities targeted by atomic bombs. Nonetheless, the American government presented a positive image of atomic energy development as benefiting the health of its own citizens. The fifth chapter examines this apparent paradox. In the late 1940s and 1950s, the AEC sought to utilize atomic energy for humanitarian purposes, above all by advancing cancer research, therapy, and diagnosis. This objective picked up on hopes articulated in the 1930s by E. O. Lawrence and others that artificial radioisotopes would transform the treatment of cancer. In addition, health physicists generally assumed that the occupational risks associated with radiation could be rendered insignificant by carefully limiting exposure. The hazards of radioactivity were understood largely in terms of acute effects triggered by a relatively high dose.

Knowledge emerging in the 1950s concerning long-term radiation effects, including documentation of leukemia incidence among Japanese survivors at some distance from the atomic detonations, revealed the hazards of low-level exposures. In a classic 1957 paper in *Science*, Edward B. Lewis showed that the probability of leukemia attributable per dose of exposure was roughly the same across four exposed populations of doctors, patients, and Japanese survivors.[37] He surmised that radioactivity from continued atomic weapon tests could increase the leukemia incidence in the US population by as much as 10%. The increasing clinical reliance on

37. Lewis, "Leukemia and Ionizing Radiation" (1957).

radioisotopes in the 1950s developed in this context of changing perceptions of radiation's hazards. Radioisotopes, particularly from radioactive fallout, began to be seen as causes rather than cures of cancer. This complicated the agency's plans for advancing the other dividend of atomic energy, namely nuclear power.

The partially public, partially private nature of nuclear industry in the 1940s and early 1950s reflected the contradictions of government policy that promoted "free enterprise" while stringently guarding materials and technologies related to national security. Chapter 6 stresses the uneasy relationship between the US government and industry in developing the civilian uses of atomic energy, a problem that the 1954 revision of the Atomic Energy Act was aimed at redressing. The construction of nuclear reactors outside of the AEC's facilities changed the government's role in radioisotope production, as marked by the closing of the original Oak Ridge production reactor in 1963. By that time, users obtained most radioactive materials from companies that prepared radiolabeled compounds and radiopharmaceuticals, based on reactor-generated radioisotopes purchased in bulk from either the AEC or, increasingly, nongovernment suppliers. At the same time, the emergence of a civilian reactor industry expanded the AEC's responsibility for regulating radiological protection to the private sector, which in turn impacted the oversight of radioisotope buyers.

If the US government's decision to use part of the infrastructure of the bomb project to produce radioisotopes is judged by the volume sold, it was wildly successful. Figure 1.2 shows the cumulative curies shipped from the AEC's facility in Oak Ridge. The head of the Isotopes Division estimated that there were 50,000 purchases of radioisotopes, radiolabeled compounds, and radiopharmaceuticals in 1956 alone.[38] To give a sense of what this meant in one field, the percentage of publications in the *Journal of Biological Chemistry* that employed radioactive isotopes rose from 1% in 1945 to 39% in 1956.[39] The Commission regarded its provision of radioisotopes in economic terms, as seen by the note on the graph ascribing the small dip in 1952 to a new policy that charged 20% production costs on

38. This estimate includes sales from the secondary retailers. Paul C. Aebersold, Outline of Isotope Production and Licensing, 9 Mar 1956, Aebersold papers, box 2, folder 2-4 Gen Corr Jan–Mar 1956, p. 1.

39. Broda, *Radioactive Isotopes* (1960), p. 2. By contrast, the increase in papers using radioactive isotopes was not as high in leading British, Russian, or German journals of biochemistry.

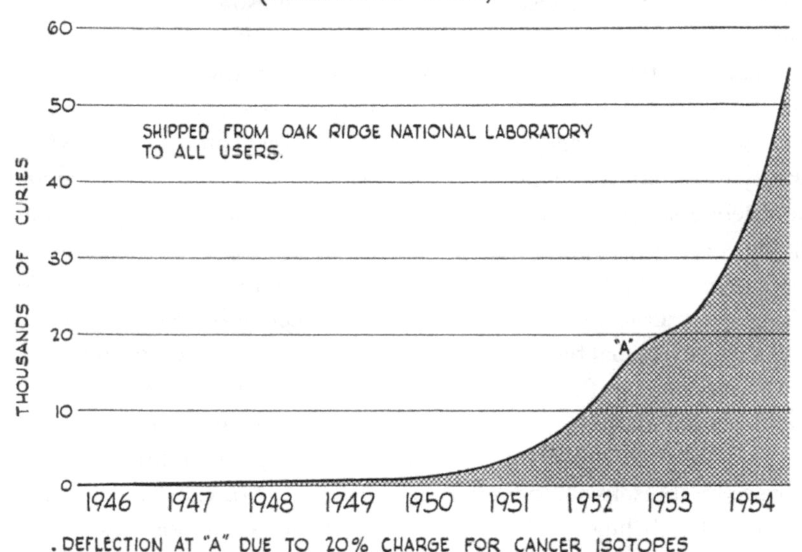

FIGURE 1.2. Graph of cumulative curies shipped from the AEC's Oak Ridge National Laboratory to all users, 1946 to 1955. From US Atomic Energy Commission, *Eight-Year Isotope Summary*, vol. 7 of *Selected Reference Material, United States Energy Program* (Washington, DC: US Government Printing Office, 1955), p. 79.

previously free shipments for use in cancer research, therapy, and diagnosis. Price apparently mattered to these customers. But despite the involvement of the private sector in secondary distribution, this was a heavily subsidized economy, not a free market.

What were the consequences of the AEC's promotion and provision of radioisotopes in biomedicine? If the first half of *Life Atomic* is about the establishment of a technological system for radioisotope production and consumption, the second half of the book focuses on some representative users and on how the technology mattered.[40] These chapters examine specific applications of radioisotopes in biochemistry and molecular biology, clinical investigation, nuclear medicine, and ecology. As these samplings illustrate, the availability of radioisotopes shaped not only experimental

40. Hughes, *Networks of Power* (1983); Cowan, "Consumption Junction" (1987).

methods, but also the ways in which life and disease were conceptualized. These episodes show as well how the transition from cyclotron-based production to reactor-based production of radioisotopes played out on the ground.

Chapter 7 follows radioisotopes into the laboratories of biochemists and molecular biologists. By focusing on how radioisotopes illuminated metabolic pathways and genetic transmission, one can also see how the supply of radioisotopes from the AEC shaped the questions asked and knowledge sought by researchers in the early postwar period. Radioisotopes reinforced a preoccupation with elucidating chemical pathways and processes, by labeling molecules with radioactive atoms so as to make visible the transformations they undergo, in a cell or in an organism. The metabolic maps that biochemists drew represented chemical changes over time as movement through space, along the pathway. Gene transfer experiments similarly traced the movement of atoms from the hereditary material of parent to that of progeny, examining how reproduction involved the transmission of molecules. These kinds of experiments established radiolabeling as a key technique that took on a momentum of its own. Carbon-14 and tritium became standard labels of substrates used in enzyme assays, a trend due in part to the development of the automated scintillation counters. Phosphorus-32 became the standard label for DNA and RNA. In this sense, by the 1960s, tagging molecules for in vitro experiments overtook tracing pathways in vivo.

Chapter 8 extends the use of radioisotopes in biomedical research to human subjects. In physiology and endocrinology, radioisotopes were used to investigate the regulation of hormones and the absorption and movement of micronutrients. The first part of the chapter examines the use of iron-59 in studies of mammalian metabolism of this element. Like the experiments discussed in chapter 7, this was the application of a radioisotope as a tracer, but the use of human subjects brought up both logistical and ethical issues not faced by most biochemists using isotopes. A direct outgrowth of this line of research was the controversial, large-scale study of iron metabolism in pregnant women that took place at Vanderbilt University Medical School in the mid-1940s. The second example concerns the development of radioimmunoassays, in which research associated with the clinical use of radioiodine in a Veterans Administration Hospital led Rosalyn Yalow and Solomon Berson to develop a diagnostic method with wide applicability. Here the boundary between laboratory research and clinical application was especially permeable, and applied knowledge generated new tools for basic research. Administering radiolabeled insulin to veterans turned up surprising results about antibodies, which were then utilized in a novel diagnostic test.[41] In both of the chapter's cases, medical research with radioisotopes relied on human patients as subjects, and the push to apply radioisotopes in these settings came at the expense of caution about their radiation exposure.

The AEC's radioisotope distribution program and its broader biomedical research policy fostered the emergence of nuclear medicine in the 1950s and 1960s, the focus of chapter 9. The quest for novel ways to use radioisotopes in cancer treatment remained largely disappointing. The most important development was the use of cobalt-60 in teletherapy machines (so-called cobalt bombs), which began to replace radium as an external radiation source in hospitals. By contrast, the growth of medical diagnostics with radioisotopes to locate tumors and observe organ function mirrored the biochemical usage of radioisotopes as tracers to study metabolism and heredity discussed in chapter 7. The growth in radioisotope-based diagnostics relied on the invention of new detection devices, such as the whole-body scintillation scanner, which prompted the search for isotopes with more suitable half-lives or decay energies. These medical applications widened the gulf, particularly in dosage, between therapeutic uses and tracer uses, the former in radiation sources of unprecedented strength ("beams"), the latter for shorter-lived and lower-energy radioisotopes ("emanations") that could be considered safe for routine diagnostics. Particularly in the development of diagnostics, one can see how the medical establishment sought to respond to both the public concern and new scientific evidence about the hazards of low-level radiation.

Ecology was as profoundly shaped by the AEC as biomedical research. Not only did ecologists throughout the country use radioisotopes as tracers, but the Commission also launched important investigations of the environmental consequences of radioactive contaminants. Chapter 10 shows how the adoption of radioisotopes as tools in ecology enabled researchers to track the flow of materials and energy in ecosystems. G. Evelyn Hutchinson and others took inspiration from how physiologists and biochemists used radioisotopes, seeking to understand the "metabolism" of entire lakes and other ecosystems. The last part of the chapter focuses on the development of radioecology at three AEC installations:

41. Sturdy, "Looking for Trouble" (2011).

Hanford, Oak Ridge, and Savannah River. Strikingly, radioactive waste itself provided unintended tracers for ecological research, yielding information about the movement of materials through aquatic and terrestrial ecosystems. In the end, radioisotopes became "model pollutants" for developing means of detecting other environmental contaminants, especially synthetic chemicals.

By the 1970s, the AEC's vision of a society transformed by atomic energy was challenged by a vocal political movement opposing the continued construction of nuclear power plants.[42] Given this context, the concluding chapter assesses the longer-term impact of the other dividend of atomic energy the Commission had promoted, radioisotopes. Even in the age of environmentalism, radioisotopes continued to be vital tools for scientific research and medical diagnosis. Particularly in molecular biology, the main techniques of the era of genetic engineering, including blots and sequencing, employed radiolabels. But since the Human Genome Project was completed, the emphasis on tracing single biochemical changes in cells and organisms, and on the role of single genes in determining biological traits, has given way to a systems approach in biology attuned to networks of molecular interactions and epigenetics. In addition, the burden of regulation for radiation exposure and radioactive waste disposal made alternative labeling technologies worth pursuing, particularly for research uses. That said, it might be premature to refer to the twilight of the atomic era in biomedicine. The reliance on radioisotopes in nuclear medicine continues to be strong, particularly the use of technetium-99m in diagnostic tests.

The historical traces of radioisotopes can be detected in the bodies of patients and human subjects, diagrams of metabolic pathways and ecosystems, countless nucleic acid sequences, as well as an approach to environmental pollution that focuses on the movement of contaminants through ecosystems. The legacy of radioisotopes also includes the emergence of government regulation of radiation exposure in the laboratory. When the AEC began supplying radioisotopes, it also regulated their uses by civilians, in hospitals and laboratories. The codification of rules for radiological protection and the increasing level of their enforcement reflected changing public expectations about the need for government oversight of scientific research.

42. Walker, *Three Mile Island* (2004), ch. 1.

War and Peace

Life Atomic grapples with an issue central to the physics-oriented historiography on the Manhattan Project: the degree to which the atomic sciences were militarized. At one level, this book challenges the idea that scientific and technological developments connected to atomic energy were dictated by the US military.[43] But if it is too reductionist to view postwar science and technology as extensions of what Eisenhower termed the "military-industrial complex," neither does this book defend the unimpeachable autonomy of civilian science and scientists.[44] Rather, radioisotopes, as part of a classic "dual-use" technology, exemplify the blurriness of the civilian-military divide. The postwar growth of the federal government's defense organizations and programs, with an increasingly elaborate security apparatus, implied a civilian counterpart, and vice versa.[45] The civilian-military boundary, although porous, was nonetheless important politically and culturally.[46]

For the AEC in particular, its viability as a civilian agency relied on being able to differentiate certain of its activities and programs as nonmilitary. It was in this respect that the radioisotope program proved so valuable to the agency, particularly during its first decade. The vast majority of radioisotope shipments went to civilian scientists or clinicians to aid them in their own research or medical practice. As Smyth attested, the AEC's display of the peaceful atom through radioisotope distribution showed that the civilian agency was fulfilling its mandate to develop peaceable applications of atomic energy. In this context, the AEC routinely represented advances in biology, agriculture, and medicine as the "peaceful" face of atomic energy. By contrast, physics was often associated with the military uses of atomic energy, not least through the credit physicists received for designing the first atomic bombs.[47] Putting on view the biomedical

43. For the debate over the effects of military and AEC patronage on the physical sciences, see Forman, "Behind Quantum Electronics" (1987) and Kevles, "Cold War and Hot Physics" (1990).

44. On the "military-industrial complex," Wolfe, *Competing with the Soviets* (2013), pp. 23–39.

45. Edgerton, *Warfare State* (2006); McEnaney, *Civil Defense* (2000).

46. On the role of boundaries in scientific credibility, Gieryn, *Cultural Boundaries of Science* (1999).

47. Schwartz, *Making of the Atomic Bomb* (2008).

benefits of the bomb project served an important political function for an agency charged with atomic energy's continuing military development. The civilian-qua-biomedical vision of the isotope program was borne out in practice: more than three-quarters of the shipments were used in medical therapy and diagnosis or biological research.[48]

Yet this image of biomedical research with radioisotopes as inherently humanitarian obscured the relevance of some of these same investigations to the military, particularly research into the biological effects of radiation exposure.[49] Several of the clinics and laboratories that developed nuclear medicine, especially new radioisotope-based diagnostics and cobalt-60 teletherapy for cancer, were also engaged in military-sponsored research on human subjects, exploring, for instance, the effects of whole-body irradiation or the metabolism of fission products. Medicine was not the only area where civilian and military interests converged. Investigations at the AEC's national laboratories of the environmental fate of radioactive waste around plutonium production plants revealed the ecological processes of nutrient cycling and the bioconcentration of contaminants. Radioisotopes used in biological warfare experiments—clearly for the military—came from the same Oak Ridge reactor that produced radioisotopes to sell to civilian scientists. That said, military applications of radioisotopes included conducting research for occupational health and safety at the government's reactors and production plants. After the Atomic Energy Act of 1954, and the beginning of a nuclear power industry, differences between civilian and military agendas, especially regarding safety, were increasingly difficult to distinguish.

The overlap between civilian and military uses of atomic energy is also reflected in the changing technologies for radioisotope supply. The early history of radioisotope production was tied to academic cyclotrons, whereas the postwar production of radioisotopes emerged out of the militarization of atomic energy and the related invention of nuclear reactors. Yet the dynamics of this shift are more complex than a neat transition from a civilian to military technology suggested by this chronology. On the one hand, most cyclotrons in the United States did not remain civilian, but

48. AEC, *Isotopes* (1949), p. 53. The remaining shipments were used in physics, chemistry, and industrial research.

49. See Welsome, *Plutonium Files* (1999); Whittemore and Boleyn-Fitzgerald, "Injecting Comatose Patients" (2003); Kutcher, "Cancer Therapy" (2003); idem, *Contested Medicine* (2009); Kraft, "Manhattan Transfer" (2009); Leopold, *Under the Radar* (2009).

were militarized as part of the Manhattan Engineer District, beginning with the Berkeley cyclotrons. The military uses of radioactive materials in human experimentation, most notoriously the postwar experiments with plutonium ingestion by cancer patients, were largely administered by the former medical physicists and physicians of the Manhattan Project. Joseph Hamilton and Robert Stone, who in the 1930s pioneered medical uses of cyclotron-produced radioisotopes in Berkeley and San Francisco, became involved during the war with classified human experiments using radioisotopes and radiation sources.[50] This legacy of military sponsorship and secret research persisted into the postwar years under the AEC. From this perspective, the early medical uses of atomic energy became enduringly militarized.

On the other hand, one may interpret the postwar government production of radioisotopes as the attenuation of military control. To be sure, reactors had been developed for the Army in the massive scientific effort to produce nuclear weapons. But many scientists were passionately committed to the demilitarization of the new technologies of atomic energy after the war, particularly nuclear reactors. Putting atomic energy development, including the production of nuclear weapons, in the hands of a civilian agency was conceived as a way to liberate atomic energy from the military and unleash its potential for civilian benefits. From this vantage point, the government's radioisotope program, launched by the Manhattan Engineer District before the AEC began, evinced the authority that these scientists wrested from the military at the beginning of peacetime. Thus, the complex push-me-pull-you of civilian and military uses of atomic energy grew out of the earlier assimilation of nuclear science into the war effort and the strong push during demobilization to "free the atom" for peaceful development.

At another level, the role of radioisotopes in postwar biology and medicine drew importantly on the pre-World War II uses of radioactive sources (particularly in therapy) and stable isotopes as tracers. In this respect, many of the notable advances enabled by the government's supply of radioisotopes—such as the use of carbon-14 in studying metabolic pathways or the development of teletherapy for cancer using cobalt-60—should not be seen as originating in the atomic age, but as continuing older technologies

50. Jones and Martensen, "Human Radiation Experiments" (2003); ACHRE, *Final Report, Supp. Vol. 1* (1995), ch. 2; Westwick, "Abraded from Several Corners" (1996).

and approaches with new materials and purposes.[51] Yet the development of nuclear reactors during the atomic bomb project fundamentally changed the scale of the production and use of radioisotopes—and mass production facilitated their commodification. The government's pricing structure, aiming to recoup only production costs and not infrastructure costs, as well as its subsidies for materials used in cancer work, made radioisotopes affordable as well as accessible (to Americans, at least). The cheapness and availability of radioisotopes—which were intimately tied to the politics of atomic energy—propelled their widespread usage, even as this built on preexisting demand from the era of cyclotron production. Assessing the degree to which these developments were continuous with pre-World War II trends rather than arising from postwar conditions is complex, and depends on the field. For medical therapy and for biochemistry, the uses of radioisotopes continued already-productive lines of research and practice; for molecular genetics and ecology, discoveries came out of the AEC's postwar supply.

In the end, the point is not to reduce all of the important postwar developments in biology and medicine to the government supply of radioisotopes. The AEC's own contributions to postwar life science and medicine went beyond that. For example, the Commission sponsored influential research on radiobiology, genetics, and nuclear medicine in its own national laboratories and in universities throughout the country.[52] Other important trends in life science, even those related to the Cold War, were not directly linked to the politics of atomic energy or its infrastructure. Yet radioisotopes provide a useful means for detecting events and reactions in scientific systems, often not as a cause but as an indicator or residue. Once you raise a historical Geiger counter and scan the last half of the twentieth century, you find the chatter of radioactivity everywhere, not only around atomic weapons facilities but also concentrated in places where life was studied and diseases were diagnosed. Radioisotopes became an essential element in thousands of medical diagnostics; enzyme assays; nucleic acid sequences; environmental studies; countless agricultural, medical, and biological experiments; and more varieties of laboratory tests than can

51. Santesmases, "Life and Death" (2010).

52. See, e.g., Beatty, "Genetics in the Atomic Age" (1991); Westwick, *National Labs* (2003); Rader, "Hollaender's Postwar Vision" (2006); Kraft, "Manhattan Transfer" (2009); Semendeferi, "Legitimating a Nuclear Critic" (2008). On British contributions to radiobiology, de Chadarevian, "Mice and the Reactor" (2006) and "Mutation in the Nuclear Age" (2010).

be named here. The spread of radioisotopes was as messy as using them often was—they left traces everywhere.[53] Their movement through the postwar landscape, both technological and natural, can be understood only in terms of the aftermath in the United States of Hiroshima and Nagasaki. Just as significantly, following radioisotopes from Oak Ridge into the many settings where they were used provides a valuable way to map the growth and regulation of postwar biomedicine.

53. I owe this felicitous phrasing to Crispin Barker, personal communication, 6 Nov 2008.

CHAPTER TWO

Cyclotrons

The nuclear physicist can now induce radioactivity in practically all of the elements, and he can harness a beam of neutrons of intense biological activity. This new wonderland for the biologist has been brought about by such events as the first successful experiments of Joliot and Curie in artificial radioactivity, the discovery of the neutron by Chadwick, the discovery of heavy hydrogen by Urey, and the development of the cyclotron by E. O. Lawrence and his associates.—John Lawrence, 1940[1]

The construction of cyclotrons in the 1930s gave physicists a new instrument with which to produce artificial radioactive isotopes, albeit often in limited quantities. Life scientists and physicians wishing to use these radioisotopes relied on physicists and chemists to provide them in a moral economy of shared material and credit.[2] This chapter focuses on developments in Ernest O. Lawrence's Radiation Laboratory (Rad Lab) to illustrate the cyclotron-based system of radioisotope production that existed in the United States before World War II. Unlike the market for radium, whose clinical and industrial uses were well established, the early market for artificial radioisotopes was not commercial.[3]

1. Lawrence, "Some Biological Applications" (1940), p. 125.

2. Several historians of science have adapted E. P. Thompson's notion of a moral economy (see "The Moral Economy of the English Crowd in the Eighteenth Century" and "The Moral Economy Revisited" in Thompson, *Customs in Common* [1991]); Shapin, *Social History of Truth* (1994); Daston, "Moral Economy of Science" (1995); and especially Kohler, "Moral Economy" (1999) and *Lords of the Fly* (1994).

3. Landa, "Buried Treasure" (1987); idem, "First Nuclear Industry" (1982); Badash, *Radioactivity in America* (1979), chapters 9 and 10; Rentetzi, *Trafficking Materials* (2008). Note that MIT also had its own wartime Rad Lab for developing microwave radar technology, completely separate in purpose from that at Berkeley.

In Berkeley, biological research with radiotracers was closely related to—and sometimes intertwined with—therapeutic experimentation with radioisotopes, an effort largely overseen by John H. Lawrence (brother of Ernest). Physicians as well as researchers began obtaining phosphorus-32 and other radioisotopes through Ernest and John Lawrence once they became available. The circulation of these cyclotron-generated radioisotopes relied on scientific networks of patronage and was not regulated by the state. Distribution was also nonmonetary, despite—or perhaps because of—Lawrence's unsuccessful attempt to patent his production method with an eye on the emerging radiopharmaceutical market.[4] This supply system may be described in terms of gift exchange, insofar as the recipient scientists and physicians remained personally obligated to E. O. Lawrence for his provision of radioactive materials.[5] Not every request was fulfilled. Moreover, the noncommercial character of these transactions did not automatically extend to the bedside; patients did not necessarily receive radioisotope treatment gratis.

After the war, the US Atomic Energy Commission trumpeted the inauguration of their program allowing foreigners to purchase radioisotopes for use in biology and medicine.[6] Yet this program was not as novel as it appeared: a private international distribution of American-produced radioisotopes existed before World War II. The Rad Lab became the main clinical supplier of phosphorus-32 in California by 1940, and shipped radioisotopes elsewhere in the United States and abroad. The supply of radioisotopes from Berkeley was suspended by the involvement of its cyclotron enterprise in the Manhattan Project. Distribution of radioisotopes to foreign scientists resumed, under quite different terms as a formal government-run program, only in 1947. Immediately after the war, E. O. Lawrence was not free to send radioisotopes to his friends (and clients) abroad; the Rad Lab continued to be a military installation and subject to the US government's embargo.

The mobilization of E. O. Lawrence's laboratory for the American war effort in the early 1940s posed serious constraints on the availability of radioisotopes to any physicians and scientists outside the Rad Lab, though

4. Heilbron and Seidel, *Lawrence and His Laboratory* (1989), p. 192 ff.

5. The classic work on gift exchange is Mauss, *The Gift* (1954). See also Findlen, "Economy of Scientific Exchange" (1991); Biagioli, *Galileo Courtier* (1993); idem, *Galileo's Instruments of Credit* (2006); and especially Anderson, "The Possession of Kuru" (2000).

6. See chapter 4.

some domestic clinical distribution continued. In addition, new military priorities shaped the nature of ongoing human experiments conducted by Berkeley scientists and physicians. Joseph Hamilton began a classified program of research under contract to the Manhattan Engineer District investigating the toxicity and metabolism of fission products. As it transpired, some of the most criticized of the human radiation experiments were conducted as part of this effort.[7] Human experiments with radioisotopes did not begin with the war effort, but the bomb project changed their orientation from serving clinical to military ends.

Cyclotrons, Isotopes, and Vaudeville

The availability of artificial radioactive isotopes in the 1930s was closely related to the emergence of cyclotron-based research programs in physics aimed at studying fast particles. In 1930, Ernest O. Lawrence and his graduate student M. Stanley Livingston built the "proton merry-go-round," only five inches in diameter.[8] It was followed by an 11-inch version in 1932, which produced million-volt protons.[9] Curiosity about nuclear transformations spurred Lawrence's research program, but just as vital were technical breakthroughs in the electrical power industry, developed for handling high-voltage generation and transmission to meet the growing demand for electrical power.[10] The Research Corporation, a philanthropic organization dedicated to scientific research, gave Lawrence a $5,000 grant and patented his machine. Lawrence and his collaborators subsequently adapted a massive, discarded Federal Telegraph magnet to construct a 27-inch cyclotron. In December 1932, this machine could produce 4.8 million electron volt hydrogen ions.[11]

Even with this technological edge, it was not cyclotroneers in Berkeley who discovered that particle bombardment could be used to generate radioactive elements. On New Year's Day, 1934, Frédéric Joliot and

7. ACHRE, *Final Report, Supp. Vol. 1* (1995), ch. 2; Jones and Martensen, "Human Radiation Experiments" (2003).

8. Heilbron and Seidel, *Lawrence and His Laboratory* (1989), p. 87.

9. Heilbron, Seidel, and Wheaton, *Lawrence and His Laboratory* (1981), p. 15.

10. Heilbron, Seidel, and Wheaton, *Lawrence and His Laboratory* (1981), p. 7; Seidel, "Origins of the Lawrence Berkeley Laboratory" (1992).

11. Heilbron, Seidel, and Wheaton, *Lawrence and His Laboratory* (1981), pp. 11–17.

Irène Joliot-Curie first produced artificial radioactivity by bombarding aluminum with alpha particles. Soon thereafter, Enrico Fermi demonstrated that sixteen radioisotopes could be made—albeit only in minute amounts—by means of neutron bombardment from a radon-beryllium source.[12] Small sources of this kind were not uncommon in physics departments, but larger sources tended to be owned by hospitals, and inaccessible to researchers.[13] Physicists developed other high-voltage sources that could be used to generate particles for bombarding targets: a voltage-multiplier circuit by John Cockcroft in Cambridge, a "cascade transformer" by Charles Lauritsen at Caltech, and an electrostatic generator by Merle Tuve at the Carnegie Institution of Washington.[14] But the cyclotron was particularly promising for inducing radioactivity, and Lawrence seized on this opportunity.[15] In September 1934, Lawrence and his group generated sodium-24 as a result of bombarding table salt with deuterons (hydrogen ions containing one proton and one neutron).[16] (See figure 2.1.)

Lawrence was immediately attuned to the potential medical significance of the material.[17] In a letter to the Commonwealth Fund, he claimed that radiosodium had properties "superior to those of radium for the treatment of cancer," and for a fraction of the cost.[18] By 1936, the cyclotron could generate 200 millicuries of sodium-24 a day from rock salt worth less than a penny.[19] That same year, Lawrence's group expanded and became an autonomous research entity within the Physics Department.[20]

12. Brucer, *Chronology of Nuclear Medicine* (1990), p. 215; Fermi, "Radioactivity Induced" (1934).

13. Kohler, *Partners in Science* (1991), p. 378.

14. Seaborg, "Artificial Radioactivity" (1940), p. 200.

15. Livingston, "Early History" (1980), p. 32; Henderson, Livingston, and Lawrence, "Artificial Radioactivity" (1934).

16. Heilbron, Seidel, and Wheaton, *Lawrence and His Laboratory* (1981), p. 24.

17. According to an oral history with John Lawrence, when he (John) was still at Yale in 1934 he was already corresponding with Ernest about the possibility of injecting subjects with radiosodium. John H. Lawrence, MD, "Medicine Pioneer and Director of Donner Laboratory, University of California," oral history conducted in 1979 and 1980 by Sally Smith Hughes, Regional Oral History Office/History of Science and Technology Program, the Bancroft Library, University of California, Berkeley, 2000, p. 22.

18. E. O. Lawrence in letter to the Commonwealth Fund, 7 Dec 1934, as quoted by Heilbron and Seidel, *Lawrence and His Laboratory* (1989), p. 189.

19. The salt was in fact donated by the Myles Salt Company of Louisiana. Heilbron and Seidel, *Lawrence and His Laboratory* (1989), p. 189.

20. Heilbron, Seidel, and Wheaton, *Lawrence and His Laboratory* (1981), p. 26.

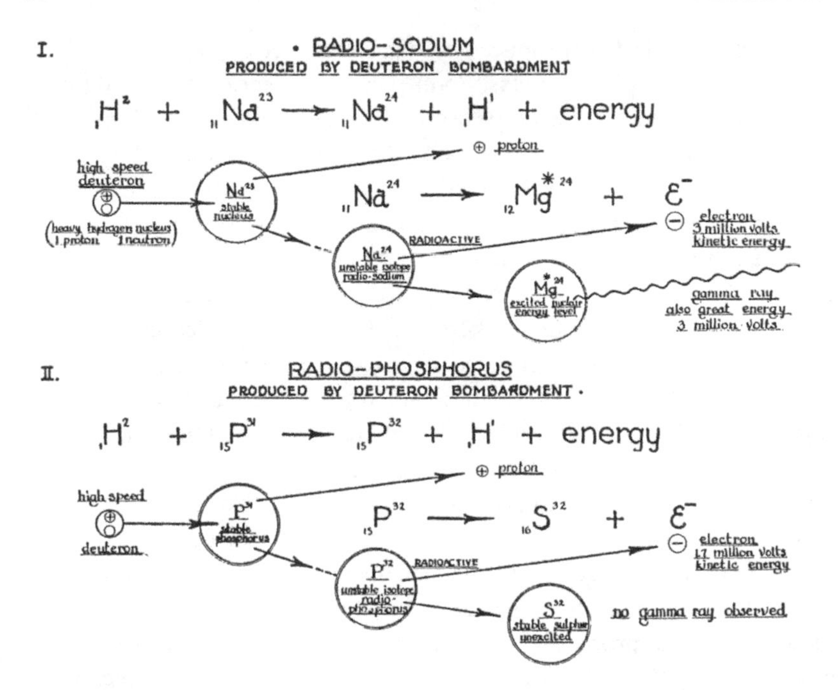

FIGURE 2.1. Schematic diagram depicting the reactions taking place when radiosodium and radiophosphorus are made in the cyclotron through bombardment with high-speed deuterons, as well as their decay reactions. Image and caption from John H. Lawrence, "Artificial Radioactivity and Neutron Rays in Biology and Medicine," in *Handbook of Physical Therapy,* 3rd edition (Chicago, IL: American Medical Association, 1939), pp. 438–55, on p. 443.

The Crocker Radiation Laboratory was broadly oriented to nuclear science, including biological and medical applications in addition to particle physics.[21] Lawrence received support from the Research Corporation, Chemical Foundation, and Macy Foundation that totaled about $50,000 from 1932 to 1936.[22] A 37-inch cyclotron became operational on August 18, 1937.[23]

To explore the clinical uses of cyclotron-produced isotopes, Lawrence recruited medically trained personnel to the Radiation Laboratory. Jo-

21. Ibid.

22. "History of the University of California Radiation Laboratory," Jones papers, box 2, folder UCB-Lawrence Berkeley Lab, History, p. 5.

23. Letter from Herbert Childs to John Lawrence, 24 Jun 1966, JHL papers, carton 10, folder 23.

seph Hamilton joined the group in 1936, having trained in chemistry as a Berkeley undergraduate before taking his medical degree in San Francisco. In addition, Ernest's brother and physician John left Yale Medical School in 1936 to build up the biomedical program in Berkeley. Both John Lawrence and Joseph Hamilton obtained teaching appointments at the University of California's medical school in San Francisco (UCSF), giving them access to clinical research subjects.[24]

John Lawrence's initial efforts at the Rad Lab involved not radioisotopes but a cyclotron-based radiation source, neutron beams. He studied the biological effects of neutron radiation with an eye toward developing new cancer therapies.[25] Seeking a physicist with whom to collaborate, John Lawrence recruited Paul C. Aebersold, a physics graduate student with his brother Ernest.[26] Aebersold obtained a medical-research fellowship with the help of his advisor, and spent two years at UCSF in the mid-1930s studying radiation biology.[27] Aebersold developed a device to focus the neutron beam for therapeutic application, first used on a patient in September 1938. Whereas radioisotopes might be imagined as surrogates for expensive radium, neutrons were more comparable with x-rays, though potentially more efficient in producing biological effects. Thus applications of the neutron beam developed alongside experimental therapies with radioisotopes, one key difference being that radioisotopes were portable.[28]

Joseph Hamilton and radiologist Robert Stone undertook the first clinical administration of artificial radioisotopes in 1936, administering sodium-24 to two patients at the UCSF hospital.[29] One received an

24. Westwick, "Abraded from Several Corners" (1996); ACHRE, *Final Report, Supp. Vol. 1* (1995), p. 603.

25. Lawrence, "Biological Action" (1937). On the history of neutron therapy, see Svensson and Landberg, "Neutron Therapy" (1994); Kutcher, "Fast Neutrons" (2010).

26. Davis, *Lawrence and Oppenheimer* (1968), p. 68; Paul C. Aebersold, "Professional History," appendix to "Application for Federal Employment," 13 Jun 1946, Aebersold papers, box 1, folder 1-1 Biographical Materials.

27. Heilbron and Seidel, *Lawrence and His Laboratory* (1989), pp. 230–31.

28. Lawrence and Lawrence, "Biological Action" (1936); Lawrence, Aebersold, and Lawrence, "Comparative Effects" (1936); Axelrod, Aebersold, and Lawrence, "Comparative Effects" (1941).

29. Heilbron and Seidel, *Lawrence and His Laboratory* (1989), p. 395. According to Jones and Martensen ("Human Radiation Experiments" [2003], p. 87), the date on which Joseph Hamilton first administered radiosodium to a leukemic patient was Christmas Eve 1936. But

intravenous dose of 13 mCi and the other 15 mCi.[30] These represented substantial doses of radioactivity, even for the time, although no exposure limits for so-called internal emitters (ingested or inhaled radioactive sources) were set until 1941, and they did not apply to therapy.[31] In effect, Hamilton and Stone performed an old-fashioned physiology "intake-output" study, following the excretion of the radioactivity in urine, feces, and sweat.[32] Another patient was treated with an order of magnitude more radiosodium.[33] Although these patients did not improve, neither did they appear to suffer ill effects.

In their paper, Hamilton and Stone cited the intravenous administration of radium chloride to patients two decades earlier by Frederick Proescher, who reported improvement in patients with hypertension and arthritis with "no immediate toxic manifestations."[34] Apparently, the subsequent recognition of the long-term dangers of radium did not undermine its general role as a therapeutic precedent for radioisotopes.[35] In any case, radiosodium seemed advantageous over radium on account of both its much briefer half-life (15 hours compared with 1600 years, meaning it would irradiate a patient's body from within for a much shorter period of time) and the fact that it "does not tend to become fixed in the body tissues."[36]

Stone claimed the motivation behind giving these patients radioisotopes was therapy, not research. "I, personally, would not have been a party to giving radiosodium to these patients if it had not been for the purpose of treatment."[37] By contrast, John Lawrence viewed these treatments as a "stunt," because "there was no selective localization of sodium [in

---

Robert Stone agrees with the date given by Paul Aebersold of March 23, 1936. Robert S. Stone to Paul C. Aebersold, 25 Apr 1957, Aebersold papers, box 2, folder 2-11 Gen Corr Apr 1957.

30. Hamilton and Stone, "Excretion of Radio-Sodium" (1937); Aebersold, "Development of Nuclear Medicine" (1956), p. 1029. The publications do not indicate whether patients paid for this experimental therapy.

31. Hacker, *Dragon's Tail* (1987), p. 25. The standard limit for occupational exposures published in 1941 in National Bureau of Standards handbook 27 was 0.1 μCi of radium.

32. Holmes, "Intake-Output Method" (1987).

33. As calculated from information given in Hamilton, "Rates of Absorption" (1937), pp. 523–24. Heilbron and Seidel, *Lawrence and His Laboratory* (1989), p. 395.

34. Hamilton and Stone, "Intravenous and Intraduodenal Administration" (1937), p. 178.

35. Evans, "Radium Poisoning" (1933).

36. Hamilton and Stone, "Excretion of Radio-Sodium" (1937), p. 595.

37. Robert S. Stone to Paul C. Aebersold, 25 Apr 1957, Aebersold papers, box 2, folder 2-11 Gen Corr Apr 1957.

bone] nor possible effect at the dosage which they used." That is, Hamilton "simply wanted to rush into using an isotope on a sick patient."[38] In part, priority was at stake; if not for this application of radiosodium, John Lawrence's administration of radiophosphorus the following year would have counted as the first therapeutic use of an artificial radioisotope.

On the basis of the apparent safety (or, at least, nontoxicity) of radiosodium in these clinical experiments, Hamilton launched a broader investigation of the rate of absorption of sodium in humans, feeding healthy subjects smaller amounts of sodium-24.[39] His initial publication reported results from eight individuals, including two women; most of these received 80 μCi to 200 μCi radiosodium by mouth.[40] The subject would put one hand around a Geiger-Müller counter encased in a lead cylinder, and use the other hand to drink the radioactive salt solution. The appearance of radioactivity in the shielded hand, registered by the counter after ingestion, was used as an "'indicator' of absorption."[41] Hamilton detected the radiosodium two and a half to ten minutes after ingestion; absorption appeared to be complete after as little as three hours in some subjects, although "in others equilibrium was not reached at the end of 10 hours."[42]

In the meantime, E. O. Lawrence touted the physiological significance of sodium-24 in public lectures, including live demonstrations. Drawing on the experimental set-up that Hamilton had devised, Lawrence had a volunteer swallow a solution of the isotope, so that he could track it through the subject's body.[43] (See figure 2.2.) These subjects included members of his own laboratory—J. Robert Oppenheimer, Luis Alvarez, and Joseph Hamilton. As Oppenheimer recounted the experience,

38. John H. Lawrence to Herbert Childs, 13 Jul 1966, JHL papers, carton 10, folder 23. See also Jones and Martensen, "Human Radiation Experiments" (2003).

39. Some sources claim that Hamilton himself was the first experimental subject; others claim these experiments began as early as 1935. Brucer, *Chronology of Nuclear Medicine* (1990), p. 215; "Donner Laboratory—Developing Atomic Medicine," JHL-LBL Technical, box 5, folder 12 History Donner Laboratory; "History of Donner Laboratory and the Division of Medical Physics," JHL-LBL Technical, box 5, folder "Firsts" in Biology and Medicine at Donner Laboratory.

40. One subject received a larger dose of 2 mCi before Hamilton realized that 5%–10% of this amount of radioactivity gave results just as satisfactory. Hamilton, "Rates of Absorption" (1937), p. 524.

41. Hamilton, "Rates of Absorption" (1937), p. 523.

42. Ibid., p. 527.

43. Heilbron, Seidel, and Wheaton, *Lawrence and His Laboratory* (1981), p. 25.

FIGURE 2.2. Picture of Robert Marshak drinking a solution of radiosodium with his hand around a Geiger-Müller counter encased in a lead cylinder. Joseph Hamilton is at right in the background. Credit: University of California, Lawrence Berkeley National Laboratory.

> He got me out on the platform and used me as a guinea pig. . . . He had me put my hand around a Geiger counter and gave me a glass of water in which part of the salt had radioactive sodium in it. For the first half minute all was quiet, but about fifty seconds after I drank, there was a great clattering of the Geiger counter. This was supposed to show that in at least one complex physiochemical system, the salt had diffused from my mouth through my bloodstream to the tip of my fingers and that the time scale for this was fifty seconds.[44]

This demonstration became a stock part of Lawrence's presentations on the national lecture circuit; he referred to it as "vaudeville."[45]

When traveling, Lawrence had his laboratory members ship him fresh samples of radiosodium for each demonstration. Lawrence and his patron, the Research Corporation, tried to file a patent on the production of radiosodium, in anticipation of a lucrative radiopharmaceutical industry.[46] This was unsuccessful, for the patent examiner claimed that the isotope had been produced previously even if it were not identified. Moreover, continuing experiments on the localization of sodium-24 in healthy subjects suggested this radioelement was better suited to analyze the vascular system and investigate water balance than to treat cancer.[47] After all Lawrence's publicity, radioactive sodium did not supersede costly radium. However, other isotopes beckoned.

Radiophosphorus, Radioiodine, and Beyond

Also medically promising was phosphorus-32, which Joliot and Curie had identified in 1934. The following year, Otto Chievitz and George Hevesy first used phosphorus-32, obtained from Niels Bohr, in a biological experiment. They demonstrated that ingested radiophosphorus concentrated in the bones and, to a lesser degree, muscles of rats.[48] By 1936,

44. As quoted in Davis, *Lawrence and Oppenheimer* (1968), p. 68; see also Welsome, *Plutonium Files* (1999), p. 25.

45. Heilbron and Seidel, *Lawrence and His Laboratory* (1989), p. 191.

46. Ibid., p. 192.

47. Ibid., p. 396; Joseph G. Hamilton, Dec 1938, untitled report on research, p. 3. EOL papers, series 1, reel 13, folder 8:25 Hamilton, Joseph G. Reports.

48. Chievitz and Hevesy, "Radioactive Indicators" (1935); Brucer, *Chronology of Nuclear Medicine* (1990), p. 215. Hevesy's continuing work applying radioactive tracers (mostly in biomedical research) proved crucial to his being awarded the Nobel Prize in 1944, twenty years

Donald Cooksey could produce more radiophosphorus in the cyclotron, and of higher-specific activity, than Bohr could with his radium-beryllium source.[49] A target material would be placed for bombardment by deuterons as they exited the cyclotron's dees, or D-shaped vacuum chambers, placed side-by-side to form a circular path for the particles as they were accelerated by the surrounding magnet. A paste of red phosphorus, a substance unpleasant to work with, was pressed into a knurled copper platc that was, in turn, covered with gold foil. The 6 MeV beam dissipated so much heat at the copper target as to require a water-cooling system and dispersed radioactive phosphorus everywhere. The bombardment produced a radioactive stew of metals such as copper and zinc as well as a few millicuries of radiophosphorus, which had to be purified out and chemically prepared for use. This entailed the conversion of elemental phosphorus-32 into phosphoric acid and often subsequently into dibasic sodium phosphate.[50]

Medical researchers in San Francisco and in Berkeley were keen to put this hard-won phosphorus-32 to use as soon as it became available in 1936.[51] In the Division of Physiology in San Francisco, K. G. Scott and S. F. Cook fed phosphorus-32 to chicks of different ages to look for biological effects, particularly in blood cells. Radiophosphorus suppressed the number of polymorphonuclear leucocytes, a type of white blood cell. Scott and Cook attributed this effect to the phosphorus-32 deposited in the bone, where the bone marrow was bombarded with beta rays.[52] Because this is a portion of the body hard to reach with external radiation sources (x-rays), they thought that radiophosphorus might be used to treat blood cell diseases, particularly leukemia. In addition, other features of phosphorus-32 made it an attractive candidate for therapeutic use, namely its brief half-life (14 days), the relatively low penetrating power of its beta rays (2 to

---

after he was first nominated for his contributions to chemistry. See Pallo, "Scientific Recency" (2002).

49. Lawrence and Cooksey, "Apparatus for Multiple Acceleration" (1936).

50. Kamen, *Radiant Science* (1985), pp. 80–81; Heilbron and Seidel, *Lawrence and His Laboratory* (1989), p. 279; Warren, "Therapeutic Use" (1945), p. 702.

51. In Copenhagen, Chievitz and Hevesy conducted similar experiments administering phosphorus-32 to patients and rats, but they had access only to microcuries of the radioisotope as compared with millicuries at Berkeley. Chievitz and Hevesy, "Studies on the Metabolism" (1937); Brucer, *Chronology of Nuclear Medicine* (1990), p. 222.

52. Scott and Cook, "Effect of Radioactive Phosphorus" (1937).

4 millimeters of tissue), and the fact that its reaction product, sulfur, was neither radioactive nor toxic.[53]

John Lawrence put radiophosphorus to work in cancer research, having induced lymphatic leukemia in an inbred mouse strain (Strong A) that was particularly susceptible to the implantation of tumors.[54] He and K. G. Scott found that these leukemic mice concentrated more radiophosphorus in their lymph glands and spleens than did healthy mice after both groups received tracer doses.[55] This finding stoked hopes that radioisotopes would be selectively absorbed and localized in cancer patients, where they could serve to irradiate tumors.

John Lawrence did not wait for the completion of these mice studies to commence clinical trials.[56] He administered phosphorus-32 to a patient with chronic lymphatic leukemia on December 14 and again on December 26, 1937.[57] In 1938, he gave radiophosphorus to a woman suffering from polycythemia vera, with apparent success.[58] That same year, John Lawrence's request for 1,520 mCi of radiophosphorus tied up the 37-inch cyclotron for a day and a night. He divided the material between his cancerous mice and a patient with chronic leukemia of the bone marrow, who received 70 mCi of phosphorus-32 over two months. After treatment, the patient's blood picture improved, which encouraged further clinical experimentation.[59]

Other animal experiments were aimed at establishing the safety of radiophosphorus. Radiation safety in the 1930s focused on "tolerance," the

53. Warren, "Therapeutic Use" (1945), p. 702.

54. Lawrence and Gardner, "Transmissible Leukemia" (1938); Strong, "Establishment of the A Strain" (1936).

55. Lawrence and Scott, "Comparative Metabolism" (1939).

56. Lawrence, Tuttle, Scott, and Connor, "Studies on Neoplasms" (1940), p. 271.

57. There is disagreement about whether Lawrence first gave radiophosphorus to a leukemia patient in December 1936 or December 1937. Retrospective documents in JHL-LBL Technical, box 5, folders 12 and 13, give the Christmas 1937 date, as does Robert S. Stone (see his letter to Paul C. Aebersold, 25 Apr 1957, cited above), who was correcting Aebersold's recently published "Development of Nuclear Medicine" (1956). Herbert Childs concurs with the 1937 date in *American Genius* (1968), p. 280. For the late December 1936 dating, see John Lawrence to Edward B. Silberstein, 13 Oct 1978, JHL-LBL Technical, box 2, folder S.

58. Polycythemia vera is a bone marrow disorder in which blood cells are overproduced, leading to symptoms such as nosebleeds, an enlarged spleen, and risk of blood clots. On the dating, see Lawrence, "Early Experiences" (1979).

59. Heilbron and Seidel, *Lawrence and His Laboratory* (1989), pp. 279 and 399.

dose of radiation living systems could absorb without irreversible damage.[60] Accordingly, John Lawrence and Kenneth Scott injected radiophosphate into the abdominal cavities of four young rhesus monkeys to establish the "maximum dosage" that could be tolerated.[61] All four exhibited a marked decline in the number of white blood cells, and one monkey died of radiation poisoning. This unfortunate monkey received over one millicurie of radiophosphorus per pound of body weight, equivalent to 114 mCi dosage in a 150-pound human. The researchers sacrificed another monkey, which received the smallest dose (0.45 mCi per pound of body weight). The other two monkeys, which received up to 0.76 mCi per pound of body weight (three-quarters of the fatal dose), survived.[62] One year later these two monkeys were reported to be "essentially the same as they had been at the beginning of the experiments."[63] Lawrence and Scott concluded that the tolerance level of radiophosphorus was about ten-fold higher than the usual therapeutic dose.[64]

The induced bone cancers of radium dial painters had already demonstrated that radiation's damaging effects could appear years after exposure. These workers, almost entirely young women, painted luminous dials by hand to meet demand for watches and instruments with self-lit figures, particularly during World War II. Exposure to radium in the paint, ingested as the women licked their brushes, led by the mid-1920s to numerous cases of jaw necrosis and, over time, high cancer mortality.[65] However, John Lawrence believed that these dangers would never be seen with the cyclotron-generated radioisotopes he was using, as they were not alpha particle emitters and would not become permanently deposited in the bone.[66] In addition, the relatively short half-lives of therapeutic isotopes such as radiophosphorus limited long-term exposure. At this stage, possible genetic consequences from exposure did not attract medical consideration.[67]

In 1937 Ernest Lawrence assigned radiochemist Martin D. Kamen the task of processing requests and preparing radioisotopes for users at

60. Hacker, *Dragon's Tail* (1987), p. 2.
61. Scott and Lawrence, "Effect of Radio-Phosphorus" (1941), p. 155.
62. Scott and Lawrence, "Effect of Radio-Phosphorus" (1941).
63. Hamilton, "Radioactive Tracers" (1942), p. 549.
64. Scott and Lawrence, "Effect of Radio-Phosphorus" (1941), p. 158.
65. Hacker, *Dragon's Tail* (1987), pp. 20–23; Clark, *Radium Girls* (1997).
66. Lawrence, "Early Experiences" (1979), p. 562.
67. On the genetic effects of radiation, see chapter 5.

Berkeley and elsewhere. As he recollected, "There was an almost insatiable need for radiophosphorus (^{32}P) to implement experimentation on therapy of cancer and various blood diseases, particularly polycythemia vera."[68] The increase in clinical demand was matched by improvements in production method: in mid-1938 the cyclotroneers developed a more efficient way of producing phosphorus-32 that simultaneously generated iron-59.[69] Clinical administration required more intensive preparation of the materials than had experiments on mice and monkeys. The chemical processing to make phosphorus-32 safe for human consumption was nontrivial—both radioactive contaminants and pyrogens had to be removed, and the salt prepared in neutral solution.[70]

By the summer of 1939, John Lawrence had treated over a dozen patients, each taking 20 or 25 mCi of phosphorus-32 per year.[71] Of six patients with leukemia (one lymphatic, one monocytic, and four myelogenous) treated with radiophosphorus, none recovered and four died. Nonetheless, Lawrence concluded that "since at present there is no completely satisfactory method for the treatment of this disease, it seems justifiable to use this material cautiously in therapeutic attempts, in addition to its use as a tracer of the metabolism of phosphorus in this disease."[72] From his point of view, radiophosphorus was at least as effective as x-ray therapy in treating leukemia and polycythemia vera. Hamilton noted that the dose of phosphorus-32 administered to leukemic patients was not accompanied by the same level of radiation sickness as x-ray radiation.[73] But not everyone agreed with this positive assessment. For Robert Stone at UCSF, who was developing the Sloan X-Ray Tube for radiation treatments, radioisotopes represented competition rather than merely opportunity. In general, radiologists were slow to embrace radioisotopes as an alternative source of radiation therapies.[74]

In tandem with ongoing clinical experiments, local biochemists used radiophosphorus as a tracer to study metabolism. Through labeling with phosphorus-32, Israel L. Chaikoff from Berkeley's Department of

68. Kamen, *Radiant Science* (1985), p. 80.

69. Heilbron and Seidel, *Lawrence and His Laboratory* (1989), p. 280.

70. Kamen, *Radiant Science* (1985), p. 81.

71. Heilbron and Seidel, *Lawrence and His Laboratory* (1989), p. 400.

72. Lawrence, Scott, and Tuttle, "Studies on Leukemia" (1939), p. 57.

73. Hamilton, "Use of Radioactive Tracers" (1942), p. 550.

74. See Erf to Lawrence, 27 Aug 1945, JHL papers, series 3, reel 4, folder 4:10 Correspondence E 1945.

Physiology found unusually high phospholipid turnover in the tumors of John Lawrence's cancerous mice.[75] For that matter, even in normal mice, the uptake of radiophosphorus into the phospholipids of active tissues was rapid.[76] Similarly, Berkeley biochemist David Greenberg and his graduate student Waldo Cohn analyzed the assimilation of phosphorus-32 into the tissues of rats from both absorption through the digestive system and intraperitoneal injection. All organs except the brain showed a rapid uptake of the labeled phosphate.[77] Radioactive phosphorus brought into view the dynamics of the synthesis, transport, and breakdown of key biological molecules.[78]

Plant scientists also investigated radiophosphorus uptake. Daniel Arnon and his coworkers in Berkeley's Division of Truck Crops obtained phosphorus-32 from the Rad Lab in the form of sodium biphosphate, which they added to unlabeled ammonium phosphate in the nutrient solution for tomato plants. The radiophosphorus was rapidly absorbed by the seven-foot tall plants, making its way into the leaves and fruit. (See figure 2.3.) Phosphorus-32 accumulated most in the foliage and fruit in the upper portion of the plants, the region of most active growth. In addition, there was differential absorption among the tomatoes: the smaller the tomato, the more radioactivity it took up.[79] As in the case of mouse tumors, tissues growing most rapidly concentrated more phosphorus-32 than slower-growing tissues. These kinds of tracer studies generally required about a thousand-fold less radioactivity than most therapeutic uses, and these results reinforced the value of the cyclotron for the life sciences.

In the late 1930s, phosphorus-32 was the most widely used radioisotope in therapy and research. In a 1939 review, Hevesy treated the uses of radiophosphorus as exemplary of the "type of problems which can be successfully attacked by the use of tagged elements."[80] That said, other radioelements being produced by physicists at the Rad Lab were similarly utilized in tracer studies.[81] Those that were physiologically significant invariably made their way into research with clinical subjects at Berkeley:

75. Jones, Chaikoff, and Lawrence, "Radioactive Phosphorus" (1939).
76. Perlman, Ruben, and Chaikoff, "Radioactive Phosphorus" (1937).
77. Cohn and Greenberg, "Studies in Mineral Metabolism" (1938).
78. Bennett, "I. L. Chaikoff" (1987), p. 367; chapter 7.
79. Arnon, Stout, and Sipos, "Radioactive Phosphorus" (1940).
80. Hevesy, "Application of Radioactive Indicators" (1940), p. 641.
81. Heilbron, Seidel, and Wheaton, *Lawrence and His Laboratory* (1981), p. 25.

FIGURE 2.3. A contact radiograph of the young leaf of a tomato plant 36 hours after $^{32}PO_4$ was added to the nutrient solution. Notice the concentration of radioactivity (the lighter areas) in the parts of the plant that are growing. That said, the light areas indicated by letters were caused by folds in the leaves (a–d) or the bunching of several small leaflets (e). Image and caption from D. I. Arnon, P. R. Stout, and F. Sipos, "Radioactive Phosphorus as an Indicator of Phosphorus Absorption of Tomato Fruits at Various Stages of Development," *American Journal of Botany* 27 (1940): 791–98, on p. 794.

Hamilton performed human absorption experiments with isotopes of potassium, chlorine, and bromine, as well as sodium and iodine.[82]

The other radioisotope that physicians keenly sought in the late 1930s was radioiodine. Robley D. Evans, a physicist at MIT, used a radium-beryllium neutron source (devised from discarded and donated medical radon sources) to make iodine-128. Evans's collaborators, Saul Hertz and

82. Hamilton, "Rates of Absorption" (1938); and, for animal studies, Hamilton and Alles, "Physiological Action" (1939).

Arthur Roberts of the Thyroid Clinic at Massachusetts General Hospital, conducted the first biological tracer experiments with this isotope. In a 1938 paper, they demonstrated the rapid, selective concentration of this isotope in the thyroid of 48 rabbits that had been injected with iodine-128.[83] Under conditions of thyroid stimulation, even more radioiodine was localized to the thyroid. This suggested that iodine-128 might be used for diagnosis or even treatment of thyroid conditions such as hyperthyroidism and cancer, despite its brief half-life of 25 minutes.[84]

Later that year in Berkeley, John J. Livingood and Glenn T. Seaborg announced the discovery of a new, longer-lived radioisotope of iodine with a half-life of eight days, iodine-131.[85] Shortly thereafter, Joseph Hamilton collaborated with Mayo Solley of the medical school in San Francisco to administer iodine-131 orally to patients. Those with overactive thyroids took up more than ten times the amount of radioiodine that healthy individuals did.[86] This finding laid the groundwork for the subsequent widespread use of iodine-131 as a therapy for hyperthyroidism—which was already being treated with x-ray radiation.[87] One group reported that metastatic carcinoma of the thyroid gland accumulated radioiodine.[88] Unfortunately, only some thyroid cancers proved to selectively concentrate radioiodine.[89] For both radiophosphorus and radioiodine, the most successful clinical applications were for nonmalignant diseases—polycythemia vera for phosphorus-32, and hyperthyroidism for iodine-131.

Not all biologically relevant radioisotopes were sought for medical application. The isolation of an isotope of carbon in 1938, carbon-11,

83. Hertz, Roberts, and Evans, "Radioactive Iodine" (1938).

84. Ibid., p. 513.

85. Livingood and Seaborg, "Radioactive Isotopes" (1938). Iodine-126, with a half-life of thirteen hours, was also discovered in 1938, but the longer-lived iodine-131 took precedence in medicine. Tape and Cork, "Induced Radioactivity" (1938).

86. Hamilton and Soley, "Studies in Iodine Metabolism" (1939); Hamilton and Soley, "Studies in Iodine Metabolism" (1940); Heilbron and Seidel, *Lawrence and His Laboratory* (1989), pp. 396–98.

87. Hamilton and Lawrence, "Recent Clinical Developments" (1942); Hertz and Roberts, "Application of Radioactive Iodine" (1942); Adelstein, "Robley Evans" (2001). On x-ray treatment of thyroid disorders: Means and Holmes, "Further Observations" (1923); Simpson, "X-Ray Treatment" (1924).

88. Keston, Ball, Frantz, and Palmer, "Storage of Radioactive Iodine" (1942); Marinelli, Foote, Hill, and Hocker, "Retention of Radioactive Iodine" (1947).

89. According to Ross, about 15% of thyroid cancers would take up radioiodine; "Radioisotope Division" (1951), p. 39. See also Hamilton, "Use of Radioactive Tracers" (1942), p. 556; Aebersold, "Development of Nuclear Medicine" (1956), p. 1030.

inspired a host of innovative tracer experiments with humans, animals, plants, and microbes. John Lawrence recalled that "several of us inhaled radioactive carbon monoxide and measured the possible conversion of the CO to CO_2 in our bodies with a Geiger counter placed next to a soda-lime canister."[90] Initial results were disappointing, failing to show conversion or oxidation. Martin Kamen collaborated with Berkeley chemist Sam Ruben and biochemist Zev Hassid to trace the assimilation of ^{11}C-labeled CO_2 by plants. Their efforts laid the groundwork for postwar use of radiocarbon to elucidate the photosynthetic pathway, but experiments with carbon-11 were hampered by its short half-life of 21 minutes.[91] In 1940 Kamen and Ruben identified carbon-14, a much more promising radiolabel, but were unable to utilize it in further research before their work was interrupted by World War II.[92]

Networks of Distribution

When Kamen began preparing isotopes for biological and medical use, the Rad Lab was supplying about half a dozen Berkeley biologists with radioisotopes. Despite the precedent of using stable isotopes in metabolic tracing, biochemists at Berkeley did not clamor en masse for radioisotopes; the main early adopters have already been mentioned. As Kamen put it, "E.O.L. had the bad luck to be on a campus that did not house a faculty of biologists by and large ready to seize the opportunities presented by the cyclotron's production of radioisotopes imaginatively."[93] The AEC similarly found after the war that demand for isotopes had to be cultivated, and in both cases visible success nurtured wider emulation.

90. John Lawrence, "Isotopes and Nuclear Radiations in Medical Research, Diagnosis and Therapy," talk prepared for Nuclear Education Symposium, Sacramento, California, 17 Apr 1962, JHL-LBL Technical, box 11, folder Seattle, Washington Am. Nuclear Society, Annual National Meeting.

91. Ruben, Hassid, and Kamen, "Radioactive Carbon" (1939); Ruben, Kamen, and Hassid, "Photosynthesis with Radioactive Carbon, II." (1940); Ruben, Kamen and Perry, "Photosynthesis with Radioactive Carbon, III." (1940); Ruben and Kamen, "Photosynthesis with Radioactive Carbon, IV." (1940).

92. Kamen, "Early History" (1963). On photosynthesis research with radiocarbon, see chapter 7.

93. Kamen, *Radiant Science* (1985), p. 80.

The Rad Lab also supplied physicians beyond the Bay Area with radioisotopes for use in therapy, beginning in the late 1930s and continuing during World War II. John Lawrence supplied only physicians who he knew had some experience with radioisotopes, and he provided radiophosphorus only for use in treating five diseases for which therapy had been shown effective.[94] As one example among many documented in his personal correspondence, in 1943, John Lawrence sent Dr. F. E. Jacobs of San Diego four millicuries of radiophosphorus with directions on dosages and dates.[95] And four days after the atomic blast over Hiroshima, Dr. Frank H. Bethell of the Thomas Henry Simpson Memorial Institute of Ann Arbor, Michigan, wrote John Lawrence thanking him for the sample of radioactive phosphorus and for "continuing to supply us with this material."[96] This last shipment was in keeping with John Lawrence's decision to use most of the limited radiophosphorus produced during wartime (400 millicuries every six weeks) to supply five collaborating centers with the material: Jefferson Medical College of Philadelphia, University of Pennsylvania, Ohio State University College of Medicine, University of Michigan, and the Mayo Clinic of Rochester, Minnesota.[97] By comparison, before the war John Lawrence had sent radioisotopes to clinicians as far away as South America.[98] Those requiring smaller amounts for tracer research from the Rad Lab were more likely to have their requests met by Hamilton, who handled much of the radioisotope production from the sixty-inch cyclotron.[99]

The Rad Lab did not institute a formal protocol for obtaining radioisotopes nor did they charge researchers or physicians, at least not until

94. The diseases were polycythemia vera, chronic myelogenous leukemia, chronic lymphatic leukemia, lymphosarcoma, and some cases of Hodgkin's disease. John H. Lawrence to George W. Corner, 13 Feb 1945, JHL papers, series 3, reel 4, folder 4:9 C Correspondence 1945.

95. John Lawrence to F. E. Jacobs, 3 Apr 1943, JHL papers, series 3, reel 4, folder 4:6 Correspondence 1943.

96. Frank H. Bethell to John H. Lawrence, 10 Aug 1945, JHL papers, series 3, reel 4, folder 4:8 B Correspondence 1945.

97. John H. Lawrence to George W. Corner, 13 Feb 1945, JHL papers, series 3, reel 4, folder 4:9 C Correspondence 1945.

98. John H. Lawrence to The Honorable Edward R. Stettinius, 8 Jun 1945, JHL papers, series 3, reel 4, folder 4:20 S Correspondence 1945.

99. Joseph G. Hamilton to John E. Christian, 10 March 1945, JHL papers series 3, reel 4, folder 4:9 C Correspondence 1945.

the laboratory was funded by the government for war-related work.[100] As E. O. Lawrence explained to fellow cyclotroneer Lee A. DuBridge in 1939,

> At this stage we are not willing to undertake to prepare radioactive materials for any research project which in our opinion is not very much worthwhile, as there are obviously so many that are worthwhile. If we should charge for the materials, we might find it difficult to refuse requests from some quarters. Moreover, charging for the materials has an unfortunate, undesirable psychological effect on the production of the materials in the laboratory. When voluntarily giving the materials, all of us in the laboratory have the feeling that our efforts are in the direction of furthering important work, while if we should sell the materials I am sure that some of the boys would have a little bit the feeling that they were functioning as routine technicians.[101]

Rather than placing a purchase order, individuals who knew E. O. Lawrence, John Lawrence, or Joseph Hamilton approached them personally to ask if they could obtain surplus radioelements. This gave the leadership of the Rad Lab discretion over who had access to the scarce materials. Of 156 enumerated requests E. O. Lawrence received between 1934 and 1946, around thirty were denied.[102]

Most requests were biomedical in orientation. George Whipple and his colleagues at University of Rochester Medical School used more than twenty shipments of iron-59 produced in the Berkeley cyclotron for tracer

100. See Hamilton to Christian, ibid. After accepting a contract with the government for war work, E. O. Lawrence did begin charging users a rate of $25/hour. See E. O. Lawrence to D. M. Yost, 5 Dec 1941, and Martin Kamen to A. Seligman, 18 Dec 1941, both in EOL papers, series 1, reel 14, folder 10:10 Kamen, Martin D. Like the Berkeley Rad Lab, the Harvard cyclotron provided radiomaterials to researchers without charge, although they still had to be chemically processed. Oral history of Joseph F. Ross by Eric Hoffman, 11–12 Jun 1986, Columbia University Oral History Library.

101. Ernest O. Lawrence to Lee A. DuBridge, 14 Jun 1939, EOL papers, series 1, reel 23, folder 15:26A Rochester, University of, 1939.

102. "Special Materials," 7-pg. typescript, undated [1946?], EOL papers, series 3, reel 32, folder 21:22 Special Materials. This list does not seem to include John Lawrence's distribution of phosphorus-32 to physicians at the five medical centers, and it is also missing some shipments from E. O. Lawrence to researchers. The number of denials given is approximate because some of the listed requests are marked with a "?" In addition, some researchers requested several radioisotopes and were sent only one.

experiments with dogs; they found that it was absorbed only when the body's store of iron had been depleted. In recognition of the important contribution of iron-59, which was difficult to provide in the quantities Whipple desired, the resulting publications listed E. O. Lawrence and Kamen as coauthors.[103] In supplying Whipple, who had received the 1934 Nobel Prize in Physiology or Medicine for his work on anemia, E. O. Lawrence could enhance his own reputation.[104]

Others at University of Rochester Medical School, a center for experimental work with new radiation sources, also received Rad Lab radioisotopes. Stafford Warren, future head of health and safety for the Manhattan Project, obtained radiophosphorus to treat patients with leukemia.[105] Warren was interested in using this isotope in laboratory experiments as well as clinical treatment. In 1939, he requested 40 millicuries of phosphorus-32 for experimentation with laboratory tumors, to see if radiophosphorus could affect the tumor growth of a Brown-Pearce rabbit epithelioma. This was a large request for an experiment John Lawrence doubted would work; he encouraged Warren to first try tracer work with a much smaller quantity.[106]

The Rad Lab shipped radioisotopes beyond the continental United States. A professor at University of Hawaii asked Lawrence for phosphorus-32 to study the use of fertilizer in growing pineapple. The radiophosphorus was transported via Pan Am clipper, a seaplane that cut the time for transpacific delivery from six weeks to six days, an important consideration for a short-lived material. On account of its mode of transport, the "University Explorer" radio program featured this shipment on May 25, 1939.[107] Even more strikingly, if less publicly, the Rad Lab became a supplier of radioisotopes to researchers abroad, most notably at Niels Bohr's Institute for Theoretical Physics in Copenhagen. Following Bohr's

103. Hahn, Bale, Lawrence, and Whipple, "Radioactive Iron" (1938); idem, "Radioactive Iron" (1939); Hahn, Bale, Hettig, Kamen, and Whipple, "Radioactive Iron" (1939). A full account of these experiments is offered in chapter 8.

104. Kohler, *Partners in Science* (1991), p. 384.

105. John Lawrence to Stafford Warren, 28 Nov 1939; Lawrence to Warren, 2 Dec 1939, JHL papers, series 3, reel 4, folder 4:4 Correspondence 1938–39.

106. Warren to Lawrence, 26 Dec 1939, JHL papers, series 3, reel 4, folder 4:4 Correspondence 1938–39; Lawrence to Warren, 4 Jan 1940, JHL papers, series 3, reel 4, folder 4:5 Correspondence 1940.

107. "Radio-Active Fertilizer Goes to Hawaii," 25 May 1939, EOL papers, series 14, reel 60, folder 40:15 Radio and Television Talks 1934–56; Heilbron and Seidel, *Lawrence and His Laboratory* (1989), p. 405.

visit to E. O. Lawrence's laboratory in the spring of 1937, George Hevesy wrote Lawrence to ask if the Rad Lab might send him "a strong active phosphorus preparation."[108] Even before his letter was received in Berkeley, apparently, Hevesy had received a sample of radiophosphorus, and he thanked Lawrence for "the magnificent gift."[109] Lawrence's cyclotron-produced radiophosphorus had a specific activity a thousand-fold higher than what Hevesy could obtain from the local radium-beryllium source.[110]

This package was the first of many such shipments to Hevesy through the fall of 1940; these continued for several months into the German occupation of Denmark. Each package from Berkeley was followed by a gracious thank-you note from Hevesy for the gift, sometimes with a reprint or book included.[111] In response to one packet that arrived in December, Hevesy wrote, "That was a fine Christmas present you sent me."[112] After two shipments, Hevesy suggested that Lawrence send the radiophosphorus as solid sodium phosphate, so that the preparations could be sent through ordinary postal airmail.[113] These provisions enabled Hevesy to stay at the cutting edge of radiotracer work; he used the radioisotope to follow phosphorus exchange in hen's eggs and in milk, and gave it to human subjects to analyze the role of phosphorus in the formation of tooth enamel.[114]

The radioactivity was so valuable to Hevesy that when a small amount remained on one letter, he dissolved the paper to extract the residual phosphorus-32. As he told Lawrence, "When historians will once describe your life history I hope they won't omit this incident showing how precious

108. George Hevesy to E. O. Lawrence, 6 May 1937, EOL papers, series 1, reel 13, folder 9:7 Hevesy, George C. de.

109. George Hevesy to E. O. Lawrence, 14 May 1937, EOL papers, series 1, reel 13, folder 9:7 Hevesy, George C. de.

110. George Hevesy to E. O. Lawrence, 21 Jun 1937, EOL papers, series 1, reel 13, folder 9:7 Hevesy, George C. de.

111. Hevesy sent reprints and other reports of his work to Lawrence regularly; on his gift to Lawrence of a book (a new edition of his *Manual of Radioactivity*), see George Hevesy to E. O. Lawrence, 5 Oct 1938, EOL papers, series 1, reel 13, folder 9:7 Hevesy, George C. de.

112. George Hevesy to E. O. Lawrence 14 Jan 1938, EOL papers, series 1, reel 13, folder 9:7 Hevesy, George C. de.

113. George Hevesy to E. O. Lawrence, 8 Nov 1937, EOL papers, series 1, reel 13, folder 9:7 Hevesy, George C. de.

114. Hevesy to Lawrence, 21 Jun 1937; Lawrence to Hevesy, 4 Dec 1937; Hevesy to Lawrence, 1 Nov 1940, all EOL papers, series 1, reel 13, folder 9:7 Hevesy, George C. de. Aten and Hevesy, "Formation of Milk" (1938); Hevesy, Levi, and Rebbe, "Origin of the Phosphorus Compounds" (1938).

your letters were."[115] (Lawrence responded: "It certainly is true that one's letters are usually not so completely devoured."[116]) For his part, Lawrence paid homage to Hevesy's pioneering role in using isotopes:

> When Professor Bohr was here he told us about your work, and since then I have read reprints of your very beautiful experiments. Although we are supplying some of our colleagues here with radioactive samples for similar biological experiments, we will be more than glad to send you radioactive material from time to time, since it is so clear that there is no one in the world who could make better use of it.[117]

Lawrence supplied isotopes only for projects he deemed worthwhile, but this was no anonymous meritocracy. The highly cordial correspondence between Hevesy and Lawrence conveys the language of generosity, gratitude, and deferred obligations characteristic of networks of personal patronage.[118]

As it turned out, Hevesy was competing with Lawrence's own brother for the precious radiophosphorus, and in March 1938 John's administration of phosphorus-32 to leukemia patients took precedence.[119] In principle, there was an end in sight to Hevesy's dependence on Lawrence for radioisotopes. In 1935, the Rockefeller Foundation gave Bohr's Institute in Copenhagen a grant to build a cyclotron.[120] (The Rockefeller Foundation also funded the construction of a cyclotron in Joliot's center in Paris.) Endless technical problems delayed the completion of the cyclotron in Copenhagen, however, and by the time it was running, the outbreak of war

115. Hevesy to Lawrence, 14 Jan 1938, EOL papers, series 1, reel 13, folder 9:7 Hevesy, George C. de.

116. E. O. Lawrence to George Hevesy, 15 Mar 1938, EOL papers, series 1, reel 13, folder 9:7 Hevesy, George C. de.

117. E. O. Lawrence to George Hevesy, 26 May 1937, EOL papers, series 1, reel 13, folder 9:7 Hevesy, George C. de.

118. See correspondence between Lawrence and Hevesy, 1937–1940, EOL papers, series 1, reel 13, folder 9:7 Hevesy, George C. de. As a characteristic example, on 18 Feb 1938, Hevesy wrote Lawrence, "If you should have occasionally some phosphorus to spare I would be most obliged to you."

119. E. O. Lawrence to George Hevesy, 15 Mar 1938, EOL papers, series 1, reel 13, folder 9:7 Hevesy, George C. de.

120. RAC, RF 1.1, series 713, box 4, folders 46 and 47; Pais, *Niels Bohr's Times* (1991), pp. 388–94.

in Europe threatened its operation.[121] Hevesy wrote Lawrence on September 11, 1939, that scarcity of coal, and hence limited electricity, constrained operation of the cyclotron: "This state induces me to ask you once more if, should you have some active phosphorus remaining, you would kindly let us have it occasionally."[122] In fact, Kamen sent the next sample of radiophosphorus to Copenhagen in several small shipments, "inasmuch as present conditions in Europe entail some risk of any given sample's being lost."[123] Only the entry of the United States into the war ended the shipments of radioisotopes from Berkeley to Copenhagen.[124]

A 1962 report on Donner Laboratory described the Rad Lab as having been for years "virtually the sole source of radioisotopes distributed free to scientists all over the world."[125] Not only Hevesy's research, but also that of other Europeans, relied on Lawrence's supply.[126] Recipients included physicists Emilio Segrè in Palermo, John Cockcroft in England, and Lisa Meitner after she arrived in Stockholm.[127] Lawrence's provision of radioisotopes enabled him to participate as a benefactor in the international network of nuclear scientists. He was also a recipient within these patronage networks; early in 1939 Niels Bohr recommended E. O. Lawrence to the Nobelkommittén for that year's prize in physics.[128]

The construction of cyclotrons at many institutions in the United States and Europe meant that Berkeley's near-monopoly on radioisotope

121. Kohler, *Partners in Science* (1991), pp. 375–81, esp. p. 378.

122. George Hevesy to E. O. Lawrence, 11 Sep 1939, EOL papers, series 1, reel 13, folder 9:7 Hevesy, George C. de.

123. Martin Kamen to George Hevesy, 13 Sep 1939, EOL papers, series 1, reel 13, folder 9:7 Hevesy, George C. de. Kamen writes in his autobiography, "I sent the samples to him divided into three batches so that if any of the ships carrying them were torpedoed, the chances of at least one getting through would be greater." *Radiant Science* (1985), p. 80.

124. See Hevesy to Lawrence, 2 Oct 1945, EOL papers, series 1, reel 13, folder 9:7 Hevesy, George C. de.

125. Regents' Meeting, November 16, 1962, Appendix C, Report on Donner Laboratory, JHL-LBL Technical, box 5, folder 12 History Donner Lab, p. 3.

126. Hardin Jones, "Donner Laboratory: Summary of Major Scientific Accomplishments Over the Period 1936–66. Report to the Donner Foundation," Jones papers, box 2, folder UCB–Donner Laboratory, History and Reports of Activities, p. 4; Hevesy, "Application of Radioactive Indicators" (1940), p. 642.

127. These radioisotope shipments date from 1937 to 1939: "Special Materials." See also the acknowledgment in Artom, Sarzana, Perrier, Santangelo, and Segrè, "Rate of 'Organification'" (1937).

128. Heilbron and Seidel, *Lawrence and His Laboratory* (1989), p. 489.

production was short-lived.[129] By 1940, two dozen cyclotrons had been constructed in the United States, and five cyclotrons were running in Europe.[130] Cyclotrons at MIT and Washington University were constructed specifically to produce isotopes for medical use.[131] Even so, the demand for radioisotopes made physicists anxious about retaining adequate machine time for fundamental research. As the cyclotron at the Carnegie Institution of Washington neared completion, Vannevar Bush expressed his concern that the demand for radioisotopes for biological investigation and medical therapy "may become so large as to swamp facilities."[132] John Lawrence responded:

> We consider our cyclotron a research tool and not a machine for the commercial manufacture of radioactive substances and neutrons. I believe that it will be five years and possibly ten years before the radioactive substances will be generally used throughout the country in therapy. By that time organizations such as the American Cyanamid Corporation will undoubtedly have the various radioactive substances available for sale.[133]

As he went on to explain, the Rad Lab's 60-inch cyclotron was operated only five hours a week to produce sufficient radiophosphorus to treat a large number of patients. He estimated that treating all the leukemia patients in California with phosphorus-32 could be done with ten hours a week of cyclotron time, but insisted that he did not want to put radioisotope production "purely on a commercial basis." The Rad Lab's nonmonetary distribution policy allowed its leadership considerable discretion over which requests to answer. But a commercial radioisotope supply did

129. Kohler, *Partners in Science* (1991), p. 372.

130. The Rockefeller Foundation funded many of these. Heilbron and Seidel, *Lawrence and His Laboratory* (1989), p. 310. In Europe, there were cyclotrons in Cambridge, Copenhagen, Liverpool, Paris, and Stockholm. Heilbron, "First European Cyclotrons" (1986).

131. A. L. Hughes, "The Washington University Cyclotron (July 23, 1942)," report attached to letter from Hughes to Hugh Wilson, 15 Nov 1950, with *The Contribution Made by Washington University in the Study and Development of Atomic Energy*, undated pamphlet, courtesy of Michael Welch; Brucer, "Nuclear Medicine Begins" (1978), p. 594.

132. Vannevar Bush to John Lawrence, 4 September 1940, JHL papers, series 3, reel 4, folder 4:5 Correspondence 1940.

133. John Lawrence to Vannevar Bush, 10 September 1940, JHL papers, series 3, reel 4, folder 4:5 Correspondence 1940.

not materialize in the way John Lawrence expected despite the proliferation of cyclotrons; war intervened.

The Militarization of Cyclotrons

Over the next three years, E. O. Lawrence's scientific ambitions became entangled with the American mobilization for war and the Manhattan Project. In September 1939, E. O. Lawrence announced plans to build the largest cyclotron yet—100 million electron volts.[134] He anticipated that this machine would enable his group to discover new radioisotopes, particularly those generated by the fission of uranium and other heavy elements.[135] That same month, the Nazis invaded Poland, precipitating war in Europe. Two months later Lawrence received the Nobel Prize in physics, and that spring he received a commitment of $1,150,000 from the Rockefeller Foundation for his new machine.

By the end of 1940, E. O. Lawrence began receiving requests for special radioactive materials for war-related research in the United States and the United Kingdom.[136] British physicist John D. Cockcroft wrote to R. H. Fowler in Washington, DC, to ask if Lawrence could produce enough element 94 (later called plutonium) to investigate its potential for fission; this was followed by a request for uranium-235. (In 1937 Cockcroft had received radioactive phosphorus and vanadium from Lawrence.) As Fowler's letter to Lawrence put it, "obviously Cockcroft's mouth is watering at the thought of the strength of the source of 94 which you ought to be able to produce for experiment."[137] In conjunction with groups at Columbia and Chicago exploring the military application of atomic fission, the Rad Lab began studying the properties of both uranium and plutonium, and demonstrated that plutonium could be produced from uranium-238 via neutron capture.[138] Plutonium, in turn, would undergo fission just as

134. Heilbron, Seidel, and Wheaton, *Lawrence and His Laboratory* (1981), p. 30.

135. "History of the University of California Radiation Laboratory," Jones papers, box 2, folder UCB-Lawrence Berkeley Lab, History, p. 22.

136. Ibid., pp. 28–30.

137. R. H. Fowler, Central Scientific Office, British Purchasing Commission, to E. O. Lawrence, 28 Jan 1941, as cited in ibid., p. 32.

138. "History of the University of California Radiation Laboratory," p. 32; Jones, *Manhattan* (1985), p. 21.

uranium-235 did. The fission properties of neptunium-237 and uranium-237 were also being explored.[139]

Mobilization enhanced the resources available for the Rad Lab. What had been a rather informal coordination of Lawrence's laboratory with defense work became formalized on June 16, 1941, when the secretary of the National Defense Research Committee, Irwin Stewart, wrote the Regents of the University of California to propose a contract for the production of unnamed elements by the cyclotron for other laboratories doing military research. The Rad Lab's production of plutonium was critical for the early scientific work of the Manhattan Project. Soon government contracts were paying the salaries of Rad Lab personnel.[140]

In turn, the designation of Lawrence's work as important for the military became crucial to the laboratory's expanding infrastructure. In order to obtain the steel needed to build the 100 million electron volt cyclotron in 1941, Lawrence relied on an A-1-a procurement rating from the newly created Office of Scientific Research and Development (OSRD).[141] The massive cyclotron was built on the hill overlooking the University of California campus. The morning after the Japanese attacked Pearl Harbor, Ernest Lawrence announced to the Rad Lab employees that all their work would be henceforth directed to the war effort.[142] The 37-inch cyclotron was converted to serve as a huge mass spectrograph, to be used in separating isotopes of uranium. By March 1942 Lawrence had achieved a five-fold enrichment of uranium-235. (His brainchild was dubbed the calutron after *Cal*ifornia *U*niversity cyclo*tron*.) Up the hill, the magnet for the new 184-inch cyclotron was used in testing the alpha calutron tanks, a model for the huge electromagnetic isotope separation complex built in Oak Ridge beginning in 1943.[143]

Nearly coincident with the mobilization of the Rad Lab was the completion of Donner Laboratory, a new privately funded center for John Lawrence's group in medical physics.[144] This left Joseph Hamilton as direc-

139. "History of the University of California Radiation Laboratory," pp. 37–38.

140. "History of the University of California Radiation Laboratory," pp. 35–36; Kamen, *Radiant Science* (1985), pp. 140–41.

141. Heilbron, Seidel, and Wheaton, *Lawrence and His Laboratory* (1981), p. 32.

142. Kamen, *Radiant Science* (1985), p. 148.

143. Heilbron, Seidel, and Wheaton, *Lawrence and His Laboratory* (1981), pp. 32–34; "History of the University of California Radiation Laboratory," p. 42.

144. Westwick, "Abraded from Several Corners" (1996); Williams, "Donner Laboratory" (1999). Conflict between physicians at UCSF and John Lawrence erupted over whether clini-

tor of Crocker Laboratory and its 60-inch medical cyclotron. Wealthy Californian William Donner, who had lost a son to cancer, donated $165,000 to Lawrence's venture in the spring of 1941. After the United States entered the war, Donner Laboratory was the last building completed on the Berkeley campus. There John Lawrence's group engaged in war-related research on high-altitude physiology and decompression sickness.[145] The radioisotope work also continued, as physicians in the Bay Area referred patients to Donner for treatment.[146]

These reconfigurations at Berkeley were part of the broader mobilization of nuclear physics for the bomb project. The Manhattan Engineer District (MED) was established on August 13, 1942, in New York City. That same month Robert Stone left San Francisco to become head of the Health Division at the University of Chicago's Metallurgical Lab. On September 17, the US Army appointed (and then promoted to general) Colonel Leslie R. Groves to head this top-secret organization. In February 1943, the MED contracted with the University of California to administer Los Alamos Laboratory (Contract 36). A few months later, Contract 48 between the University of California and the Army enlisted the Rad Lab as one of the central MED facilities.[147]

Not only the physicists and chemists at the Rad Lab were mobilized. Joseph Hamilton had already begun investigating the metabolism and biological effects of plutonium and other fission products in experimental animals for the military through a contract with the OSRD. A component of the new Army contract with the Rad Lab, 48A, assimilated this line of investigation. Hamilton's project was aimed at establishing the occupational dangers for Manhattan Project employees of working with plutonium and the dozens of isotopes produced as uranium fission products.[148] Only one primary fission product—radioiodine—had yet been relatively well studied. Moreover, the order of magnitude of radioactivity to which workers would be exposed was over a million times higher than that from

---

cal experiments should be allowed at Donner Laboratory. Jones and Martensen, "Human Radiation Experiments" (2003).

145. Hardin Jones, "Donner Laboratory: Summary of Major Scientific Accomplishments Over the Period 1936–66. Report to the Donner Foundation," undated transcript, Jones papers, box 2, folder UCB–Donner Laboratory–History and Reports of Activities, p. 5.

146. ACHRE, *Final Report, Supp. Vol. I* (1995), p. 603.

147. Ibid., p. 604.

148. See Stone, *Industrial Medicine* (1951).

radium use worldwide.[149] Many of the over 200 products of uranium fission were radioactive isotopes of rare earths, and knowledge of how these elements were metabolized (even when nonradioactive) was virtually nonexistent.[150]

Hamilton's wartime study relied on use of the 37-inch and 60-inch cyclotrons to make specific fission products through bombardment of a uranium target. All of his experiments were considered to be tracer studies due to the small amounts utilized. There was some overlap with his earlier research. In particular, Hamilton had already become interested in the possible use of radioactive strontium (also a fission product) for the clinical treatment of bone diseases.[151] But the war work on fission products was larger in scale and more systematic, involving the testing of each of eighteen fission products on twelve rats, exposed in groups of three at various time-points before sacrifice and analysis. By 1943, the rates of accumulation in various organs of fourteen of the radioisotopes had been determined, as well as their rates of elimination. Beyond assessing occupational hazards for Manhattan Project workers, Hamilton was interested in the use of fission products in radiological warfare.[152]

As the Manhattan Project progressed, Oppenheimer was increasingly concerned about the safety of plutonium. In February 1944 Hamilton was provided eleven precious milligrams of plutonium for biological studies. (Only milligram quantities of plutonium existed by the end of 1943, with only gram quantities being made a few months later.) Hamilton's group rapidly ascertained that plutonium, like radium, was a bone seeker and could be expected to cause cancer. Although its risk of absorption from ingestion was less than that of radium, when inhaled it persisted longer in the lungs.[153] Given plutonium's danger and the difficulty in extrapolating from rat to man, the MED leadership decided to embark on research with human subjects.[154] In the spring of 1945, deliberate injection of plutonium into a human subject occurred at the University Hospital in San

149. Stannard, *Radioactivity and Health* (1988), vol. 1, p. 299.

150. Hacker, *Dragon's Tail* (1987), p. 43.

151. Hamilton, "Use of Radioactive Tracers" (1942), p. 566.

152. Joseph G. Hamilton, "A Report of the Past, Present, and Future Research Activities for Project 48-A-1" [c. 1948], DC-LBL files, box 4, folder 49 Medical Physics J. H. Lawrence's Group, General; Stannard, *Radioactivity and Health* (1988), vol. 1, p. 305. On his interest in radiological warfare, see chapter 3.

153. Hacker, *Dragon's Tail* (1987), pp. 53 and 63.

154. ACHRE, *Final Report, Supp. Vol. 1* (1995), p. 605.

Francisco, as part of Hamilton's Contract 48A. Albert Stevens, thought to be suffering from stomach cancer, was given 0.932 micrograms of a mixture of plutonium-238 and plutonium-239.[155] He was later found to have a gastric ulcer rather than cancer. Stevens was designated CAL-1; he was among 18 patients (including two more in San Francisco) who received injections of plutonium between April 1945 and July 1947.[156] Many were never informed of their exposure to plutonium or their status as research subjects.[157]

In the 1990s, reporter Eileen Welsome disclosed the identities and life circumstances of these eighteen patients, making the plutonium injection experiments a new focus of public outrage.[158] How did leading researchers in the medical application of radioisotopes end up conducting these experiments for the military? In part, it was because these experiments built on nonmilitary research that preceded them—the plutonium injection experiments, like those of sodium-24 in 1937, were tracer studies of the metabolism of a radioelement. Clearly, these experiments differed in crucial ways—the selection of terminally ill patients as subjects signaled their potential danger. Yet the earlier pattern of human experimentation at Berkeley facilitated the subordination of research there to the emerging occupational health and safety requirements of the military. Even after the war, Hamilton was convinced that "under appropriate and suitable circumstances, it is highly desirable to conduct human tracer studies with certain of the fission products and fissionable elements."[159] In fact, Contract 48A with its human experiments continued uninterrupted in peacetime.[160]

155. Ibid.

156. A table of these subjects appears in Stannard, *Radioactivity and Health* (1988), vol. 1, p. 352.

157. ACHRE, *Final Report* (1996), pp. 156–57.

158. Welsome, *Plutonium Files* (1999). Her original articles were published beginning in Nov 1993 in the *Albuquerque Tribune*. This was a catalyst for the formation of the Advisory Committee on Human Radiation Experiments. See ACHRE, *Final Report* (1996), pp. xxi–xxiii.

159. Hamilton, "A Report of the Past, Present, and Future Research Activities for Project 48-A-1" [c. 1948], p. 9.

160. Joseph G. Hamilton to Colonel E. B. Kelly, Subject: Summary of Research Program for Contract #W-7405-eng-48-A, JHL papers, series 3, reel 5, folder 5:30 Correspondence H 1946. See also Jones and Martensen, "Human Radiation Experiments" (2003), pp. 93–96.

As the Advisory Committee on Human Radiation Experiments noted in 1995, the militarization of biomedical research in the Rad Lab also had significant institutional repercussions. The new regime of secrecy meant that university administrators, including President Robert Sproul and Secretary-Treasurer of the Regents Robert Underhill, were often not informed about what was happening in the laboratories they were managing. For instance, these university officials were unaware of the purpose of the work at Los Alamos until it was well underway. The Army's decisions about which experiments to classify were driven by concerns about public relations and liability as well as by the strictures of national security. The MED, and later the AEC, insisted on classifying some human radiation experiments even when the findings were not militarily sensitive. The "unclear lines of authority" created by the military funding of a large scientific enterprise at a public university also meant that researchers could avoid oversight (for instance, of human subjects research) or withhold findings.[161]

Another effect of the broader mobilization during the war was that radioisotope distribution, increasingly concentrated on the MIT cyclotron, began to involve monetary transactions. The MIT cyclotron had been built between 1938 and 1940 with a $30,000 grant from the John and Mary R. Markle Foundation to physicist Robley D. Evans, a leader in the establishment of occupational standards for radium exposure. The grant stipulated that the cyclotron was "to be used exclusively for cooperative medical research and therapy."[162] Unlike the Berkeley cyclotrons, the MIT cyclotron was not directly involved in the Manhattan Project. Evans endeavored to keep his team of physicists and radiochemists together as the scientific mobilization proceeded by offering their services to aid defense-related projects. He corresponded with leadership of the Committee on Medical Research about this prospect:

> I also emphasized the concept that the service which we offer is that of a complete applied radioactivity group involving all of the radiochemical features, detection work, cyclotron services, and a personnel particularly adapted by

161. ACHRE, *Final Report, Suppl. Vol. 1* (1995), p. 602 and p. 624n1. For more on the history of radioisotope research on human subjects, see chapter 8.

162. Letter from Robley D. Evans to Archie S. Woods, Vice President, The John and Mary Markle Foundation, 20 May 1938, MIT President's papers, box 81, folder 16 Robley Evans 1938.

> experience and inclination to cooperative work on medical problems. It must always be borne in mind that this is a very much bigger thing than the use of a cyclotron alone, which is the only well-organized service which would be available from other institutions.[163]

Evans calculated a charge of $25 per hour for target irradiation and radiochemical purification.[164] Most of his initial clients were doing war-related research, so charges were made to their government contracts. But the MIT Radioactivity Center began supplying civilian scientists and physicians as well, and this continued after the war.[165] By 1945, the sale of radioactive materials had brought in $88,359.45.[166] As Evans put it, "radioactive tracer work has represented our main source of operating funds."[167]

The supply of radioisotopes was only incompletely commercialized through the circumstances of wartime mobilization. The Berkeley cyclotron continued to distribute radioisotopes gratis for medical use through John Lawrence's contacts, and, more importantly, within the Manhattan Project, particularly the isotope studies in Chicago under Seaborg (who had moved there in the spring of 1942). The Rad Lab continued to receive other requests for radioisotopes. On July 21, 1944, W. L. Davidson, in the Physical Research department at B. F. Goodrich Company, wrote the Rad

163. Robley D. Evans, memorandum, 19 Nov 1941, MIT President's papers, box 81, folder 18 Robley Evans 1940-1941.

164. The Berkeley Rad Lab instituted a similar rate; as Kamen explained to A. Seligman, who had previously received sulfur-35 gratis: "We are being forced to charge for these samples now as we have a contract with the defense council." Martin D. Kamen to A. Seligman, 18 Dec 1941, EOL papers, series 1, reel 14, folder 10:10 Kamen, Martin D. See also E. O. Lawrence to D. M. Yost, 5 Dec 1941, same folder.

165. By March 1945, the MIT cyclotron began providing radiophosphorus to clinicians throughout New England and New York. See Lowell A. Erf to John H. Lawrence, 26 Jun 1945, JHL papers, series 3, reel 4, folder 4:10 Correspondence E 1945; B. E. Hall to John H. Lawrence, 10 Mar 1945, JHL papers, series 3, reel 4, folder 4:14 Correspondence H 1945. MIT became another key site for clinical experimentation with radioisotopes: Using radiophosphorus from the MIT cyclotron, Shields Warren undertook experimental treatment of leukemia along the same lines John Lawrence had pioneered in Berkeley. See John Lawrence to Archie Woods, 28 Dec 1940, JHL papers, series 3, reel 4, folder 4:5 Correspondence 1940.

166. Robley D. Evans, "Radioactivity Center, 1934-1945," unpublished history, 28 Jun 1945, Evans papers, box 1, folder Radioactivity Center 1934–1945, p. 32. Appendix IX, dated 24 Mar 1945, gives a list of wartime government projects that relied on the Radioactivity Center and a separate list of individuals who were receiving radioisotopes for research or therapy.

167. Ibid., p. 23.

Lab's Donald Cooksey, asking if he might be able to provide one to five millicuries of phosphorus-32 and 0.01 millicurie of sulfur-35. Cooksey declined; "at the present time we are not able to furnish any radio-active materials for outside usage."[168] He told Davidson to try asking scientists at three other institutions with cyclotrons, Washington University Medical School, MIT, and the Bio-Chemical Research Foundation in Delaware. It appears from archival records that the Rad Lab did not send any radioisotopes to requesters in 1944.[169]

An even more dramatic illustration of the new restrictions at Berkeley involved Kamen, the radiochemist who prepared radioisotopes for outside users. Kamen was a devoted amateur violist and played chamber music with various groups in the Bay Area. Through his musical activities he became friends with the violinist and conductor Isaac Stern. In 1941, Kamen attended a cocktail party at the home of Stern, in honor of the latter's recent return from a USO tour. There Stern introduced Kamen to the Soviet consul and vice-consul, and mentioned that Kamen worked at the Rad Lab. The vice-consul, Gregory Kheifetz, then asked if Kamen knew Dr. John Lawrence. Kamen said that he saw him nearly every day. Kheifetz had been trying unsuccessfully to reach John Lawrence about whether he could treat an official with leukemia at the Soviet consulate in Seattle with phosphorus-32. Kamen offered to inquire with Lawrence, who said he would need a complete case history to consider treatment; Lawrence followed up with Kheifetz about the matter from that point.[170] Kheifetz contacted Kamen to express his gratitude and offered to take

168. W. L. Davidson to Donald Cooksey, 21 Jul 1944, and Donald Cooksey to W. L. Davidson, 28 Jul 1944, EOL papers, series 3, reel 31, folder 21:7 Administration, Job Orders.

169. "Special Materials." There were seven requests listed for 1942, and eight for 1943, of which many were denied or the response is not known. No requests are even listed for 1944, though the exchange cited between Davidson and Cooksey indicates that they did not entirely cease.

170. The individual was a Commander Kalinin of the Russian Navy who was being treated in the United States Navy Hospital in Seattle, Washington. As Lawrence wrote, "In the case of Commander Kalinin the prognosis would seem to be very serious and I doubt whether any form of therapy would be very effective. If you wish to try radioactive phosphorus in combination with x-ray, we can send it to you. When the next sample becomes available and that probably will not be for another week, we shall send you approximately four doses with directions for administration. . . . I assure you that we would consider it a great privilege to be of assistance to you in the treatment of this serious problem and regret that we are so helpless." John H. Lawrence to Captain J. P. Brady, 5 May 1944, JHL-LBL Administrative, box 1, folder 11 Correspondence A-B 1944.

him out to dinner, along with another consular official. The three of them were photographed together by Manhattan Project intelligence agents.[171] This incident occurred after the Rad Lab was mobilized for the war, and unbeknownst to Kamen, the Army's G-2 security agents had developed an elaborate file on him, tracking his associations with leftist musicians and others. In July 1944, Kamen was called into Donald Cooksey's office, where he was informed that the Army had ordered his dismissal from the Rad Lab as a security risk.[172] Kamen's meeting with the Soviet consular officials was a lynchpin of the case against him, and even after he left the Rad Lab the Federal Bureau of Investigation (FBI) tracked his every move. In this case, the personal cost of brokering the laboratory's radioisotope supply was especially high.[173]

Conclusions

The production of artificial radioisotopes in the Berkeley cyclotrons in the late 1930s enabled many of the early biological tracer and therapy experiments. Initially, production and consumption occurred locally, with the participation of physical scientists in preparing materials sometimes reflected in coauthorship. As the supply network grew, in response to requests to E. O. and John Lawrence from scientists and physicians, isotopes continued to be granted on a case-by-case basis. A more comprehensive account of cyclotron-based radioisotope production and distribution would include other sites, such as MIT, Washington University, the Carnegie Institution of Washington, and the University of Copenhagen.[174] However, the sum total of these pioneering research programs, even with generous Rockefeller Foundation backing, changed the practices of few

171. The photograph was later released by Senator Bourke Hickenlooper in the *Chicago Herald Tribune* on July 8, 1951 (in which the caption purported that Kamen was giving Soviet agents secret papers), in conjunction with allegations about American scientists divulging nuclear secrets to Soviet spies. See clippings in EOL papers, series 1, reel 4, folder 10:10 Kamen, Martin D. For more on Kamen's troubles with such accusations see Kamen, *Radiant Science* (1985), esp. ch. 12 and 13.

172. The communication from Colonel K. D. Nichols that Kamen was to be discharged is in the Lawrence papers: Memorandum for E. O. Lawrence from H. A. Fidler, 11 July 1944, EOL papers, series 1, reel 4, folder 10:10 Kamen, Martin D.

173. Kamen, *Radiant Science* (1985), pp. 164 and 167.

174. Brucer, "Radioisotopes in Medicine" (1966), p. 60.

life scientists. Cyclotrons produced small amounts of isotopes at high cost, and access to these scarce resources was limited. Even so, the Rad Lab's circulation of materials extended to Europe, such as the regular shipments of phosphorus-32 to Copenhagen by mail.

Human experiments were part of the patterns of radioisotope use from the outset. The same language of "tracers" was used to describe biochemical experiments using radioisotopes to follow metabolic processes in animals or plants and those in which small amounts of radioelements were administered to human subjects to track their absorption and localization. The results of these largely nontherapeutic human experiments in turn fed into ongoing clinical experiments, which generally used much larger doses of radioactivity in order to irradiate pathological (usually tumorous) tissue.

How did concerns about safety inform these applications? Rad Lab scientists liked to recall the mouse experiment that John Lawrence and Paul Aebersold performed in 1935.[175] Their aim was to expose the animals to a short burst of neutrons to see if their estimates of a tolerable dose were correct. The machine shop built a special four-inch cylinder to hold the mouse, with rubber tubing to enable an air supply from an outside pump. The rodent-holding cylinder was then placed inside the cyclotron, and the beam turned on for three minutes.[176] After the beam was turned off, the crew crowded around the chamber, only to find that the mouse was dead! Inferring that the animal had died from radiation exposure, Ernest Lawrence immediately ordered that the cyclotron be shielded with a wall of water. Only after an autopsy did the scientists realize that the animal had perished of suffocation, because Aebersold had forgotten to turn on an air pump.[177]

This anecdote is recounted frequently in the literature, with two valences. On the one hand, it serves to illustrate how cavalier the cyclotroneers were about radiation exposure before they began collaborating with physicians. The physicists were enchanted by the cyclotron beam;

175. In his own accounts, John Lawrence refers to the subject sometimes as a rat, and sometimes as a mouse. Subsequent accounts are also inconsistent on this point. Given the size of the set-up, I am calling the animal a mouse.

176. According to Brucer (*Chronology of Nuclear Medicine* [1990], p. 219), the exposure time was set to be a fraction of what Paul Aebersold's team had calculated as the neutron equivalent of an LD_{50} roentgen exposure.

177. Brucer, *Chronology of Nuclear Medicine* (1990), pp. 219–20.

according to John Lawrence, they would even "go in and look at this beautiful purple deuteron beam to get a glance at it."[178] The surprise of finding the mouse dead after such a short exposure "intensified concern over radiation hazards" and spurred the establishment of stricter laboratory safety practices.[179] At the same time, the incident is used to poke fun at the irrational fear of radiation—after all, radiation did not kill the rodent. Apparently, John Lawrence did not publicize that piece of information until twenty years after the event.[180] If so, this disclosure occurred in the midst of the fallout debates, when many scientists who used radioactivity felt that public concern had become hysterical.[181]

It was the militarization of atomic energy in the early 1940s rather than safety concerns that brought the US government into the picture. Under the veil of war-related work, not only did developments become secret, but also the patterns of human experimentation that had already developed around the cyclotron in the 1930s took on a new and more ominous dimension. Rather than using radioisotopes to treat diseases, most notably cancer, physicians and collaborating physicists extended nontherapeutic human experiments from tracer experiments with elements such as radiosodium and radiopotassium to include fission products that were often chemically toxic elements. These studies were aimed at assessing the occupational dangers that workers experienced in the Manhattan Project facilities as well as exploring the uses of fission products in radiological warfare. At the same time, the military development of atomic energy, with its complex implications for biomedicine, resulted in the much greater availability of radioisotopes. The US government replaced E. O. Lawrence as the purveyor of isotopes.

178. "The History of the Donner Laboratory," interview of John Lawrence by Daniel Wilkes, 1 Jul 1957, p. 5, OpenNet Acc NV0714926.

179. Hardin Jones, "Donner Laboratory: Summary of Major Scientific Accomplishments Over the Period 1936–66. Report to the Donner Foundation," undated typescript, Jones papers, box 2, folder UCB–Donner Laboratory, History and Reports of Activities, p. 3.

180. Stannard, *Radioactivity and Health* (1988), vol. 1, p. 291; he refers to Lawrence, "Early Experiences" (1979), which was written (but not published) in 1956. See also J. H. Lawrence, "Isotopes and Nuclear Radiations in Medicine or a Quarter Century of Nuclear Medicine," typescript of talk presented in Mexico City in 1961, p. 4, OpenNet Acc NV0722058.

181. See chapter 5.

CHAPTER THREE

Reactors

Production of tracer and therapeutic radioisotopes has been heralded as one of the great peacetime contributions of the uranium chain-reacting pile. This use of the pile will unquestionably be rich in scientific, medical, and technological applications.—Headquarters, Manhattan Project, 14 June 1946[1]

In 1943, the US Army built a top-secret nuclear reactor (the second in the world, with Chicago being first) in the hills of Tennessee. The graphite "piles" at Clinton Engineer Works in Oak Ridge were constructed for two purposes: to serve as a pilot plant for the Hanford plutonium-producing reactors under design and to produce small amounts of plutonium.[2] The site of the reactor with its adjacent processing plant (also a pilot for Hanford) was dubbed X-10, the early code name for plutonium. By the beginning of 1945, with the Hanford reactors and separation plants in full operation, the X-10 pilot plant had fulfilled its initial purpose.[3] The fate of this installation, and the continuing employment of its personnel, remained uncertain for months after the end of World War II.

Scientists involved in the Manhattan Project put forward and subsequently championed the proposal that the federal government use its new nuclear facilities to produce and distribute radioisotopes for civilian use after the war. This plan was part of a broader political agenda to "free the atom" from military domination for peacetime development. In this sense, the enterprise reflected a tension between the military and some scien-

1. "Availability of Radioactive Isotopes" (1946), p. 697.

2. US Army Corps of Engineers, *Manhattan Project* (1976), reel 6, book IV, vol. 2, part II, pp. S4–S8; Hewlett and Anderson, *New World* (1962), p. 364; Smyth, *Atomic Energy for Military Purposes* (1945), pp. 111–12.

3. Quist, "Classified Activities" (2000), p. 13.

tists that existed throughout the Manhattan Project. In 1944, researchers at the Chicago Metallurgical Laboratory, who were already proposing that atomic energy should be demilitarized after the war ended, touted radioisotopes as a peacetime benefit of the new technology. For their part, researchers at Clinton Laboratories in Oak Ridge argued that the X-10 reactor and chemical processing facility would provide the perfect base of operations for a government radioisotope distribution program. Such a program would serve the dual aims of benefiting civilian science and justifying the establishment of a postwar national laboratory in Tennessee.

The inauguration of the US radioisotope program on August 2, 1946, represented a departure from the government's single-minded pursuit of an atomic bomb to the development of peacetime uses of atomic energy. However, the program also bore important continuities with the wartime effort, rhetoric about civilian benefits notwithstanding. First, the program was developed, launched, and administered by the military Manhattan Engineer District before the civilian Atomic Energy Commission came into existence. Like other aspects of the Manhattan Project, its early establishment reflected the decisions and priorities of General Leslie Groves as well as those of nuclear scientists lobbying for the nonmilitary development of atomic energy.

Second, the selection of the X-10 graphite piles at Oak Ridge to generate radioisotopes continued the use of this reactor late in the war as a special-purpose radioisotope production facility. During the war, X-10 supplied not only plutonium but also other special radioactive materials to Manhattan Project scientists, particularly for weapons design and testing. Occasionally isotopes were for civilian use: Phosphorus-32 prepared at X-10 was shipped to the Berkeley Rad Lab, where it was distributed to clinicians in place of cyclotron-generated material. After the war, the government began to ship isotopes produced at Oak Ridge to outside users as well as to Manhattan District laboratories. While nearly all of the radioisotopes used by "off-Project" scientists were for unclassified work, many (but not all) "on-Project" applications of radioisotopes were for classified research and weapons development.

Third, and relatedly, the continuing design and production of nuclear weapons after the war meant that, at a practical and material level, weapons-related activities existed alongside the isotope program at Oak Ridge, and there were areas of overlap. The capacious X-10 reactor was being used simultaneously for classified research and radioisotope production. In particular, experiments on the feasibility of radiological warfare utilized materials prepared in the Oak Ridge reactor. In addition, the government offered stable isotopes as well as radioisotopes for sale, in part because producing them was useful for weapons-related research.

This chapter focuses on the events at and around site X-10 in Oak Ridge that led to the launching of national radioisotope distribution in the summer of 1946. It fell to the leadership of the Manhattan Project, which had operated its nuclear weapons facilities in complete secrecy, to recognize and communicate the civilian benefits of atomic energy. Oak Ridge provides an underappreciated vantage point for perceiving how the transition from the wartime bomb project to postwar atomic energy development played out—at the level of infrastructure as well as institutions. The chapter follows the isotope distribution program through the transition to the Atomic Energy Commission (AEC) and into the late 1940s. On the ground at Oak Ridge, the transition from Manhattan Project to AEC management was not dramatic. However, over time the civilian leadership expanded and further subsidized the program in ways that Groves, who felt the program should be financially self-sufficient, had resisted. The expansion of the isotope program by the AEC intersected with Congressional and Cold War politics in potent ways; this is taken up in the next two chapters, as the geographical focus shifts from Oak Ridge, Tennessee, to Washington, DC.

The Chain-Reacting Pile in Oak Ridge

The first large-scale nuclear reactors were built as a means for mass-producing one particular artificial radioisotope, plutonium, for the Manhattan Project.[4] The development of this technology under the auspices of the military had consequences for why and how scientists used reactors to generate other artificial radioisotopes. The feasibility of controlled nuclear fission was demonstrated on December 2, 1942, by Enrico Fermi and his collaborators at the University of Chicago in an experimental graphite pile (later termed reactor). The Manhattan Engineer District (MED) embarked on developing two different kinds of atomic bombs, one

4. There are many historical accounts of the Manhattan Project, e.g., Smyth, *Atomic Energy for Military Purposes* (1945); Hewlett and Anderson, *New World* (1962); Rhodes, *Making of the Atomic Bomb* (1986); Hoddeson et al., *Critical Assembly* (1993); Norris, *Racing for the Bomb* (2002); Rotter, *Hiroshima* (2008).

FIGURE 3.1. Aerial view of the X-10 graphite reactor building (tall black structure) with the chemical processing plant behind it. March 10, 1944, Clinton Engineer Works, Tennessee. National Archives, RG 434-OR, box 22, notebook 65, photograph no. 206-28.

using uranium-235 and the other, newly discovered plutonium-239.[5] Separation of uranium isotopes was critical for both weapons, because plutonium-239 was produced from neutron bombardment of uranium-238.

In the early months of 1943, Du Pont, working under an MED contract, began constructing a plutonium semiworks, a manufacturing plant at which large-scale industrial methods could be tested and developed. Clinton Engineer Works, named after the nearby town of Clinton, Tennessee, sat on 112 acres between two ridges that ran along the small creek in Bethel Valley. The semiworks, dubbed X-10, consisted of a graphite pile reactor to generate plutonium from uranium fission, and a chemical separation plant to extract and purify it.[6] (See figure 3.1.) The X-10 reactor was vastly larger than Fermi's experimental apparatus. In terms of the generation of power, the scale-up was from a level of 0.2 kilowatts in

5. One of each type of bomb was used against Japan: the uranium-235 bomb, "Little Boy," was dropped over Hiroshima, and the plutonium-239 bomb, "Fat Man," was exploded over Nagasaki.

6. Jones, *Manhattan* (1985), p. 204.

Chicago to 1,000 kilowatts in Clinton.[7] When completed, the "piles" consisted of seventy-three layers of graphite that formed a cube 24 feet per side, surrounded by seven-foot-thick concrete walls for radiation shielding. The reactor contained 1,248 channels that allowed for cooling and held the 60,000 slugs of uranium, canned by the Aluminum Company of America.[8] It went critical at 5 a.m. on November 4, 1943.[9] (See figure 3.2.)

X-10 was the first of four major production facilities constructed at Oak Ridge; the others, located at other sites in the broad valleys of the Clinch River, were built to separate uranium isotopes by three different methods: Y-12 (the electromagnetic plant), K-25 (the gaseous diffusion plant), and, after 1944, S-50 (the liquid thermal diffusion plant, built alongside K-25). These plants in Tennessee were among numerous installations throughout the country that composed the MED. In addition to laboratories in Chicago (the Met Lab) and Berkeley (the Rad Lab), there were three major MED reservations: Site X in Tennessee, with the plutonium semiworks and the three uranium separation plants; Site W in Hanford, Washington, on the Columbia River, built for large-scale production of plutonium; and Site Y, the laboratory at Los Alamos where design and construction of the bombs took place under the scientific direction of J. Robert Oppenheimer.[10]

Oak Ridge also became the administrative center of the Manhattan Project in August 1943, when the District headquarters moved there from New York City.[11] Both the plants and the adjacent town were designed and built rapidly; even at the blueprint stage, the anticipated population grew from 5,000 to 13,000.[12] After the eviction of 1,000 families by the Army, the influx of workers, scientists, military personnel, and families overwhelmed the growing city. Skidmore, Owings & Merrill, the firm contracted to build the town and its infrastructure, tried to expand development to accommodate the total population forecast of 42,000. In reality, telephone service, food, and coal could not keep pace with the population, and the housing and commuting conditions remained dismal for the thousands of workers at the plants, especially African Americans placed in

7. Smyth, *Atomic Energy for Military Purposes* (1945), pp. 108–9.

8. Oak Ridge National Laboratory, *Swords to Plowshares* (1993), p. 3.

9. Johnson and Schaffer, *Oak Ridge National Laboratory* (1994), p. 22.

10. Johnson and Jackson, *City Behind a Fence* (1981), pp. xix–xx.

11. Jones, *Manhattan* (1985), p. 201.

12. Johnson and Jackson, *City behind a Fence* (1981), p. 14.

FIGURE 3.2. Workers pushing uranium fuel slugs into the loading face of the X-10 graphite reactor. Clinton Engineer Works, Tennessee. Department of Energy Photography, Oak Ridge, photograph 7576-1.

segregated "hutments."[13] Attrition was high. From March 1, 1943, to June 30, 1945, the University of Chicago managed Clinton Laboratories. Compared with the large MED uranium-separation plants at Oak Ridge, Clinton Laboratories was smaller scale and research oriented, and remained closely connected to the Chicago Metallurgical (Met) Lab.[14]

In addition to being a pilot plant for Hanford, the X-10 semiworks provided small amounts of plutonium to various Manhattan Project laboratories. On December 30, 1943, 1.5 milligrams of plutonium were shipped from Oak Ridge to the Met Lab; over the next year, 326 grams of purified plutonium were produced at Clinton.[15] In addition to converting uranium-238 into plutonium, the reactor generated radioisotopes of many other elements as by-products of uranium fission and neutron bombardment. These various fission products could be isolated for further use. In addition, the neutron flux of the reactor could be harnessed to irradiate target materials. It was chemically simpler to obtain pure radioisotopes this way than from the stew of fission products in reactor waste. Some radioisotopes besides those of uranium and plutonium were useful in weapons development, either for military research or in radiological warfare (see below).

Due to its relatively high neutron flux, the X-10 reactor was suitable for producing radioactive materials besides plutonium—and as Hanford Engineer Works became operational, the pilot plant in Tennessee could be spared for such tasks.[16] For example, initiating a chain reaction quickly and accurately was one of the technical challenges faced by designers of the plutonium bomb. Robert Serber suggested using a neutron initiator, which could be made from radium-beryllium or polonium-beryllium. J. Robert Oppenheimer informed Groves in June 1943 of the need for polonium, and suggested they produce it in the X-10 reactor. In the summer of 1943, Los Alamos scientists arranged to have 440 pounds of bismuth slugs irradiated in the Clinton reactor for 100 days. The irradiated bismuth was

13. Hewlett and Anderson, *New World* (1962), pp. 117–20; Hales, *Atomic Spaces* (1997).

14. Johnson and Schaffer, *Oak Ridge National Laboratory* (1994), p. 21.

15. Quist, "Classified Activities" (2000), p. 13.

16. US Army Corps of Engineers, *Manhattan Project* (1976), reel 6, book IV, vol. 2, p. 4.10; W. E. Thompson, "Oak Ridge National Laboratory Research and Radioisotope Production," Jan 1952, MMES/X-10/Vault, CF-5-1-212, DOE Info Oak Ridge, ACHRE document ES-00226, pp. 3–4, 12–13.

shipped to a Monsanto facility in Dayton, Ohio, for chemical isolation of the polonium. The first batch arrived in Los Alamos in March 1944.[17]

Work on the implosion device for the plutonium bomb also required a special radioisotope. In November 1943, Serber came up with a novel diagnostic material for testing the device: lanthanum-140, nicknamed RaLa (for radiolanthanum), which emitted easily detectable gamma rays. If RaLa were placed at the center of the spherical implosion assembly, its dispersion upon explosion would provide needed information about "density changes in the collapsing sphere of metal."[18] Lanthanum-140 was an irradiation product of barium-140, and could be produced in the Oak Ridge reactor. At the request of Oppenheimer, the supply of RaLa to Los Alamos became a major effort of scientists and engineers associated with the X-10 reactor and chemical processing facility in 1944 and 1945. Clinton Laboratories had to construct a special laboratory and plant to make this material; the chemist in charge of its separation at Clinton referred to it as "the first production of a radioisotope on a large scale."[19]

The facility for producing RaLa also exposed workers to higher levels of radiation than had cyclotrons or industry. One Los Alamos chemist remarked: "No one ever worked with radiation levels like these before, ever, anywhere in the world. Even radium people normally deal with fractions of grams, fractions of a curie."[20] The MED formed its Health Division to take charge of occupational safety issues arising from the radiation and chemical toxicity of materials at its installations—not to conduct research. But as the biological effects of plutonium and the many reactor fission products were entirely unknown, the Manhattan Project sponsored investigations on them. As previously noted, when the MED contracted with the University of California in April 1943 to manage its operations at the Rad Lab, one part of their agreement (Contract 48A) covered Joseph Hamilton's research on the metabolism of fission products.[21]

Biomedical research within the Manhattan Project, limited as it was, also drew on the resources of the X-10 reactor. Members of the Biology Section of the Medical Division of Clinton Laboratories wanted to be able

17. Hoddeson et al., *Critical Assembly* (1993), pp. 119–25.

18. Ibid., p. 148; Serber, *Peace and War* (1998), p. 89.

19. Miles Leverett, as quoted in Johnson and Schaffer, *Oak Ridge National Laboratory* (1994), p. 24; Hoddeson et al., *Critical Assembly* (1993), p. 150.

20. Rod Spence, as quoted in Hoddeson et al., *Critical Assembly* (1993), p. 150.

21. ACHRE, *Final Report, Supp. Vol. 1* (1995), p. 604; chapter 2.

to purify radioisotopes from separation wastes to conduct research, and were unable to do so. They were, however, using the X-10 reactor as a radiation source for biological experiments.[22] Employing a diverse group of experimental organisms, including bacteria, guinea pigs, mice, rabbits, and flies, the Clinton group investigated the relative lethal doses of fast neutrons, gamma rays, and various combinations of the two.[23]

On at least one occasion, the X-10 reactor produced radioisotopes for use outside the Manhattan Project. In the spring of 1944, Hamilton made an urgent plea that Clinton Laboratories begin supplying Berkeley with phosphorus-32 for clinical distribution. This need arose from the anticipated shut-down of the Berkeley cyclotron, which provided the usual supply. Hamilton requested 500 millicuries a month, which his group would be willing to process chemically from the irradiated target material. They asked for shipments beginning in late June, and continuing through at least October.[24] Stone appealed to Clinton Laboratories' director, Martin Whitaker, to approve this request, even though it was not explicitly part of the war effort:

> I would suggest the following as reasons why we should do this. First, the cyclotron operates as a sub-contract to the University of Chicago both in regard to certain physical and chemical experiments and in certain biological experiments. As a convenience to our sub-contractor we should help them out of a difficulty. Second, the cyclotron must be repaired from time to time as part of normal operations and these repairs have been rendered more urgent because

22. Biological Research section of Report for Month Ending 4 Sep 1943, Metallurgical Project, A. H. Compton, Project Director, S. K. Allison, Laboratory Director, Health Division, R. S. Stone, M.D. Division Director, Health, Radiation and Protection, MMES/X-10/Vault, CF Met Lab Report #CH-908, ORF24155, DOE Info Oak Ridge, ACHRE document ES-00157.

23. Clinton Laboratories, Medical Division, Biology Section, Report for Month Ending 29 Feb 1944, MMEX/X-10/Vault, CF Met Lab Report #CH-1470, DOE Info Oak Ridge, ACHRE document ES-00143; H. J. Curtis, "Biological Work at Clinton Laboratories," 24 Mar 1945, MMES/X-10/Vault, CF-45-3-343, ORG24197, DOE Info Oak Ridge, ACHRE document ES-00199; K. S. Cole, "Experimental Biology," Rough Draft, 21 Apr 1943, [for] Stone, S. Warren, and Cole, MMES/X-10/Vault, CF-43-4-33, DOE Info Oak Ridge, ACHRE document ES-00443.

24. The request was evidently met. Raymond E. Zirkle, "Possible Use of Chain-Reacting Pile for Radiotherapy & Associated Chemical & Biological Investigations," 2 Sep 1944, MMEX/X-10/Vault, CF-44-9-506, DOE Info Oak Ridge, ACHRE document ES-00192, p. 2.

> of the almost continuous and hard operation required by the work being done for the University of Chicago. The cyclotron management has certain obligations for P-32 that have been incurred before the war and should be fulfilled if at all possible in spite of the war.[25]

Stone's language of "obligations" called up the moral economy that had characterized the distribution of cyclotron-produced isotopes to physicians and researchers. As he pointed out, the Berkeley scientists were altruistic in meeting the clinical demand for radiophosphorus: "the University of California in distributing this material makes no charge to anyone for the material."[26] Yet the highly secret provenance of this radiophosphorus necessitated some subterfuge in distribution. Having received the phosphorus-32 from Oak Ridge, Hamilton passed it on to medical centers "under the pretense of being cyclotron-produced."[27] Some of this radiophosphorus even made its way outside US borders. In 1944, phosphorus-32 was shipped from Berkeley to Brisbane, through the intermediary of Major Paul McDaniel, stationed there with the United States Army. It was used to treat nineteen patients in the Queensland Radium Institute, the first therapeutic use of a radioisotope in Australia.[28]

Except in these unusual cases of secondary distribution, Clinton Laboratories did not supply radioactive materials to those outside the reach of the Manhattan Project, who would not have known of the reactor's existence anyway. During the war and continuing afterwards, the MIT cyclotron became the main radioisotope supplier, especially to scientists on other wartime projects.[29] However, the media coverage of the dropping of

25. Memorandum from Robert S. Stone to M. D. Whitaker re: "P-32 for Dr. Hamilton," 26 May 1944, MMES/X-10/Vault, CF-44-5-379, DOE Info Oak Ridge, ACHRE document ES-00459.

26. Ibid.

27. Myers and Wagner, "How It Began" (1975), p. 10.

28. Broderick, "History" (1988), p. 117; Korszniak, "Review" (1997), p. 212. In the fall of 1946, when scientists at Brisbane requested radioisotopes from Clinton Laboratories, Aebersold regretfully denied their request. Aebersold to A. G. S. Cooper, 23 Oct 1946, JHL papers, series 3, reel 4, folder 4:27 Correspondence C 1946. See also John Lawrence to Cooper, 16 Aug 1946, same folder. On the basis of another letter in the same file, it appears that the shipments from Berkeley did not cease completely until March 1946.

29. Robley D. Evans, Appendix IX to "Radioactivity Center, 1934–1945," unpublished history, 28 Jun 1945, Evans papers, box 1, folder Radioactivity Center 1934–1945. MIT supplied both government contractors and private organizations and individuals, including a few recipients in Brazil and Canada.

the atomic bombs over Hiroshima and Nagasaki, in conjunction with the release of the Smyth report on August 12, made public the existence of the government's nuclear reactors, fueling speculation about how they might be used after the war.

Redeploying the X-10 Reactor

The idea of using the Clinton reactor to produce radioisotopes for civilians originated with Manhattan Project scientists, who anticipated the postwar uncertainty surrounding atomic energy with a variety of initiatives. By 1944, scientists at the Chicago Met Lab, who were not as involved as their Los Alamos colleagues in the final phases of research and development of the first atomic bombs, focused attention on what would happen to the MED facilities and scientists after the war.[30] Zay Jeffries, a Met Lab metallurgist, headed a committee to address postwar nuclear research and development. Other members of the committee included Robert S. Mulliken, as secretary, Enrico Fermi, James Franck, Thorfin R. Hogness, Robert S. Stone, and Charles A. Thomas. Arthur Compton received their recommendation on November 18, 1944, in a "Prospectus on Nucleonics," also called the Jeffries Report.[31]

The report highlighted the potential usefulness of reactor-produced radiation and radioactive isotopes for research and therapy. Tracer uses of isotopes were singled out as "even more important for biology and medicine than . . . for physics and chemistry."[32] Radiotracers, as the authors observed, would shed new light on photosynthesis, metabolism, immunology, and cell growth, as well as advance medical diagnosis and agri-

30. Hewlett and Anderson, *New World* (1962); Price, "Roots of Dissent" (1995).

31. Smith, *Peril and a Hope* (1965), pp. 19–24. Many of the same individuals were involved in the scientists' movement against the continuing development of atomic weaponry; the Jeffries Report also included one of the early versions of what came to be called the Franck Report.

32. "Prospectus on Nucleonics (Jeffries Report)," reprinted (in abridged form) as Appendix A in Smith, *Peril and a Hope* (1965), pp. 539–59, quote p. 544. A slightly earlier report, from the head of the Met Lab radiobiology group, proposed that the government built a new Clinton-type reactor for biomedical applications: Raymond E. Zirkle, "Possible Use of Chain-Reacting Pile for Radiotherapy & Associated Chemical & Biological Investigations," 2 Sep 1944, MMEX/X-10/Vault, CF-44-9-506, DOE Info Oak Ridge, ACHRE document ES-00192.

cultural research. Alongside radioisotopes, the committee pointed to the potential harnessing of atomic energy for power plants, submarines, and battleships.

The Jeffries Report also suggested that the federal government establish national laboratories to continue atomic energy research after the war. This idea was echoed in other reports and recommendations through the summer of 1945, but which facilities should be included in such a postwar scheme remained a point of dispute.[33] Groves initially felt that the task of selecting national laboratories should fall to the postwar agency for atomic energy, but legislation authorizing its formation was held up in Congress as politicians debated the merits of civilian versus military control of atomic energy.[34] The uncertainty contributed to significant attrition of the scientific staff of MED facilities, especially at Los Alamos and Clinton. Late in 1945, Groves and Oppenheimer began to authorize continuing scientific research at Berkeley, Los Alamos, and Clinton in the absence of a comprehensive framework for postwar organization.

At Clinton Laboratories, whose rural location inspired the nickname "Dogpatch" during the war, scientists had particular reasons to worry about the future.[35] To the extent that converting wartime facilities into national laboratories relied on the support of academic scientists and institutions (like the universities that rallied around the establishment of Argonne and Brookhaven), Oak Ridge was at a distinct disadvantage.[36] Prominent physicists expressed skepticism that a laboratory in the hills of Tennessee could attain scientific distinction in peacetime.[37] Yet like their counterparts at the Met Lab, scientists at Oak Ridge lobbied for the postwar furtherance of their facility's resources, particularly the X-10 reactor. Under the circumstances, Clinton Laboratories' survival could not be linked to aspirations for brilliant nuclear science; rather, it was the facility's "avowedly semi-industrial" character that informed strategies for

33. Westwick, *National Labs* (2003), ch. 1.

34. Ibid.; Balogh, *Chain Reaction* (1991), ch. 2.

35. Monsanto employees referred to Oak Ridge as "Dogpatch," a stereotypical hillbilly town in Al Capp's "Li'l Abner" comic strips, even in official telegrams. See Johnson and Schaffer, *Oak Ridge National Laboratory* (1994), p. 29.

36. See US Army Corps of Engineers, *Manhattan Project* (1976), reel 1, book I, vol. 4, ch. 1; Hewlett and Duncan, *Atomic Shield* (1969), p. 35.

37. For this reason the AEC transferred reactor development to Argonne, a major setback for physicists and engineers at Oak Ridge. Johnson and Schaffer, *Oak Ridge National Laboratory* (1994), p. 75.

its continuing existence.[38] Clinton's industrial image derived in part from the location of the massive MED uranium-separation plants in the same area, and in part from its contractor at the war's end. The University of Chicago managed Clinton Laboratories from its origin in 1943 until July 1945, when Monsanto Chemical Company became the contractor. Monsanto, the first industrial firm to manage one of the Manhattan District's large research laboratories, reorganized the administration of Clinton Laboratories and expanded the staff to 2,141 people.[39]

Late in 1945, the Director of Clinton Laboratories appointed a Radioisotope Committee to "study the possibilities of distributing radioisotopes for general and cancer research."[40] The reference harkened back to the long precedent of radioactive sources being used in medical therapy, but also tapped into strong public sentiment in favor of mobilizing science to fight cancer.[41] During World War II, the American Cancer Society raised large sums of money, and by the summer of 1945 the society called on the National Research Council (an arm of the National Academy of Sciences) to help disburse research grants through a newly created Committee on Growth.[42] This committee also focused attention on the inadequacy of cyclotron-produced radioactive materials for cancer research. In order to "establish evidence of the demand which exists," Merle Tuve, head of the Physics Panel for the Committee on Growth, sent an application for radioactive materials to about a dozen prominent researchers so that they could document their needs.[43] Prompted by these activities, Frank Jewett, president of the National Academy of Sciences (NAS), wrote to Secretary of War Robert Patterson in October 1945 advocating that the radioisotope

38. Johnson and Schaffer, *Oak Ridge National Laboratory* (1994), p. xi.

39. Westwick, *National Labs* (2003), p. 35; Johnson and Schaffer, *Oak Ridge National Laboratory* (1994), p. 31.

40. US Army Corps of Engineers, *Manhattan Project* (1976), reel 1, book I, vol. 4, p. 3.10. According to this source, M. D. Whitaker (Director of Clinton Laboratories) appointed this Radioisotope Committee on the advice of a Committee on Growth at Clinton, though this latter group was apparently not related to the National Research Council's committee by the same name. See also M. D. Whitaker to K. Z. Morgan, 14 May 1946, NARA Atlanta, RG 326, OROO Files Relating to K-25, X-10, Y-12, Acc 67A1309, box 14, folder Press Releases & Background Dope.

41. Creager, "Mobilizing Biomedicine" (2008) and chapter 5.

42. Eugene P. Pendergrass to Robert S. Stone, 31 July 1945, AEC Records, NARA Atlanta, RG 326, OROO Lab & Univ Div Official Files, Acc 68A1096, box 13, file Isotopes–3; National Research Council, *Research Attack* (1946).

43. M. A. Tuve to Andrew H. Dowdy, 10 Dec 1945, NARA Atlanta, RG 326, OROO Lab & Univ Div Official Files, Acc 68A1096, box 13, folder Isotopes–3.

by-products of the "chain-reaction piles of the Manhattan District" be made available immediately to medical researchers.[44] Patterson referred Jewett's letter to Groves, who was at the same time being petitioned on the issue by MED scientists.[45]

Subsequently, the Oak Ridge–based Radioisotope Committee, composed of Waldo Cohn, J. R. Coe, C. D. Coryell, and Arthur H. Snell, circulated a memorandum on January 3, 1946, recommending that Clinton Laboratories begin supplying research institutions with reactor-generated radioisotopes and offering a step-by-step plan for organizing production and distribution.[46] The report pointed out that publicity surrounding the release of the Smyth Report a few months earlier resulted in public awareness "that many radioisotopes are available from pile operation at this time." In addition, there had developed "a persistent and ever-increasing demand" among scientists for radioisotopes, "with the result that a distribution system must be set up at the earliest possible moment."[47]

The memorandum was an argument on behalf not only of government-organized radioisotope distribution, but also of Clinton Laboratories as the best facility at which to base such a program. Only two other facilities possessed sufficient piles for producing radioisotopes—Argonne and Hanford. Argonne's reactor was an experimental unit, whose low power made it unsuitable for large-scale production. The Hanford piles by contrast

44. Frank B. Jewett, President of the National Academy of Sciences, to Robert P. Patterson, Secretary of War, 18 October 1945, copy in records of the NARA Atlanta, RG 326, OROO Lab & Univ Div Official Files, Acc 68A1096, box 13, folder Isotopes–3. Jewett also volunteered the Committee on Growth to undertake "controlled and intelligent distribution [of radioisotopes] to competent workers in the field."

45. Jewett had previously been the first president of Bell Labs. On referring letter to Groves, see Secretary of War to Frank B. Jewett, 13 Nov 1945, NARA Atlanta, RG 326, OROO Lab & Univ Div Official Files, Acc 68A1096, box 13, folder Isotopes–3. On scientists lobbying Groves, see J. R. Dunning to Major General L. R. Groves, 29 Oct 1945, NARA Atlanta, RG 326, OROO Lab & Univ Div Official Files, Acc 68A1096, box 13, folder Isotopes–3.

46. US Army Corps of Engineers, *Manhattan Project* (1976), reel 1, book I, vol. 4, p. 3.10, and Oak Ridge Operations Office correspondence from 1945 and 1946 in NARA Atlanta, RG 326, OROO Lab & Univ Div Official Files, Acc 68A1096, box 13, folder Isotopes–3.

47. W. E. Cohn (for the Radioisotope Committee of Clinton Laboratories), "The National Distribution of Radioisotopes from the Manhattan Engineer District," 3 Jan 1946, NARA Atlanta, RG 326, OROO Files Relating to K-25, X-10, Y-12, Acc 67A1309, box 14, folder Press Releases & Background Dope, p. 2. For inquiries from scientists outside the Manhattan Project into the availability of reactor-produced isotopes, see NARA Atlanta, RG 326, OROO Lab & Univ Div Official Files, Acc 68A1096, box 13, folder Isotopes–2.

had the highest neutron flux of any in existence, but they were not as easy to use for irradiating target materials, the most common procedure for producing radioisotopes. Clinton's graphite reactor, on the other hand, was highly suitable in terms of both power and flexibility. Oak Ridge had processing facilities as well. Radioisotope preparation entailed two steps: irradiation in the reactor (either of uranium or target material), followed by chemical separation, usually with an elaborate remote-control apparatus to limit radiation exposure. Clinton Laboratories already possessed a radiochemical processing facility as well as personnel for this second task.[48]

The plan from the Radioisotope Committee of Clinton Laboratories gave lower priority to medical research and human uses than other scientific applications of radioisotopes. As the memorandum put it,

> Research involving the use of human material (here called clinical research) offers less overall gain to society per unit of radioisotope material than does research in the basic sciences. The principal reasons for this are (a) the size of the experimental animal, requiring larger amounts, (b) the complexity and inaccessibility of the human organism (e.g., sacrifice is prohibited), (c) the lack of standardization of research material, requiring the use of larger numbers of the already very large organisms, (d) the non-existence of bona-fide clinical uses for radioisotopes in present-day medicine other than as palliatives in certain special forms of cancer. Furthermore, legal responsibilities that involve purity and constancy of supply must be avoided. Hence, if rationing is in order, it is both wise and safe to encourage research in the basic sciences at the expense of clinical investigation.[49]

Given the postwar emphasis on fighting cancer and civilian benefits, it is striking that these authors saw clinical research as offering "less overall gain to society" than fundamental research. In fact, medical uses of radioisotopes received greater attention and priority once the system of production was established under the new agency. But the AEC was equally troubled by the "legal responsibilities" involved with having the government charge for radioisotopes. In the end, purchasers had to waive the li-

48. Cohn, "National Distribution of Radioisotopes," p. 4.
49. Ibid., p. 7.

ability of the US government and the contractor of Oak Ridge for "injury to persons or other living material" in order to obtain radioisotopes.[50]

By February 1946 an "Isotopes Branch" had been set up in the Manhattan Engineer District Office, the headquarters building located in Oak Ridge a few miles from X-10.[51] Paul C. Aebersold was given leave by the University of California to become its chief. Although based in Berkeley since the late 1930s, where he had earned a PhD with E. O. Lawrence, Aebersold had worked as a technical consultant in Oak Ridge during the war and in Los Alamos on health protection measures for the Almagordo atomic bomb test in 1945.[52] According to Cohn, he and Aebersold had provided the driving force for the formation of the Radioisotope Committee and the resulting plan of action.[53] In fact, in December 1945, Aebersold had drawn up a draft purchase application for Stafford Warren, director of the MED's Medical Section, to consider. He argued that requests should be judged by scientists in Oak Ridge rather than by bureaucrats in DC.[54] In the next months, Aebersold continued to agitate for the MED to begin making radioisotopes available to scientists outside the Project. He wrote Colonel K. E. Fields that seven months after V-J Day, "impatience with the District by non-project members on the matter of isotope distribution, has mounted to the tolerance limit."[55] Accumulating correspondence confirmed the growing, if still modest, demand; by March 1946, the Manhattan District had received thirty-five requests for radioisotopes from

50. The Agreement and Conditions for Order and Receipt of Radioactive Materials, as well as the Certificates issued, are archived in NARA Atlanta, RG 326, Files Relating to K-25, X-10, Y-12, Acc 67A1309. See chapter 6 for more on the regulatory apparatus for radioisotopes.

51. Committees were set up in accordance with the Memorandum from Radioisotope Committee of Clinton Laboratories, 3 Jan 1946, to Colonel S. L. Warren, NARA Atlanta, RG 326, OROO Files Relating to K-25, X-10, Y-12, Acc 67A1309, box 14, file Press Releases & Background Dope.

52. Paul C. Aebersold, "Professional History," appendix to "Application for Federal Employment," 13 Jun 1946, Aebersold papers, box 1, folder 1-1 Biographical Materials.

53. Human Radiation Studies: Remembering the Early Years, Oral History of Biochemist Waldo E. Cohn, Ph.D. Conducted January 18, 1995 through the Department of Energy by Thomas Fisher, Jr. and Michael Yuffee, and published at http://www.hss.energy.gov/HealthSafety/ohre/roadmap/histories/index.html.

54. Paul C. Aebersold to Col. S. L. Warren, 19 Dec 1945, NARA Atlanta, RG 326, OROO Lab & Univ Div Official Files, Acc 68A1096, box 13, folder Isotopes–3.

55. Paul C. Aebersold to Col. K. E. Fields, 8 Mar 1946, NARA Atlanta, RG 326, OROO Lab & Univ Div Official Files, Acc 68A1096, box 13, folder Isotopes–3.

researchers at nineteen universities, two federal government agencies, and six companies.[56]

There was some resistance to the proposal to use the X-10 reactor as a production facility for outsiders. Manhattan Project physicists and chemists wished to use the reactor for their own experimental purposes, the military was concerned about the dissemination of nuclear "secrets," and radiologists worried about the hazards of radiation exposure to laboratory researchers and the public.[57] In the end, these countervailing forces were not enough to derail the plan for an Oak Ridge–based radioisotope supply. However, the military did act to protect its own interests in isotopes. In February 1946, the Army proposed a "Manhattan Project Isotope Committee" composed of representatives from each of the major Project facilities, to equitably allocate "scarce isotopes, stable and radioactive," and facilitate the exchange of information about their uses.[58] Aebersold was appointed as the group's Secretary.

During the fall of 1945, a flow chart had circulated among leaders within and beyond the Manhattan Project that laid out a proposed arrangement for radioisotope distribution.[59] This radioisotope distribution plan was officially endorsed on March 8 and 9, 1946, by the MED's Advisory Committee on Research and Development.[60] Groves, in response

56. List of Requests and Estimated Needs for Radioisotopes, with memorandum from Paul C. Aebersold to Colonel K. E. Fields, 30 Apr 1946, NARA Atlanta, RG 326, OROO Lab & Univ Div Official Files, Acc 68A1096, box 13, folder Isotopes–2.

57. Cohn, "Introductory Remarks" (1968), pp. 8–9.

58. Proposed Agenda for First Meeting of a Committee on Project Distribution of Isotopes (Stable & Radioactive), attached to memorandum from A. V. Peterson, Lt. Col., Corps of Engineers, to Col. F. E. Fields, 21 Feb 1946, NARA Atlanta, RG 326, OROO Lab & Univ Div Official Files, Acc 68A1096, box 13, folder Isotopes–3. See also Memorandum (unsigned) to The Area Engineer, 21 Feb 1946, Appointment of "Manhattan Project Isotope Committee," as well as H. A. Fidler to N. E. Bradbury, 6 Mar 1946, both in NARA Atlanta, RG 326, OROO Lab & Univ Div Official Files, Acc 68A1096, box 13, folder Isotopes–3.

59. Proposed Interim Arrangement for National Distribution of Declassified Isotopes, NARA Atlanta, RG 326, OROO Lab & Univ Div Official Files, box 13, folder Isotopes–3.

60. The Advisory Committee on Research and Development for the Manhattan District consisted of seven men who had been prominent in the scientific mobilization during the war: Robert F. Bacher (who was appointed as a Commissioner of the AEC seven months later), Arthur H. Compton, Warren K. Lewis, John R. Ruhoff, Charles A. Thomas, Richard C. Tolman, and John A. Wheeler. (This was not the same as the Tolman Committee appointed by Groves in 1944.) US Army Corps of Engineers, *Manhattan Project* (1976), reel 1, book I, vol. 4, p. 2.3; Hewlett and Anderson, *New World* (1962), pp. 632–36.

to this endorsement, formally asked Jewett to nominate a committee of outside scientists to advise the headquarters of the Manhattan Project on relevant policy issues.[61] At Aebersold's urging, the NAS-appointed group was predominantly composed of civilian scientists rather than participants in the Manhattan Project.[62] This Interim Advisory Committee on Isotope Distribution Policy (hereafter Interim Committee) was chaired by Lee DuBridge and included fellow physicist Merle Tuve, chemists Linus Pauling and Vincent du Vigneaud, medical researchers Cornelius Rhoades and Cecil J. Watson, biologists Raymond Zirkle and A. Baird Hastings, as well as Zay Jeffries and L. F. Curtiss, representing applied sciences, and nonvoting secretary Paul Aebersold.[63] The group required only one meeting (on April 20, 1946, at the NAS) to come up with their recommendations.[64] They urged the MED to go ahead and begin distributing the isotopes already available to civilian scientists.[65]

The Interim Committee advocated charging radioisotope purchasers at no higher than "out-of-pocket" costs to the Manhattan District. Groves subsequently requested that Monsanto (the contractor for Clinton) "estimate costs of production and distribution, exclusive of pile costs, plant

61. L. R. Groves to Frank B. Jewett, 22 May 1946, NARA Atlanta, RG 326, OROO Lab & Univ Div Official Files, Acc 68A1096, box 6, folder Radioisotopes–National Distribution; Memo from Col. K. E. Fields to T. F. Trowbridge, 14 Jan 1947, NARA College Park, RG 326, E67A, box 45, folder 3 Distribution of Radioisotopes–Domestic. On publicity, "Beneficial Isotopes Available from Atom Bomb Project," press release for 14 Jun 1947, NARA Atlanta, RG 326, OROO Lab & Univ Div Official Files, box 13, folder Isotopes–2.

62. Paul C. Aebersold to M. D. Whitaker, Director, Clinton Laboratories, 6 Apr 1946, NARA Atlanta, RG 326, OROO Lab & Univ Div Official Files, Acc 68A1096, box 13, folder Isotopes–3.

63. The Interim Advisory Committee on Isotope Distribution Policy continued their work until Dec. 31, 1947, by which time the AEC had appointed a permanent Committee on Isotope Distribution. "Committee on Isotope Distribution: Report by the Manager of the Office of Oak Ridge Directed Operations in Collaboration with the Directors of the Division of Research and the Division of Biology and Medicine," Dec 1947, AEC General Secretary Records, RG 326, E67A, box 25, folder 7 Isotope Distribution, Committee on. See also Paul C. Aebersold, "Status of Program Involving Distribution of Radioisotopes," 15 Jan 1947, NARA Atlanta, RG 326, OROO Files Relating to K-25, X-10, Y-12, Acc 67A1309, box 14, file Press Releases & Background Dope.

64. Paul C. Aebersold to Members of Interim Advisory Committee on Isotope Distribution Policy, 26 Apr 1946, NARA Atlanta, RG 326, OROO Lab & Univ. Div., Acc 68A1096, box 13, folder Isotopes–3.

65. L. A. DuBridge to Major General L. R. Groves, 1 May 1945, NARA Atlanta, RG 326, OROO Lab & Univ Div Official Files, Acc 68A1096, box 13, folder Isotopes–2, p. 2.

rentals, or costs of existing laboratories."[66] According to Waldo Cohn, Groves first wanted the cost of constructing the reactor to be amortized into the price schedule, but scientists objected that "you put that price on carbon-14 and you won't sell a microcurie of it. You can't [put] it on the back of biological researchers."[67] The scientists prevailed. What this in fact meant was that outside purchasers would not have to pay for the wartime infrastructure that enabled the radioisotopes to be produced this way in the first place.

However, pricing radioisotopes according to the "cost of production" was still unwieldy.[68] The auditing that was part of government appropriations—the Army's auditing of Monsanto, and Congress's General Accounting Office's auditing of the Army—resulted in "red tape and auditors, with more time and money being wasted in being certain that no penny is unaccounted for than the entire produce being distributed will gross."[69] As a result, initial prices for Oak Ridge radioisotopes were, in the eyes of advocates for distribution, unduly high. Moreover, the reactor capacity was not utilized to the full extent possible.[70]

After the system had been agreed upon, the stalling of atomic energy legislation in Congress remained an obstacle. As Colonel K. D. Nichols wrote to the head of Monsanto on May 24, 1946, "The passage of legislation has been so long delayed that in the national interest it is desired to distribute such radioisotopes to non-project users as can be made available."[71] The McMahon Bill finally passed the Senate on June 1, but faced a

66. US Army Corps of Engineers, *Manhattan Project* (1976), reel 1, book I, vol. 4, p. 3.14.

67. Human Radiation Studies: Remembering the Early Years, Oral History of Biochemist Waldo E. Cohn, Ph.D. Conducted January 18, 1995 through the Department of Energy by Thomas Fisher, Jr. and Michael Yuffee, and published at http://tis.eh.doe.gov/ohre/roadmap/histories, p. 19.

68. L. F. Curtiss, Member, Interim Advisory Committee on Isotopic Distribution Policy, to Lee DuBridge, Chairman, 9 Dec 1946, copy in Aebersold papers, box 1, folder 1-11 Gen Corr July–Dec. 1946.

69. Waldo E. Cohn to Paul C. Aebersold, 1 Dec 1946, Aebersold papers, box 1, folder 1-11 Gen Corr July–Dec 1946, p. 6.

70. See Curtiss to DuBridge, 9 Dec 1946.

71. K. D. Nichols to C. A. Thomas, Monsanto Chemical Co., 24 May 1946, NARA Atlanta, RG 326, OROO Files Relating to K-25, X-10, Y-12, Acc 67A1309, box 15, folder 441.2 Isotopes.

tortuous summer in the House.[72] In the meantime, Nichols directed Monsanto to begin setting up the radioisotope program at Clinton.[73]

Distribution Begins

The first catalog of US government–produced radioisotopes was published in *Science* magazine on June 14, 1946, under the title "Availability of Radioactive Isotopes: Announcement from Headquarters, Manhattan Project." The unnamed authors were Waldo Cohn, R. T. Overmann, and Aebersold.[74] About one hundred isotopes were listed as available and purchasers were instructed to send requests to the Isotopes Branch at Oak Ridge, where each would be reviewed. The procedures followed the Interim Committee's recommendations quite closely. Isotopes would be provided only to individuals at "qualified institutions," and secondary distribution would not be permitted (beyond administering radioisotopes to patients). Each "qualified institution" was required to establish a local radioisotope committee, "of persons familiar with the biological effects of radiation, qualified to advise on proper selection and measurement of isotopes, and competent to advise on the chemical problems of preparation and of extraction for measurement of isotopes."[75] Thus local committees would oversee health and safety issues associated with employing radioisotopes.

Highest priority would go to supplying radioisotopes for fundamental and publishable research. Allocation for "routine commercial application" or nonpublishable research was considered beyond the scope of

72. See Hewlett and Anderson, *New World* (1962), chapter 14.

73. US Army Corps of Engineers, *Manhattan Project* (1976), reel 1, book I, vol. 4, p. 3.14; K. D. Nichols to C. A. Thomas, Monsanto Chemical Co., 24 May 1946.

74. "Availability of Radioactive Isotopes" (1946). Subsequently catalogs were published by the AEC as pamphlets. According to Cohn, Aebersold wrote the first part of the *Science* article (on administration), and Cohn wrote the second (technical) part. Cohn, "Introductory Remarks" (1968), p. 10; Human Radiation Studies: Remembering the Early Years, Oral History of Biochemist Waldo E. Cohn, Ph.D. Conducted January 18, 1995 through the Department of Energy by Thomas Fisher, Jr. and Michael Yuffee, and published at http://tis.eh.doe.gov/ohre/roadmap/histories, p. 17.

75. Memo from Col. K. E. Fields to T. F. Trowbridge, 14 January 1947, NARA College Park, RG 326, E67A, box 45, folder 3 Distribution of Radioisotopes–Domestic, p. 2.

distribution.[76] The emphasis on furthering unclassified research reflected the commitment of Manhattan Project scientists to realize civilian benefits of the technologies developed to build the bomb.[77] The announcement also indicated that research uses of radioisotopes would have higher priority than therapeutic uses, in part because therapy tended to require greater quantities.

All paperwork was to be handled by the Isotopes Branch, located at the Manhattan Project headquarters, Oak Ridge Operations. (The Isotopes Branch was renamed the Isotopes Division in 1947.) This office had to approve both the specific request and the institution, on the basis of the adequacy of facilities and oversight for handling radioactivity. Domestic applicants were to file an "Application for Radioisotope Procurement" along with two other forms: an "Acceptance of Terms and Conditions for Order and Receipt of Byproduct Materials (Radioisotopes)," and a Certificate to meet Section 505(i) of the Federal Food, Drug and Cosmetic Act that indicated whether the material was to be administered to humans, either as a drug or experimentally. Applicants deemed "adequately equipped" and who proposed to use the requested radioisotope in acceptable ways were issued an "Authorization for Radioisotope Procurement" that permitted the purchase.[78]

Any requests for use of radioactive materials in humans first had to be cleared through the Subcommittee on Human Application. This committee, which met for the first time on June 28, 1946, was concerned principally with medical experiments but also addressed industrial exposure. Perhaps reflecting prior experience with cyclotron-produced radioisotopes, the committee viewed its ethical responsibilities mainly in terms of dealing with scarcity—setting priorities among possible recipients of radioisotopes. In retrospect, allocation appears less significant than the ethics of safety and informed consent. Although the AEC acknowledged

76. Those doing classified research within the Manhattan Project continued to be supplied with Oak Ridge radioisotopes. The AEC altered the program to promote industrial usage of isotopes.

77. L. A. DuBridge to Major General L. R. Groves, 1 May 1945, NARA Atlanta, RG 326, OROO Lab & Univ Div Official Files, Acc 68A1096, box 13, folder Isotopes–2, p. 4.

78. See AEC memo 398, "Regulations for the Distribution of Radioisotopes," 22 Jan 1951, NARA College Park, RG 326, E67A, box 45, folder 1 Regulations for the Distribution of Radioisotopes. A copy of the forms used when the program began in 1946 is in NARA Atlanta, OROO Lab & Univ Div Official Files Acc 68A1096, box 6, folder Radioisotopes–National Distribution.

the importance of consent from the spring of 1947, the agency did not develop a consent requirement for experimental volunteers until the late 1950s. The local radioisotope committees at institutions that received AEC radioisotopes were expected to both monitor radiation hazards and safeguard patients.[79] Radioisotopes would be sent not to physicians in private practice, but only to institutions "capable of using them in integrated research programs and with suitable instruments for measuring the radiations."[80]

The catalog in *Science* included a chart indicating, for every radioactive isotope available for sale, what target material was used and what kind of nuclear transmutation reaction was involved in pile production. Some elements absorbed neutrons and released a gamma ray, undergoing what was termed a neutron-gamma (photon) (n,γ) reaction to produce a radioactive isotope of the same element. For example, pile irradiation of phosphorus-31 causes some atoms to undergo transmutation to become phosphorus-32. The resulting radioactive phosphorus cannot be chemically separated from the stable phosphorus. The stable element, phosphorus-31, is called the "carrier" for the small amount of radioactive isotope, phosphorus-32.[81] Hence these are referred to as non–carrier-free radioisotopes. Other elements experienced transmutation, either through neutron-proton (n,p) or neutron-alpha particle (n,α) reactions, yielding isotopes of chemical elements different from the parents. These radioisotopes were referred to as carrier-free, since all of the element present in the sample was radioactive—it was not mixed with the nonradioactive form. For instance, a carrier-free form of phosphorus-32 could be produced from sulfur-32. In addition, other radioisotopes frequently requested by biologists and physicians, such as carbon-14, sulfur-35, and calcium-45, could be produced carrier-free. In general, separation of such a carrier-free element from its chemically distinct target material ("parent") yielded a radioactive isotope of high purity and high specific activity. Lastly, in certain cases, the (n,γ) reaction produced an unstable isotope, whose radioactive decay produced "daughters" that were nonisotopic with the original element.

Radioisotope production at X-10 built on the "experimental-lot production" of specific isotopes late in the war—most notably RaLa and

79. ACHRE, *Final Report* (1996), pp. 46–47, 173–74, 189.

80. S. Allan Lough to W. H. Bergen, 18 Nov 1947, NARA Atlanta, RG-326, MED CEW Gen Res Corr, Acc 67B0803, box 145, folder AEC 441.2 (R–Bergen's Pharmacy).

81. US Army Corps of Engineers, *Manhattan Project* (1976), reel 1, book I, vol. 4, p. 3.4

polonium for Los Alamos, and phosphorus-32 for Berkeley.[82] Target materials were placed in the reactor in small aluminum containers, inserted into holes in the graphite blocks referred to as "stringers." The graphite block could then be "pushed into the interior of the pile through an opening in the surrounding thick concrete shield."[83] Irradiation could take place over a period of days or months, depending on the materials and nuclear reactions involved. (See figure 3.3.) Removal of the target material was a delicate, even dangerous, operation requiring careful coordination and monitoring:

> When the reactor is shut down for the removal of irradiated samples, each member of the team of workers must know his assignment precisely and carry it out quickly and without error. Geiger counters and other radiation-detection equipment must be used constantly to check radiation. In subsequent chemical treatment of materials, special remotely controlled equipment, located inside "hot cells" must be operated by a skilled operator who, in many cases, can see what he is doing only by looking in a mirror, through a periscope, or through a transparent shield.[84]

There were physical limits to the amount and kind of isotopes that could be produced by irradiating target materials in the X-10 reactor.[85] But as it turned out, the system of production proved adequate for filling approved requests, the majority of which came from biologists and clinicians.

On August 1, 1946, President Harry S. Truman signed into law the Atomic Energy Act, providing the necessary legal framework for the Manhattan Project to distribute reactor "byproduct materials."[86] The following day the government delivered the first radioisotope shipment, with "appropriate formalities in front of the Clinton pile."[87] This was a carefully orchestrated event with full press coverage. Fifty journalists

82. "Availability of Radioactive Isotopes" (1946), p. 699.

83. *Isotopes Catalog and Price List No. 3, July 1949* (Oak Ridge, Tennessee: Isotopes Division, United States Energy Commission, 1949), p. 1, copy in NARA College Park, RG 326, E67A, box 46, folder 6 Foreign Distribution of Radioisotopes vol. 2.

84. W. E. Thompson, "Oak Ridge National Laboratory Research and Radioisotope Production," Jan 1952, MMES/X-10/Vault, CF-52-1-212, DOE Info Oak Ridge, ACHRE document ES-00226, p. 13.

85. "Availability of Radioactive Isotopes" (1946), p. 697.

86. "Atomic Energy Act of 1946" (1946).

87. US Army Corps of Engineers, *Manhattan Project* (1976), reel 1, book I, vol. 4, p. 3.20.

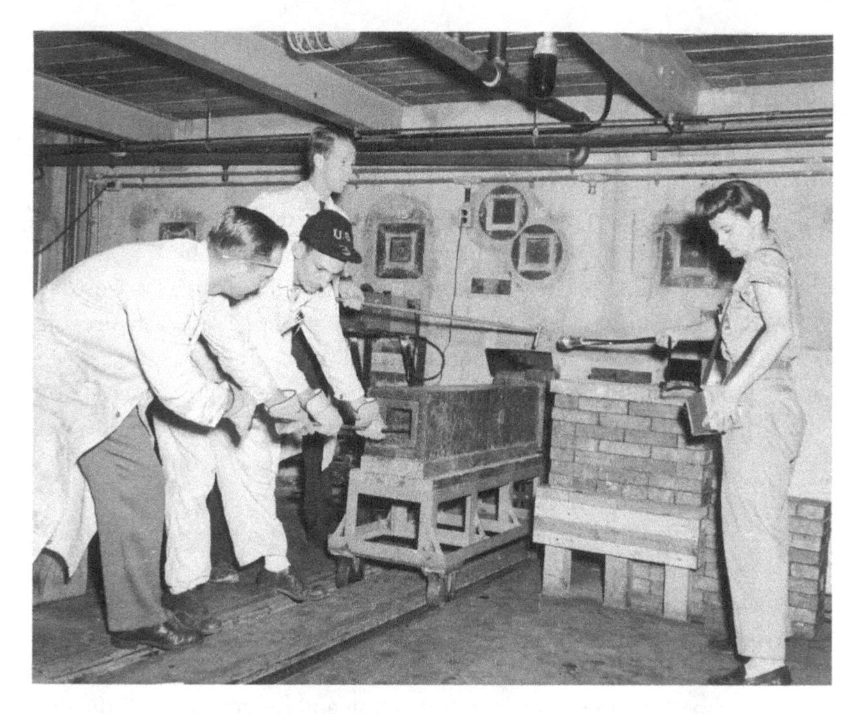

FIGURE 3.3. Target material being unloaded from the X-10 reactor, with radioactivity being monitored by a hand-held counter at right. June 14, 1946, Clinton Engineer Works, Tennessee. National Archives, RG 434-OR, box 22, notebook 65, photograph MED-308.

and photographers were brought in from Washington, DC, for a press conference and a tour of unrestricted portions of the laboratories, especially the radioisotope production facility.[88] Finalizing the invitation list proved challenging, since several science writers who covered the beat were planning to be in the Pacific for the Bikini tests.[89] The photographs and captions selected for press release were carefully vetted for any unintentional disclosure of classified information. In addition, there was debate about whether the journalists would be allowed to visit security-restricted areas. Nichols expressed to Groves his concerns that "if we permit newspapermen in the hot lab we will be under pressure to approve every request to

88. Schedule of Events, Radio and Press Conference, Monsanto–Clinton Laboratories, 2 Aug 1946, NARA Atlanta, RG 326, OROO Files Relating to K-25, X-10, Y-12, Acc 67A1309, box 14, folder Press Releases & Background Dope.

89. Memorandum to Colonel K. E. Fields regarding Publicity Concerning Distribution of Radioisotopes by the Manhattan Project, 27 May 1946, NARA Atlanta, RG 326, OROO Lab & Univ Div Official Files, Acc 68A1096, box 13, folder Isotopes–2.

take scientists thru in the same fashion."[90] In the end, the itinerary did include a stop at the Chemistry Building or "hot lab," as well as at the Pile Building and shipping deck.

The climax of the day was the staging of the first "shipment" in a "transfer ceremony," scheduled to begin at 11:40 a.m. As photographers snapped their cameras and Colonel Elmer E. Kirkpatrick (Deputy District Engineer) looked on, Eugene Wigner (Director of Clinton Laboratories) handed a millicurie of carbon-14, sold for $367, to E. V. Cowdry and William L. Simpson, research director and associate research director of the Barnard Free Skin and Cancer Hospital, St. Louis, Missouri. (See figure 3.4.)

Wigner's remarks for the occasion of the staged shipment focused on the shift from using atomic energy to destroy to using it to heal:

> We hope . . . that this day will mark a turning point in the history of atomic energy and that which has been used so effectively for the purpose of destruction will henceforth be used with even greater efficiency for the saving of lives, and the increase of our knowledge for the benefit of mankind.[91]

Cowdry echoed the theme of transformation and redemption: "If the development of a terrible weapon of war opened the door of a new scientific age, then this is truly the first step across the threshold."[92] Not only was this first Oak Ridge shipment dedicated to cancer research, but also the recipient was from St. Louis, home to the headquarters of Clinton's contractor, Monsanto.[93]

By the end of 1946, the Isotopes Branch at Oak Ridge had received 306 requests for radioisotopes and Clinton Laboratories had processed 125 shipments. Sales were just under $30,000. This volume of requests was

90. Memo Routing Slip from "N." to General Groves and Col. Fields, in Folder "Press Release on Isotopes," NARA Atlanta, RG 326, OROO Lab & Univ Div Official Files, Acc 68A1096, box 13, folder Isotopes–2.

91. Text of remarks by Dr. Eugene Paul Wigner, Director of Research and Development, Clinton Laboratories, at transfer ceremony, NARA Atlanta, RG 326, OROO Files Relating to K-25, X-10, Y-12, Acc 67A1309, box 14, folder Press Releases & Background Dope.

92. Text of remarks by Dr. E. V. Cowdry, Research Director of Barnard Free Skin and Cancer Hospital, at transfer ceremony, NARA Atlanta, RG 326, OROO Files Relating to K-25, X-10, Y-12, Acc 67A1309, box 14, folder Press Releases & Background Dope.

93. Hewlett and Anderson, *New World* (1962), p. 636.

FIGURE 3.4. (A, B) Staged sale of first radioisotope shipment from Oak Ridge reactor, August 2, 1946. (A) Eugene Wigner, director of Oak Ridge National Laboratory, handing one millicurie of carbon-14 to E. V. Cowdry, cancer researcher, in front of the reactor. To the left of Wigner is Colonel Elmer E. Kirkpatrick; to the right of Cowdry is William L. Simpson. (B) Newspapermen and photographers reporting the shipment. Credit: James E. Westcott, Oak Ridge. National Archives, RG 434-OR, box 21, notebook 22, photographs no. 430-OR-58-1870-4 and 430-OR-58-1870-13.

lower than Manhattan Project planners anticipated. A survey of research institutions to find out why orders were not more numerous yielded three main reasons: "(a) lack of trained personnel, (b) lack of electronic detection instruments, and (c) the speed with which radioisotopes were made available found researchers with plans incompleted [*sic*] for their utilization."[94] The Isotopes Branch set out to overcome these obstacles. Under the framework of civilian control, the mindset of scarcity turned into a culture of promotion.

Isotope Alley

Each of the isotopes on sale had its own production requirements and constraints. Early on, the production of carbon-14 took about half of the total flux capacity of the graphite reactor that was being used for isotope production. Clinton scientists were also working to extract carbon-14 from beryllium nitride that had been irradiated at Hanford, to free up more reactor space at Oak Ridge.[95] The 5,730-year half-life of this isotope meant that little activity was lost in cross-country transit. By contrast, the half-life of one of the other isotopes in high demand, iodine-131, was only eight days, making the loss of activity due to transportation to customers a serious concern. The two radioisotopes requested most often for therapeutic use were also the ones in shortest supply. Initially, the Oak Ridge reactor was not able to keep up with the domestic demand for iodine-131 and phosphorus-32.[96] However, the inadequacy of supply was short-lived: by June 1947, as noted in one AEC memo, "production capacity now surpasses demands."[97]

Special arrangements had to be made for the transport of radioactive materials from Oak Ridge to laboratories and hospitals around the country

94. US Army Corps of Engineers, *Manhattan Project* (1976), reel 1, book I, vol. 4, p. 3.20.

95. "Pile Capacity and Rate of Production at Oak Ridge National Laboratory," Annex A to Appendix D of "Study of Wider Use of Isotopes," Info Memo 48-92, 29 July 1948, RG 326, E67A, box 45, folder 6 Study of Wider Use of Isotopes, p. 20.

96. Memorandum from Colonel K. D. Nichols to Carroll L. Wilson, 15 Jan 1947, AEC Records, RG 326, E67A, box 46, folder 3 Foreign Distribution of Radioisotopes vol. 1.

97. Memorandum from Walter J. Williams to Carroll L. Wilson, "Availability of Radioisotopes for Export," 19 Jun 1947, NARA College Park, RG 326, E67A, box 46, folder 3 Foreign Distribution of Radioisotopes vol. 1, p. 1.

(and later, around the world).[98] In the short term, the government simply extended existing regulations for the shipping of radium by rail to radioisotopes.[99] However, the short half-lives of many of the radioisotopes made more rapid delivery desirable.[100] Through the Civil Aeronautics Board, airlines developed regulations to enable air transportation of radioactive materials from Oak Ridge, although not all the airlines were willing to accept shipments.[101] One memorandum detailed their concerns: "Would radioactive materials affect delicate aircraft instruments or radio communication? Would the welfare of the pilots and the safety of the passengers be endangered? What cargo might be damaged?"[102] Fortunately, all three of the airlines based in Knoxville, namely Delta, Capital, and American, agreed to accept shipments of radioisotopes from Clinton Laboratories. Scientists at Oak Ridge dismissed concerns from the airline industry that transporting radioisotopes would affect navigation instruments: "The radiation intensity at the surface of an isotope container is approximately the same as that at the radium dial of the pilot's wrist watch."[103] Many radioisotope shipments consisted, in fact, of small amounts of radioactivity. Still, the shipping containers could weigh as much as a ton, due to lead shielding; the average container for shipping iodine-131 was over 100 pounds. The staff at the Isotopes Branch packed instructions for safely removing radioisotopes from their sealed containers, but could not control whether the agency's procedures would be followed by those handling the materials once shipped.[104]

98. US Army Corps of Engineers, *Manhattan Project* (1976), reel 1, book I, vol. 4, pp. 3.18–3.19.

99. H. A. Campbell, Chief Inspector, Bureau of Explosives, 20 Apr 1946, NARA Atlanta, RG 326, OROO Lab & Univ Div Official Files, Acc 68A1096, box 13, folder Isotopes–3.

100. This became even more of an issue for foreign shipments. Because radioactive iodine being shipped to Australia lost so much of its activity in transit, the Australian ambassador appealed to the Commission to "consider relaxing its more stringent requirements with regard to the transportation of isotopes of this [short-lived] type, by air." N. A. Whiffen to T. E. Jones, 15 Jan 1948, NARA Atlanta, RG 326, MED CEW Gen Res Corr, box 145, folder AEC 441.2 (R–Australian Embassy). See also Broderick, "History" (1988), p. 117.

101. US Army Corps of Engineers, *Manhattan Project* (1976), reel 1, book I, vol. 4, p. 3.19.

102. "Background Material on Activity in First Year of Distribution of Pile-Produced Radioisotopes," Press Release, 2 Aug 1947, AEC Records, NARA College Park, RG 326, E67A, box 45, folder 3 Distribution of Radioisotopes–Domestic, p. 7.

103. Ibid., p. 6.

104. Ibid., p. 5.

The legacy of cyclotrons shaped the early sales of reactor-produced radioisotopes in two striking ways. First, the isotopes already in common use for research and therapy—namely phosphorus-32 and iodine-131—were those most in demand during the early years of the government distribution program. They comprised about two-thirds of domestic shipments and, later, a higher proportion of foreign shipments.[105] Cyclotron production for clinical use had cultivated the demand for these particular isotopes. Second, those institutions that possessed cyclotrons also purchased the most radioisotopes from the government early on.[106] Paul Aebersold observed at the end of 1947:

> The greatest use of isotopes has thus far been made by those sections of the country in which isotopes were previously used as a consequence of association with one of the larger cyclotrons, such as the University of California, Massachusetts Institute of Technology, Carnegie Institution of Washington, Columbia University, University of Chicago, Biochemical Research Foundation in Newark, Delaware, Washington University in St. Louis, and Harvard University.[107]

The names of those who first applied to purchase isotopes bear out Aebersold's statement, including David Rittenberg at Columbia, Raymond Zirkle, W. F. Libby and James Franck at the University of Chicago, Martin Kamen at Washington University, A. K. Solomon at Harvard, John Lawrence and Cornelius Tobias at Berkeley, and Robley Evans at MIT.[108] This was the network of scientists at institutions that already possessed either cyclotrons or heavy isotopes before World War II, above all Berkeley's Rad Lab.

The transition from a decentralized regime of cyclotron-based radioisotope production to a national government-controlled distribution system was not completely smooth. For one thing, those who had been able

105. Estimates based on information up to June 30, 1949 presented in AEC, *Isotopes* (1949), pp. 5–6, 22, 54, 59.

106. Aebersold, "Isotopes and Their Application" (1948), p. 151.

107. Paul Aebersold, draft of "Isotopes and Their Application to Peacetime Use of Atomic Energy," 29 Dec 1947, presented at the 114th meeting of the American Association for the Advancement of Science, Appendix B to "Study of Wider Use of Isotopes," Info Memo 48-92, 29 July 1948, NARA College Park, RG 326, E67A, box 45, folder 6 Study of Wider Use of Isotopes, p. 4.

108. Certificates filed with Isotopes Branch, Aug through Dec 1946, in NARA Atlanta, RG 326, OROO Files Relating to K-25, X-10, Y-12, Acc 67A1309, box 14 (not in folders).

to send radioisotopes to colleagues or clinicians did not wish to relinquish that right. In the fall of 1945, John Lawrence sought in vain to get airlines to transport radiophosphorus to physicians who had requested it in South America and England.[109] For another, expert users proved to be the most fastidious customers at the government's shop. John Lawrence complained of problems with both impurities and measurement in radioisotopes he received: "it now appears that the material separated at Oak Ridge is unsatisfactory for our purposes."[110] Lawrence's solution was to ask that Oak Ridge ship him unseparated isotope material—in the case of phosphorus-32, the can of irradiated sulfur—from which his laboratory would extract material to his specifications. Aebersold noted that other consumers, too, had to reprocess radioisotopes after receiving shipments from Oak Ridge, particularly those used in therapy: "Many users of separated P 32 employ radiochemists to further process the material before use in human beings if the material is to be employed intravenously."[111]

There were similar concerns about the purity and therapeutic safety of radioiodine that the US government made available. The agency's position was that "Clinton Laboratories is in no position to assume legal responsibility or to make commitments for the furnishing of material in a form suitable for human use."[112] However, most of the iodine-131 requested was for medical application, which placed the burden of additional purification on radioisotope purchasers. The alternative for clinicians was to continue obtaining this isotope from collaborators at cyclotron facilities. Robley Evans at MIT, a major supply site for iodine-131 before the commencement of the government's program, expressed a willingness to continue to chemically purify reactor-produced radioiodine from Oak Ridge to supply

109. See John H. Lawrence to Joseph G. Hamilton, 16 Nov 1945, JHL papers, series 3, reel 4, folder H correspondence 1945; John Lawrence to Sir Cecil Kisch, 27 Aug 1945, series 3, reel 4, folder I-K correspondence 1945. It is not clear from these letters whether the decision to refuse foreign shipments originated with the airlines or with the US government.

110. John Lawrence to Area Engineer, undated [c. Jan. 1947], NARA Atlanta, RG 326, OROO Files Relating to K-25, X-10, Y-12, 1943-1948, box 57, folder AEC 400.32 Isotopes vol. 1.

111. Paul C. Aebersold to M. E. Hubbard, 2 Jun 1947, NARA Atlanta, RG 326, MED CEW Gen Res Corr, Acc 67B0803, box 177, folder AEC 441.2 (R–Veterans Administration). Aebersold pointed out that for oral administration, many users were simply diluting the Oak Ridge preparation with water.

112. Paul C. Aebersold to Edgar J. Murphy, 7–8 Aug 1946, NARA Atlanta RG 326, OROO Files Relating to K-25, X-10, Y-12, Acc 67A1309, box 14, folder Press Releases & Background Dope, p. 2.

physicians.[113] For his part, Aebersold at the Isotopes Branch felt that even if the government did not want to certify the safety of isotope preparations for human consumption, they could well furnish material "in such a form that the final adaption of the material for human injection would be simple," such as dilution.[114] Within a few years, industry stepped into the void, as Abbott Laboratories constructed a radiopharmaceutical plant at Oak Ridge.[115]

Other safety concerns arose specifically with iodine-131. This radioelement was nearly unique among isotopes most widely used in biology and medicine in that it was actually a by-product of the uranium fission reactions sustained in the reactor. This meant that iodine-131 could be extracted chemically from the stew of reactor fission products. In this sense, radioiodine fit the exact wording of the Atomic Energy Act, which authorized the Commission to "distribute, with or without charge, byproduct materials to applicants seeking such materials for research or development activity, medical therapy, industrial uses, or such other useful applications as may be developed."[116] (The agency interpreted "byproduct" liberally so as to encompass any artificial radioisotope produced in the reactor.) The chemical extraction of radioiodine from spent reactor fuel was difficult enough that during the early days of isotope distribution, the Oak Ridge team instead prepared iodine-131 by irradiating tellurium.[117] When the Commission was able to begin producing iodine-131 from Hanford's abundant plutonium by-products, this change brought new safety concerns—especially the possible presence of plutonium as a contaminant. This was worrisome since most iodine-131 was used in medical diagnosis and therapy. The AEC tried to be reassuring that the level of impurity

113. F. R. Keating, Jr., Mayo Clinic, to Paul C. Aebersold, 26 Jul 1946, NARA Atlanta, RG 326, OROO Files Relating to K-25, X-10, Y-12, Acc 67A1309, box 14, folder Press Releases & Background Dope, p. 2.

114. Aebersold to Murphy, 7–8 Aug 1946, p. 2.

115. See chapter 6.

116. "Atomic Energy Act of 1946" (1946), quote on p. 20.

117. Minutes of Initial Meeting, Advisory Sub-Committee on Human Applications of the Interim Advisory Committee on Isotope Distribution Policy, NARA Atlanta, RG 326, OROO Lab & Univ Div Official Files, Acc 68A1096, box 6, folder Radioisotopes–National Distribution.

was vanishingly small: "in the case of I-131, human injection is possible without undue risk of injury due to plutonium."[118]

There were also issues of standardization for those users shifting from obtaining radioiodine from a cyclotron to purchasing it from Oak Ridge. By November 1946, customers reported to scientists at the Isotopes Division that their own measurements of radioactivity in separated phosphorus-32 did not agree with those provided by the Manhattan District at the time of purchase. As one member of the Isotopes Division noted, "most of our customers have, in the past, obtained their material from M.I.T. where it was prepared on a cyclotron, and accordingly, they have standardized their measurements with M.I.T."[119] They requested permission from the District Engineer to transfer ten millicuries of phosphorus-32 to the cyclotron group at MIT so they could determine its radioactivity using their methods and instruments.[120]

The regulatory oversight that the AEC set in place, particularly for human applications, also caused problems with users accustomed to obtaining their radioisotopes from cyclotrons. Aebersold at the Isotopes Branch insisted that all human uses of radioisotopes had to be approved by the Subcommittee on Human Applications.[121] He interpreted this rule to apply even for those laboratories whose applications of radioisotopes had preceded the establishment of the AEC (e.g., the Berkeley Rad Lab), as well as those in the Commission's own facilities. Those who had been administering radioisotopes clinically since the late 1930s found the new approval process cumbersome and unnecessary. As Aebersold noted in one memo, "this procedure has not been uniformly followed in the past."[122]

118. "Proposed Shipments of Isotopes to UK and Canada under the Technical Cooperation Program," Report by the AEC Member of the CPC Subgroup of Scientific Advisers, 10 Jan 1950, NARA College Park, RG 326, E67A, box 46, folder 5 Foreign Distribution of Stable Isotopes, p. 6.

119. J. A. Cox to The District Engineer, 25 Nov 1946, NARA Atlanta, RG 326, OROO Files Relating to K-25, X-10, Y-12, Acc 67A1309, box 14, folder Reports to Aebersold on Shipments and Shipping Memo.

120. The Oak Ridge activity measurements, at least for phosphorus-32, had been checked against those at both MIT and the Bureau of Standards by June 1947. See Paul C. Aebersold to M. E. Hubbard, 2 Jun 1947, NARA Atlanta, RG 326, MED CEW Gen Res Corr, Acc 67B0803, box 177, folder AEC 441.2 (R–Veterans Administration).

121. Paul C. Aebersold to A. H. Holland, Jr., Director of Research and Medicine, 5 Oct 1949, NARA Atlanta, RG 326, OROO Lab & Univ Div Official Files, Acc 68A1096, box 33, folder Isotopes Program 1–General Policy.

122. Ibid.

He wrote John Lawrence at the Donner Laboratory in Berkeley and Herman Lisco, Director of the Medical Division at Argonne, about their need to seek approval from the Subcommittee on Human Applications even when the radioisotopes did not originate in Oak Ridge. "It should be emphasized that the instruction applies even though the radiomaterial is produced in the laboratory where it is to be used."[123] Similarly, Aebersold took researchers at thc Oak Ridge Cancer Research Hospital to task for not seeking clearance from the Subcommittee on Human Applications for their use of isotopes. One wrinkle in the regulatory framework was that AEC's initial regulations for radioisotope usage did not extend to cyclotron-produced isotopes, unless they were produced in Commission facilities.[124]

There were infrastructural limits on radioisotope production at Oak Ridge. When the neutron flux of the X-10 reactor was not high enough to generate certain high specific activity radioisotopes (e.g., of iron, cobalt, nickel, and zinc), high-flux reactors at Argonne and Hanford were used.[125] In general, however, the limitations at Oak Ridge were related to inadequate workspace for the chemical processing after irradiation, particularly as production levels increased.[126] Crowding resulted in problems with cross-contamination of samples, as well as the need to move some work outdoors.[127] Ultimately, these inadequacies were resolved only as Clinton Laboratories was put on longer-term footing as a national laboratory. This took some time; the early postwar years were turbulent ones

123. Memorandum by Paul C. Aebersold to A. Tammaro, Manager, Chicago, Use of Radioisotopes in Human Subjects, 5 Oct 1949, AEC Records, NARA Atlanta, RG 326, OROO Lab & Univ Div Official Files, Acc 68A1096, box 33, folder Isotopes Program 1–General Policy.

124. "Regulations for the Distribution of Radioisotopes," AEC398, 22 Jan 1950, Report by the Division on Research and Isotopes Division, NARA College Park, RG 326, E67A, box 45, folder 1 Regulations for the Distribution of Radioisotopes.

125. "Pile Capacity and Rate of Production at Oak Ridge National Laboratory," Annex A to Appendix D of "Study of Wider Use of Isotopes," Info Memo 48-92, 29 Jul 1948, NARA College Park, RG 326, E67A, box 45, folder 6 Study of Wider Use of Isotopes.

126. Paul Aebersold, "Isotopes and Their Application to Peacetime Use of Atomic Energy," 29 Dec 1947, presented at the 114th meeting of the American Association for the Advancement of Science, Appendix B to "Study of Wider Use of Isotopes," Info Memo 48-92, 29 Jul 1948, RG 326, E67A, box 45, folder 6 Study of Wider Use of Isotopes, p. 3.

127. "Production Limitations," Appendix D to "Study of Wider Use of Isotopes," Info Memo 48-92, 29 Jul 1948, NARA College Park, RG 326, E67A, box 45, folder 6 Study of Wider Use of Isotopes.

at Oak Ridge, in part due to turnover of contractors. Morale was low, and retention of scientists was a major concern. Early in 1948, the AEC gave the contract for Clinton Laboratories to Carbide, in part because the firm also managed the uranium gaseous diffusion plant (K-25). This, however, generated labor problems. There were differentials between X-10 and K-25 with respect to wages and fringe benefits, and, not surprisingly, the American Federation of Labor, which represented workers at X-10, objected to these differences. Despite a Taft-Hartley injunction invoked over the deadlock in labor negotiations, the X-10 plant almost shut down before the conflict was resolved.[128]

Nonetheless, the designation of the X-10 site as a National Laboratory in 1947, and its renaming as Oak Ridge National Laboratory (ORNL) in 1948, signaled the AEC's long-term commitment to research in Tennessee. The portfolio of activities at Oak Ridge diversified to encompass a wide range of basic and applied research and development, including a radiobiology program.[129] The construction of permanent facilities began in 1949, including a new complex of ten buildings for processing, packaging, and shipping radioisotopes.[130] This came to be known as "Isotope Alley."[131] By 1949, there were seventy-two personnel working in pile operation or processing for isotope production.[132] In 1950, the AEC further expanded the facilities with a new isotope processing area, with remote control equipment and an assembly line set-up.[133] (See figure 3.5.)

Organizationally, the Isotopes Division did not fit neatly within the structure of the AEC. While the plant was at Oak Ridge National Laboratory, the office that processed authorizations and sales was located a few miles away in the Oak Ridge Operations building. This worked well on the ground, but the relationship to Washington was more complex. As one

128. Atomic Energy Labor Relations Panel, Report for Period 1 Jun–31 Oct 1949; Dec 1949, NARA College Park, RG 326, E67A, box 3, folder AEC Labor Policy and the Labor Situation (General), vol. 1.

129. Quist, "Classified Activities" (2000), p. 18; Rader, "Hollaender's Postwar Vision" (2006).

130. Johnson and Schaffer, *Oak Ridge National Laboratory* (1994), p. 57; Quist, "Classified Activities" (2000), p. 18.

131. Fred Strohl, ORNL, personal communication, 25 Jul 2006.

132. A. M. Weinberg, "Research Program at ORNL," 22 Mar 1949, MMES/X-10/Vault, CF-49-3-233, DOE Info Oak Ridge, ACHRE document ES-00497.

133. Press Release, 1 Feb 1950, NARA College Park, RG 326, E67A, box 47, folder 1 Foreign Distribution of Radioisotopes vol. 3.

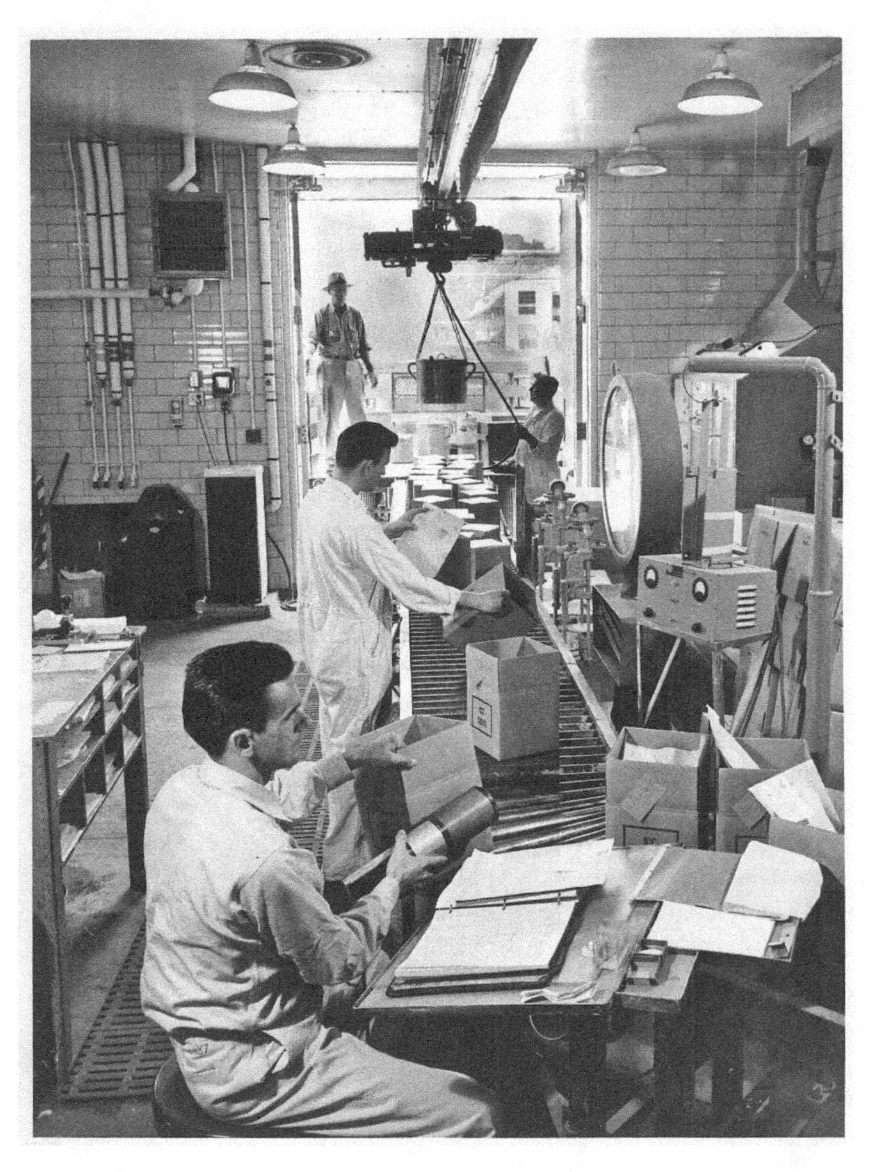

FIGURE 3.5. The isotope shipping room at Oak Ridge National Laboratory, where isotopes were packaged, labeled, weighed, and checked for radiation leakage prior to shipment. National Archives, RG 434-SF, box 25, folder 2, photograph no. AEC-55-5353.

memo appraised, the Isotopes Division could not be "related solely to one division or office since it is intimately concerned with both the Division of Research and the Division of Biology and Medicine, and has frequent business with the Division of Production."[134] The sales activities also posed new accounting issues, not only for isotopes being shipped to outsiders but also those used in-house, as radioisotopes had to be placed "in the category of accountable property."[135]

These bureaucratic complexities notwithstanding, the Commission welcomed the radioisotope distribution program it inherited from the MED. As the agency stated in its midyear report to Congress in 1947, "The cost of producing these radioactive and nonradioactive [stable] isotopes is a very small fraction of the cost of the atomic energy program as a whole; but the value of the benefits is incalculable."[136] The one politically difficult issue facing the newly appointed Commissioners was whether to allow radioisotopes to be shipped to foreign laboratories, over the objections of conservative Congressional representatives who did not want any materials sent abroad that might conceivably aid the development of nuclear technology elsewhere.[137] But the AEC's domestic program supplying isotopes to universities, hospitals, and companies was extolled by scientists and politicians alike.

One-Stop Isotope Shop

The transfer of the radioisotope program from the MED to the AEC, which occurred officially at midnight on December 31, 1946, is barely visible in the correspondence of the Isotopes Branch. From the perspective of individuals on the ground at Oak Ridge, the system that had been set up under Groves simply continued under new management. However, the establishment of a civilian agency for atomic energy did bring a new political value to the "public interest" aspects of radioisotope distribution.

134. "Administration and Administrative Controls," Appendix F to "Study of Wider Use of Isotopes," Info Memo 48-92, 29 Jul 1948, NARA College Park, RG 326, E67A, box 45, folder 6 Study of Wider Use of Isotopes, p. 29.

135. Memorandum to Lt. Col. W. P. Leber, Operations Officer, X-10, 29 July 1946, NARA Atlanta, RG 326, OROO Lab & Univ Div Official Files, Acc 68A1096, box 15, folder Accountability–Isotopes.

136. AEC, *Second Semiannual Report* (1947), p. 27.

137. See chapter 4.

This transition is illustrated by comparing the ways in which managers viewed costs associated with the program as well as by the AEC's interest in expanding the program to include isotopes not obtainable from the X-10 reactor. The Atomic Energy Act authorized the agency to distribute "radioactive byproduct material *with or without charge*."[138] In setting up the program, the MED decided to charge for radioisotopes as a "brake on the indiscriminate use of material."[139] During the first twenty-one months of the program, the AEC recovered $256,449 for isotope purchases that cost the agency $705,227 in operating expenses. However, the income included less than $200,000 from outside consumers—about $60,000 of the sales moved within the AEC's accounts covering radioisotopes for "Project" use.[140] Groves, who saw the isotope program as beyond the purview of the Army's work, was bothered that the radioisotope program did not even cover costs.[141] This was part of his larger anxiety that Congress would investigate the enormous costs associated with the Manhattan Project.[142]

The AEC did not face the same pressures, and treated the radioisotope program as inexpensive relative to its political benefits. According to AEC documents, prices were set for radioisotopes so as to recover approximately 60% of the cost of production.[143] The second Radioisotopes Catalog and Price List, effective March 1, 1947, reflected lowered prices for the three radioisotopes most widely used in biological research and

138. "Costs and Prices," Appendix C to "Atomic Energy Commission, Study of Wider Use of Isotopes," Info Memo 48-92, 29 Jul 1948, NARA College Park, RG 326, E67A, box 45, folder 6 Study of Wider Use of Isotopes, p. 9. Emphasis in original.

139. Ibid., p.10.

140. "Radioactive Isotopes Sales, Oak Ridge National Laboratory" and "Radioactive Isotope Production Costs, Oak Ridge National Laboratory," Annex B and Annex C to Appendix C of "Study of Wider Use of Isotopes," Info Memo 48-92, 29 Jul 1948, NARA College Park, RG 326, E67A, box 45, folder 6 Study of Wider Use of Isotopes.

141. Groves, *Now It Can Be Told* (1962), p. 385. Cohn predicted the radioisotope program would stop losing money as carbon-14 production took off; this proved overly optimistic. Waldo E. Cohn to Major General L. R. Groves, 1 Oct 1946, NARA Atlanta, RG 326, OROO Lab & Univ Div Official Files, Acc 68A1096, box 13, folder Isotopes–3.

142. Groves, *Now It Can Be Told* (1962), p. 70; Rader, "Hollaender's Postwar Vision" (2006), p. 692.

143. For information on the subsidization of reactor-produced radioisotopes, see AEC memo 195, "Program for Production and Distribution of Cyclotron-Produced Isotopes," NARA College Park, RG 326, E67A, box 45, folder 3 Distribution of Radioisotopes–Domestic.

medical therapy: carbon-14, phosphorus-32, and iodine-131.[144] Sales from these three isotopes represented "the major portion of the income from non-Project distribution."[145] In its publicity, the Commission frequently touted the vast savings realized by using nuclear reactors rather than cyclotrons to produce radioisotopes. An oft-repeated example was that a millicurie of carbon-14, which Oak Ridge sold for $50, "would cost about $1,000,000 to produce using a cyclotron."[146]

The AEC sought to make other isotopes, including nonradioactive ones, available. The electromagnetic separation plant (Y-12) at Oak Ridge that had been built during the war to separate uranium-235 and uranium-238 generated several stable isotopes as by-products. From December 1945 to December 1948, 129 stable isotopes of 30 elements were produced this way in the Y-12 facility.[147] By the end of 1946, physicists were lobbying the AEC to release enriched stable isotopes to off-Project users.[148] In response, the AEC announced in May 1947 that stable isotopes, like radioactive isotopes, would be available to researchers outside the Commission's laboratories.[149] However, because the supply was limited and cost was high, most stable isotopes were available not by sale but on a "loan basis" (with a per-isotope handling fee of $50).[150] Accordingly, the

144. *Radioisotopes, Catalog and Price List*, NARA College Park, RG 326, E67A, box 46, folder 3 Foreign Distribution of Radioisotopes vol. 1, p. 10.

145. Memorandum from A. V. Peterson to Col. K. E. Fields, 23 Dec 1946, Revised Cost Estimate for Radioisotopes, NARA Atlanta, RG 326, OROO Lab & Univ Div Official Files, Acc 68A1096, box 13, folder Isotopes–3.

146. AEC, *Second Semiannual Report* (1947), p. 25; also "Background Material on Activity in First Year of Distribution of Pile-Produced Radioisotopes," Press Release, 2 Aug 1947, NARA College Park, RG 326, E67A, box 45, folder 3 Distribution of Radioisotopes–Domestic, p. 3.

147. AEC Info Memo 163, 6 Apr 1949, NARA College Park, RG 326, E67A, box 45, folder 7, Production of Stable Isotopes by Electromagnetic Processes, p. 1.

148. See Robley D. Evans to Robert F. Bacher, 31 Dec 1946, NARA College Park, RG 326, E67A, box 46, folder 3 Foreign Distribution of Radioisotopes vol. 1.

149. This followed an announcement seven months earlier that the AEC would distribute heavy water and deuterium gas, manufactured by the Stuart Oxygen Company of San Francisco, to qualified users. Press Release, "United States Atomic Energy Commission Announces Distribution of 'Heavy Water,'" 1 May 1947, NARA College Park, RG326, E67A, box 45, folder 13 Distribution of Stable Isotopes Domestic.

150. A list of the stable isotopes available for sale (some as compounds) is given in Isotopes Division, US Atomic Energy Commission, Oak Ridge, Tennessee, *Isotopes Catalogue and Price List No. 3, July 1949*, NARA College Park, RG 326, E67A, box 46, folder 6 Foreign Distribution of Radioisotopes, vol. 2, p. 35.

AEC prioritized research, such as nuclear measurements, that would not use up the material.[151] In 1948, the Isotopes Division made 98 shipments of 75 stable isotopes representing 25 different elements. However, when the Commission decided to discontinue electromagnetic separation of uranium-235, the future of the Y-12 production facility, including its stable isotope program, became uncertain. It was certainly costly. Production of stable isotopes in 1948 cost $2,000,000, including $500,000 in research costs.[152] Carbide proposed a reorganization to cut cost by half, and by 1950, the Isotope Research and Production Division of Y-12 became part of Oak Ridge National Laboratories' Isotopes Division.[153]

Part of the problem was that the demand for stable isotopes was not as great as that for radioactive isotopes. Stable isotopes could not be used in medical treatment, which was a main source of demand for AEC-produced radioisotopes (particularly iodine-131 and phosphorus-32). Biologists could obtain the two stable isotopes of greatest interest as tracers—carbon-13 and nitrogen-15—from commercial suppliers.[154] In addition, the equipment for detecting stable isotopes, namely mass spectrometers, was costly. To address this obstacle, the Isotopes Branch explored the possibility of providing services for mass detection at Y-12, but this did not materialize.[155] In the end, the Commission continued the expensive stable isotopes program because it benefited nuclear science at its own laboratories, especially weapons research.[156] American researchers could piggy-back on this agency priority. This had a side benefit for the AEC as well: the fact that university scientists employed these isotopes for their nonclassified research gave a program driven by military priorities a civilian function.

In the interests of becoming a one-stop shop, in 1947 the AEC's Iso-

151. Press Release, "United States Atomic Energy Commission Announces Distribution of Stable Isotopes," 4 Dec 1947, NARA College Park, RG 326, E67A, box 45, folder 13, Distribution of Stable Isotopes Domestic, p. 3; Memorandum from Walter J. Williams to Carroll L. Wilson, 29 Aug 1947, NARA College Park, RG 326, E67A, box 45, folder 13, Distribution of Stable Isotopes Domestic, p. 4.

152. AEC Info Memo 163, 6 Apr 1949, NARA College Park, RG 326, E67A, box 45, folder 7, Production of Stable Isotopes by Electromagnetic Processes, p. 2.

153. Quist, "Classified Activities" (2000), p. 21.

154. Aebersold, "Isotope Distribution" (1947). Eastman Kodak Company sold carbon-13 and nitrogen-15.

155. AEC Info Memo 163, 6 Apr 1949, NARA College Park, RG 326, E67A, box 45, folder 7, Production of Stable Isotopes by Electromagnetic Processes, p. 5.

156. Ibid., p. 4.

topes Division proposed a cyclotron program to supplement the reactor-based program.[157] Not all radioisotopes in demand could be produced in a reactor; some could be made only in a cyclotron. Moreover, as agency officials noted, university cyclotron laboratories were not equipped "to handle the voluminous administrative detail involved in correspondence related to distribution, screening of formal requests, shipment problems, and legal arrangements connected with sales."[158] Moreover, handling outside requests would interfere with cyclotron research programs. The AEC plan allowed the contractor at Oak Ridge to purchase cyclotron time from various institutions (MIT, the University of Pittsburgh, the University of California, the Carnegie Institution of Washington, and Washington University).[159] The isotopes were then shipped to Clinton Laboratories for processing and sale. In this way, the AEC sought to consolidate cyclotron-produced isotopes under its regulatory framework for safe handling, disposal procedures, and human uses.[160]

Supplying cyclotron-produced radioisotopes required heavier subsidies than reactor-generated isotopes; the AEC felt that they could charge no more than one-third of the actual production and processing costs.[161] Because cyclotrons had been constructed in many foreign countries, the proposed program restricted purchase to the United States. In addition, the initial program included only those radioisotopes with at least a thirty-day half-life. Even so, these included many that were valuable to

157. On outside support for this idea: L. F. Curtiss, Chair, Committee on Nuclear Science, National Research Council, "Proposal to the Atomic Energy Commission for the Distribution of Cyclotron-Produced Radioisotopes," 28 Sep 1948, Appendix "B"; and Walter J. Williams (Manager, Field Operations, Oak Ridge) to Carroll Wilson, 12 Jun 1947, "Proposal to Distribute Cyclotron-Produced Isotopes," Appendix "D" to AEC 195, "Program for Production and Distribution of Cyclotron-Produced Isotopes," 30 Mar 1949, all in NARA–College Park, RG326, E67A, box 45, folder 3 Distribution of Radioisotopes–Domestic.

158. "Program for Production and Distribution of Cyclotron-Produced Radioisotopes," NARA College Park, RG 326, E67A, box 45, folder 3 Distribution of Radioisotopes–Domestic, p. 2.

159. "AEC Announces Distribution Program for Cyclotron-Produced Radioisotopes," Press Release, 24 Jul 1949, NARA College Park, RG 326, E67A, box 45, folder 3 Distribution of Radioisotopes–Domestic.

160. "Program for Production and Distribution of Cyclotron-Produced Radioisotopes," NARA College Park, RG 326, E67A, box 45, folder 3 Distribution of Radioisotopes–Domestic, p. 6.

161. "Program for Production and Distribution of Cyclotron-Produced Radioisotopes," AEC 195, 18 Mar 1949, NARA College Park, RG 326 E67A, box 45, folder 3 Distribution of Radioisotopes–Domestic, p. 8.

biomedical research, including sodium-22, iron-59, zinc-65, arsenic-63, and iodine-125.[162] On March 23, 1949, the Commission approved the Isotopes Division's plan, with one condition. The agency required the staff to "insure [*sic*] that wherever possible, as the Commission's reactor program develops, isotopes distributed by the Commission are produced in reactors rather than cyclotrons."[163] There were technical trade-offs. Some isotopes that could be prepared in the reactor, such as iron-59, were more radiochemically pure when coming from a cyclotron.[164] The Commission's stated preference was not scientific, but political—only reactor-produced isotopes represented the civilian benefits of atomic energy.[165]

In a 1949 press release, the AEC referred to the cyclotron program as instigated to "round out a comprehensive program of distribution of radioactive materials."[166] The consequence of this one-stop isotope shop was the government-subsidized commodification of all these research materials. In part, the shift was necessary as consumption of radioisotopes dramatically increased; cyclotrons could not produce the quantities of carbon-14, phosphorus-32, iodine-131, and sulfur-35 demanded by researchers and physicians. But the fact that the new infrastructure grew out of the manufacturing sites for atomic weaponry brought radioisotopes under the umbrella of governmental nuclear surveillance (concerning both safety and security). Moreover, the federal government served many functions with respect to radioisotope consumption: supplying, subsidizing, promoting, regulating. The multiplicity of responsibilities grew out of the Atomic Energy Act, which mandated a government nuclear monopoly, but even-

162. "AEC Announces Distribution Program for Cyclotron-Produced Radioisotopes," Press Release, 24 Jul 1949, NARA College Park, RG 326, E67A, box 45, folder 3 Distribution of Radioisotopes–Domestic.

163. Extract from Status Report, 1–15 Jan 1950, "Program for Production and Distribution of Cyclotron-Produced Isotopes (AEC 195)," NARA College Park, RG 326, E67A, box 45, folder 3 Distribution of Radioisotopes–Domestic.

164. "A Review of the Possibility of Reactor Production of Isotopes Currently Produced by Cyclotron Bombardment," report by the Manager, Oak Ridge Operations Office, 20 Feb 1950, NARA College Park, RG 326, E67A, box 45, folder 3 Distribution of Radioisotopes–Domestic, p. 4. On the uses of iron-59 in medical research, see chapter 8.

165. "Procedures to Insure Distribution of Reactor-Produced Rather than Cyclotron-Produced Radioisotopes Wherever Possible," AEC 195/3, 26 May 1950, NARA College Park, RG 326, E67A, box 45, folder 3 Distribution of Radioisotopes–Domestic.

166. Press release, "AEC Distributes 8,363 Shipments of Radioactive and Stable Isotopes in Three Years," 3 Aug 1949, NARA College Park, RG326, E67A, box 45, folder 13 Distribution of Stable Isotopes Domestic. This program was discontinued in 1955; see chapter 6.

tually, the fact that the AEC both promoted and regulated atomic energy use was to be seen as a irremediable conflict of interest.[167]

Radioisotopes for War

The AEC made the argument that simply making radioisotopes available to civilian scientists would lead to greater military preparedness in the event of a nuclear war: "It has also been mentioned that should an atomic war occur, it would be essential that as many scientists as possible be trained in the technique of working with radioactive material."[168] More immediately, the radioisotopes manufactured at X-10 supplied both classified and unclassified research in other "Project" facilities. Oak Ridge also supplied various Army research units with radioisotopes. While these users filed the same forms as civilian purchasers, they were not required to disclose the planned uses of the research. As noted on an application from the Army's research center on biological warfare, "paragraphs 9, 10, and 11 of this form have been deleted inasmuch as compliance therewith is impossible due to the fact that the work performed at this installation is of a highly classified nature."[169]

Another military application of reactor-produced radioactive materials was their use as direct agents in warfare.[170] Even before the Manhattan Project got underway, scientists in the United States considered the prospect that the Germans might use "radioactive poisons" against Allied troops. Physicists Eugene Wigner and Henry DeWolf Smyth calculated that "fission products produced in one day's run of a 100,000 kw chain-reacting pile might be sufficient to make a large area uninhabitable."[171] Subsequent secret correspondence focused largely on defensive measures

167. Walker, *Containing the Atom* (1992).

168. "Background Material on Activity in First Year of Distribution of Pile-Produced Radioisotopes," Press Release, 2 Aug 1947, NARA College Park, RG 326, E67A, box 45, folder 3 Distribution of Radioisotopes–Domestic, p. 10.

169. John D. M. Shaw to Military Liaison Committee, 27 May 1948, NARA Atlanta, RG 326, MED CEW Gen Res Corr, Acc 67B0803, box 148, folder AEC 441.2 (R–Camp Detrick). After 1956 this facility was called Fort Detrick.

170. See Bernstein, "Radiological Warfare" (1985); Hacker, *Dragon's Tail* (1987), pp. 46–48; de la Bruheze, "Radiological Weapons" (1992).

171. Smyth, *Atomic Energy for Military Purposes* (1945), p. 65. Their report dated to 10 Dec 1941.

against radiological warfare.[172] A Military Section of the Met Lab's Health Division, set up to study radiological warfare, was short-lived.[173] Nonetheless, this agenda surfaced in the Manhattan Project's biological research programs. The investigations of fission products done in the name of occupational health and safety by Joseph Hamilton at Berkeley inspired him to advocate consistently for the military use of these materials as poisons.[174] In addition, wartime research at Clinton Laboratories included study of radiological warfare, both defensive and offensive. Director Kenneth Cole summarized his group's work along three lines:

> The group on experimental biology aims to investigate the unknown factors in the effects of radio-activity on organisms to provide a basis for (1) protection to personnel and the public in the process, (2) an estimate of the possible damage from enemy action, and (3) results of military use of the product.[175]

Oppenheimer and Fermi corresponded in 1943 about the development of radioactive agents from fission products, particularly the possible use of radiostrontium to poison food.[176] However, the momentum to build an atomic bomb soon overshadowed the quest for radiological weapons.

Discussions of radiological warfare reemerged after the war. The unexpected level of radioactive contamination caused by the underwater nuclear weapons test at Bikini (Test Baker) on July 23, 1946, revived interest in military uses of radioactive materials in warfare.[177] Preliminary tests conducted at Oak Ridge provided the basis for an AEC report in October 1947 on possible research and development programs for radiological warfare.[178] In the spring of 1948, a Joint AEC-National Military

172. See memorandum from J. C. Stearns to A. H. Compton and R. L. Doan, 25 Jul 1942, MMES/X-10/Vault, CF-42-7-11, DOE Info Oak Ridge, ACHRE document ES-00442.

173. Westwick, *National Labs* (2003), p. 242.

174. Grover, "All the Easy Experiments" (2005). Hamilton's interest in this topic continued after the war: Memorandum from Joseph G. Hamilton to Professor Ernest O. Lawrence, 28 Apr 1948, LBL Archives, ARO-998, folder Joseph G. Hamilton Radioactive Warfare (Project 48A) Reports, 1946–1948.

175. K. S. Cole, "Experimental Biology," rough draft, 21 Apr 1943, [for] Stone, S. Warren, and Cole, MMES/X-10/Vault, CF-43-4-33, DOE Info Oak Ridge, ACHRE document ES-00443.

176. Bernstein, "Oppenheimer" (1985); idem, "Radiological Warfare" (1985).

177. ACHRE, *Final Report* (1996), pp. 325–26.

178. de la Bruheze, "Radiological Weapons" (1992), p. 213.

Establishment Panel, chaired by chemist W. A. Noyes of the University of Rochester, began meeting.[179] By that fall, the group had submitted a report recommending the military and AEC pursue both offensive and defensive aspects of radiological warfare.[180] This became a major focus of the Armed Forces Special Weapons Project, established in 1949, which brought together representatives from all branches of the military to work with the AEC on the operational level, particularly on the development of atomic weapons.[181] Groves was the initial chief of this group; Nichols succeeded him in 1948.

Oak Ridge figured prominently in radiological warfare research, in part due to its special facilities and personnel. Like radioisotope distribution, radiological warfare was viewed as a suitable project for ORNL, whose research priorities were still being set.[182] Indeed, the same infrastructure—the X-10 reactor and radioisotope processing facilities—could be used to generate materials for both projects. There were several radioelements being considered as radiological warfare agents, such as zirconium and tantalum; others, such as lanthanum, could be useful in field tests.[183]

In a June 1948 meeting at Oak Ridge, four lines of radiological warfare investigation were planned at ORNL, two chemistry studies and two field tests.[184] The chemical projects concerned first, developing a process for recovering materials to use as radiological weapons from Hanford's plutonium production waste, and, second, studying the feasibility of irradiating tantalum on a large scale in the Hanford reactors. The two field tests used radioactive materials prepared in the X-10 reactor. These radiation sources were to be placed in specific locations so that the resulting dosage levels could be measured and mapped. First was a single-source field test, conducted using radioactive lanthanum (RaLa) in Oak Ridge on July 23,

179. The National Military Establishment was created in 1947, and renamed the Department of Defense in 1949.

180. ACHRE, *Final Report* (1996), p. 326; K. D. Nichols, "Conduct of Research and Development in Radiological Warfare. Includes Draft Motion on Radiation Warfare Policy," 13 Sep 1948, OpenNet Acc NV9757161.

181. Nichols, *Road to Trinity* (1987), p. 253.

182. "Minutes of the Discussions That Took Place in Oak Ridge June 28 and 29, 1948—Re: RW Program," MMES/X-10, Director's Files CF-48-7-110, DOE Info Oak Ridge, ACHRE document ES-00420.

183. For a discussion of various fission products as possible radiological warfare agents, see Ridenour, "Radioactive Poisons" (1950).

184. "Minutes of the Discussions."

1948.[185] The second was a grid test using a uniformly distributed source of radioactive tantalum. This second test took place on July 2, 1948, on a cleared plot of land near the K-25 plant, with five by seven rows of radioactive sources laid out on the 275 by 300 yard rectangle. Readings of the radiation levels were taken on the edges and centers of each grid square, as well as up to 100 yards outside the gridded area. According to Karl Z. Morgan (head of Health Physics at ORNL), the readings indicated that "1 megacurie per square mile would give a dose of about 3.5 roentgens/hr near the center."[186] Suitable radiological warfare agents would be expected to give someone a daily dose of 10–100 roentgens.[187] Hence achieving a dose high enough for injury would require very large radiation sources.

Both of these tests involved sealed sources of radioactivity, which, once removed, did not leave any radioactive contamination on the fields.[188] After these initial field tests at Oak Ridge, the Army Chemical Corps conducted sixty-five radiological warfare tests from 1949 to 1952 at Dugway Proving Ground, Utah, with radioactive tantalum prepared in Oak Ridge at the X-10 plant.[189] The administration at Oak Ridge determined that 700 curies of radioactive tantalum could be generated per month without interfering with the "irradiations of other materials now in progress."[190] In these tests, the radioactive tantalum was not in a sealed source but was dispersed in particulate form.

In 1952, the Chemical Corps proposed a larger test of 100,000 curies, but the following year such field tests were suspended. One reason that the military did not expand testing was that existing facilities, presumably the

185. This may be the same as a planned test for July 15, which apparently used three different single sources, of 1000, 100, and 10 curies respectively. Memorandum on AHRUU Project to Karl Z. Morgan, 14 Jul 1948, MMES/X-10, CF-48-7-171, DOE Info Oak Ridge, ACHRE document ES-00095.

186. Letter from Karl Z. Morgan to Carl B. Marquand, re: RW tests, 21 Jun 1949, MMES/X-10/Vault, CF-49-6-250, DOE Info Oak Ridge, ACHRE document ES-00453. On the layout, see Karl Z. Morgan, "Uniformly Distributed Source, ARUU Program," MMES/X-10, Director's Files ORNL-126, DOE Info Oak Ridge, ACHRE document ES-00422. Since 1928 the roentgen has been defined as the quantity of ionizing radiation (originally understood as x-rays) that would produce one electrostatic unit of charge in a cubic centimeter of air under standard conditions. Hacker, *Dragon's Tail* (1987), p. 16.

187. de la Bruheze, "Radiological Weapons" (1992), p. 219. This estimate is from 1950.

188. ACHRE, *Final Report* (1996), p. 326. A third test was apparently conducted in Oak Ridge in 1948.

189. Ibid.

190. "Minutes of the Discussions."

X-10 reactor, were insufficient to provide the greater amounts of radioactive sources needed. There were also safety concerns having to do with the hazards of radioactive dust being released in the tests in Utah. Instead, investigations of radiological warfare would have to rely on other activities as sources of information, particularly atmospheric nuclear tests.[191]

Conclusions

The wartime entry of the government into the production of radioisotopes did not make other suppliers immediately obsolete. Rather, the production of radioisotopes from cyclotrons and from reactors was concurrent and these systems were briefly entangled. On the one hand there were material exchanges, particularly between the reactor at Oak Ridge and the cyclotron at Berkeley. On the other hand, immediately after the war there was an uneasy relationship between cyclotron-associated scientists and physicians, who were used to controlling the supply and setting the rules about radioisotope usage, and the new government agency, with its own regulatory apparatus. The tensions were not between government and nongovernment entities (the Berkeley cyclotron laboratory had become a central part of the Manhattan Project and subsequently an AEC facility), but struggles over authority within the Manhattan Project-turned-civilian agency.

The US government's incursion into the economy of radioactive materials was not only as a regulator but also as a producer. In contrast with the cyclotron-based gift exchange of radioisotopes, the US government's system was industrialized and commodified. As a lieutenant colonel observed, "The initiation of the non-project isotope distribution program has placed a monetary value on isotopes produced."[192] In part this reflected the imprint of General Groves, who did not believe that radioisotopes should be provided free of charge. As he put it in his autobiography, "I did not think that it was in my province to give away material belonging to the United States; and, secondly, I thought the material would be much

191. ACHRE, *Final Report* (1996), pp. 326–27.

192. Lt. Col. Walter P. Leber to Mr. Prescott Sandidge, 31 Jul 1946, NARA Atlanta, RG 326, OROO Files Relating to K-25, X-10, Y-12, Acc 67A1309, box 14, folder Approvals. The quote continues, "It becomes necessary and desirable that all declassified isotopes produced, distributed and used be picked up on suitable accountable property records."

more carefully used if it were paid for than if it cost nothing."[193] Robley Evans, who ran the MIT cyclotron, concurred: "In all cases I believe that a charge should be made for radioactive materials supplied. Even surplus Government property is sold, not given away."[194] MIT set a precedent for the government sales of radioisotopes by charging for their shipments, most of which were to scientists working on wartime contracts. Lawrence's patent on the cyclotron would have affected postwar commercialization of radioisotope production using that technology, whereas the Atomic Energy Act forbade patenting most nuclear technologies.[195]

The Atomic Energy Act also guaranteed a government monopoly over reactor-produced radioisotopes. The new economy of radioisotopes may have been monetized, but it was not a free market. The AEC subsidized radioisotope production specifically to encourage purchasing. Such government policy, which verged at times on propaganda, met with little resistance—there was in fact a growing demand for these materials before the end of the war. In effect, the AEC regarded radioisotope provision as a kind of public service, and did not expect sales to be revenue-generating for the agency. Yet some aspects of the gift exchange system remained, insofar as the AEC expected its provision of affordable radioisotopes to have political benefits. However, concerns about potential military applications of radioisotopes complicated the agency's efforts to garner public support, not only in the United States but also abroad, through export of government-controlled isotopes to foreign scientists and physicians.

193. Groves, *Now It Can Be Told* (1962), p. 386.

194. Robley D. Evans to Lee Dubridge, 10 Apr 1946, Evans papers, box 1, folder 1 Isotopes–Clinton Lab.

195. Wellerstein, "Patenting the Bomb" (2008).

CHAPTER FOUR

Embargo

There is nothing secret or evil about radioisotopes in the forms in which they are sold in this country and abroad. While their utilization cannot significantly advance the atomic energy programs of nations, they can contribute, and are contributing, significantly to advancements in basic science, medicine, agriculture and industry. As of today, isotopes constitute the single most important contribution of atomic energy to peacetime welfare.—AEC Press Release, 1951[1]

In the initial years of the Atomic Energy Commission, radioisotopes became political instruments, used by the US government in its promotional efforts and by critics of the civilian control of atomic energy. Congress established a civilian agency for atomic energy, at the urging of scientists, with the expectation that peacetime benefits would soon materialize. But the controversies the AEC faced in the immediate postwar years, particularly whether to ship radioisotopes to foreign scientists, revealed the agency's political vulnerabilities. Scientists outside the United States found access to the bountiful radioisotopes generated through the Manhattan Project infrastructure entangled in the politics of national security, in which the ideals of internationalism in science vied with American suspicions of communists abroad.[2] National security concerns effectively resulted in a year-long embargo on US-produced isotopes to foreign institutions. Anxieties that sending radioisotopes abroad might threaten American nuclear supremacy continued to reverberate in Congressional politics and the national press through the early 1950s.

1. Quote from Atomic Energy Commission in Proposed Press Release, "AEC Enlarges Radioisotope Export Program," Appendix E to AEC 231/16, NARA College Park, RG 326, E67A, box 47, folder 1 Foreign Distribution of Radioisotopes vol. 3, p. 16.

2. Smith, *Peril and a Hope* (1965); Wang, *American Science* (1999); Slaney, "Eugene Rabinowitch" (2012).

The Atomic Energy Act of 1946 was aimed at protecting (or at least prolonging) the American atomic monopoly; to this end it prohibited the export of fissionable materials. At the same time, the bill charged the new agency with promoting civilian uses of atomic energy, authorizing the distribution of so-called by-product materials (namely, radioisotopes) for peaceful uses. The original wording of the announcement of the availability of pile-produced radioisotopes referred to "national distribution," and there was also a general understanding at Oak Ridge that domestic needs were to be filled first.[3] But foreign purchasers were not merely at the back of the line; they could not purchase the AEC's isotopes at all. As one physicist reported to the agency at the end of 1946, "Although no one in this country knows of any regulations against sending isotopes to foreign users, the conviction is wide spread abroad that scientists in this country are unwilling to share their materials."[4] For European scientists such as George de Hevesy, who had received radiophosphorus from Berkeley throughout the late 1930s and early 1940s, the US government effectively disrupted the availability of American-generated isotopes.

Restrictions against sending radioisotopes abroad did not come from the leadership of the Manhattan Project; the Army was inclined to share the fruits of government reactors, at least with scientists in the British and Canadian atomic energy programs. The Truman administration, however, took a dim view of the prospect of continuing any nuclear exchanges after the war. In the summer of 1947, the issue of foreign distribution came to a head among the five AEC Commissioners, who voted—without achieving unanimity—to allow export. They justified their decision by appealing to the Marshall Plan, not Anglo-American military cooperation. In announcing the program, Truman spoke of the foreign shipments as securing "greater international cooperation in the field of medical and biological research."[5] The first shipments of radioisotopes reached foreign hospitals and laboratories in the fall of 1947.

3. "Availability of Radioactive Isotopes" (1946). On the filling of domestic needs first, see "Foreign Distribution of Radioisotopes," Appendix A to Memorandum by J. B. Fisk, Director, Division of Research, to Carroll L. Wilson, General Manager, 13 Aug 1947, NARA College Park, RG 326, E67A, box 46, folder 3 Foreign Distribution of Radioisotopes vol. 1, p. 8.

4. Robley D. Evans to Robert F. Bacher, 31 Dec 1946, NARA College Park, RG 326, E67A, box 46, folder 3 Foreign Distribution of Radioisotopes vol. 1.

5. Telegram from President Truman to E. V. Cowdry, President of the Fourth International Cancer Research Congress, 3 Sep 1947, printed with Press Release, 4 Sep 1947, NARA College Park, RG326, E67A, box 46, folder 6 Foreign Distribution of Radioisotopes vol. 2.

As the Cold War intensified, conservative Congressional watchdogs of the AEC monitored the agency's foreign radioisotope shipments with suspicion. In 1949, they alleged that certain shipments of isotopes to Norway and Finland were undermining national security. The Commission never altered its policy—in fact, the agency expanded the purview of exports to include industrial shipments in the early 1950s—but these shipments featured in the 1949 Congressional Investigative Hearings of the agency as purported evidence of the AEC's national security lapses. Just a few months later, the explosion of the first Soviet atomic bomb shattered the US monopoly on nuclear weapons.[6] In addition, the governments of Britain and Canada began selling radioisotopes to foreign purchasers, with fewer restrictions than the AEC. The US policy based on denying radioactive materials and nuclear technology to other nations had become pointless. In the early 1950s, President Eisenhower took a new approach to American nuclear supremacy in his Atoms for Peace program, shifting the emphasis from guarding secrets to sharing technology. His depiction of radioisotopes as tools of international diplomacy, in conjunction with the 1954 revisions to the Atomic Energy Act permitting greater (though controlled) access to nuclear materials and technology, finally quelled political suspicions about foreign shipments.[7]

Two perceptions about radioisotopes gave them their political salience in the early Cold War. First, civilian uses were distinguished from military uses along disciplinary lines, with biology and medicine seen as inherently civilian and physics and engineering as intrinsically military. The popular view that nuclear physics research was unavoidably related to atomic weapons development led the AEC to emphasize medical therapy and biological research in the export program, which the agency represented as a strictly humanitarian endeavor.[8] Notably, the most controversial radioisotope exports went to physical scientists in Finland and Scandinavia. Congressional critics alleged that these radioisotopes could end up in the hands of the Soviet military, undermining American national security. Second, critics of these shipments, especially dissenting Commissioner Lewis Strauss, insinuated that the sharing of nuclear materials in the form

6. Gordin, *Red Cloud at Dawn* (2009).

7. Krige, "Atoms for Peace" (2006) and, on the earlier period, idem, "Politics of Phosphorus-32" (2005).

8. Creager and Santesmases, "Radiobiology" (2006); Santesmases, "Peace Propaganda" (2006).

of isotopes was equivalent to the dissemination of nuclear information to foreign powers in a way prohibited under the 1946 Atomic Energy Act. Conservatives used anxieties about the loss of the US atomic monopoly to denounce the international circulation of the AEC's radioisotopes.

The Politics of Foreign Distribution of "American" Isotopes

From the fall of 1945 to the spring of 1946, legislation over atomic energy stalled in Congress. One of the sticking points concerned the level of scientific and technical exchange that the United States would have with its former allies.[9] Roosevelt and Churchill had negotiated the so-called Quebec Agreement in 1943, enabling exchanges of technical information among the American, British, and Canadian participants in the bomb project.[10] The British government hoped to continue Anglo-American cooperation in the postwar period. Truman's administration, however, viewed the sharing of technical information as antithetical to the American aim of maintaining a nuclear monopoly. The US Congress overwhelmingly agreed: the McMahon Bill that President Truman signed into law on August 1, 1946, forbade the government "to share information about nuclear technology with any other power."[11]

At the same time that the bill limited the sharing of information, it also authorized distribution of so-called by-product material. The predominant by-products of nuclear fission were radioactive isotopes, both those generated from uranium-235 decay and target materials exposed to the neutron flux of the reactor. This clause enabled the Manhattan Engineer District (MED) to commence their radioisotope distribution program. In early August, radioisotope shipments that had been pending all summer began being sent out.

In the meantime, the MED was receiving requests for isotopes from foreign institutions. The military leadership of the Manhattan Project

9. See Mallard, "Quand l'expertise se heurte" (2006); idem, *Atomic Confederacy* (2008).

10. Hewlett and Duncan, *Atomic Shield* (1969), chapter 8.

11. Ball, "Military Nuclear Relations" (1995), p. 440. The law stated "there shall be no exchange of information with other nations with respect to the use of atomic energy for industrial purposes," though strictly scientific information exchange was permitted. This inconsistency was itself a problem for the AEC. "Atomic Energy Act of 1946" (1946); Paul, *Nuclear Rivals* (2000).

also received direct appeals for shipments from British and Canadian researchers affiliated with the atomic energy programs in those countries. These radioisotope requests drew on the wartime precedent of technical cooperation. On September 9, 1946, J. D. Cockcroft wrote to General Leslie Groves about outstanding requests for isotopes from British scientists. One such request, for 10 millicuries of phosphorus-32 for biological research, had been transmitted over five months earlier. Groves told Cockcroft that the MED could provide radioactive isotopes to the United Kingdom and Canada prior to the formal establishment of the AEC.[12] He sent an identical letter to W. B. Lewis at Chalk River, Canada, in response to requests there.[13] As it developed, the purchase orders were not received in time to permit shipment before the new year, so the Army deferred action to the new agency.[14]

At midnight on December 31, 1946, most of the infrastructure of the MED was legally transferred to the civilian AEC. The agency was to be headed by a panel of five Commissioners: David J. Lilienthal (chair), Robert Bacher, Lewis Strauss, Sumner T. Pike, and William W. Waymack. The five Commissioners met for the first time on January 2, 1947.[15] Colonel Kenneth D. Nichols, the Army liaison to the AEC selected by Groves, updated them on the civilian radioisotope program launched by the MED. He explained that domestic orders were already being filled, but action on foreign requests had been postponed. He emphasized the special status of requests from Harwell and Chalk River, as they originated from scientists who participated in the atomic bomb project:

12. Letters from L. G. Groves to J. D. Cockcroft at Harwell, England and W. B. Lewis at Chalk River, Canada, both 24 Oct 1946, NARA Atlanta, RG 326, OROO Lab & Univ Div Official Files, Acc 68A1096, box 13, folder Isotopes–3. The request for phosphorus-32 had been transmitted by Cockcroft on behalf of the UK's Medical Research Council.

13. At this stage, the Canadians had requested various preparations of phosphorus-32, carbon-14, and sulfur-35; their colleagues at Harwell requested these same isotopes and also zinc-65 and calcium-45. Memorandum from Colonel K. D. Nichols to Carroll L. Wilson, 15 Jan 1947, NARA College Park, RG 326, E67A, box 46, folder 3 Foreign Distribution of Radioisotopes vol. 1.

14. Memorandum re: Distribution of Radioisotopes Abroad, from E. E. Huddleson, Jr., Deputy General Counsel, to Carroll L. Wilson, General Manager, 5 Mar 1947, NARA–College Park, RG 326, E67A, box 46, folder 3 Foreign Distribution of Radioisotopes vol. 1.

15. On November 1, 1946, Truman gave the five Commissioners recess appointments; their confirmation hearings lasted through March 1947. Hartmann, *Truman and the 80th Congress* (1971), p. 32.

> In view of the past collaboration in atomic energy matters among Canada, Great Britain, and the United States, it is recommended that the AEC approve continuance of the policy of making available for sale to Canada and Great Britain isotopes currently on sale for unclassified off-Project use in this country and surplus to U.S. requirements.[16]

Most of the foreign requests, however, were unconnected to the wartime alliance. By March 1947, twenty nations were represented among requests for isotopes from the US government.[17]

As Carroll Wilson, the AEC's General Manager, saw it, the agency's position in the forefront of the radioisotope field had been "forced upon it more or less by accident," as a nonmilitary application of their nuclear reactors. Thus the Commission found itself "custodian of these peculiar resources," radioactive isotopes.[18] Most scientists, especially those who had worked in the Manhattan Project, felt that the isotopes produced by the AEC should be made available to researchers as readily as possible. Paul Aebersold, who had directed the isotope distribution program since its establishment within the MED, strongly advocated opening it to foreign scientists. But the Commission hesitated to authorize international distribution, even though a memorandum prepared on "the case against sale of radioisotopes abroad" rebutted the most obvious objections.[19]

The Commission's indecisiveness on the question of isotope distribution was partly attributable to political battles on other fronts. The confirmation hearing of Lilienthal, former head of the Tennessee Valley Authority (TVA), proved contentious, dragging out until March 1947. Republicans had gained a majority in the November 1946 elections, and the hearings

16. Memorandum from Colonel K. D. Nichols to Carroll L. Wilson, 15 Jan 1947, NARA College Park, RG 326, E67A, box 46, folder 3 Foreign Distribution of Radioisotopes vol. 1.

17. The countries listed by MED officials include Argentina, Australia, Belgium, Bolivia, Brazil, Canada, Chile, Cuba, England, France, Holland, Iceland, Mexico, New Zealand, Peru, Portugal, Russia (not listed as the USSR), Spain, Sweden, and Switzerland. Colonel K. D. Nichols to Carroll L. Wilson, 21 Jan 1947, and Colonel C. G. Haywood to Bennett Boskey, both in NARA College Park, RG 326, E67A, box 46, folder 3 Foreign Distribution of Radioisotopes vol. 1.

18. "Foreign Distribution of Radioisotopes," 13 Aug 1947, Appendix A to memorandum from Carroll L. Wilson to J. B. Fisk, of same date, NARA College Park, RG 326, E67A, box 46, folder 3 Foreign Distribution of Radioisotopes vol. 1, p. 4.

19. Memorandum from W. A. Shurcliff to Bennett Boskey, 14 May 1947, subject: The Case Against Sale of Radioisotopes Abroad, copy in Strauss papers, AEC Files, folder Isotopes Jan–Aug 1947.

gave them a chance to object to Lilienthal's New Deal career and raise once again the question of whether atomic energy should be under military control.[20] While the hearings were underway, conservative syndicated columnist Drew Pearson disclosed the existence of a network of nuclear spies in Canada that allegedly included government officials.[21] Critics of Lilienthal fixed on the specter of atomic espionage. Senator Robert A. Taft of Ohio described Lilienthal as "too 'soft' on issues connected with communism and Soviet Russia."[22] That spring, reports surfaced in the press that "secret files" had been lost or stolen from the AEC's laboratories, necessitating that the Commission, and especially Lilienthal, defend the very legitimacy of a civilian atomic agency just a few months after its creation.

Congressman J. Parnell Thomas, who presided over the House Un-American Activities Committee, was especially critical of Lilienthal. He published two incriminating magazine articles on the AEC in June. The first, appearing in *American* magazine, claimed that atomic energy patents, including information held secret during the war, were now available from the US Patent Office to the Soviets or anyone else who wanted to build an atomic bomb.[23] The second article, printed in *Liberty* magazine and leading to further newspaper investigations and stories, attacked the security system at Oak Ridge, citing missing and possibly stolen classified documents.[24] The FBI soon reported to Lilienthal that other classified materials had been missing from Los Alamos for a year and only recently retrieved. Critics of the AEC called for military takeover of the agency on account of its ineptitude in managing security—despite the fact that some of the alleged thefts had taken place while the labs were under control of the Army.[25]

20. Hartmann, *Truman and the 80th Congress* (1971), pp. 31–35; Wang, *American Science* (1999), pp. 160–61.

21. Craig and Radchenko, *Atomic Bomb* (2008), p. 121.

22. As quoted in Hewlett and Duncan, *Atomic Shield* (1969), p. 11.

23. Wellerstein, "Patenting the Bomb" (2008); idem, *Knowledge and the Bomb* (2010).

24. Thomas claimed that the article was based on information gleaned when he visited the facility with Robert E. Stripling, an investigator for the House Un-American Activities Committee (Hewlett and Duncan, *Atomic Shield* [1969], p. 89). Thomas, "Russia Grabs Our Inventions" (1947); Thomas and Jones, "Reds in Our Atom-Bomb Plants" (1947). Lilienthal, for his part, was appalled by the misrepresentations, writing: "Such a disgraceful performance does make me a bit vomity." Lilienthal, 7 Jun 1949, *Journals* (1964), vol. 2, p. 190.

25. Hewlett and Duncan, *Atomic Shield* (1969), pp. 88–95, 324.

Lilienthal quite reasonably approached the prospect of radioisotope export with skepticism. So far as he could see, sending isotopes abroad would not promote "international accord on control of atomic energy."[26] The benefits of sharing radioisotopes with foreign scientists seemed meager compared with the prospect that American openness might speed the development of atomic weaponry elsewhere. "Radioactive isotopes, which could aid fundamental research in France and Sweden, say, and perhaps add a bit to a restoration of the international fraternity of knowledge, set off [my] train of thought that leads to a bitter isolation, and something more aggressive than isolation."[27] However, countering this isolationist inference was the fact that many of the uranium deposits that the AEC needed for its plutonium production were located in other countries.[28] Moreover, Lilienthal recognized that the issue of radioisotope distribution, like other aspects of atomic energy control, revealed contradictions in US foreign policy: the aim of slowing the Russian development of the bomb was in tension with "a doctrine of aiding and assisting in the restoration of Europe, which Truman and Marshall are pressing these days."[29] Over time, he came to view the export of radioisotopes as part of a larger system of exchange that would benefit US interests.

Lilienthal's evolving views were informed by the agency's trusted scientists. At the beginning of June, the General Advisory Committee (GAC) of the AEC discussed whether the agency should make isotopes available to researchers abroad. The GAC, headed by J. Robert Oppenheimer, was composed entirely of physical scientists; the group wielded substantial influence on the Commission in its early years. They strongly supported foreign distribution of the AEC's radioisotopes, arguing it would "prove that this democratic country will do all it can, consistent with its own defense and security, to improve the public welfare and raise the standard of living throughout the world."[30]

26. Lilienthal, 7 Jun 1947, *Journals* (1964), vol. 2, p. 190. As author of the Lilienthal-Acheson report, he had thought substantially about the issue of the international control of atomic energy.

27. Ibid., p. 191.

28. Colonel C. G. Haywood to Bennett Boskey, 31 Jan 1947, NARA College Park, RG 326, E67A, box 46, folder 3 Foreign Distribution of Radioisotopes vol. 1.

29. Lilienthal, 7 Jun 1947, *Journals* (1964), vol. 2, p. 191.

30. Draft of proposed public statement on release of isotopes abroad, prepared from discussions of the General Advisory Committee, 1 Jun 1947, Oppenheimer papers, box 176,

On June 5, 1947, the Commission took up the GAC's recommendation that it release radioisotopes to foreign scientists. Cited in the meeting's minutes is a variety of "pro" and "con" factors. The main perceived risk was that sharing isotopes might compromise American military supremacy.[31] None of the radioisotopes under the proposed export program could directly assist in the development of atomic weapons. However, obtaining experience with isotopes would help scientists in other nations develop atomic energy and might contribute to the pursuit of radiological warfare. There was also the problem of controlling secondary distribution. Strauss had already observed that if the agency was unwilling to ship radioisotopes to the Soviet Union, it should not let them outside American borders at all, since it could not control their ultimate destination.[32]

On the other side of the ledger, there were many advantages to a policy of controlled sharing. First, publications from foreign scientists using radioisotopes would benefit American researchers. Second, making radioisotopes available outside the United States, as a step away from American isolationism, would generate political goodwill among foreign nations. It would show the world that the AEC was not dominated by the military. As Lilienthal saw it, restoring a sense of the "international fraternity of knowledge" was crucial to shoring up alliances with European "friends," in the spirit of the Marshall Plan.[33] Third, sharing radioisotopes might help the United States negotiate access to uranium ores, needed for continued nuclear weapons production, that were located in other countries or their colonial possessions.[34] Fourth, the radioisotope program could be used to gain information about the nuclear programs in other countries, bolstering national security through intelligence. As one memorandum put it:

---

folder GAC–Radioisotopes, Foreign Distribution. On the GAC, Sylves, *Nuclear Oracles* (1987).

31. Minutes from 62nd AEC Meeting, 5 Jun 1947, NARA College Park, RG 326, E67A, box 46, folder 3 Foreign Distribution of Radioisotopes vol. 1.

32. Lewis L. Strauss, memorandum to Robert F. Bacher, 23 May 1947, Strauss papers, AEC Files, folder Isotopes Jan–Aug 1947. Interestingly, there is also a copy in the Hickenlooper papers, Senate Committee Files, folder JCAE–Isotopes Jan–Aug 1947, suggesting that Strauss was passing information to him about radioisotope exports.

33. Lilienthal, 7 Jun 1949, *Journals* (1964), vol. 2, pp. 190–91. On the significance of science in American-European relations, see Krige, *American Hegemony* (2006).

34. Haywood to Boskey, 31 Jan 1947.

> It will presumably be useful to the United States to know that certain foreigners are busy with no more mischievous work than radioisotope research; conversely, the absence of certain expected names from the list of applicants might suggest inquiry into the possibility that these men are employed against the interests of the United States.[35]

There was, in addition, an element of urgency: the political benefits would be greatest if the United States acted before other nations developed production-scale nuclear reactors with which to supply radioisotopes.

During the summer of 1947, while the Commissioners equivocated about radioisotope export, scientists became increasingly frustrated with the AEC's restrictive policy. American physicist Robley Evans described British biologists, who were relying on the Cavendish cyclotron for minute supplies of phosphorus-32 and sodium-24, as "materially handicapped by the almost complete lack of isotopes."[36] British and European researchers complained bitterly to their American colleagues about the refusal of the US government to share the fruits of their atomic reactors:

> It illuminates the ill feeling of foreign scientists at the failure of the United States, bountiful as it is in radioisotope production and development, to share even a small fraction of these useful and unclassified by-products with other nations. Many of us regret to find that scientists of countries, which normally are friendly to the United States, have become impatient and view with much disfavor our continued exercise of scientific monopolies in this and other ways.[37]

The AEC's isotope embargo gave the impression that the military controlled American science. Caltech physicist Charles Lauritsen wrote Commissioner Bacher, "It is quite generally believed that most laboratories here are financed and controlled by the military and that there is little or no freedom of research and publication."[38] Aebersold echoed this sense, asserting that the American refusal to share its radioisotopes "goes as far

35. "Foreign Distribution of Radioisotopes," 13 Aug 1947.

36. Robley D. Evans to Robert F. Bacher, 15 Jul 1947, NARA College Park, RG 326, E67A, box 46, folder 3 Foreign Distribution of Radioisotopes vol. 1.

37. Memorandum from Paul C. Aebersold to Walter J. Williams, 6 Aug 1947, NARA College Park, RG 326, E67A, box 46, folder 3 Foreign Distribution of Radioisotopes vol. 1.

38. Charles C. Lauritsen to Robert F. Bacher, 25 Jun 1947, NARA College Park, RG 326, E67A, box 46, folder 3 Foreign Distribution of Radioisotopes vol. 1.

as to class us in somewhat the same light as Russia on scientific and political matters."[39]

The discontent of scientists spilled into the press on July 21, 1947, through an editorial entitled "Scientific Monopoly" in the *New York Herald Tribune* (whose internationalist orientation reflected its liberal Republican ownership). The piece criticized the AEC for changing national policy from openness in sharing radioisotopes to withholding them. "Before the war, when isotopes were made in minute quantities by cyclotrons, America was magnanimous enough to ship tiny amounts to foreign scientists; today, when the supply is comparatively huge, the nation holds on to it grimly."[40] Even small amounts of radioisotopes would be materially significant to European scientists, without in any way endangering the American nuclear monopoly. As a correspondent of Niels Bohr wrote the AEC,

> Bohr says, and I agree, that one of the most useful, convincing, and friendly things we can do is to send now to Europe some small quantities of the biologically useful isotopes, tracer elements, etc. We both know (and Bob Evans, recently here, will underscore this) that *even the bottle-washings* we throw away can be used literally for months of research over here. And nothing, obviously, can possibly be adduced regarding fissionable material.[41]

Even cyclotron-produced isotopes from American laboratories were less available to foreign scientists. George Hevesy, who had received small amounts of radiophosphorus from the Berkeley Rad Lab from 1937 through 1940, began asking E. O. Lawrence again in 1946 for a few isotopes. Lawrence was obliged to reply, "The whole of our laboratory now is being supported by the Army and those in authority have informed me that until all demands for phosphorus in this country are reasonably met we should not undertake to send the material abroad."[42]

39. Memorandum from Paul C. Aebersold to Walter J. Williams, 6 Aug 1947, NARA College Park, RG 326, E67A, box 46, folder 3 Foreign Distribution of Radioisotopes vol. 1.

40. "Scientific Monopoly" (1947). There is a copy in NARA College Park, RG 326, E67A, box 46, folder 3 Foreign Distribution of Radioisotopes vol. 1. On the *New York Herald Tribune*, see Kluger, *The Paper* (1986).

41. Albert Stone to US Naval Research attaché, 1 Jul 1947, NARA College Park, RG 326, E67A, box 46, folder 3 Foreign Distribution of Radioisotopes vol. 1. Emphasis in original.

42. E. O. Lawrence to George Hevesy, 10 Aug 1946, EOL papers, series 1, reel 13, folder 9:7 Hevesy, George C. de. Paul Aebersold advised another scientist from Copenhagen who

By mid-1947, Oak Ridge had ninety-six unfilled foreign requests for radioisotopes, of which seventy-three were for medical research and therapy. Nearly half were from England and continental Europe, including researchers in Belgium, Denmark, France, Holland, Italy, Portugal, Spain, and Sweden.[43] The agency's Division of Research, which oversaw isotope sales, had drawn up a provisional policy to allow distribution of twenty-eight different radioisotopes of nineteen elements to foreign scientists. The list included those isotopes of greatest interest to biomedical researchers.[44] Foreign distribution of naturally radioactive elements—those with atomic number higher than 83—was not authorized, and the only fission product that could be exported was iodine-131, on account of its importance in medical therapy. The memorandum asserted that the export program created "no hazard of possible use in some form of poison warfare."[45] Export of radioisotopes would "help overcome a hostile feeling of foreign scientists toward our own scientists and assist in reestablishing the internationalism of science."[46] Production of the four isotopes of greatest biological interest—carbon-14, iodine-131, phosphorus-32, and sulfur-35—was already sufficient at Oak Ridge to supply foreign as well as domestic demand.[47]

---

wanted sodium-22 to request it from one of the cyclotrons at MIT, Carnegie Institution, or Ohio State. "These are privately owned and operated machines and are not under the supervision of the United States Atomic Energy Commission." Aebersold to Lorin J. Mullins, 17 Jun 1947, NARA Atlanta, RG 326, MED CEW Gen Res Corr, Acc 67B0803, box 158, folder AEC 441.2 (R–Institute for Theoretical Physics).

43. Appendix, "List of Foreign Countries from which Requests for Isotopes Have Been Received," NARA College Park, RG 326, E67A, box 46, folder 3 Foreign Distribution of Radioisotopes vol. 1. See also "Foreign Distribution of Radioisotopes," memorandum with letter from David E. Lilienthal to George C. Marshall, 27 Aug 1947, Hickenlooper papers, Senate Committee Files, folder JCAE–Isotopes, 1947–48.

44. Namely carbon-14, calcium-45, iodine-131, phosphorus-32, sodium-24, and sulfur-35. Memorandum from J. H. Manley to Carroll L. Wilson, General Manager, 24 Jun 1947, NARA College Park, RG 326, E67A, box 46, folder 3 Foreign Distribution of Radioisotopes vol. 1.

45. "Memorandum to the Department of State," Appendix B to Memorandum by J. B. Fisk, Director, Division of Research, to Carroll L. Wilson, General Manager, 13 Aug 1947, NARA College Park, RG 326, E67A, box 46, folder 3 Foreign Distribution of Radioisotopes vol. 1, p. 5.

46. J. H. Manley to Carroll L. Wilson, 24 Jun 1947, p. 4.

47. Memorandum from Walter J. Williams to Carroll L. Wilson, General Manager, 19 Jun 1947, NARA College Park, RG 326, E67A, box 46, Folder 3, Foreign Distribution of Radioisotopes vol. 1.

On August 19, when the Commissioners voted on the pending proposal, the tally was four-to-one in favor of radioisotope export.[48] Strauss, the holdout, did not believe that the safeguards proposed by the AEC would prevent radioisotopes from being used to military advantage by other nations. This risk trumped all other benefits in his assessment. He regarded his fellow Commissioners as naive in presuming "that scientists are practically all on our side of the international political argument."[49] Strauss was suspicious that foreign scientists (including Communist sympathizers in Europe) would channel these resources to aid the Soviet Union and its client states.[50] He also questioned whether radioisotopes would enable the United States "to buy the good will of foreign scientists." In the end, he viewed the distribution of radioisotopes to foreigners as "a serious breach of security comparable with that of the publication of the Smyth report."[51]

In the eyes of the other four Commissioners, the damage to the credibility of the United States if it failed to export radioisotopes outweighed the risk. In addition, since their last discussion of the issue, the first Canadian reactor at Chalk River had become operational. For the Commission majority, this made action urgent:

> The United States should take the leadership in this matter rather than reluctantly follow the actions of Canada or Britain. By denying the foreign distribution of radioisotopes at this time the United States will be giving unfriendly countries a propaganda weapon that might be more hurtful to national security

48. Hewlett and Duncan, *Atomic Shield* (1969), p. 109–10.

49. Memorandum from Lewis Strauss to Carroll L. Wilson, 25 Aug 1947, on Foreign Distribution of Radioisotopes, NARA College Park, RG 326, E67A, box 46, folder 6 Foreign Distribution of Radioisotopes vol. 2, p. 1.

50. Atomic Energy Commission, Minutes of Meeting No. 95 at Bohemian Grove, 19 Aug 1947, NARA College Park, RG 326, E67A, box 46, folder 6 Foreign Distribution of Radioisotopes vol. 2.

51. Ibid., p. 201. The Smyth Report was the first official account of the Manhattan Project, which many conservatives regarded as providing too much technical detail about atomic weaponry. Smyth, *Atomic Energy for Military Purposes* (1945). Strauss claimed he did not oppose sending isotopes for therapy, and even approved of export of radioisotopes for the purpose of "basic scientific research or instruction." However, he felt that radioisotope export for any military or industrial use should be prohibited. The AEC policy was not restrictive enough for his liking. See Strauss, *Men and Decisions* (1962), pp. 258–59.

> than would be any possible harm in the release of radioisotopes under appropriately safeguarded conditions.[52]

Waymack observed that radioisotope shipments would accord with the goals of the Marshall Plan, "the number one key to the policy of the United States."[53] The Department of State approved the AEC's policy at the end of August.[54]

At the opening of the Fourth Annual International Cancer Research Congress in St. Louis on September 3, 1947, President Truman announced that the US AEC's radioisotopes would be made available to foreign scientists "principally for medical and biological research." The decision was framed as enabling the "open, impartial, and truly international character of medical research [to] carry over into the realm of other problems of world concern."[55] Not only could the AEC reap the political benefit from being able to associate their announcement with cancer research, but also E. V. Cowdry, who had received the agency's first official radioisotope shipment on August 2, 1946, headed this conference. Accordingly, the setting highlighted continuities with the domestic distribution program. The press release suggested that the US government was making isotopes available to foreigners at this time as a consequence of an increase in production at Oak Ridge rather than a change in policy.[56] European observers greeted the announcement as the lifting of a "ban against the exportation

52. Minutes of the Atomic Energy Commission on Foreign Distribution of Radioisotopes, 19 Aug 1947, copy in Oppenheimer Papers, box 186, folder Isotopes–Miscellaneous Information, p. 3.

53. Ibid.

54. Summary of actions at Commissioners' meetings re: foreign distribution of radioisotopes, NARA College Park, RG 326, E67A, box 46, folder 6, Foreign Distribution of Radioisotopes vol. 2.

55. Telegram from President Truman to E. V. Cowdry, President of the Fourth International Cancer Research Congress, 3 Sep 1947, printed with Press Release, 4 Sep 1947, NARA College Park, RG326, E67A, box 46, folder 6 Foreign Distribution of Radioisotopes vol. 2.

56. Press Release, "Radioisotopes for Medical and Biological Research Available to Users Outside United States," 4 Sep 1947, NARA College Park, RG 326, E67A, box 46, folder 6 Foreign Distribution of Radioisotopes vol. 2. See also "Radioisotopes for International Distribution Catalog" (same folder); Memorandum from E. E. Huddleson, Jr., Deputy General Counsel, to Carroll L. Wilson, General Manager, 5 Mar 1947, NARA College Park, RG 326, E67A, box 46, folder 3 Foreign Distribution of Radioisotopes vol. 1.

of radioactive isotopes to scientists of foreign countries."[57] In fact, though the public presentation stressed the medical and biological uses of radioisotopes, the AEC did not prohibit shipments to foreign scientists working outside of those fields. But neither did they advertise this, and some scientists in Europe remained under the impression that the AEC would export radioisotopes only for medical use.[58]

Truman's wording suggested that radioisotope exports could be a successful element in the contentious diplomatic arena of international control of atomic energy. Some officials in the AEC had gone a step further, explicitly comparing the foreign distribution program with proposed plans for international control of nuclear weapons. An agency memorandum rebutting "The Case Against Distribution" asserted, "There is a parallel between this proposal and the Lilienthal-Baruch proposal for the international control of atomic energy. Both require a degree of openness of activities to individuals associated with the agency which owns the raw material."[59] The linkage proved inopportune in the weeks following Truman's announcement: In September, negotiations at the United Nations with the Soviets over the terms of international control reached a standstill.[60] Strauss drafted a memorandum to his fellow Commissioners asking them to consider suspending all foreign shipments "until the United Nations has come to an agreement on a satisfactory plan for the international control of atomic energy."[61] His plea was ineffective.

57. Lt. Col. F. G. Camino, Military Attaché, Spanish Embassy, Washington, to T. Raymond Jones of Isotopes Branch, Oak Ridge, 4 Sep 1947, AEC Records, NARA Atlanta, RG 326, MED CEW Gen Res Corr, Acc 67B0803, box 173, folder AEC 441.2 (R–Spanish Embassy). See also Vladimir Houdek to Isotopes Branch, 11 Sep 1947, AEC Records, NARA Atlanta, RG 326, MED CEW Gen Res Corr, Acc 67B0803, box 151, folder AEC 441.2 (R–Czechoslovak Embassy).

58. See K. T. Bainbridge to Paul C. Aebersold, 7 Nov 1947, NARA Atlanta, RG 326, MED CEW Gen Res Corr, Acc 67B0803, box 151, folder AEC 441.2 (R–Harvard University).

59. Memorandum from J. H. Manley to Carroll Wilson, 24 Jun 1947, NARA College Park, RG 326, E67A, box 46, folder 3 Foreign Distribution of Radioisotopes vol. 1.

60. Hewlett and Duncan, *Atomic Shield* (1969), p. 272.

61. Draft memorandum from Lewis Strauss to Robert Bacher on question "Has a mistake been made in deciding to export isotopes to foreign countries?" 23 Sep 1947, with a note that the draft was shown to the three other Commissioners that week; quote from p. 9. Strauss Papers, AEC Files, folder Isotopes Sep–Dec 1947. See also memorandum from Lewis L. Strauss to David E. Lilienthal, Robert F. Bacher, Sumner T. Pike, W. W. Waymack [undated but c. Aug 1947], Strauss papers, AEC Files, folder Isotopes Jan–Aug 1947.

Although the majority of Commissioners had set policy, Strauss would not let the issue rest. Determined to monitor the program, he asked the AEC's manager to begin sending him a monthly listing of "the isotopes that we export, with a description of the isotope, the amount in units of radioactivity or weight, or both, the country of destination, the consignee, and the purpose for which requested."[62] He began scouring the records for shipments whose recipients or countries were of questionable loyalty to the United States, so that he could pass this information to the AEC's Congressional critics to stoke concern about radioisotope exports.[63] Some others shared his discontent. The head of the AEC's Military Liaison Committee, General Brereton, did not feel the issue had been properly vetted with them, and complained to the Commissioners at their joint September 24 meeting.[64]

The AEC's protocol for sales of radioisotopes to foreigners differed substantially from that for domestic purchasers. Recipient countries had to obtain the Department of State's explicit approval. This involved going through the usual diplomatic channels to make the request to the Secretary of State, and designating a US-based representative to act on the country's behalf in handling requests. This agent, which could be a diplomatic official, a company, or an individual, was expected to take care of a wide variety of tasks: "arrangements for shipments, payment, forwarding of technical circulars to interested scientists in his country and [submitting] progress reports."[65] This designated representative was also responsible for obtaining a license for export of radioisotopes from the Department of Commerce. The agency required that all correspondence be in English.[66] Other, less bureaucratic issues came into play for certain foreign applicants. Before action could be taken on "requests from Russia

62. Memorandum from Lewis L. Strauss to Carroll L. Wilson, 18 Sep 1947, Strauss papers, AEC Files, folder Isotopes 1947 Sep–Dec.

63. There are several copies of internal AEC documents, such as Lewis Strauss's memorandum to Robert F. Bacher, 23 May 1947 (cited above) in the Hickenlooper papers, Senate Committee Files, folder JCAE–Isotopes Jan–Aug 1947.

64. Excerpt of AEC-MLC meeting, 24 Sep 1947, NARA College Park, RG326, E67A, box 46, folder 6 Foreign Distribution of Radioisotopes vol. 2.

65. Press Release, "27 Nations Qualify to Receive Radioisotopes from Atomic Energy Commission," 3 Feb 1949, NARA College Park, RG 326, E67A, box 46, folder 6 Foreign Distribution of Radioisotopes vol. 2.

66. *Radioisotopes for International Distribution, Catalog and Price List*, Sep 1947, Isotopes Branch, NARA College Park, RG 326, E67A, box 46, folder 6 Foreign Distribution of Radioisotopes vol. 2.

or Russian-dominated countries," the application had to be referred to the AEC's General Manager for approval.[67]

The AEC required foreign recipients to report semiannually on the results obtained with the radioisotopes, to use the reagents only for the purposes specified in the application, to abide by the same laboratory safety guidelines as domestic users, and to "permit qualified scientists of all nations to visit their institutions and freely obtain information about the work."[68] As it happens, the issue of visitation and inspection had also been a sticking point in international atomic energy control. For countries interested in receiving AEC isotopes, the policy had a coercive edge—it meant opening up laboratories to American visitors, including the scientific attachés the US government was appointing in various European countries.[69] In other words, it meant acquiescing to US intelligence-gathering.[70]

Destinations and Complications

The first thirteen foreign shipments of AEC radioisotopes went to Australia in the fall of 1947. Most of these initial shipments were destined for use in cancer and thyroid disease treatment, although some were employed in physiological and metabolic research, including studies of plant viruses. The next foreign country was Argentina, as the National Academy of Medicine in Buenos Aires received 10 millicuries of phosphorus-32 for medical therapy in early December. A shipment of phosphorus-32 also was sent to the National Institute for Medical Research at Hampstead, London, to be used in physiological studies. The first shipment to reach

67. "Procedure for Handling Foreign Requests for Radioisotopes," 26 Sep 1947, memorandum from John C. Franklin, Manager, Oak Ridge Operations, to Carroll L. Wilson, General Manager, AEC Records, NARA College Park, RG 326, E67A, box 46, folder 6 Foreign Distribution of Radioisotopes vol. 2, p. 4.

68. AEC, *Fourth Semiannual Report* (1948), p. 15.

69. In the meeting on December 15, 1948, at which the Commissioners discussed the first application from Finland for radioisotopes, Sumner Pike pointed out that the US scientific attaché in Sweden could check on the uses of the radioisotopes in Finland. Excerpted minutes, attached to memorandum from T. O. Jones, Acting Secretary of the Commission, to Edwin E. Huddleson, Jr., Acting General Counsel, 16 December 1948, NARA College Park, RG 326, E67A, box 47, folder 6, Foreign Distribution of Radioisotopes vol. 2.

70. Krige, "Atoms for Peace" (2006); Doel and Needell, "Science, Scientists, and the CIA" (1997). To be clear, US scientific attachés were not explicitly intelligence agents, so this information-gathering was in addition to their official diplomatic duties.

continental Europe was iodine-131 sent for medical therapy to the Radium Center in Copenhagen, Denmark, on December 30; a shipment of carbon-14 to the Naples Zoological Station followed in January 1948, for use in metabolic studies of invertebrate eggs.[71]

From the fall of 1947 to the end of 1948, the AEC sent out 356 shipments of radioisotopes to various laboratories and treatment centers around the world. (See figure 4.1.) Nearly 70% went to Europe and the British Commonwealth. Sweden was the largest consumer, having received sixty-two shipments, followed by England with fifty-eight. Beyond Australia and New Zealand, the other non-European countries whose institutions received shipments were Argentina, Peru, and South Africa.[72] Approximately 90% of the uses of isotopes by foreign recipients were in the fields of medical therapy or physiological research. The other 10% of research uses was accounted for by basic research in the physical sciences and agriculture.[73] The slant toward applications in biology and medicine was stronger still in the foreign distribution program than in the domestic program.[74] This orientation was important to allay fears that these radioisotopes might contribute to the development of military technologies—especially nuclear weapons—elsewhere.

The official list of exported isotopes is incomplete in one important way. Beginning on October 1, 1948, the AEC authorized shipments of stable and radioactive isotopes to the Canadian and British atomic energy installations at Chalk River and Harwell. These shipments were part of the Technical Cooperation Program, as specified by the modus vivendi agreements among the United States, the United Kingdom, and Canada, hammered out in the final months of 1947.[75] The American motivation for this agreement was clear. Half of the uranium ore from the Belgian Congo was going to the British; the competition for raw materials would soon hamper the US nuclear weapons program. As more policy makers became aware of this reality, they better understood the disadvantages

71. Monthly Reports of Foreign Shipments–Radioisotopes, Reports 1–5, NARA College Park, RG 326, E67A, box 46, folder 4 Reports of Foreign Shipments.

72. Monthly Reports of Foreign Shipments–Radioisotopes, Report 16, Dec 1948, NARA College Park, RG 326, E67A, box 46, folder 4 Reports of Foreign Shipments.

73. Ibid., p. 24.

74. Press Release, "AEC Sends Radioisotopes to 22 Nations for Research and Therapy," NARA College Park, RG 326, E67A, box 46, folder 6 Foreign Distribution of Radioisotopes vol. 2.

75. Hewlett and Duncan, *Atomic Shield* (1969), chapters 9 and 10. On these agreements from the British side, see Gowing, *Independence and Deterrence* (1974), vol. 1, ch. 8.

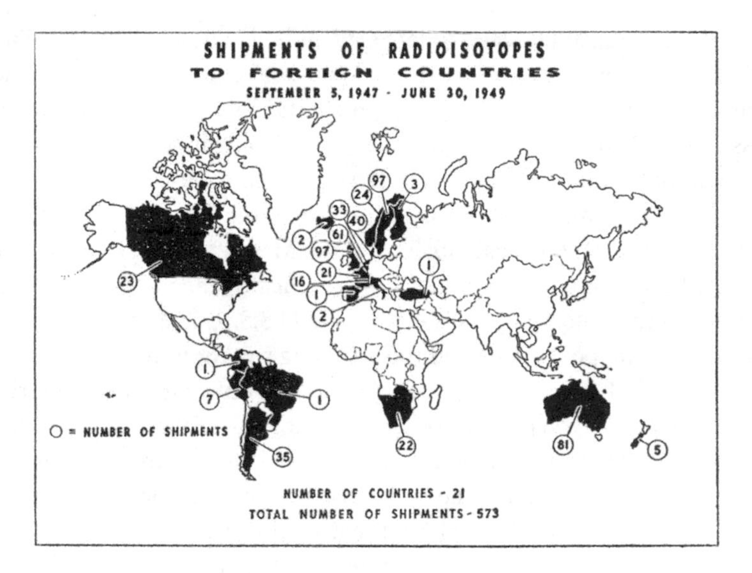

FIGURE 4.1. A world map indicating the destinations of foreign shipments of radioisotopes from the US AEC during the first twenty-two months of the program. Reproduced from US Atomic Energy Commission, *Isotopes* (1949), p. 59.

of the 1946 Atomic Energy Act's restrictions. Nonetheless, the Technical Cooperation program, like radioisotope exports, remained a sore point with Lewis Strauss and other national security watchdogs. The radioisotope shipments to Chalk River and Harwell were not numerous—three to the United Kingdom and fifteen to Canada between October 1, 1948, and August 1, 1949. They included shipments of boron-10, tritium, oxygen-18, carbon-14, phosphorus-32, iron-58, helium-3, and actinium-227. These materials were available to the British and Canadian researchers on the understanding that they would not be "transferred outside the official atomic energy projects or to unauthorized persons within those projects."[76] These isotopes, which were not part of sales to civilian institutions, do not appear on the monthly reports of foreign shipments.[77]

76. "Proposed Shipments of Isotopes to UK and Canada Under the Technical Cooperation Program," AEC 43/215, 10 Jan 1950, NARA College Park, RG 326, E67A, box 46, folder 5, Foreign Distribution of Stable Isotopes, p. 1 and box 47, folder 1 Foreign Distribution of Radioisotopes vol. 3. On the development of the modus vivendi agreements in late 1947 with the British and Canadian governments, see Hewlett and Duncan, *Atomic Shield* (1969), pp. 273–84.

77. See Monthly Reports of Foreign Shipments–Radioisotopes, NARA College Park, RG 326, E67A, box 46, folder 4 Reports of Foreign Shipments.

The AEC was, at the same time, participating in developing guidelines for compliance with export control of industrial equipment and supplies connected with atomic energy development. On December 19, 1947, Congress passed legislation enlarging the export control powers of the Department of Commerce. The bill authorized a blanket control plan for all shipments to countries in continental Europe, the British Isles, Iceland, Turkey, the USSR, Portugal, Spain, and the Mediterranean islands.[78] Despite its broad nature, the policy was specifically aimed at "controlling exports which could be of assistance to the U.S.S.R. in an atomic energy program."[79] The policy reflected the changing situation since the passage of the Atomic Energy Act in 1946. Originally, only export of materials directly related to the production of fissionable fuel for nuclear weapons was prohibited.[80] The expanded program of blanket control also forbade the export of general industrial and construction equipment, now recognized "as being of considerable indirect importance to Russia's atomic energy program."[81] The new policy reflected not only the changed status of the USSR as a potential nuclear rival, but also the role of industrial-scale production facilities in the manufacture of atomic weapons.

In this environment of political sensitivity about exports, the AEC was at pains to emphasize that its foreign shipments of radioisotopes could not aid the development of atomic energy programs abroad. Strauss, on the other hand, was seeking any evidence he could find of suspicious shipments. In August 1948, he had his assistant inquire with the head of the Isotopes Division, Paul Aebersold, to ask explicitly whether any foreign

78. AEC 23, Export Control Program, NARA College Park, RG 326, E67A, box 44, folder 1 AEC Export Policy vol. 1, p. 3. Colonies of these countries were generally included. On the history of the export control program, see Berman and Garson, "United States Export Controls" (1967) and Funigiello, *American-Soviet Trade* (1988).

79. Draft Letter to Secretary of Commerce from the Chairman, AEC, Appendix A to "Export of General Industrial Equipment," 10 Dec 1947, included with AEC 23, Export Control Program, NARA College Park, RG 326, E67A, box 44, folder 1 AEC Export Policy vol. 1, p. 5.

80. Memo from Walker L. Cisler to Carroll L. Wilson, General Manager, 2 Jul 1947, re: Export of General Industrial Equipment, Appendix C to "Export of General Industrial Equipment," 10 Dec 10, 1947, included with AEC 23, Export Control Program, NARA College Park, RG 326, E67A, box 44, folder 1 AEC Export Policy vol. 1. The activities of the Industrial Advisory Group are discussed in chapter 6.

81. Draft Letter to Secretary of Commerce from the Chairman, AEC, Appendix A to "Export of General Industrial Equipment," 10 Dec 1947, included with AEC 23, Export Control Program, NARA College Park, RG 326, E67A, box 44, folder 1 AEC Export Policy vol. 1.

shipments had been made for other than biological or medical projects.[82] He was convinced that even those shipments being used in medical research and therapy were not achieving the desired ends. An article published in the British socialist journal *The Scientific Worker* criticized the restrictions on the US radioisotope export program. In a gesture of red-baiting, Strauss forwarded the article to Senator Bourke Hickenlooper, Republican from Iowa and chair of the Joint Committee on Atomic Energy, as evidence against the "argument that was made at the time to the effect that the export of isotopes to foreign countries would make friends for us among their scientists."[83] In effect, Strauss's anticommunism and overriding concern with national security meant that he and the other Commissioners were talking past each other when it came to the issue of radioisotope export.

In December 1948, the Government of Finland applied to purchase radiophosphorus for use in medical therapy. The Commissioners were wary of approving the request.[84] The agency consulted extensively with the Department of State, whose stated policy was to encourage Finland's independence from Soviet influence. Despite the fact that Finland had recently negotiated a mutual defense pact with the USSR, the State Department emphasized that "Finland is not behind the Iron Curtain."[85] Including Finland in the radioisotope program signaled its place among the European democracies that the United States sought to assist—and influence. The State Department's view reflected the logic of the Marshall Plan, which integrated European countries into an American-led

82. Memorandum from Lewis Strauss to William T. Golden with request to query Aebersold, 6 Aug 1948, Strauss papers, AEC Files, folder Isotopes 1948.

83. Note from Strauss to Bourke B. Hickenlooper, 9 Mar 1948, Strauss papers, AEC Files, folder Isotopes 1948.

84. The AEC was cognizant of the publicity that would likely attend this decision: "the fact that Finland will have qualified to receive Commission-produced radioisotopes will be the subject of considerable legitimate interest on the part of the press." Memorandum from Morse Salisbury to Roy B. Snapp, Secretary to the Commission, 14 Dec 1948, NARA College Park, RG 326, E67A, box 46, folder 6 Foreign Distribution of Radioisotopes vol. 2.

85. Finland Policy Statement, Department of State, 2 Sep 1948, in NARA College Park, box 46, folder 6 Foreign Distribution of Radioisotopes, vol. 2, p. 5. Finland and the USSR signed the mutual defense pact on May 4, 1948; the Soviets had first broached the idea with Finland in 1938 and the pact, once obtained, compromised Finland's aspirations for neutrality. Jakobson, *Finland in the New Europe* (1998), p. 57.

capitalist world system, not by military force but by political persuasion.[86] On December 21, the Commissioners voted four-to-one to include Finland in the export program. Again Lewis Strauss dissented.[87] In preparation for likely criticism, the Commissioners asked the Department of State for a written confirmation that "Finland is not regarded as a satellite of the USSR and that it is US policy to provide Finland with non-military aid and support."[88]

Despite the Commission's efforts to guard against criticism of the isotope program, the agency's credibility on national security was weakened by a host of other troublesome issues and allegations. First, the Commission decided in 1948 that its agency's postdoctoral fellowships in the biomedical and physical sciences would be open to applicants irrespective of their political affiliations. This gave the AEC greater credibility with scientists but became difficult to defend to the public and key Congressional patrons. In particular, the awarding of one AEC fellowship to a scientist with a record of membership in the Communist Party became a lightning rod for criticism.[89] Hickenlooper used the confirmation hearings for Henry DeWolf Smyth as an AEC Commissioner on May 12 to attack Lilienthal on this issue, arguing that the FBI should be asked to clear all government fellowship holders.[90] Second, the Joint Committee on Atomic Energy raised concerns about the publication of the Commission's fifth semiannual report to Congress, which contained detailed information about their facilities and programs—too much for politicians eager to see the country safeguard its atomic knowledge.[91] Third, on May 17 re-

86. Craig and Radchenko, *Atomic Bomb* (2008), p. 128; Krige, *American Hegemony* (2006).

87. Foreign Distribution of Radioisotopes [A Summary of Commission Actions], NARA College Park, RG 326, box 46, folder 6 Foreign Distribution of Radioisotopes vol. 2, p. 2. Hickenlooper also raised concerns about the size of the shipment requested: 300 millicuries of phosphorus-32. He had his assistant William Golden look into the amount of other requests, and Golden found out that it was the largest request to date, although there had been several requests for 100 millicuries each. William T. Golden to Lewis L. Strauss, 22 Dec 1948, Strauss papers, AEC Files, folder Isotopes 1948.

88. Foreign Distribution of Radioisotopes [A Summary of Commission Actions], NARA College Park, RG 326, box 46, folder 6 Foreign Distribution of Radioisotopes vol. 2.

89. See Kaiser, "Cold War Requisitions" (2002), especially pp. 140–41.

90. Hewlett and Duncan, *Atomic Shield* (1969), p. 356; Lilienthal, 14, 15, 17, and 19 May 1949, *Journals* (1964), vol. 2, pp. 528–31.

91. This despite the fact that much of this information, including locations of facilities, had previously been published in the Smyth Report. Lilienthal, 20 Mar 1949, *Journals* (1964), vol. 2, pp. 488–89; Hewlett and Duncan, *Atomic Shield* (1969), pp. 289, 340, 352.

ports surfaced of another alleged security breach—that some fissionable uranium was missing from Argonne National Laboratory. This allegation spurred the Congressional Joint Committee on Atomic Energy to launch an investigation of the AEC, to "make a complete inquiry into the grave charges which have been made."[92] Making his position clear, Hickenlooper called for Lilienthal's resignation on May 22.[93]

Strauss's close surveillance of the foreign shipment records had turned up exactly the sort of fishy export he had hoped to find. On April 28, 1949, a shipment of iron-59 was shipped to the Norwegian Defense Establishment at Kjeller for metallurgical research on high-temperature steel, which might have had application for jet engines.[94] As Strauss wrote about this shipment in a memorandum to the other Commissioners a week before the hearings began, "It would require quite a stretch of the imagination to regard that as furthering the 'beneficent use' of isotopes."[95] Given Strauss's close connections to Hickenlooper, this was a warning signal of trouble ahead. A few days later, on May 24, 1949, the AEC sent out a press release setting out "the facts concerning the development of the program for foreign distribution of radioisotopes." Hickenlooper marked up his own copy, underlining the emphasis on exports for "medical and biological research," a slant that he knew sat uneasily with the Norwegian radio-iron shipment.[96]

Critics made use of various hearings in progress to raise questions about the AEC's reliability on issues of national security, including isotope exports. On May 24, the Senate Appropriations Subcommittee focused

92. JCAE, *Investigation* (1949), Part I, 26 May 1949, p. 1. I have compressed a more complex sequence of events: Hickenlooper charged the AEC with "incredible mismanagement" in a May 1949 letter. When the AEC demanded an opportunity to respond, the chair of the JCAE, Brien McMahon, held hearings. Balogh, *Chain Reaction* (1991), p. 70.

93. Hewlett and Duncan, *Atomic Shield* (1969), p. 358. The same concerns had surfaced privately months earlier: Senator Hickenlooper wrote Lilienthal on January 24, 1949, expressing his concern that shipments of radioisotopes to Finland might well be diverted to the Soviet Union and used for unauthorized purposes. NARA College Park, RG 326, E67A, box 37, folder 6 Foreign Distribution of Radioisotopes vol. 2.

94. This was shipment 508, one millicurie of high specific activity iron-59. See Monthly Reports of Foreign Shipments–Radioisotopes, Report 20, Apr 1949, NARA College Park, RG 326, E67A, box 46, folder 4 Reports of Monthly Shipments. Another copy with this shipment highlighted is in the Strauss papers, AEC Files, folder Isotopes 1949.

95. Lewis L. Strauss, Memorandum to the Commissioners, 18 May 1949, Strauss papers, AEC Files, folder Isotopes 1949.

96. "Foreign Distribution of Radioisotopes," AEC Press Release, 24 May 1949, copy in Hickenlooper papers, Senate Committee Files, folder JCAE–Isotopes 1949, p. 5.

attention on the week-old disclosure that three bars of uranium, removed from Hanford, were not missed for three months.[97] Senator Joseph O'Mahoney of Wyoming took the occasion to draw attention to the dissent of Commissioner Strauss on the policy of the AEC to sell radioisotopes abroad.[98] Lilienthal defended the agency's radioisotope shipments, including those to scientists and hospitals in Norway and Sweden (worrisome for their proximity to the Soviet Union).[99] He argued strenuously that the radioisotopes in question could not aid in atomic bomb research or development. However, Strauss countered him, testifying that the United States had no guarantee that the "isotopes would be used only in research on benign matters . . . once they left our hands."[100] Moreover, just weeks after the AEC approved Finland's participation in the isotope export program, the Finnish purchaser canceled the first order without explanation.[101] Critics of the AEC's inclusion of Finland found this worrisome; Strauss forwarded this information to Hickenlooper.[102] The press's scrutiny of the issue was handled poorly by the AEC. As the *New York Times* relayed, "Commission officials did not know . . . whether any isotopes had actually been sent to that Russian-dominated country."[103]

The Joint Committee on Atomic Energy soon amplified the alarm about isotope exports in their own Investigative Hearings. Strauss expressed his concerns in two executive sessions with the committee from which Lilienthal was excluded, and at which he likely provided information

97. US Senate, *Independent Offices Appropriation* (1949), 24 May 1948, p. 588.

98. Ibid., pp. 584–85.

99. US Senate, *Independent Offices Appropriation* (1949), 24 May 1948, p. 577 ff.

100. Ibid., p. 578. This was reported in the *New York Times* by Morris, "Two Uranium Bars Taken" (1949).

101. See letter from David Lilienthal to Brien McMahon, 9 Jun 1949, NARA College Park, RG 326, E67A, box 46, folder 6 Foreign Distribution of Radioisotopes vol. 2. Lilienthal explained that the Finnish university was able to obtain the radiophosphorus they ordered more quickly from the British government. Finland was approved for the isotope export program by the Department of State on January 28, 1949. The Legation of Finland designated the Barr Shipping Company as their agent on August 31, 1948, and this company submitted a request for 300 millicuries of radiophosphorus on September 29. See AEC 173/3, "Inclusion of Finland in the Program for Foreign Distribution of Radioisotopes," NARA College Park, RG 326, E67A, box 46, folder 6, Foreign Distribution of Radioisotopes, vol. 2.

102. Cover note from Lewis L. Strauss, 4 Feb 1949, on memorandum to AEC Commissioners from Morse Salisbury regarding cancellation of order for radiophosphorus by University of Helsinki. Hickenlooper papers, Senate Committee Files, folder JCAE–Isotopes 1949.

103. Morris, "Two Uranium Bars Taken" (1949), p. 15.

on the most suspicious foreign shipments.[104] On June 8, Hickenlooper charged that the AEC's policy "clearly violates the scope and the limitations of the isotope distribution program as announced by the President on September 3, 1947."[105] As he pointed out, the April shipment of radioactive iron to the Norwegian Defense Research Establishment in Kjeller did not fit with the principle that "isotopes would be made for biological and medical research almost exclusively."[106] Adding to the sense of urgency, the *New York Times* reported shortly before the hearings that Norway was pressing to build a nuclear reactor at Kjeller.[107] Hickenlooper raised concerns about other shipments as well, three to Finland for research in the physical sciences, and others that had been sent to Frédéric and Irène Joliot-Curie's laboratory in France, viewed as suspicious because of the physicists' connections to the French Communist party. The senator warned, "Once these isotopes leave our possession we lose control of the actual use to which they may be put and we lose control over the destination of either the isotopes or the information gained therefrom."[108] In his view, these shipments ran afoul of the Atomic Energy Act's ban on exchange of "information with other nations with respect to the use of atomic energy for industrial purposes" until international safeguards had been established by Congress.[109]

Strauss publicly aligned himself with Hickenlooper. As he stated while testifying that same day, "The question of whether indiscriminate dissemination of knowledge on atomic energy ought to be made during the period of a cold war is the real point at issue."[110] By referring to the dissemination of "knowledge," rather than isotopic materials, Strauss insinuated that exporting radioisotopes abrogated the Atomic Energy Act's prohibition

104. Lilienthal, 9 Jun 1949, *Journals* (1964), vol. 2, p. 541.

105. JCAE, *Investigation* (1949), Part 5, 8 Jun 1949, p. 204.

106. Ibid. Hickenlooper dictated a memorandum highlighting this shipment to Norway and one other to the Swedish Royal Institute of Technology; 31 May 1949, Hickenlooper papers, Senate Committee Files, folder JCAE–Isotopes 1949.

107. "Norwegian Defense Board Pressing an Atomic Pile," *New York Times*, 29 May 1949, clipping in Hickenlooper papers, Senate Committee Files, folder JCAE–Isotopes 1949. The concerns about Norway's nuclear developments were not misplaced; the nation's first reactor at Kjeller (JEEP) went critical on 30 Jun 1951. This was the first reactor outside of the United States, Britain, France, Canada, and the Soviet Union.

108. JCAE, *Investigation* (1949), Part 5, 8 Jun 1949, p. 207.

109. Ibid.

110. JCAE, *Investigation* (1949), Part 6, 9 Jun 1949, p. 236.

against sharing nuclear information. Smear tactics blurred distinctions among the alleged violations: Finland was regarded with suspicion on account of its pact with the Soviet Union, France owing to Joliot being an outspoken communist, and Norway because of its military research—this despite the fact that Norway was a US ally and a founding member of the North American Treaty Organization (NATO).[111] In their loathing of Lilienthal, Strauss and Hickenlooper simply saw red.

Hickenlooper's allegation was reported in newspapers the next day under headlines such as "U.S. Isotope Export Held Dangerous."[112] The *New York Times* reported that the accusation "threw the Joint Congressional Committee on Atomic Energy into severe dispute at the resumption of hearings in Mr. Hickenlooper's 'incredible mismanagement' case against David E. Lilienthal."[113] The AEC had its strong defenders, not least Senator Brien McMahon, head of the Joint Committee and sponsor of the 1946 act, and Henry DeWolf Smyth, newly appointed as a Commissioner. J. Robert Oppenheimer testified on June 13, defending the radioisotope export decisions of the AEC, but Congressmen and Senators asked him repeatedly whether the radioisotopes being sent to Europeans might find their ways into the hands of Soviets and speed their development of atomic weapons.[114] Oppenheimer retorted:

> No one can force me to say that you cannot use these isotopes for atomic energy. You can use a shovel for atomic energy; in fact, you do. You can use a bottle of beer for atomic energy. In fact, you do. But to get some perspective, the fact is that during the war and after the war these materials have played no significant part, and in my own knowledge, no part at all. That is not true of all isotopes. Plutonium is a good one; that played a big part. It is true of the group of isotopes falling under the Commission's export policy.[115]

Efforts by Oppenheimer and other defenders of the AEC to differentiate between the radioisotopes being exported and those actually useful in developing nuclear weaponry proved fruitless—and perhaps beside

111. On Norway and NATO, see Krige, *American Hegemony* (2006).

112. "U.S. Isotope Export" (1949).

113. Morris, "Isotopes Shipment" (1949), p. 1.

114. E.g., J. Robert Oppenheimer, testimony, JCAE, *Investigation* (1949), Part 7, 13 Jun 1949, p. 284.

115. JCAE, *Investigation* (1949), Part 7, 13 Jun 1949, p. 282.

the point. For Strauss, even the use of radioisotopes in developing conventional military technologies was damning enough.[116] In any event, just three months after the hearings concluded, the Soviets detonated their first atomic bomb, reinforcing suspicions that disloyal Americans scientists, and perhaps even the government itself, had given away the country's nuclear secrets.[117]

Radioisotope Distribution in a Competitive Foreign Market

The Congressional Hearings of the summer of 1949 exacted a toll on the agency's leadership. Lilienthal resigned from leading the AEC, to be replaced by Gordon E. Dean, former law partner of Senator McMahon. While several aspects of the Commission were subject to change under Dean's leadership, the small radioisotope export program continued unabated.[118] The AEC did begin vetting foreign requests for isotopes for nonbiological or nonmedical uses with the Commissioners.[119] Disagreement among them persisted. Strauss made clear his guiding principle: "If the requests are for medical or basic scientific research or instruction purposes I should like to be voted in approval. If for military or industrial use I should like to be voted against granting the requests."[120]

116. See Lewis L. Strauss to Senator Brien McMahon, 24 Jun 1949, NARA College Park, RG 326, E67A, box 46, folder 6 Foreign Distribution of Radioisotopes, vol. 2.

117. See Kaiser, "Atomic Secret" (2005); Gordin, *Red Cloud at Dawn* (2009).

118. For example, procedures for Congressional appropriations changed. Hewlett and Duncan, *Atomic Shield* (1969), pp. 442–84.

119. See AEC 231/5, "Swedish Request for Radioisotopes for Non-Medical or Biological Uses," 5 October 1949, and other similar requests, NARA–College Park, RG 326, E67A, box 46, folder 6 Foreign Distribution of Radioisotopes vol. 2. In January 1950 the Commission proposed an amendment to the Atomic Energy Act that would clarify its authority for distributing isotopes outside the United States. The changes cleared the Bureau of Budget in May 1950, but stalled there. Gordon Dean, Acting Chairman, to Senator Brien McMahon, Chairman of the JCAE, Enclosure "A" to AEC 236, "Revision of the Language of the Atomic Energy Act Relating to Foreign Distribution of Radioisotopes," 23 May 1950, NARA College Park, RG 326, E67A, box 47, folder 1 Foreign Distribution of Radioisotopes vol. 3.

120. Note in Strauss's hand, 24 Aug 1949, appended to memorandum AEC 231/2, Requests from Foreign Countries for Radioisotopes for Non-Medical Uses, 30 Aug 1949, NARA College Park, RG 326, E67A, box 46, folder 6 Foreign Distribution of Radioisotopes vol. 2. A duplicate copy is in the Strauss papers, AEC files, folder Isotopes 1949.

Improvements to the radioisotope processing facility in Oak Ridge as well as increased demand drove up the number of shipments going out from Tennessee. By late 1950, the AEC had sent out more than 14,500 shipments of radioisotopes, of which around 5,000 were shipped in the last year alone. Foreign laboratories received 975 through November 1950; 563 of those went to continental Europe and England. Medical uses, including therapy, diagnosis, and research, continued to dominate uses among foreign recipients: 436 of the 975 foreign shipments were of phosphorus-32, 220 were of iodine-131, 106 were of carbon-14, and 54 were of sulfur-35.[121] Even as the agency touted its radioisotope export program in its publications, disparities persisted between the policies for domestic and foreign sales. In some cases even pricing was different. Whereas radioisotopes used in cancer research were free for purchasers in the United States, the Commissioners declined to give foreigners this discount.[122]

In the spring of 1950 the Commission considered how it might expand the isotope export program in response to the rival distribution programs in the United Kingdom and Canada.[123] Nuclear capabilities were also being developed outside the Anglo-American network. By this time France possessed its own experimental reactor and the United States had evidence of a reactor operating in the Soviet Union. Several other countries, including Sweden, Norway, Belgium, India, and Switzerland, had plans for reactor projects.[124] The British and Canadian programs did not restrict shipments to use in scientific research and medical therapy, but allowed them for industrial applications. The AEC recognized that these supply programs would lead American researchers to turn the tables and request permission to import foreign isotopes.[125]

121. AEC, *Ninth Semiannual Report* (1951), p. 31; AEC, *Isotopes* (1949), pp. 23–24.

122. See Conclusions and Recommendations, "Study of Wider Use of Isotopes," Info Memo 92/1, 9 Oct 1949, NARA College Park, RG 326, E67A, box 45, folder 6 Study of Wider Use of Isotopes.

123. The day after Truman announced the commencement of the US isotope export program, the Canadian Atomic Energy Board announced that its reactor was operational and that it would sell isotopes to both domestic and foreign purchasers. See Memo from Carroll L. Wilson to the Commissioners, Progress and Procedure on Foreign Distribution of Isotopes, 25 Sep 1947, NARA College Park, RG 326, E67A, box 46, folder 6 Foreign Distribution of Radioisotopes vol. 2.

124. Note by the Secretary, 17 May 1950, AEC 231/12, NARA College Park, RG 326, E67A, box 47, folder 1, Foreign Distribution of Radioisotopes vol. 3, p. 3.

125. For example, the British offered phosphorus-32 of higher specific activity than Oak Ridge. See p. 249.

Having already relaxed reporting requirements for foreign recipients, the Commission voted to expand the radioisotope export program on January 30, 1951.[126] The only isotope that was available to domestic purchasers that would not be available to foreigners was tritium, on account of objections from the Military Liaison Committee to shipping even small amounts abroad. (Tritium was a critical component of hydrogen bombs.) The expanded program would also permit foreigners to purchase isotopes for industrial uses, as the British and Canadian programs allowed. Relatedly, American companies would be allowed to market certain isotope-containing instruments, such as beta-ray thickness gauges, abroad. The AEC appealed to their counterparts in Canada and Great Britain to address remaining disparities. On April 25, 1951, representatives from American, Canadian, and British atomic energy programs reached some agreement on pricing and access to isotopes for foreign purchasers.[127]

Announcing its expanded program for foreign radioisotope distribution in May 1951, the AEC assured its main patron in Congress, Senator McMahon, that the newly broadened program "is wholly consistent with the paramount objective of assuring the common defense and security."[128] The press release for the policy was also carefully worded to allay public worries about the exports (see the chapter epigraph for the paragraph preceding this quote):

126. In the spring of 1950, while waiting on responses by the Bureau of Budget and Joint Committee to the proposed amendment, the Commission went ahead and enacted changes to the reporting requirements such that purchasers could file reports every twelve months, rather than every six months, and there was no obligation to publish results. See "International Distribution of Radioisotopes," 29 Mar 1950, AEC 231/10, and letter from Sumner T. Park to Brien McMahon, 10 May 1950, NARA College Park, RG 326, E67A, box 47, folder 1 Foreign Distribution of Radioisotopes vol. 3. The reporting requirement was finally dropped on October 4, 1954, when Strauss was chair of the Commission. At that point, the agency acknowledged that most of these uses were entirely routine and "conditions for export of radioisotopes from Great Britain and Canada are less stringent that those . . . from the United States." See AEC 231/24, International Distribution of Radioisotopes, 1 Sep 1954, and Lewis L. Strauss to Sterling Cole, 1 Oct 1954, NARA College Park, RG 326, E67B, box 29, folder 1 Isotopes Program 3 Foreign, p. 10.

127. "International Distribution of Radioisotopes," AEC 231/15, 11 May 1951, NARA College Park, RG 326, E67A, box 47, folder 1 Foreign Distribution of Radioisotopes vol. 3.

128. Sumner T. Pike, letter to Senator Brian McMahon, 7 May 1951, Appendix B to AEC 231/16, NARA College Park, RG 326, E67A, box 47, folder 1 Foreign Distribution of Radioisotopes vol. 3.

> Enlargement of our isotope export program is, we feel, in keeping with the foreign policy of the United States, which calls for aid to foreign nations in peacetime development, and, even in the absence of international control of atomic energy, constitutes a field in which international cooperation can be increased.[129]

The press release also emphasized the role of training programs for foreign scientists in increasing the demand for isotopes abroad. The Isotope School of the Oak Ridge Institute of Nuclear Studies at Oak Ridge, Tennessee, had accepted a limited number of foreign nationals in the late 1940s through 1951.[130]

Even with these changes, the upshot of the restrictions that the AEC placed on radioisotope exports was that the Americans lost substantial market share to the British and Canadian governments. Within a year of the commencement of the British export program, radioactive shipments abroad from Harwell exceeded those from Oak Ridge.[131] An AEC memo from June 1951 conceded that the three governments had become regional, not global, isotope suppliers: "the British can, and for the most part do, furnish short-lived radioisotopes to Western European countries; we distribute these materials to Latin American countries."[132] Moreover, the Soviet Union was rapidly developing a radioisotope supply for the Eastern bloc countries.[133] If the United States had an edge, it was in the production of radiolabeled compounds, although the gap was being narrowed by the British national company, Amersham, which began offering radiolabeled compounds in 1949.[134] This outfit became a major competitor to American companies, and—to their consternation—was not under the same profit pressure.[135]

129. Proposed Press Release, "AEC Enlarges Radioisotope Export Program," Appendix E to AEC 231/16, NARA College Park, RG 326, E67A, box 47, folder 1 Foreign Distribution of Radioisotopes vol. 3, p. 16.

130. Ibid. John Krige has analyzed the importance of such training programs in association with Eisenhower's Atoms for Peace initiative in "Techno-Utopian Dreams" (2010). See also Herran, "Isotope Networks" (2009).

131. Herran, "Spreading Nucleonics" (2006).

132. AEC 231/17, 19 June 1951, NARA College Park, RG 326, E67A, box 47, folder 1 Foreign Distribution of Radioisotopes vol. 3.

133. Medvedev, *Soviet Science* (1978), pp. 48–49.

134. Chapter 6 addresses the emergence of US suppliers of radiolabeled compounds. On Amersham: Kraft, "Between Medicine and Industry" (2006).

135. Interview with Paul McNulty, New England Nuclear sales manager in 1960–1984, 29 Mar 2002, Newton, Massachusetts. The United Kingdom and Canada tended to follow the

Because all three Anglo-American governments were subsidizing isotope sales, the United States could not actually compete on a cost basis—from the outset, the Canadian and British governments had set their prices to match those of Oak Ridge. As the agency admitted, "Such differences in prices as do exist would not in themselves be the determining factor in choosing one country over another as a supplier. Differences in shipping costs are a more important economic consideration."[136] Beyond price, the US AEC was at a disadvantage on account of the complex diplomatic controls it exercised in the name of national security.

Atoms for Peace

The political association that conservatives crafted between foreign distribution of isotopes and lax national security was finally severed by a Republican. President Dwight Eisenhower delivered his "Atoms for Peace" speech to the United Nations at the end of 1953, with its core proposal that the United States and the USSR both contribute fissionable material to a program aimed at peaceful atomic energy.[137] The amount of fissionable material was fixed to a figure the Americans could manage much more easily from their stockpile than the Soviets, and in fact the joint atomic pool never materialized.[138] Other aspects of the President's vision did come to fruition, including the establishment of an International Atomic Energy Agency.[139] Eisenhower's speech marked a strategic policy shift from denying foreign nations nuclear materials and technologies to sharing these

US AEC in their pricing policy; see Sumner Pike to Brien McMahon, 28 Jun 1951, NARA College Park, RG 326, E67A, box 47, folder 1 Foreign Distribution of Radioisotopes, vol. 3. Information on the Radiochemical Centre, which was a British government–owned company until its privatization in 1982 under Margaret Thatcher, appears on the Amersham web site: http://www.amersham.com/about/heritage.html.

136. Sumner Pike, Acting Chair, to Senator Brien McMahon, 28 June 1951, NARA College Park, RG 326, E67A, box 47, folder 1 Foreign Distribution of Radioisotopes vol. 3, p. 7. Pike observes: "Actually most of the orders for routinely-produced isotopes for Western European countries are being met by the United Kingdom."

137. See Eisenhower, *Atoms for Peace* (1990).

138. Krige, "Atoms for Peace" (2006), p. 164; Weart, *Nuclear Fear* (1988), p. 158; and, for a more complete account, Soapes, "Cold Warrior Seeks Peace" (1980).

139. The International Atomic Energy Statute was unanimously approved by the 81 countries of the United Nations in October 1956, leading to the creation of the International Atomic Energy Agency in 1957. Hewlett and Holl, *Atoms for Peace and War* (1989), p. 225; AEC, *Eighteenth Semiannual Report* (1955), p. 5.

resources as a display of American beneficence and a means to secure diplomatic ties to the developing world.[140] His proposal highlighted the long-standing emphasis on civilian benefits that had been emblematized by the radioisotope distribution program.

The subsequent passage of the Atomic Energy Act in 1954 relaxed the tight security restrictions of the 1946 legislation, permitting companies to patent atomic technologies and license fissionable materials. The law aimed at encouraging commercial development of civilian atomic power, to keep the United States on a competitive basis with other nations that already possessed this technology.[141] To enable industrial participation in reactor development, a massive declassification effort was set into motion. The act also contained provisions for the sharing of technical nuclear information outside of the United States, through the framework of bilateral agreements with "friendly nations."[142] These agreements permitted countries to access equipment and materials from the United States, particularly reactor fuel. For American companies entering the newly open area of civilian nuclear power, these agreements also offered a foreign market for commercial reactors.[143]

Atoms for Peace sought to showcase American capitalism in the nuclear sphere while diverting resources from atomic weapons development abroad.[144] The bilateral agreements that controlled the sharing of reactor technology were designed to restrict the nuclear development of other countries to the purely civilian sphere.[145] Not that the United States intended to slow its own military development of atomic energy. When Eisenhower was elected in 1952, the US stockpile of nuclear weapons was 841; by the time John F. Kennedy was elected in 1960, it had grown to 18,638.[146] The shift to thermonuclear weapons further increased their lethal capacity. Yet Eisenhower understood that the use of nuclear weapons would have devastating consequences for the United States, and that

140. Krige, "Techno-Utopian Dreams" (2010).

141. Chapter 6 addresses the AEC's attempt to foster a civilian nuclear industry.

142. AEC, *Seventeenth Semiannual Report* (1955), p. vii.

143. Medhurst, "Atoms for Peace" (1997).

144. Krige, "Atoms for Peace" (2006); Osgood, *Total Cold War* (2006), ch. 5.

145. On civilian nuclear power programs that were not necessarily incompatible with military ones, see Hecht, *Radiance of France* (1998). John Krige has described a similar US strategy for diverting the resources of other countries from military to civilian program in the space sector: Krige, "Technology, Foreign Policy and International Cooperation" (2006).

146. Krige, "Atoms for Peace" (2006), p. 162.

threat of force was not sufficient to win allies in the global Cold War. Atoms for Peace provided a distraction from the intensifying arms race, to win "hearts and minds" of those in developing nations who might otherwise focus on the seeming militarism of the United States.[147]

In this new phase of the Cold War, technical information and fissionable materials joined radioisotopes as scarce items to be bartered for political influence.[148] In the first six months of 1955, twenty-seven bilateral agreements were signed between the United States and other nations, allowing the dissemination of technical information on reactor design and the transfer of not more than six kilograms of 20% uranium-235 to fuel foreign reactors.[149] The diffusion of reactor technology and fuel would eventually make radioisotopes less scarce—in principle, countries would no longer depend on the major atomic powers of the United States, United Kingdom, Canada, or the USSR to supply isotopes, but could generate them domestically with their own civilian reactors. But such infrastructure was slow and expensive to develop, especially since the United States offered most countries only small research reactors.

In fact, Atoms for Peace included plans for "accelerating and facilitating the distribution of radioisotopes to foreign countries."[150] The AEC continued to hold up radioisotopes to represent the humanitarian face of the country's atomic energy enterprise, an image unchanged since the onset of the export program in 1947. In reality, it remained harder for most foreigners to purchase isotopes from the American government than from the United Kingdom or Canada. As Shields Warren observed to the Advisory Committee for Biology and Medicine in 1956, "There has been much criticism of the isotopes program because of the amount of red tape and forms as compared to the much simpler method of obtaining British isotopes."[151] The one region of the world where the United States retained

147. Krige stresses the winning of "hearts and minds" in "Atoms for Peace" (2006); Medhurst focuses on the civilian program as a "diversion" ("Atoms for Peace" [1997]). I discuss the effects on world opinion of US nuclear weapons testing, particularly the disastrous Bravo shot, in chapter 5.

148. On the ramifications of the 1954 Atomic Energy Act for industrial development, see Palfrey, "Atomic Energy" (1956).

149. Hewlett and Holl, *Atoms for Peace and War* (1989), p. 236; AEC, *Eighteenth Semiannual Report* (1955), pp. 6–7.

150. Minutes, 57th ACBM Meeting, 21–22 Sep 1956, Washington, DC, OpenNet Acc NV0411750, p. 8.

151. Ibid., p. 20.

FIGURE 4.2. Hand-drawn sketch of Paul C. Aebersold, trotting around the globe to promote radioisotopes, prepared for his farewell party in Oak Ridge before moving to the AEC headquarters in Washington, DC, April 1957. The "DCA" on his briefcase below his initials stands for Division of Civilian Application, AEC. Reproduced with permission from the Aebersold papers, box 2, folder 11, Cushing Memorial Library and Archives, Texas A&M University.

an advantage in supplying radioisotopes was Latin America, and this was also the region most impacted by the Atoms for Peace initiative.[152]

In June 1956, the AEC in collaboration with the Fund for Peaceful Atomic Development sent an "isotope mission" to visit Venezuela, Brazil, Argentina, and Uruguay.[153] Paul Aebersold, who headed the Isotopes Division from 1947 to 1957, went on this mission, and his notes from the trip

152. Alonso, "Impact in Latin America" (1985).

153. Memo to Fields from Hall, Re: Radioisotope Mission to South America, 25 May 1956, NARA College Park, RG 326, E67B, box 29, folder 1 Isotopes Program 3 Foreign.

emphasize the political value of radioisotopes in combating "propaganda by anti-US groups."[154] (See figure 4.2.) He had attended the Sixth International Cancer Congress in São Paulo, Brazil, two years earlier, and noted in the summer of 1956 that some improvements had taken place in terms of space and equipment for radioisotope usage. But the increase in isotope users in each country was "not great," and he saw only four or five laboratories in all of South America that were comparable with an "average isotope lab" in the United States.[155] The Atoms for Peace objectives could be partly realized by training more personnel in handling radioactive materials, providing labs with US equipment, and exporting more radioisotopes. As far as the AEC was concerned, cultivating the biomedical utilization of radioisotopes in Latin America would advance American foreign policy as well as prepare these countries to develop (with help from the United States) their own civilian nuclear infrastructure.[156]

Conclusions

The summer 1949 Hickenlooper hearings, in conjunction with the first Soviet atomic test, ended a period of hope after World War II that peaceful applications of atomic energy would predominate over military uses.[157] This changed the symbolism of radioisotopes in two ways. First, to the degree that the radioisotope program exemplified the new era of peaceful uses of atomic energy, it began to appear a mirage in light of the intensifying Cold War. As the successor to the Manhattan Project, the AEC had never ceased being the supplier of nuclear weapons, but as the prospect of economical nuclear power faded, so did the vision that the agency might develop into an atomic TVA. The nuclear arms race with the Soviet Union

154. Paul C. Aebersold, handwritten notes from "Atoms for Peace Mission to SA," Jun 1956, Aebersold papers, box 2, folder 2–6 General Correspondence, June 1956.

155. Ibid.

156. In reality, civilian nuclear technology was slow to arrive in most Latin American countries. Claims by a government-sponsored scientist in Argentina of building the first working nuclear reactor in South America turned out to be fraudulent. To see this as a US policy failure would be too simple. As John Krige has observed ("Techno-Utopian Dreams" [2010]), Atoms for Peace was about the cultivation of desire for nuclear modernity in developing countries as much as implementation. On Argentina, see de Mendoza, "Autonomy, Even Regional Hegemony" (2005).

157. Hewlett and Duncan, *Atomic Shield* (1969).

resulted in new policy conflicts: Debates over whether the United States should pursue building a "super" hydrogen bomb exacerbated rifts within the AEC and between the agency and members of the Congressional committee for atomic energy.[158] Second, the new reality that the United States no longer held an atomic monopoly made it harder for conservatives to portray the export of radioisotopes as jeopardizing the country's supremacy. The establishment of foreign radioisotope distribution programs by Britain and Canada eased the political acceptability of this expanded program in the United States by competing with it. By 1950 the window of opportunity for the American government to show itself generous with reactor-produced materials was closing, and AEC officials sought to capture its declining market share.

Despite—or rather, because of—the intensification of the Cold War, the peaceful development of atomic energy remained politically useful. Eisenhower's Atoms for Peace initiative reinvigorated the use of radioisotopes as tools of foreign policy—extended to include nuclear information and technology, whose dissemination the Atomic Energy Act of 1946 had previously restricted. In this sense, not only radioisotopes but also nuclear "secrets" were divested of their perceived national security risk. This reflected the changed geopolitical situation since the end of World War II. The United States had lost its nuclear monopoly and the Soviet Union was already sharing the "peacetime" benefits of atomic energy with nations to secure political alliances, especially among what soon came to be called the nonaligned nations.[159] The Atomic Energy Act of 1954, with its provisions to encourage both the circulation of previously classified information and private participation in development of atomic energy, brought existing legislation into accord with this new global reality. With the anticipated commercialization of nuclear power, radioisotopes lost their status as the exemplary peacetime benefit of atomic energy. Yet radioisotopes remained potent emblems of the humanitarian promise of the atom, both for American citizens and in the world. One reason was the enduring hope that radioisotopes would heal, particularly those afflicted with cancer.

158. Galison and Bernstein, "In Any Light" (1989).

159. Some non aligned countries, such as India, were especially adept at playing the superpowers off each other to maximize the assistance received. On India, see Abraham, *Making of the Indian Atomic Bomb* (1998).

CHAPTER FIVE

Dividends

We were surprised at first to find so little interest in the Bikini tests. But we really had no right to be. Atomic energy was an uncomfortable subject. Things like John Hersey's *Hiroshima* were rough. How much more pleasant to consider the coming miracles of healing, the prolongation of life, the days of sunny leisure which people were everywhere promising.—David Bradley, 1948[1]

The terrible human toll of the atomic bomb detonations over Hiroshima and Nagasaki did not lessen hopes that artificial sources of radioactivity would revolutionize medicine. If anything, the new scale of atomic energy heightened the level of expectation. As Albert Q. Maisel stated in a May 1947 article for *Collier's*, "The first benign results of atomic bomb research have been new tools for medical scientists which promise cures for hitherto incurable diseases." A striking image illustrated this article, a composite photograph of a mushroom cloud and a man in pajamas arising from a wheelchair, presumably healed. (See figure 5.1.) In Maisel's telling, these "medical dividends"—symbolized by this atomic transfiguration—counteracted the fearsome new possibilities represented by atomic weapons:

> From the Oak Ridge laboratories that gave us the most painful problem of our age now flows a small but steady stream of radioactive isotopes. These strange by-products of atomic energy production may yet prove as important to mankind as the atomic bomb or the cheap atomic power that is still only a promise on the horizon. For to scientists, radioactive isotopes are tools of power in the eternal fight against pain and death; they may be even a means of prying open

1. Bradley, *No Place to Hide* (1948), pp. 167–68.

FIGURE 5.1. Illustration for Alfred Q. Maisel's article "Medical Dividend" in the "Man and the Atom" feature of the May 3, 1947, issue of *Collier's*.

> the once tightly shut door to an understanding of the inner processes of life itself.[2]

Maisel's key examples were radioactive phosphorus and radioactive iodine, whose uses already demonstrated "the two uses of atomic energy—as tracer detectives and as interbody medical bullets."[3] At that time, phosphorus-32 and iodine-131 were precisely the isotopes that Oak Ridge shipped most frequently. They composed about two-thirds of domestic shipments and a higher proportion of foreign shipments, and in the early postwar years they were the predominant isotopes used in therapy.[4]

Maisel's article echoed the AEC's publicity, which emphasized the promise rather than the perils of radioactivity. Many scholars have examined the redemptive hope for atomic medicine in the early Cold War, usually interpreting it as either a misplaced, futuristic optimism, or a cynical attempt by the US government to compensate for the devastation and ongoing costs of atomic weaponry.[5] Yet after World War II there was some concrete basis for the widespread enthusiasm about the curative power of the atom. The AEC's projection of civilian benefits was hitched to the potential of radioisotopes to revolutionize medicine. This hope, as we have seen, was not new. Rather, the government perpetuated the expectation, articulated by E. O. and John Lawrence in the late 1930s, that artificial radioisotopes would transform the treatment of cancer. That said, the health hazards of excessive exposure to radiation had been known for decades, and the suffering among Japanese in Hiroshima and Nagasaki provided new evidence. Over the 1950s the growing public concern over radioactive fallout from atomic weapons testing undermined the image of radioisotopes as therapeutic. In addition, radioisotopes proved less useful than imagined initially for internal therapy (though other medical uses were found). Consequently, nuclear medicine emerged against a background of growing public anxiety—and scientific concern as well—about the dangers of occupational, environmental, and even clinical exposure to radioactivity.[6]

2. Maisel, "Medical Dividend" (1947), p. 14.

3. Ibid., quote on p. 43.

4. Estimates based on information up to June 30, 1949, presented in AEC, *Isotopes* (1949), pp. 5–6, 22, 54, 59. The brief half-lives of iodine-131 (8 days) and, to a lesser degree, phosphorus-32 (14 days), is one reason there were so many shipments of these two; they had a short shelf life. See Aebersold, "Isotopes for Medicine" (1948).

5. E.g., Boyer, *By the Bomb's Early Light* (1985); Leopold, *Under the Radar* (2009).

6. Boudia, "Radioisotopes' 'Economy of Promises'" (2009).

The AEC's evolving understanding of the health risks and benefits associated with radioactivity can be traced through the activities of its Division of Biology and Medicine, established in 1948, and the Advisory Committee for Biology and Medicine that involved outside experts in this unit's work. The agency's radioisotope program was not in this division, but rather was managed through the Division of Production, whose major responsibility was to produce fissile materials for atomic weapons. Nonetheless, the agency's initial commitment to biomedical research grew out of its provision of radioisotopes, an early component of the AEC's cancer program. In addition, the Division of Biology and Medicine oversaw radiological safety at agency installations.[7] Its policies regarding radiation safety had implications for the handling and disposal of radioisotopes produced and distributed by the AEC.

At the beginning of the 1950s, official concerns about radioactivity surfaced mostly in the management of occupational safety at the AEC's plants and in civil defense planning. But over the course of that decade, emerging evidence that low-dose radiation could induce leukemia and other malignancies raised questions about whether the government-adopted standards for permissible dose were adequate to protect the population. While the media tended to focus on the potential dangers of radioactive fallout from atomic weapons testing, the advent of a civilian nuclear power industry, facilitated during the Eisenhower administration through the 1954 Atomic Energy Act, posed an even greater source of environmental radioactivity.[8] The emergence of a nuclear power industry would also increase occupational exposure to ionizing radiation. By 1954, the government was firmly committed to both continued weapons testing and civilian development of atomic energy, insisting that these activities posed no hazard to human health. Many scientists, particularly geneticists and ecologists, began to challenge publicly the government's assertions that the radiation exposures associated with these operations were safe.

As these debates were being reported in the public media, the AEC itself was reassessing radiation safety. During and after World War II, radiological protection standards had been set somewhat arbitrarily, on the basis of both pragmatic considerations and incomplete knowledge

7. Statement of Shields Warren, JCAE, *Investigation* (1949), Part 22, p. 876.

8. See chapter 6 for more on these industrial developments and their regulation.

about the effects of newly created radioactive materials.[9] Health physicists charged with overseeing occupational safety worked under a toxicological framework, assuming the existence of a threshold below which radiation would not be harmful. Radiation levels that were significantly below those that induced acute injury were considered "tolerable." But over the course of the 1950s, new evidence of long-term effects of low-level radiation called these assumptions into question. This had implications not only for safety at the AEC's plants and the emerging civilian nuclear industry, but also for how the agency represented the safety of the radioisotopes it shipped from Oak Ridge—and the safe disposal of the radioactive waste their large-scale consumption generated. How could the agency assure that exposure levels were harmless when the geneticists it funded were showing that there was no lower limit to mutational damage caused by radiation exposure?

There was a related shift in popular perceptions of radioactivity.[10] By the late 1950s, the fallout debates had fixed attention on the dangers of a few particular fission products of atomic detonations, such as strontium-90 and iodine-131. The prospect of radioactively contaminated meat and milk, from strontium-90 deposited on grasslands and eaten by livestock, changed the symbolic valence of radioisotopes. Once seen as curative, radioisotopes were increasingly viewed as contaminants, even carcinogens. The perception of danger associated with radioactivity grew stronger over the 1960s and 1970s, eclipsing the government's emphasis on the healthful dividends of the atom.

Targeting Cancer

Cancer was an especially important motif in the domestic politics of atomic energy and provided the impetus for the AEC's program in biology and medicine. Harnessing atomic energy to combat cancer suited the general ambivalence about the age of nuclear weaponry—the conjoined danger and promise of the atom. Moreover, radiation itself had a double-edged relationship to cancer, recognized long before the crash program to build atomic weapons: radiation exposure could cause cancer. Both the

9. Caufield, *Multiple Exposures* (1990), esp. preface and ch. 8; and Whittemore, *National Committee on Radiation Protection* (1986).

10. Boyer, *By the Bomb's Early Light* (1985), and Weart, *Nuclear Fear* (1988).

risks and the therapeutic possibilities of new radiation sources dated to observations with x-rays and then radium.[11] These hopes and fears around cancer were recast around the power of the atom. One popular article published by an American biophysicist in the Japanese occupation was entitled "Atomic Energy: Cancer Cure . . . or Cancer Cause?"[12]

The particular hope that isotopes could cure cancer derived from the notion that they would localize to specific tumors and deliver internal radiation. One sensational case, reported in the media in 1947, reinforced these expectations. A patient diagnosed with adenocarcinoma of the thyroid was treated with iodine-131. The radioactive iodine localized to several metastatic tumors, both making them detectable and, over time, shrinking their size.[13] As Alfred Maisel put it in *Collier's*: "The case of Mr. B is one of the most hopeful things that have hit the medical world in a long, long time. For it demonstrates two uses of atomic energy—as tracer detectives and as interbody medical bullets—that are finding increasing application in scores of ways in laboratories and hospitals all over the United States."[14] Some newspapermen were less cautious in heralding the breakthrough; one headline read "Cancer Cure Found in the Fiery Canyons of Death at Oak Ridge."[15]

This representation of radioisotopes as elixirs of health drew on an older image of radioactivity as a source of rejuvenation. Some patent medicine included radium in the 1920s.[16] Radioisotopes, generated in the "atomic furnace" by nuclear fission, appeared more potent than natural sources of radiation used to destroy cancer cells. Just as significantly, the therapeutic use of radioisotopes countered, and perhaps even redeemed, the more destructive uses of atomic energy. As medical physicist Robley Evans asserted in 1946: "The sober truth is that through medical advances alone, atomic energy has already saved more lives than were snuffed out at Hiroshima and Nagasaki."[17]

11. Serwer, *Rise of Radiation Protection* (1976); Lavine, *Cultural History of Radiation* (2008).

12. Henshaw, "Atomic Energy" (1947).

13. Seidlin, Marinelli, and Oshry, "Radioactive Iodine Therapy" (1946).

14. Maisel, "Medical Dividend" (1947), p. 43.

15. As quoted by Brucer, "Nuclear Medicine" (1978), p. 595. On changing cancer therapeutics, Pickstone, "Contested Cumulations" (2007); Cantor, *Cancer in the Twentieth Century* (2008).

16. Weart, *Nuclear Fear* (1988), p. 50; Campos, *Radium and the Secret of Life* (2006).

17. Evans, "Medical Uses of Atomic Energy" (1946), p. 68. It is not clear how Evans calculated the number of lives saved; despite his "sober" assertion, it seems a bald exaggeration.

The AEC's role in medicine was, however, unclear at the outset. As the successor to the Manhattan Project, the Commission was more heavily oriented to the physical sciences and engineering than the life sciences. Its General Advisory Committee was composed entirely of physical scientists. In addition, the sole Commissioner who was a scientist, Robert Bacher, was a physicist.[18] The Atomic Energy Act of 1946 specified that the new agency should undertake research related to "utilization of fissionable and radioactive materials for medical, biological, health, or military purposes" and "the protection of health during research and production activities."[19] Two programs the AEC inherited from the MED were in line with Congress's directives. First was obviously the radioisotope program, which had been shipping materials for four months when the Commission was formally established. Second, the MED had announced in July 1946 that 15% of the medical program budget at the research-oriented project sites (e.g., Berkeley and Chicago) could be allocated to basic research "in the treatment of malignant tissues."[20] This small program represented a departure from wartime activities; under the Manhattan Project, health physics was configured to manage occupational safety, not to conduct biomedical research.[21] The MED's Medical Advisory Committee, chaired by Stafford Warren, strongly endorsed this move when they first met in September 1946.[22] When reconvened by the AEC a few months later, this group pointed to the bounty of opportunities for medical research and technology related to atomic energy.[23] However, the Commission did little to set policy in this arena during its first six months.

A new mandate from Congress to address the cancer problem impelled the AEC to build up a biomedical research program. From April to June of 1947, the House Appropriations Committee held hearings on the AEC's proposed fiscal year 1948 budget of $500,000,000. On May 15, Representative Everett M. Dirksen contended that the AEC should be involved in finding effective treatment for cancer, which was for him "something of

18. Westwick, *National Labs* (2003), p. 245.

19. Sec. 3 (a) (3) and (5), "Atomic Energy Act of 1946" (1946), p. 19; Westwick, *National Labs* (2003), p. 247.

20. As quoted in Westwick, *National Labs* (2003), p. 244.

21. Westwick, *National Labs* (2003), pp. 242–43; Hacker, *Dragon's Tail* (1987), pp. 36–44; Malloy, "'A Very Pleasant Way to Die'" (2012).

22. Westwick, *National Labs* (2003), p. 245.

23. Transcript of the Discussion at the First Meeting of the Medical Review Board, AEC, Washington, DC, 16 Jun 1947, OpenNet Acc NV0709599.

a crusade."[24] Cancer killed "one person every 3 minutes"; this amounted to "Seventy-Two Pearl Harbors Every Year."[25] Dirksen observed that the Atomic Energy Act gave the agency authority to utilize "fissionable and radioactive materials for medical, biological, health, or military purposes." Radioactive materials, he asserted, composed "the essential point in this whole cancer business."[26] Cancer, being "a virus disease," Dirksen claimed, was especially susceptible to scientific attack. Radioactive materials, he suggested, could be the equivalent of penicillin for the treatment of cancer—and penicillin itself had been developed industrially by the US government as part of the wartime mobilization. One Congressman asked, "Is not radioactivity the only hope of the scientists for finding a cure for cancer?" Dirksen replied, "That is right, and these are the people who probably have the greatest background of information in that whole radioactive field."[27]

It seemed especially fitting to Dirksen that the successor to the Manhattan Project should take on the cancer problem: "If we are going to spend a few hundred million dollars in the atomic-energy field to perfect an instrumentality of death, then let us take a little of that money to develop an instrumentality to preserve life."[28] (See figure 5.2.) Or, as a writer for the *Science News Letter* put it: "Control of cancer instead of manufacture of bombs is the alternative program offered research scientists."[29] Dirksen pointed out that Congress could tackle the problem without even passing new legislation if it would "earmark not less than $25,000,000 for cancer research" to be carried out by the AEC, about 5%–10% of the agency's budget.[30] The House committee approved Dirksen's proposal.[31]

The Senate Appropriations Committee resisted Dirksen's bold plan, as did the agency's own Commissioners. There was the problem of duplicating existing cancer-related programs, both public and private. For instance, the US Public Health Service already supported cancer research

24. Statement of Everett M. Dirksen, 15 May 1947, in US Congress, House, *Independent Offices Appropriation* (1947), p. 1539.

25. Ibid., p. 1539.

26. Ibid., p. 1538–59.

27. Ibid., p. 1542.

28. Ibid., p. 1540. See Keller, "From Secrets of Life" (1992).

29. "Control of Cancer Instead of Atomic Bombs" (1946), p. 213.

30. Statement of Everett M. Dirksen, p. 1539. For an expression of Lilienthal's consternation: Transcript of the Discussion, pp. 20–21.

31. Feffer, "Atoms, Cancer, and Politics" (1992), esp. 256–58.

FIGURE 5.2. Political cartoon from the *Dallas Morning News*, 12 August 1945, depicting the use of atomic energy to destroy cancer, only days after atomic weapons were detonated over Hiroshima and Nagasaki. Reprinted with permission of the *Dallas Morning News*.

to the tune of $12,500,000 per year. The American Cancer Society was also raising vast sums of money, $10 million in 1946 alone. Of that sum, $3.5 million was designated for medical research grants, to be awarded through the newly established "Committee on Growth" of the National Research Council.[32] As one senator said: "I wonder why drag the Atomic Energy Commission into cancer research."[33] The Commissioners expressed their own reservations. Lewis Strauss doubted that the new agency could effectively spend such a large sum on cancer research.[34] David Lilienthal made clear that the AEC had not asked for this responsibility, and he did not want to see the agency involved in either cancer therapy or awarding research grants.[35] As he saw it, the AEC's principal relevance to the issue rested on its provision of "isotopes, which will further the fundamental science on which cancer control, in the view of many, depends."[36] Despite these misgivings, a sizeable appropriation remained in place. In a budget generally austere toward science funding, Congress earmarked $5 million for the AEC's efforts in cancer research.[37]

The AEC turned to its Medical Review Board, inherited from the MED, to advise them on how to spend a large sum of money on cancer research.[38] This funding-driven approach made several of its members uneasy, reflecting a more widespread sentiment among elite scientists that "directed" research would corrode the freedom and creativity of scien-

32. As mentioned in chapter 3, the Committee on Growth also lobbied the federal government to make radioisotopes more available to researchers and physicians. Eugene P. Pendergrass to Robert S. Stone, 31 Jul 1945, NARA Atlanta, RG 326, OROO Lab & Univ Div Official Files, Acc 68A1096, box 13, folder Isotopes–3; Shaughnessy, *Story of the American Cancer Society* (1957); Patterson, *Dread Disease* (1987); Bud, "Strategy in American Cancer Research" (1978); Gaudillière, "Molecularization of Cancer Etiology" (1998).

33. Senator Clyde M. Reed, in US Congress, Senate, *Independent Offices Appropriation* (1949), p. 52; Westwick, *National Labs* (2003), pp. 252–57.

34. Strauss in US Congress, Senate, *Independent Offices Appropriation* (1949), p. 56.

35. Lilienthal in US Congress, Senate, *Independent Offices Appropriation* (1949), p. 58.

36. Ibid., p. 53.

37. Feffer, "Atoms, Cancer, and Politics" (1992), p. 258.

38. The members of the Medical Review Board were A. Baird Hastings of Harvard Medical School; Detlev Bronk of the University of Pennsylvania Medical School; Ernest Goodpasture of Vanderbilt Medical School; George Beadle of Caltech; Karl Meyer of UCSF; Joseph Wearn of Western Reserve, Cleveland; and E. C. Stakman of the University of Minnesota. AEC Press Release, "Statement by Medical Board of Review Advising United States Atomic Energy Commission on Programs and Policies in the Field of Medical Research," 22 Jun 1947, NARA College Park, RG 326, E67A, box 26, folder 9 Medical Board of Review.

tists. For many Americans, large-scale federal support of science smacked of socialism.[39] Herbert Gasser raised concerns about whether the prospect of funding would lure universities to seek AEC support even if it were "detrimental to their primary purpose and objectives."[40] Moreover, safeguarding security for the AEC might require a "partition" from the "free investigation" that characterized universities.[41] Yet the Medical Review Board urged the AEC to involve itself in research by providing reactor-generated materials: "Of equal importance, of course, is the continuing supply of radioisotopes from the Commission's facilities. No phase of the atomic energy program offers greater promise of benefit to mankind."[42] The group recommended that the AEC add an isotope "consulting service" for interested researchers and urged the Commission to extend distribution to foreign investigators.[43]

Although the AEC had not asked for the responsibility, the Congressional cancer appropriation gave the agency an immediate opportunity to demonstrate the medical benefits of atomic energy. The political value of this became evident as early as the summer of 1947, as the General Advisory Committee informed the Commission that their expectations for the quick inception of domestic atomic power were completely unrealistic.[44] To provide expert guidance on applications of atomic energy to health and life science, the Commission replaced the Medical Review Board with a permanent Advisory Committee for Biology and Medicine (ACBM), chaired by Alan Gregg of the Rockefeller Foundation.[45] In addition, the AEC created a Division of Biology and Medicine, whose director would report directly to the Commission and General Manager.[46] Shields Warren (no relation to Stafford Warren) was appointed Director in October 1947.

39. Reingold, "Vannevar Bush's New Deal" (1987).

40. Transcript of the Discussion, Record #4, p. 10.

41. Ibid., p. 9. In its final report, the Board recommended that "in so far as it is compatible with national security, secrecy in the field of biological and medical research be avoided." Report of the Medical Board of Review, 20 Jun 1947, NARA College Park, RG326, E67A, box 26, folder 9 Medical Board of Review, p. 11.

42. AEC Press Release, "Statement by Medical Board," p. 3.

43. AEC Press Release, "The United States Atomic Energy Commission Releases Report of Medical Board of Review," 6 Jul 1947, NARA College Park, RG 326, E67A, box 26, folder 9 Medical Board of Review. On the export program, see chapter 4.

44. See chapter 6.

45. Summary of the 68th AEC Meeting, 25 Jun 1947, NARA College Park, RG 326, E67A, box 23, folder 7 Advisory Committee for Biology and Medicine.

46. Westwick, *National Labs* (2003), p. 247.

A professor of pathology at Harvard Medical School, Warren had previously served on the Interim Medical Advisory Committee and headed the Navy's medical survey team of Hiroshima and Nagasaki.[47] This new Division, which was responsible for radiological safety at AEC installations as well as biomedical research, sat rather oddly alongside the other AEC units, which were largely oriented toward the continuing development of nuclear science and manufacture of atomic weapons: Raw Materials, Production, Engineering, Military Application, and Research. As Warren saw it, the mission of the Division of Biology and Medicine was to "maintain safety and to fight against disease."[48]

By that fall, the Commission had overcome its early inertia about including biomedical research. Relying on the ACBM for input, the agency began designing a cancer program in line with the Congressional appropriation, one that did not duplicate the efforts of other government agencies.[49] Formally accepted by the Commission in January 1948, the AEC's cancer program involved three major activities, two of which involved funding for programs already in place.[50] First was supporting the work of the Atomic Bomb Casualty Commission (ABCC), whose studies of the medical consequences of the bombings of Hiroshima and Nagasaki included assessment of cancer incidence in survivors.[51] Second, the agency would make certain radioisotopes already available for sale from Oak Ridge free of charge for cancer research, diagnosis, and therapy. Beginning in April 1948, Oak Ridge offered radiosodium, radiophosphorus, and radioiodine without charge for uses related to cancer. (The only costs were for shipping.) Early in 1949, the program was expanded to include all

47. AEC Press Release No. 64, "Dr. Shields Warren Appointed Interim Director of Biology and Medicine," 24 Oct 1947, NARA College Park, RG 326, E67A, box 23, folder 8 Division of Biology and Medicine–Organization and Functions.

48. Statement of Shields Warren, JCAE, *Investigation* (1949), Part 22, p. 876.

49. This was a stipulation of the legislation authorizing the cancer research budget. On the role of the ACBM in planning the cancer program, AEC Press Release No. 55, "United States Atomic Energy Commission Names Advisory Committee for Biology and Medicine," 12 Sep 1947, NARA College Park, RG 326, E67A, box 23, folder 7 Advisory Committee on Biology and Medicine.

50. A Plan for a Cancer Research Program for the Atomic Energy Commission, Report by the Director of the Division of Biology and Medicine, Atomic Energy Commission, 5 Jan 1948, OpenNet Acc NV0702018; AEC 26, Cancer Research Program for the Atomic Energy Commission, 30 Jan 1948, NARA College Park, RG 326, E67A, box 64, folder 5 Research in Biological and Medical Science.

51. See Beatty, "Genetics in the Atomic Age" (1991); Lindee, *Suffering Made Real* (1994). The AEC was already funding the ABCC year to year.

radioisotopes; by August of that year, the cancer program accounted for over 2,000 shipments of isotopes.[52]

The third part of the cancer program involved the AEC in experimental cancer therapy, which would require new organization and infrastructure.[53] The agency opened up a clinical cancer research unit in Oak Ridge through a contract with the Oak Ridge Institute of Nuclear Studies, an affiliation of southern universities that already ran the agency's isotope training courses.[54] Plans for cancer research at Argonne were more ambitious: the agency set aside $1,750,000 to build a 50-bed hospital in Chicago dedicated to experimental therapies for cancer, scheduled to open in 1951.[55] The decision to fund the Argonne cancer hospital helped the Commission use a large chunk of the $5 million cancer budget—only two days before the fiscal year 1948 deadline for spending the appropriation.[56] The agency expected that this hospital would be "primarily for research in which radioisotopes will be used to treat cancer patients."[57]

In reality, however, radioisotopes did not turn out to be the "medical bullets" envisioned. As Shields Warren had assessed already in 1948, "Only a few isotopes among those thus far tried have been useful from the standpoint of therapy."[58] Few radioisotopes localized sufficiently to one type of tissue to be used for internal radiotherapy. Researchers sought to

52. Ninety percent of the 2,059 shipments were radiophosphorus or radioiodine. Press Release for 3 Aug 1949, "AEC Distributes 8,363 Shipments of Radioactive and Stable Isotopes in Three Years," NARA College Park, RG 326, E67A, box 45, folder 13 Distribution of Stable Isotopes Domestic. When the program was expanded to include all radioisotopes, the stress was on advancing the forefront of research and treatment. Isotopes Division, US Atomic Energy Commission, Oak Ridge, Tennessee, *Isotopes Catalogue and Price List No. 3, July 1949*, copy in NARA College Park, RG 326, E67A, box 46, folder 6 Foreign Distribution of Radioisotopes vol. 2, p. 7. In 1952 the program was modified such that users paid 20% of production costs for radioisotopes used in the treatment, diagnosis, and study of cancer. AEC, *Twelfth Semiannual Report* (1952), p. 32.

53. Lilienthal to Bourke B. Hickenlooper, 15 Dec 1947, Appendix D to AEC 26, Cancer Research Program for the Atomic Energy Commission, RG 326, NARA College Park, E67A, box 64, folder 5 Research in Biological and Medical Science.

54. ORINS went on to launch the Teletherapy Evaluation Board to develop and evaluate radioisotope-based cancer therapy; see chapter 9.

55. David E. Lilienthal to Alan Gregg, 9 July 1948, NARA College Park, RG 326, E67A, box 64, folder 5 Research in Biological and Medical Science. In the end, it was dedicated on March 14, 1953. US Congress, House, *Second Independent Offices Appropriation* (1953), Part I, p. 455.

56. Feffer, "Atoms, Cancer, and Politics" (1992), p. 259.

57. AEC, *Atomic Energy and the Life Sciences* (1949), p. 91.

58. Warren, "Medical Program" (1948), p. 233.

devise ways to guide radioisotopes to specific organs and tissues by combining them with compounds or antibodies, but research along these lines remained preliminary.[59] The one isotope that did concentrate in a single organ, radioiodine, was not effective against the majority of thyroid tumors.[60] Cornelius Rhoads, director of the Sloan-Kettering Institute for Cancer Research, referred to the therapeutic use of radioiodine as "a possibility which has been hailed with great enthusiasm by those who wish to justify the manufacture of atomic bombs by the human application of their byproducts."[61] Expectations that radioisotopes would provide radically new tools for curing cancer proved overrated; the most important advance ended up being the development of cobalt-60 as an alternative external radiation source to radium.[62] Partly in response to this dilemma, the agency began stressing the value of radioisotopes used as tracers in cancer research, rather than in therapy.

Beyond its cancer program, the AEC began receiving other proposals for biological and medical research related to radiation or atomic energy. The AEC did not yet have a grants division; its extramural research support was awarded through the Office of Naval Research. Until the Division of Biology and Medicine was fully staffed, the ACBM served as a peer review panel for such extramural proposals.[63] The AEC's regional labs also expanded their purview into biomedical research. Argonne's Biology Division under Austin Brues and the Berkeley Rad Lab's program under John Lawrence and Joseph Hamilton led the pack, with Oak Ridge National Lab quickly gaining ground.[64] However, the cancer program remained a key ingredient of the AEC's biomedical research enterprise, which represented one-third of the budget of the Division of Biology and Medicine in 1949.[65]

59. AEC, *Atomic Energy and the Life Sciences* (1949), p. 100.

60. According to Joseph Ross, head of the Radioisotope Division at Evans Memorial Hospital in Boston, about 15% of thyroid cancers would take up radioiodine. Ross, "Radioisotope Division" (1951), p. 39.

61. "Business in Isotopes" (1947), p. 158.

62. See chapter 9.

63. Draft Minutes, 5th ACBM Meeting, 9 Jan 1948, Los Alamos, New Mexico, OpenNet Acc NV0711640.

64. Westwick, *National Labs* (2003), pp. 247–48.

65. Shields Warren, JCAE, *Investigation* (1949), Part 22, p. 885; Westwick, *National Labs* (2003), p. 252.

Radiation Hazards in an Age of Atomic Weapons

The AEC had more immediate motivations, beyond its participation in the national crusade against cancer, to develop a program in biology and medicine. Foremost among them was understanding better the occupational risks faced by workers in AEC plants and laboratories. As Henry DeWolf Smyth put it, "Cancer is a specific industrial hazard of the atomic energy business."[66] The agency tried to present its occupational health requirements as creating research opportunities:

> Radiation can cause cancer as well as control it and this effect requires full study. A great number of experiments undertaken at Argonne, Oak Ridge, and the National Cancer Institute, to determine what levels of radioactive rays and isotopes may be considered safe for workers in the atomic energy field, have been useful in helping to understand how cancer arises.[67]

While the Commission's responsibility for worker safety made the carcinogenic effects of radiation highly relevant, it also gave the agency an incentive to question or downplay evidence of danger. Particularly in the early postwar years, the AEC focused issues of radiological safety on acute effects—radiation "injury"—rather than on possible long-term effects from chronic exposure.

In general, both the publicity around the AEC's radioisotope program and the general emphasis on the civilian benefits of atomic energy rested on the assumption that the health risks of radiation exposure could be straightforwardly controlled. Activities in two areas highlighted the difficulties, both actual and theoretical, with this assumption. The first was the environmental contamination that resulted from atomic weapons testing, especially as part of the 1946 Operations Crossroads tests in Bikini. This issue became even more relevant to Americans as the US government began shifting nuclear weapons testing from the Pacific atolls to the Nevada Test Site. The second was civil defense planning, which intensified

66. Henry DeWolf Smyth to Ernest Goodpasture, 28 Dec 1951, as quoted in Minutes, 30th ACBM Meeting, 10–12 Jan 1952, University of California, Davis and Berkeley, OpenNet Acc NV0711826, p. 5. Smyth was apparently quoting a phrase of Goodpasture's, approvingly, back to him.

67. AEC, *Atomic Energy and the Life Sciences* (1949), p. 96.

in the 1950s on account of both the Korean War and the development of hydrogen bombs, which gave off vastly more radioactive debris than the initial fission bombs. While neither of these activities was directly related to the AEC's radioisotope program, both shaped popular understandings of radiation, ultimately countering the agency's positive press about the dividends of atomic energy.

In the early postwar period, most Americans were not strongly opposed to the peacetime testing of atomic weaponry. The Gallup Public Opinion Polls surveyed adults in the spring of 1946 on whether "the United States should carry out the atom bomb tests on Bikini Island, or should this be given up?" Forty-three percent responded in favor of holding the tests, 37% expressed a preference to call off the tests, and 20% had no opinion. Support for the tests was highest among World War II veterans and among more educated groups.[68] Those who opposed it need not have fixed on radiation hazards; continued testing could, for instance, undermine the Baruch plan for international control of atomic energy. In addition, there were predictions that atomic explosions might rupture the earth's crust, or that they would set in motion huge tidal waves threatening the continental United States.[69] Obviously, no such natural disasters resulted. The short-term effect of the early tests was to calm public fears. Atomic weapons could be detonated in a controlled fashion, or so it seemed.

However, the extensive contamination of the ships and natural environment resulting from the underwater blast (Test Baker) in the summer of 1946 gradually corroded this sense of assurance. As Paul Boyer asserts, "It was Bikini, rather than Hiroshima and Nagasaki, that first brought the issue of radioactivity compellingly to the nation's consciousness."[70] After the Baker blast, a "moving blanket of radioactive mist" showered contamination over the anchored ships in the lagoon. Decontamination of the hottest ships proved impossible, because radiation exposures among those personnel cleaning the vessels, so far as they could be measured, exceeded permissible levels. To make matters worse, members of the clean-up crews did not want to wear the protective gloves and garments in the sweltering heat.[71]

A year after the Bikini tests, *Life* magazine ran a spread on "What Science Learned at Bikini." The feature pointed out that many of the tar-

68. Interviewing Date 3/29–4/3/46, Survey #368-K, Question #10, in Gallup, *Gallup Poll* (1972), p. 571.

69. Boyer, *By the Bomb's Early Light* (1985), p. 82.

70. Ibid., p. 90.

71. Hacker, *Dragon's Tail* (1987), pp. 140–45.

get ships for Test Baker were "still too radioactive for crews to remain aboard." The persistent dangers of radioactive contamination had implications for civilian populations in the event of nuclear war: "the huge wave of radioactive spray which shot out from the base of the water column demonstrated a terrifying new possibility of atomic warfare."[72] David Bradley's 1948 book, *No Place to Hide*, also discussed the inescapable radioactive contamination that confronted the thousands of troops involved in Operation Crossroads. (He was among them.) However, he also acknowledged how difficult it was to dislodge the popular perception of radioactivity as controllable and curative.[73]

The AEC began examining the longer-term dangers of radioactive contamination, especially in the event of nuclear war. Early in 1949, the agency launched a secret study, Project Gabriel, to "determine the long-range effects which the explosion of a number of atomic bombs might have on animal and vegetable life through radioactive contamination of air, water and soil."[74] Physicist Nicolas N. Smith Jr., of Oak Ridge National Laboratory, calculated how many atomic detonations would be needed to produce significant radiological hazard. Focusing on the potential contamination of plutonium, strontium-90, and yttrium-90, Smith calculated that it would take 3,000 blasts in a single growing season to seriously disrupt agricultural production.[75] In 1951, Smith used data from more recent test series for further calculations; these results seemed even more reassuring.[76] As Shields Warren updated the ACBM: "In light of additional data from the Greenhouse and Ranger tests . . . 100,000 nominal bombs might be detonated without undue hazard from secondary effects."[77] Needless to say, everything hinged on what counted as "undue hazard."

Reports of radiological accidents and the dreadful suffering of Japanese survivors of the atomic blasts had focused public attention on

72. "What Science Learned at Bikini" (1947), p. 74.

73. See the epigraph to this chapter, taken from Bradley, *No Place to Hide* (1948).

74. Memorandum from Shields Warren, Director, Division of Biology and Medicine to the General Advisory Committee, 13 Feb 1952, OpenNet Acc NV0403989.

75. Hewlett and Anderson, *Atomic Shield* (1969), p. 499.

76. Memorandum requesting assistance of Dr. Nicholas Smith, 19 June 1951; Memorandum on Re-Evaluation of Project Gabriel, 21 Aug 1951, OpenNet Acc NV0404822; Hewlett and Holl, *Atoms for Peace and War* (1989), p. 265.

77. Minutes, 30th ACBM Meeting, 10–12 Jan 1952, University of California, Davis and Berkeley, OpenNet Acc NV0711826, p. 10. The updated calculations were generated by an ad hoc committee appointed in November 1951 to reassess Project Gabriel. The committee contained no geneticist. Jolly, *Thresholds of Uncertainty* (2003), p. 132.

radiation sickness from acute levels of exposure. The 1951 translation of *We of Nagasaki*, with its depiction of the horrors of radiation exposure after the atomic blast, reinforced this emphasis.[78] But studies of two populations exposed to lower levels of radiation provided disturbing evidence of longer-term effects. These results were more sobering than the estimates from AEC's Project Gabriel. In 1948, the Atomic Bomb Casualty Commission (ABCC) began surveying leukemia incidence among survivors of the atomic blasts over Hiroshima and Nagasaki. Their first publication, in 1952, documented a higher rate of leukemia among these survivors. Distance from the blast correlated with leukemia incidence: Those who were less than 2,000 meters from the explosion showed a higher incidence of leukemia than those further away, though both groups of exposed individuals showed increased rates of the disease.[79] Even those who did not suffer radiation sickness were not altogether spared.

Another irradiated population was domestic and professional: radiologists. In a 1950 study, H. C. March analyzed leukemia rates among physicians over a twenty-year period, finding a nine-fold higher incidence of the disease among radiologists than in the group as a whole.[80] The relevance of these results beyond clinical workers remained unclear, as the numbers were relatively small, and the dosages of radiation experienced could not be precisely quantified. The AEC leadership clung to the view that radiation below a certain threshold would not cause cancer.[81] But these studies, especially when taken together, offered a disquieting picture of radiation's imperceptible dangers.

Radiation Hazard as an Issue of Heredity

Over the course of the 1950s, there was a gradual reframing of radiation hazards away from the acute effects to long-term consequences. At issue in assessments of danger due to peacetime atomic tests—or, for that matter, the prospect of nuclear war—was what kinds of health issues counted as harmful. Early radiation protection standards were designed to prevent

78. Nagai, *We of Nagasaki* (1951).

79. Folley, Borges, and Yamawaki, "Incidence of Leukemia in Survivors" (1952); Moloney and Kastenbaum, "Leukemogenic Effects of Ionizing Radiation" (1955).

80. March, "Leukemia in Radiologists" (1950).

81. Jolly, *Thresholds of Uncertainty* (2003), ch. 3.

any known short-term effects from acute radiation exposure. In addition, longer-term carcinogenic risks associated with ingesting radioisotopes had become clear from the cases of radium dial painters.[82] Yet radiation safety experts tended to assume that there was a threshold exposure level below which biological effects were effectively innocuous. Since 1929, a nongovernmental organization, the Advisory Committee on X-Ray and Radium Protection, had made recommendations to industry for a "tolerance dose." The atomic age, with its new array of radioactive materials, made radiation safety vastly more complex. In 1946, the advisory body, rechristened the National Committee on Radiation Protection (NCRP), replaced "tolerance dose" in their recommendations with "maximum permissible dose."[83] This change in terminology subtly acknowledged that there was no certifiably safe level of exposure.

The NCRP tended to focus on the so-called somatic effects of radiation—the direct health consequences of exposure. In addition to burns and other immediate effects, radiation had been correlated with the appearance of tumors and leukemia from early observations of the consequences of radium exposure in the 1920s and 1930s.[84] But there was another class of effects, mutations in the germ line, that would affect the exposed individual's fertility and offspring. H. J. Muller had demonstrated in 1927 that x-rays could induce mutations, and this finding was rapidly extended to other forms of radiation.[85] Geneticists saw no basis for recognizing a lower threshold for the mutational effects of radiation, which showed a linear relation to dose. Well into the 1950s, there remained a gap—partly disciplinary, partly conceptual—between this genetic perspective and that of the health physicists who formulated and implemented radiological protection for the AEC.[86]

After World War II, the question arose of whether genetic effects should influence the setting of "maximum permissible dose" levels for radiation. In 1949 Robley Evans, a physicist known for his studies of the radium dial painters, published an article entitled "Quantitative Inferences Concerning

82. Ibid., p. 36.

83. Walker, *Permissible Dose* (2000), pp. 10–11. See also Boudia, *Gouverner les risques* (2010).

84. Stone, "Concept of a Maximum Permissible Dose" (1952); Whittemore, *National Committee on Radiation Protection* (1986); Kathren, "Pathway to a Paradigm" (1996).

85. Muller, "Artificial Transmutation of the Gene" (1927).

86. Jolly, *Thresholds of Uncertainty* (2003), chapters 3 and 4.

the Genetic Effects of Radiation on Human Beings."[87] Evans's analysis suggested that exposure at or under the government's permissible dose level would not significantly increase the mutation rate beyond its spontaneous level.[88] A March 2, 1949, press release from the Science Service presented Evans's article as a challenge to Muller's public pronouncements "that human beings are now in danger of acquiring harmful hereditary changes from peacetime exposures to atomic radiations and some kinds of X-rays."[89] The key point of dispute was whether radiation exposure from atomic energy installations (or, for that matter, clinical applications of radioactive materials) significantly increased the baseline mutation rate in humans. As Evans put it in a letter, "Everyone admits that a small amount of genetic change is always induced by radiation. However, one must remember that spontaneous genetic mutations are taking place all the time anyhow. I have felt that the important point to be emphasized is the ratio between induced and spontaneous mutations."[90]

Evans's figure for the spontaneous mutation rate in humans had been extrapolated from the incidence of rare genetic diseases such as hemophilia. He then calculated a "doubling dose"—that is, the radiation level that would cause a two-fold increase of the spontaneous mutation rate, as 300 roentgens per individual per generation.[91] The maximum permissible radiation dose used by the AEC was 0.1 roentgens per workday. A worker who was exposed at this level each day might accumulate 250 roentgens over a ten-year period, still below the doubling dose.[92] In reality, few work-

87. Evans, "Quantitative Inferences" (1949).

88. Evans's estimates of radiation-induced mutation rates appear to have been based on "Genetic Effects of Irradiation with Reference to Man," by D. G. Catcheside (2 Jun 1947), and "Tolerance Doses in Relation to Genetic Effects of Radiation," a memorandum by D. E. Lea. Both of these were prepared for the Tolerance Dose Panel of the Protection Sub-Committee of the Medical Research Council, and manuscripts of them, along with Evans's notes, are in the Evans papers, box 1, folder Genetics 1.

89. Frank Thone, Science Service Biology Editor, "Long-Range Debate Stated on Genetic Effects of Radiation," 25 Mar 1949, Evans papers, box 1, folder Genetics 1. For examples of media coverage of Muller's views, see "Radioactive Rays Held Peril to Race" (1947) and "Our Defective Race" (1947). Some of Muller's numerous speeches on the topic were published, e.g., Muller, "Menace of Radiation" (1949). See also Muller, "Some Present Problems" (1950).

90. Robley D. Evans to E. E. Stanford, 3 Jun 1949, Evans papers, box 1, folder Genetics 1.

91. Evans did not originate the doubling dose; underlying it was the widespread assumption that a doubling of the natural mutation rate would not cause significant injurious effects in a population. Jolly, *Thresholds of Uncertainty* (2003), p. 95.

92. Evans, "Quantitative Inferences" (1949), p. 302.

ers were exposed to the maximum permissible levels, so Evans argued that the induced mutation rate from radiation would be low, practically negligible.

After Evans published his paper, geneticist Sewall Wright challenged its calculations.[93] Wright pointed out that Evans's benchmark for the spontaneous generation rate in humans was the occurrence of two rare but devastating hereditary medical conditions—hemophilia and epiloia. This gave a rate of 10^{-5} mutations per gamete per generation. Experiments with *Drosophila*, however, had demonstrated that not all mutations were as deleterious as these two pathologies. Consequently, Evans's estimate was likely far too high. Extrapolating instead from *Drosophila* experiments gave a spontaneous mutation rate of 10^{-7}. Reducing the spontaneous mutation rate by two orders of magnitude also reduced Evans's doubling dose from 300 roentgens to a much more worrisome level of only 3 roentgens, which personnel in atomic plants, if working near permissible exposure limits, might accumulate within a few months. Wright argued that exposures at this level could significantly alter the incidence of mutations in the person's offspring (but perhaps not detectably, as most mutations are recessive).[94]

Arnold Grobman's 1951 book, *Our Atomic Heritage*, also raised concerns about whether the AEC's occupational safety standards would protect workers against genetic damage.[95] The Commission did not want to engage Grobman's criticisms, although it prepared a public statement aimed at reassuring its workers of the adequate radiological safeguards at its facilities, "should publication of the book generate undue anxiety among . . . contractor personnel."[96] Interestingly, even Grobman's criticisms of the agency did not touch the radioisotope program. *Our Atomic Heritage* included a section on "Hot Atoms" that effused about isotopes

93. According to Evans, in a discussion at the University of Chicago late in 1948 Sewall Wright agreed with him that an exposure of 0.1 roentgens per day could be considered safe from a genetic standpoint. Robley D. Evans to J. O. Hirsenfelder, 22 Jul 1949, Evans papers, box 1, folder Genetics 1.

94. Muller was concerned about detrimental effects of mutations to a whole population, whereas Wright was focused on individual effects. Geneticists disagreed about the consequences of increased mutations in populations through the 1950s. Wright, "Population Genetics and Radiation" (1950); Jolly, *Thresholds of Uncertainty* (2003), pp. 90–93; Beatty, "Weighing the Risks" (1987).

95. Grobman, *Our Atomic Heritage* (1951); Jolly, *Thresholds of Uncertainty* (2003), p. 117–19.

96. Notes on 541st AEC meeting, 26 Mar 1951, Genetic Effects of Radiation on Human Beings, NARA College Park, RG 326, E67A, box 65, folder 8 Radiation Hazards Unclassified.

as a new tool in medicine, and drew on the AEC's publications to describe many of the research results already achieved in medicine, agriculture, science, and industry.

The ACBM conferred many times on the human genetic risks for low-level radiation exposure—which were highly uncertain—and what the Commission's policy should be. At one meeting at which the AEC's program of genetics research was reviewed, members "agreed that the knowledge of human mutation rates is extremely important." A review of progress in setting permissible dose levels for external radiation and radioisotopes followed. The committee was "gratified to learn that during the past two years there has not been a single case of radiation injury due to the activities of the AEC."[97] But this very wording evaded the issue of genetic damage. Were genetic mutations caused by exposure to artificial radioactivity to be considered "radiation injuries," particularly if the damage was to one's offspring? If so, how low would the permissible dose have to be set to assure radiological safety?

New research (much of it sponsored by the AEC) bolstered geneticists' concerns about an increased human mutation rate from low-level radiation exposure. At Oak Ridge National Laboratory, William and Lianne Russell launched their "mega-mouse" study to obtain a better estimate of the spontaneous mutation rate in mammals in order to assess more accurately the human genetic consequences from low-level radiation exposure. Their early results suggested that the mammalian spontaneous mutation rate might be an order of magnitude lower than that for *Drosophila*.[98] This would render Evans's estimates of the doubling dose off by three orders of magnitude, making the prospect of safety from genetic damage even more remote. The consensus among geneticists was that mutational damage from ionizing radiation was linear, cumulative, and deleterious, a perspective the AEC was reluctant to accept, even when endorsed by its own advisors and scientists.[99]

97. Minutes, 24th ACBM Meeting, 10–11 Nov 1950, Washington, DC, OpenNet Acc NV0711806, p. 10 and p. 11.

98. Jolly, *Thresholds of Uncertainty* (2003), p. 112, summarizing a letter from William Russell to Shields Warren, Jan 1951; Rader, *Making Mice* (2004), ch. 6; idem, "Hollaender's Postwar Vision" (2006).

99. Geneticist Bentley Glass, who joined the ACBM in the fall of 1955, was particularly insistent on this point when assessing the agency's radiological protection standards. "Dr. Glass pointed out that 'the recommendations are on the basis of what seems to be practicable, not on the basis of what radiation is considered genetically harmless. There is no

Radioisotopes, Cancer, and Fallout

Certain medical studies in the mid-1950s tarnished the curative image of atomic energy. Like the radiologists and Japanese atomic bomb survivors already mentioned, children who had been treated with x-rays as infants were found to have higher rates of leukemia and cancer.[100] And in an influential British paper, Alice Stewart and colleagues at Oxford reported that children exposed to diagnostic x-rays in utero exhibited higher rates of malignant diseases.[101]

Radioisotopes, which had become commonplace in clinical usage, were implicated as well. As I. Phillips Frohman noted in the *Journal of the American Medical Association*, use of radioiodine had increased from a few millicuries per month after the war to almost 50,000 millicuries per month in 1956. Radioactive phosphorus was being used at nearly 13,000 millicuries per month, and radioactive gold, introduced in 1950, was distributed at a rate of 50,000 millicuries per month.[102] Some patients were experiencing health problems subsequent to radioisotope treatment, problems that seemed linked with their radiation exposure. Though not numerous, reports of such cases in the medical journals were troubling. Two of fourteen thyroid cancer patients treated with iodine-131 by Samuel Seidlin (a pioneer of such therapy) died from subacute myeloid leukemia.[103] Another patient who had been treated with radiogold subsequently developed aplastic anemia.[104] Phosphorus-32, which was widely used in therapy, also seemed to produce some serious long-term effects. Two researchers reported a number of acute leukemias that arose

such amount of radiation. All geneticists, I am sure, pretty well agree on this." Minutes, 56th ACBM Meeting, 26–27 May 1956, Washington, DC, OpenNet Acc NV0411749, p. 28. There were genetic experiments that challenged the linear perspective, though they did not unsettle the consensus among leading geneticists. Calabrese, "Key Studies" (2011).

100. Simpson, Hempelman, and Fuller, "Neoplasia in Children" (1955).

101. Stewart, Webb, Giles and Hewitt, "Malignant Disease in Childhood" (1956).

102. Frohman, "Role of the General Physician" (1956). A patient receiving phosphorus-32 therapy might receive a total of 3 to 80 millicuries, depending on the disease. Reinhard, Neely, and Samples, "Radioactive Phosphorus" (1959), p. 945; Chodos and Ross, "Use of Radioactive Phosphorus" (1958).

103. Seidlin, Siegel, Melamed, and Yalow, "Occurrence of Myeloid Leukemia" (1955).

104. Frohman, "Role of the General Physician" (1956); the original report is Schoolman and Schwartz, "Aplastic Anemia" (1956).

among patients previously treated with phosphorus-32.[105] Animal studies on the appearance of tumors following administration of phosphorus-32 indicated that the "therapeutic dose for man may well be within the carcinogenic range."[106]

Government and industrial advocates of radioisotopes ignored these concerns. Abbott Laboratories, the main commercial supplier of radiopharmaceuticals, touted not only the medical usefulness of isotopes but also their safety. As the firm claimed in a 1956 pamphlet, "Experience has demonstrated that isotopes, properly used, pose no appreciable hazard for the physician or his hospital. There is no evidence that a diagnostic or moderate therapeutic dose has ever done any harm to any patient."[107] The AEC continued to promote "atomic medicine." In January 1956, a nine-person citizens' panel presented a report appraising nuclear policy to Congress's Joint Committee on Atomic Energy; their recommendations for civilian benefits again highlighted using atomic energy for medicine.[108] The report noted that new research and diagnostic tools had been the most significant contributions of atomic energy to medicine, although "public attention has been focused on radiation treatment of cancer."[109] As if to illustrate this point, coverage of the report in *Time Magazine* concluded: "Atomic diagnosis and treatment of cancer will so prolong life that the United States must dump its theory that a working life ends at 65."[110]

In reality, medical researchers who had expected radioisotopes to revolutionize the treatment of cancer were less sanguine. As John Lawrence put it, "One who has worked in the field from the beginning and who is asked to sum up his experiences can't help saying that the therapeutic achievements have been disappointing."[111] As AEC staff scientist Simeon Cantril admitted in 1955, "we would be hard put to say that there has been any significant decrease in cancer mortality as a result of funds expended

105. Hall and Watkins, "Radiophosphorus in Treatment" (1947); Brues, "Biological Hazards" (1949).

106. Furth and Tullis, "Carcinogenesis" (1956), p. 10.

107. *Nuclear Medicine for the Modern Physician and His Hospital*, pamphlet published by Abbott Laboratories, 1 Jan 1956, Opennet Acc NV0723949, p. 21.

108. Hewlett and Holl, *Atoms for Peace and War* (1989), p. 328.

109. US Congress, Joint Committee on Atomic Energy, *Report of the Panel* (1956), p. 55. For more on radioisotope-based diagnostics, see chapter 9.

110. "Atomic Energy: The Nuclear Revolution" (1956).

111. Lawrence and Tobias, "Radioactive Isotopes" (1956), p. 185; Boudia, "Radioisotopes' 'Economy of Promises'" (2009), p. 255.

by the AEC to date."[112] This lack of progress did not dampen the Commission's ambitions; in 1956 the AEC increased its budget request for "research in the use of the atom to cure cancer." But as the *New York Times* noted wryly, "The report made no mention of progress in this field."[113]

These reassessments occurred in the midst of the fallout debates, which proved highly consequential for the public image of radioisotopes irrespective of their specific clinical benefits or drawbacks. The legacy of the fallout debates is complex; this recounting will focus on two pertinent aspects. First, some of the somatic effects of radiation, particularly cancer, began to be understood in mutational terms, like the genetic effects. Muller had suggested in 1948 that radiation-induced mutations in somatic cells might be responsible for malignancies, and this idea, picked up by others, gained traction over the 1950s.[114] This interpretation subverted the AEC's categorical distinction between somatic and genetic effects, and it also focused concern on the carcinogenic potential of fallout products such as strontium-90 and, to a lesser degree, iodine-131. If there were no threshold for genetic damage, then even minute amounts of these contaminants might induce cancer-producing somatic mutations. Second, and relatedly, the focus on these two fallout products as invisible dangers corroded the image of radioactive isotopes as medical dividends. Increasingly, radioisotopes were being invoked as causes of cancer, not cures. This change in perception did not make radioisotopes any less useful to researchers or physicians, but it attenuated the political credit the AEC could accrue through their distribution.

The increased pace of nuclear tests, and the "transient increases in the radioactivity level in many communities of the United States," as two AEC officials put it, catalyzed new concerns about the civilian health costs of nuclear peace.[115] The US government began testing many nuclear devices in Nevada, which meant that radioactive fallout could reach ordinary Americans.[116] In 1951 and 1952, there were a total of twenty nuclear

112. Minutes, 53rd ACBM Meeting, 1–2 Dec 1955, Washington, DC, OpenNet Acc NV0411747, p .12.

113. "A.E.C. Adds to Funds for Disease Studies" (1956).

114. Muller, "Some Present Problems" (1950), p. 56. This paper was presented at the Symposium on Radiation Genetics, sponsored by the Biology Division of Oak Ridge National Laboratory, Oak Ridge, Tennessee, 26–27 Mar 1948.

115. Eisenbud and Harley, "Radioactive Dust" (1953), p. 141.

116. However, above-ground tests in Nevada were restricted to yields in the tens of kilotons, whereas the larger explosions occurred in the Pacific. Hacker, *Elements of Controversy* (1994), p. 82.

detonations at the Nevada Test Site; eleven more took place during the first half of 1953.[117] One May 1952 test dispersed radioactive debris as far away as Salt Lake City.[118] In the spring of 1953, in a period of heavy nuclear testing, some sheep men reported unusually high losses of lambs and ewes. An investigation of their complaints, under the auspices of the Commission, absolved radiation of the livestock deaths, though questions persisted.[119] The AEC's *Thirteenth Semiannual Report* steadfastly denied that testing imperiled anyone: "No person has been exposed to a harmful amount of radiation from fall-out. In general, radioactivity resulting from fall-out has been many times below levels which could cause any injury to human beings, animals, or crops."[120]

Atomic weapons themselves were also changing in alarming ways. So-called hydrogen bombs, first tested at the Pacific Proving Ground in 1952, released significantly higher levels of radioactivity and fission products than conventional bombs. In March 1954, the AEC's test at Enewetak of one such thermonuclear device, "Bravo," spread radioactive ash widely across the Pacific, engulfing the "Lucky Dragon," an unfortunately named Japanese fishing boat.[121] Nearly two dozen fishermen aboard suffered injuries from the radiation exposure and one died in September 1954; these casualties received extensive media coverage in the United States as well as in Japan.[122] In response, the AEC made no concessions to critics. Lewis Strauss, now Chair of the Commission, publicly denied that the fallout could have injured anyone.[123]

Geneticist A. H. Sturtevant rebutted Strauss's statements about fallout in his presidential address at a June 1954 meeting of the Pacific Division

117. Hewlett and Duncan, *Atomic Shield* (1969), pp. 672–73; Hewlett and Holl, *Atoms for Peace and War* (1989), pp. 146–47.

118. Hacker, *Elements of Controversy* (1994), p. 81.

119. Hacker, "Hotter Than a $2 Pistol" (1998).

120. AEC, *Thirteenth Semiannual Report* (1953), p. 78.

121. When it was selected as a test site, the name of the atoll was spelled Eniwetok, but the US government later altered the spelling in accordance with Marshallese sensibilities to Enewetak. I follow this convention. See Hacker, *Elements of Controversy* (1994), pp. 14, 140; "H-Bomb and World Opinion" (1954); and, on the long-range consequences, Harkewicz, *"Ghost of the Bomb"* (2010).

122. Lapp, *Voyage of the Lucky Dragon* (1958).

123. "Chairman Strauss's Statement on Pacific Tests" (1954); Kopp, "Origins of the American Scientific Debate" (1979). Strauss's statement contained several patent untruths, such as the suggestion that the fishermen's skin lesions were due to "the chemical activity of the converted material in the coral, rather than to radioactivity."

of the American Association for the Advancement of Science, and his address appeared subsequently in *Science*. In recounting the hazards of radiation exposure, Sturtevant connected mutagenicity with so-called somatic effects: "There is reason to suppose that gene mutations, induced in an exposed individual, also constitute a hazard to that individual—especially in an increase in the probability of the development of malignant growths, perhaps years after the exposure." In comparison with the kind of radiation injuries taken into account in setting a permissible dose for exposure, there was not a lower limit to the amount of radiation that might induce such somatic mutations. Sturtevant concluded that "there is, in fact, no clearly safe dosage."[124]

The initiation of an open debate between the AEC and dissenting scientists made for great media coverage, which in turn fueled public alarm. Early in the winter of 1955, the Eisenhower administration was discussing how it might defuse criticisms about the dangers of fallout by disclosing more information. The preparation of a press release stalled between the AEC and the State Department.[125] Just before it was released, Ralph Lapp, a longtime critic of the agency, published two pieces on the scale of fallout from Bravo, one in the *Bulletin of Atomic Scientists* and the other in the *New Republic*.[126] Lapp asserted that a hydrogen bomb the size of the Bravo shot would release a hot cloud of debris that could travel 200 miles in several hours, contaminating the whole area with serious-to-lethal radioactivity.[127] The AEC report, appearing on the heels of Lapp's articles, largely confirmed his grim assessment.

The press amplified these concerns. *Newsweek*, which described the AEC report as the "terrible truth," projected the horrific consequences that would result from a thermonuclear explosion over the Northeast: "Drifting down from the sky, the ash will poison everything it touches within a cigar-shaped area of 7,000 square miles. It can threaten all life in a state the size of New Jersey."[128] At a February 22 hearing, members of the Senate Subcommittee on Civil Defense of the Armed Services

124. Sturtevant, "Social Implications" (1954), p. 406. Sturtevant built on H. J. Muller's earlier suggestion that the "general biological effects of radiation" were attributable to damage to the genes in somatic cells; Muller, "Some Present Problems" (1950), p. 44.

125. Hewlett and Holl, *Atoms for Peace and War* (1989), pp. 279–87.

126. Divine, *Blowing on the Wind* (1978), pp. 36–38.

127. Lapp, "Fall-Out" (1955).

128. "To Live—Or Die" (1955), p. 19; Divine, *Blowing on the Wind* (1978), p. 38.

Committee asked AEC officials why an independent scientist publicly disclosed the radiological dangers of hydrogen bombs before the agency did.[129] For their part, the Joint Committee on Atomic Energy pursued these issues at a hearing on April 15, 1955, though this amounted to little more than an opportunity for AEC officials to reassure the public that radioactive fallout from weapons testing was being monitored and current levels would not cause any immediate or long-term health hazards. Behind the scenes, however, the AEC had launched Project Sunshine, a secret worldwide data collection effort to assess the accumulation of radioactive fission products in soil, plants, animals, and humans. The Division of Biology and Medicine arranged for the covert collection of baby bones throughout the world, by cooperating with American philanthropies and medical missions with contacts in South Asia and Latin America.[130]

During the same period, a decidedly mixed picture emerged from the ABCC's long-term studies of Japanese survivors of the atomic blasts. On the one hand, as *Time* reported, the incidence of leukemia among those who had been within 1,000 meters of the blasts was "more than 600 times the normal incidence of leukemia in Japan."[131] A higher-than-normal incidence of leukemia was observed also among those nearly two miles away from the explosion, suggesting the long-range health consequences of exposure to atomic detonations. One physician associated with the study extrapolated to the question of fallout: "Dr. Moloney expects other forms of cancer to appear later, and he suspects that the radioactive fall-out of hydrogen bombs will have even greater cancer-producing effect."[132] On the other hand, studies of genetic damage in the offspring of survivors were less alarming. In 1955, the ABCC geneticists announced their results, which were "negative"—or inconclusive.[133] Conservative-leaning publications seized on this conclusion to rebut alarmism about radiation exposure. *US News & World Report*, for example, published a story en-

129. Hewlett and Holl, *Atoms for Peace and War* (1989), pp. 287–88; Straight, "Ten-Month Silence" (1955).

130. Most of the samples collected from 1953 to 1956 were from stillborn infants. ACHRE, *Final Report* (1996), pp. 402–7; Memorandum to ACHRE, 8 Feb 1995, OpenNet Acc NV0750699.

131. "Nuclear Revolution" (1955).

132. Ibid.

133. This announcement of results was to the American Academy of General Practice, 29 Mar 1955; see Hill, "Effect of A-Bomb" (1955). A monograph was published the following year: Neel and Schull, *Effect of Exposure* (1956).

titled "Thousands of Babies, No A-Bomb Effects."[134] The ABCC's genetic study had been both scientifically difficult and politically contentious from the outset, in part because defining which phenotypes qualified as mutations in survivors' children was not straightforward.[135] Moreover, most geneticists thought it unlikely that surveying congenital abnormalities in children born to survivors would reveal even those mutations that had occurred.

The BEAR Report and the Somatic Mutation Hypothesis

In part to provide an independent assessment of radiation hazards amidst public controversy, the US National Academy of Sciences convened a panel to assess the Biological Effects of Atomic Radiation. The experts appointed to the panel were organized into six committees: genetics, pathology, agriculture and food supply, meteorology, oceanography and fisheries, and disposal of radioactive waste.[136] In June 1956 the academy issued the panel's analysis, known as the BEAR Report. The report recommended reducing the maximum cumulative radiation exposure to reproductive cells from 300 to 50 roentgens, and limiting the average exposure through age thirty in the population at large to 10 roentgens.[137]

Both the genetics and the pathology committees addressed the health-related effects of radiation, but offered strikingly different assessments. Geneticists focused on low-level doses, and emphasized "any radiation is genetically undesirable."[138] Even small increases in radiation exposure would result in deleterious mutations, or "genetic defects." "Each one of these mutants must eventually be extinguished out of the population

134. "Report on Hiroshima" (1955). This story, an interview with ABCC director Robert Holmes, contained several inaccuracies.

135. Lindee, "What Is a Mutation?" (1992); idem, *Suffering Made Real* (1994). As Lindee shows, the ABCC's guidelines reflected social realities and criteria as well as scientific ones.

136. Beatty, "Genetics in the Atomic Age" (1991); Lindee, *Suffering Made Real* (1994); Jolly, *Thresholds of Uncertainty* (2003), chapter 6; Higuchi, *Radioactive Fallout* (2011).

137. "Summary Report," in National Academy of Sciences, *Biological Effects* (1956), p. 8. On the simultaneous publication of a noncoincidentally similar report in the United Kingdom: Hamblin, "'Dispassionate and Objective Effort'" (2007).

138. "Report of Committee on Genetic Effects of Atomic Radiation," in National Academy of Sciences, *Biological Effects* (1956), p. 23.

through tragedy."[139] The committee focused particular concern on the overuse of radiation in medicine, which amounted to as much radiation exposure for most Americans as background radiation (3–4 roentgens), and much more than from radiation from fallout (0.1 roentgens). The geneticists' recommendation was to "keep all of our expenditures of radiation as low as possible."[140]

The report from the Pathology Committee addressed a number of serious health problems associated with radiation exposure, but was more reassuring on the issue of low-level radiation. Radiation under certain levels could be "harmless to individuals." In discussing "late" effects of radiation—namely leukemia—observed among Japanese atom bomb survivors as well as radiologists, they emphasized that these individuals received either a nearly fatal single dose of radiation or, for those exposed occupationally, "higher than acceptable permissible dose rates."[141] The implication was that exposure below permissible doses would not result in any long-term effects such as leukemia or shortening of life. The committee dismissed outright the somatic mutation theory of cancer.

A main task of the Committee on Pathologic Effects was to evaluate concerns about whether the level of contamination of strontium-90 from fallout might lead, through its presence in milk and food, to increased incidence of human cancer. As a RAND report commissioned by the AEC put it: "The risk is simply this: The bone-retentive and radioactive properties of Sr^{90} endow it with a high carcinogenic capability."[142] So-called internal emitters were especially dangerous because once embedded in the bone, they continued to irradiate the organs. The long half-life of strontium-90—nearly thirty years—created additional concerns, as the radioelement could persist in the environment, moving through the food chain. The BEAR Committee on Pathologic Effects acknowledged a possible link between radiation from strontium-90 and certain cancers, but concluded that current levels posed no such threat.[143] Nonetheless,

139. Ibid., pp. 25 and 26.

140. Ibid., p. 30. On the committee report, see Beatty, "Masking Disagreement" (2006).

141. "Report of Committee on Pathologic Effects of Atomic Radiation," National Academy of Sciences, *Biological Effects* (1956), pp. 36 and 39.

142. RAND Corporation, "Worldwide Effects of Atomic Weapons: Project Sunshine," 6 Aug 1953, R-251-AEC, OpenNet Acc NV0717541, p. 4, also quoted in Jolly, *Thresholds of Uncertainty* (2003), p. 154.

143. National Academy of Sciences, *Biological Effects* (1956), p. 21. The Committee on Pathologic Effects similarly acknowledged the existence of injuries from long-term, low-dose

as public discontent mounted over radioactive fallout from weapons testing, this particular radioisotope became a symbol of the "atomic poison" released by nuclear weapons.[144] An editorial in *America* published on June 15, 1957, was simply entitled "The Strontium-90 Debate," and referred to the radioisotope as a "menace to life and health."[145]

In defending the need to conduct peacetime atomic weapons testing, officials at the AEC repeatedly pointed to the low level of radiation exposure resulting from tests as compared with natural exposure to radioactivity and with normal clinical uses of radiation (particularly x-rays). For example, on May 6, 1955, Willard F. Libby wrote Linus Pauling a letter explaining why the AEC was "justified in saying that although genetic effects are unknown, the test fallout is so small as compared to the natural background and, more important, to the variations in the natural background which are customarily accepted, that we really cannot say that testing is in any way likely to be dangerous."[146] The AEC's July 1956 semiannual report to Congress, reiterating Libby's analysis on the matter, asserted that "at the present level of weapons' testing, the present and potential contribution of strontium-90 to the world ecology is not a significant factor."[147] More and more scientists, however, agreed with minority Commissioner Thomas Murray, who dissented on this issue from the view in the AEC's report. Ralph Lapp published criticisms of the official interpretation, and Linus Pauling cited the link between strontium-90 and cancer in a 1959 letter in the *New York Times*.[148]

Journalists began to focus on radioactivity entering the food supply, particularly in the vicinity of the Nevada testing site. They focused public concern on the possible contamination of milk with strontium-90, due to the grazing of cattle on contaminated Western pastures, generating a groundswell of anxiety. The AEC underestimated the symbolic importance of contaminating food, especially milk. Throughout the twentieth

---

exposures, such as leukemia and skin cancer, but insisted that "among those who have adhered to present permissible dose levels, none of these effects have been detected" (p. 34).

144. Larsen, "Midwest Center for Research" (1955).

145. "Strontium-90 Debate" (1957).

146. W. F. Libby to Linus C. Pauling, 6 May 1955, in NARA College Park, RG 326, E67B, box 49, folder 7 Medicine, Health & Safety 13 Genetics; Libby, "Radioactive Fallout" (1956).

147. AEC, *Twentieth Semiannual Report* (1956), p. 106.

148. Lapp, "Strontium-90 in Man" (1957); Pauling, "Effect of Strontium-90" (1959); Jolly, "Linus Pauling" (2002).

century in the United States the purity of milk has been a potent icon; earlier worries about milk contaminated with tuberculosis bacilli were transmuted into fears about milk contaminated with radioactivity.[149] Infants and children who assimilated strontium-90 from contaminated milk into their skeletons could suffer decades of radiation exposure. During the 1956 Presidential campaign, Democratic candidate Adlai Stevenson proposed a unilateral test ban of hydrogen bombs to protect Americans from the accumulating radioactive fallout.[150] Fear of fallout was not enough to prevent Eisenhower's reelection, but the issue had become central to American politics.

The Hearings on the Nature of Radioactive Fallout and Its Effects on Man, convened by Congress in 1957, provided an opportunity for the AEC's critics—and even its supporters—to question the agency's intransigence. Congressman Chet Holifield, who chaired the Joint Committee on Atomic Energy, berated the Commission for not disclosing more information earlier on the nature and extent of radioactive fallout from weapons testing, and for establishing a "party line" of downplaying its dangers.[151] Strauss misjudged the tenacity and political connections of the agency's scientific critics, even as he dismissed their concerns as hysteria. As observed in the minutes of an ACBM meeting, "The AEC was not prepared for the emotionalism within segments of the scientific community but felt that the results of the current hearings on fallout should be beneficial."[152] Regarding radioisotopes, the picture that emerged from the hearings is complex. While the agency continued to extol the scientific benefits of radioisotopes (and atomic energy generally), there were also references to "radioisotopes as possible contaminants in food products," as well as to the hazardous genetic and somatic consequences of exposure to all forms of ionizing radiation.[153]

A 1957 publication by Caltech biology professor Edward B. Lewis in *Science* underlined the urgency of health risks from radioactive fallout.

149. Smith-Howard, *Perfecting Nature's Food* (2007), chapter 5.

150. See Stevenson, "Why I Raised the H-Bomb Question" (1957). The Public Health Service began monitoring strontium-90 levels in milk in 1958, and the increases they reported, though levels were still below the maximum permissible dose, reinforced public alarm. See Divine, *Blowing on the Wind* (1978), pp. 263–64.

151. Hewlett and Holl, *Atoms for Peace and War* (1989), pp. 454–455.

152. Minutes, 63rd ACBM Meeting, 18 Jun 1957, Washington, DC, OpenNet Acc NV0712175, p.6.

153. US Congress, Joint Committee on Atomic Energy, *Nature of Radioactive Fallout* (1957), quote on p. 1960.

Zone	Distance from hypocenter (m)	Estimated population of exposed survivors (Oct. 1950)	Number of confirmed cases of leukemia	Percentage of leukemia
A	0– 999	1,870	18	0.96
B	1000–1499	13,730	41	0.30
C	1500–1999	23,060	10	0.043
D	2000 and over	156,400	26	0.017

FIGURE 5.3. Incidence of leukemia among the combined exposed populations of Hiroshima and Nagasaki by distance from the hypocenter (Jan 1948–Sep 1955). From E. B. Lewis, "Leukemia and Ionizing Radiation," *Science* 125 (1957): 965–72, on p. 967. Reprinted with permission from American Association for the Advancement of Science.

He compared studies of leukemia in four populations exposed to ionizing radiation: "(i) survivors of atomic bomb radiation in Japan, (ii) patients irradiated for ankylosing spondylitis, (iii) children irradiated as infants for thymic enlargement, and (iv) radiologists."[154] In each case, the dose-effect response appeared linear, and there was no evidence of a threshold below which exposure posed no hazard.[155] The risk of leukemia caused by various kinds of exposure was comparable, leading Lewis to postulate a minimum estimate of induced leukemia as 2×10^{-6} per individual per rem per year, the newer dosage designation of rem being equivalent to the older roentgen.

The leukemia incidence among exposed and unexposed Japanese populations was crucial to Lewis's analysis, as it strongly suggested there was no lower threshold for risk from ionizing radiation. The Japanese atomic bomb survivors were problematic surrogates insofar as the fallout debates hinged on the detrimental effects of radiation at low levels. Many Japanese survivors had been exposed at relatively high levels, whether they had exhibited radiation sickness or not, as the AEC was quick to point out. However, the number of cases of leukemia diagnosed in Japanese survivors, even those over 1,000 meters from the blast epicenter, provided compelling evidence that ionizing radiation induced leukemia in linear, dose-dependent fashion. (See figure 5.3.)

154. Lewis, "Leukemia and Ionizing Radiation" (1957), p. 965.

155. As Jolly points out, Lewis did not claim to have proven linearity, but argued that the evidence pointed that way and that his estimates were valid within a factor of three. *Thresholds of Uncertainty* (2003), p. 494.

Lewis attributed the linear dose-dependence of leukemia incidence on radiation exposure to the somatic mutation hypothesis for carcinogenesis. On the basis of his analysis, Lewis argued that the growing concentration of strontium-90 from fallout could be sufficient to raise the incidence of leukemia in the United States by as much as 5%–10%. This estimate relied on many assumptions; the concentration of strontium-90 in the human skeleton necessary to induce leukemia was not known.[156] Despite genuine scientific uncertainties, the geneticists' perspective on the linear, dose-dependent nature of radiological hazard was influential. A report from the United Nations Committee on the Effects of Atomic Radiation published on August 10, 1958, predicted a rise in additional deaths from leukemia worldwide due to radioactive fallout, as Lewis had. The increase in radioactivity, they pointed out, was slight—only 5% of total radiation received from natural sources. However, even small increases in radiation could lead to increases in cancer.[157]

Mutations were no longer only about the future of the human race, but also represented the cancer burden of the country's atomic energy program. Just as significantly, the clear divide between somatic and genetic effects was eroding, although many AEC researchers (particularly nongeneticists) resisted giving up the threshold concept until the mid-1960s.[158] A 1958 paper published by Miriam Finkel, a researcher at Argonne National Laboratory, presented results of a study in which mice were exposed to various doses of strontium-90. The expected effects of life-shortening and leukemia were observed, but mice receiving the lowest dose did not show any of these effects.[159] Finkel argued that a threshold for strontium-90 existed, which was well below human exposures due to fallout, an interpretation soon challenged by Linus Pauling and others. In response, Argonne's

156. Minutes, 52nd ACBM Meeting, 9–10 Sep 1955, Washington, DC, OpenNet Acc NV0411746, p. 13.

157. Divine, *Blowing on the Wind* (1978), p. 222.

158. Merrill Eisenbud, who worked as Director of the AEC's Health & Safety Laboratory and Manager of the New York Operations Office, dates the decline of this threshold view to 1963, although one 1966 agency publication, while stating that there is no threshold for genetic damage from radiation, says there is a threshold for somatic symptoms. Asimov and Dobzhansky, *Genetic Effects of Radiation* (1966), pp. 35–36; Human Radiation Studies: Remembering the Early Years, Oral History of Merrill Eisenbud. Conducted January 26, 1995 through the Department of Energy by Thomas J. Fisher, Jr. and David S. Harrell, and published at http://www.hss.energy.gov/HealthSafety/ohre/roadmap/histories/index.html.

159. Finkel, "Mice, Men and Fallout" (1958).

director Austin Brues asserted that the theory of linearity remained unproven.[160]

The somatic mutation theory also had a mixed reception, particularly among physicians and oncologists.[161] For other putative cancer-causing agents, such as viruses and hormones, there was a long history of research and a possible mechanism. How mutations caused cancer was not clear to anyone, least of all the geneticists who advocated the theory. In the late 1950s the principal technique for visualizing the mutational effect of radiation was cytogenetics, the microscopic analysis of chromosomes.[162] The effects of radiation on chromosome structure had long been studied, as had chromosomal abnormalities in cancer cells, but other types of mutations were not so easily pinpointed or visualized. Although AEC officials continued to treat the somatic mutation theory as an untested hypothesis, they could no longer relegate genetic effects to the sidelines, as if they were unrelated to health issues. In the 1960s, as DNA became an increasingly important object of study for radiobiology and chemical mutagenesis, the somatic mutation theory of cancer causation gained ground.[163] Even more importantly, the geneticists' portrayal of the effects of ionizing radiation as linear and cumulative became the consensus view.

Conclusions

In material prepared for the Hearings by the Joint Committee on Atomic Energy in February 1957, the AEC presented isotopes as the "first peaceful dividend from United States atomic investment."[164] Similarly, Walt

160. Divine, *Blowing on the Wind* (1978), pp. 223–25; Brues, "Critique of the Linear Theory" (1958).

161. As Shields Warren wrote, "This theory is not accepted by many oncologists." "You, Your Patients and Radioactive Fallout" (1962), p. 1125. On the reception of the somatic mutation in the medical community, see Jolly, *Thresholds of Uncertainty* (2003), ch. 12.

162. Following studies in the 1950s of chromosomal abnormalities in patients and experimental animals with leukemia, AEC researcher (and ultimately agency critic) John Gofman employed cytogenetics in his famous studies of low-level radiation and cancer. de Chadarevian, "Mutations in the Nuclear Age" (2010); Semendeferi, "Legitimating a Nuclear Critic" (2008).

163. For example, the topic of the Cold Spring Harbor Symposium on Quantitative Biology in 1966 was Mutagenesis.

164. "Hearings Under Section 202 of the Atomic Energy Act, February 1957, Table of Contents" NARA College Park, RG 326, E67B, box 71, folder 1 Org. & Man. 7 Joint Committee on Atomic Energy (BP 1 of 5).

Disney's 1956 movie *Our Friend the Atom* represented atomic energy as a genie that, although fearfully powerful, could grant humanity three wishes: power, food and health, and peace. It emphasized the new era in medicine that atomic energy had brought forth, echoing the assertion that "radioisotopes can heal."[165] This recurrent depiction was not merely propaganda; radioisotopes did prove immensely useful in scientific research and medicine, though not always along thc lincs imagined in the late 1940s.

However, a countervailing image of radioisotopes as harmful gained momentum over the 1950s. The fallout debates and the growing concerns about radioactive waste from nuclear power plants gradually changed public perception of the relationship of atomic energy and cancer. Radioisotopes began to be understood as poisons rather than "medical bullets," as they had been termed in 1947. Over the 1950s, physicians found disturbing evidence that even when medical treatment with radioisotopes was successful, the radiation exposure could lead to blood disorders or cancer years later. In fact, clinicians drew on these observations to calibrate therapeutic doses so as to minimize these side effects, and medical researchers developed an array of safer diagnostic methods using low-level and short-lived radioisotopes.[166] Yet public awareness that exposure to even low-level radioactivity was hazardous could not be easily undone.

Not only the symbolic valence of radioactivity but also the meaning of radiation safety changed during the period. In the late 1940s, the AEC portrayed radiation exposure that did not cause any short-term harm as safe. For example, in the agency's *Sixth Semiannual Report* to Congress, this description of an experiment with human volunteers appeared amidst a section on radioisotope research:

> Some 200 young men volunteered to receive injections of radioactive iron so weak that the radiation was below the levels which human beings can safely tolerate. Tests of these young men after 6 months showed no ill effects from the irradiation.[167]

Ten years later, such a statement about the innocuous effects of assimilated radioisotopes would have drawn scrutiny if not outright criticism.

165. Haber, *Our Friend the Atom* (1956), p. 157. The image of atomic energy as a genie was older than Disney's fable; Weart, *Nuclear Fear* (1988), p. 404.

166. See chapter 9.

167. AEC, *Sixth Semiannual Report* (1949), p. 81.

The dangers of radiation came to be understood in terms of increased risk for leukemia and other cancers in the long term, effects that might be detectable only through a statistical, population-based analysis of disease incidence.[168] The exposure to radioactivity in fallout, the AEC insisted, was much less than that from the natural environment and clinical uses of radiation. This comparison, however, ceased to be sufficiently reassuring for many Americans.

By the early 1960s, some increase in mutations and cancer incidence appeared to be a side effect of the US commitment to atomic energy. Such detriments would undoubtedly accompany nuclear war. As strategist Herman Kahn maintained: "Those waging a modern war are going to be as much concerned with bone cancer, leukemia and genetic malformations as they are with the range of a B-52 or the accuracy of an Atlas missile."[169] But even the peacetime testing of atomic weapons carried a biological toll. Strauss told the ACBM that "the results of radiation exposure from tests should always be balanced against the results which would follow from atomic war."[170] However, critics of the AEC argued that national security should not require Americans to pay this biological price, contending that even small increases in environmental radioactivity were not acceptable. The 1963 test ban treaty vindicated this point of view, and reflected the changed mentality about the dividends and dangers of atomic energy. The fear of cancer, which in the 1940s was exploited by the AEC to justify its status as a civilian agency improving the health of citizens, remained a threat to the agency's other peacetime benefit, nuclear energy.

168. In 1958, members of the ACBM spoke of the need for "whole-population radiation epidemiology." They discussed whether a human experiment might be done "to ascertain the effects of very small doses of the order of several times natural background." It would have to be carried out "on the largest possible scale" to be informative. Minutes, 70th ACBM Meeting, 17–18 Oct 1958, Germantown, OpenNet Acc NV710349, p. 5.

169. Kahn, *On Thermonuclear War* (1960), p. 24.

170. Minutes, 63rd ACBM Meeting, 18 Jun 1957, Washington, DC, OpenNet Acc NV0712175, p. 7.

CHAPTER SIX

Sales

Imagine a manufacturer whose total production for 1950 weighs less than one-tenth of an ounce, an industry which makes approximately 9,000 product shipments a year with the net weight of product in each shipment being less than the weight of pencil lead used in writing your name. — W. E. Thompson, ORNL, 1952[1]

From the point of view of a biomedical researcher in 1946, radioisotopes simply joined the stock of reagents and instruments available for purchase. The industrialization of most of these items relied on the private sector.[2] The Rockefeller Foundation had nurtured the development of many new laboratory methods such as electrophoresis, spectroscopy, and ultracentrifugation as part of its Natural Sciences program aimed at bringing techniques and approaches from the physical sciences to bear on problems in biology.[3] Companies such as Spinco, Klett, and Beckman then commercialized biophysical instruments based on these methods beginning in the 1940s.[4] The principal role played by the state, especially the US federal government, was indirect—the rapid growth in postwar government funding for biomedical research made commercial markets for these sophisticated machines viable.

1. W. E. Thompson, "Oak Ridge National Laboratory Research and Radioisotope Production," Jan 1952, MMES/X-10/Vault, CF-52-1-212, DOE Info Oak Ridge, ACHRE document ES-00226, p. 16.

2. Gaudillière and Löwy, "Introduction" (1998); Elzen, "Two Ultracentrifuges" (1986); Kay, "Laboratory Technology" (1986); Zallen, "Rockefeller Foundation" (1992); Lenoir and Lécuyer, "Instrument Makers" (1995); Rasmussen, *Picture Control* (1997); Rheinberger, "Putting Isotopes to Work" (2001); Slater, "Instruments and Rules" (2002).

3. Kohler, "Management of Science" (1976); idem, *Partners in Science* (1991); Abir-Am, "Discourse of Physical Power" (1982); Kay, *Molecular Vision of Life* (1993).

4. Kay, "Laboratory Technology" (1988); Elzen, *Scientists and Rotors* (1988); Creager, *Life of a Virus* (2002), ch. 4.

The industrialization of radioisotopes displayed a different trajectory: the postwar mass production of radioisotopes was undertaken by the state itself. By using the graphite nuclear reactor at Oak Ridge for manufacture of artificial radioisotopes, the US federal government turned radioisotopes into commodities, albeit highly subsidized commodities. The AEC aimed, through its low pricing and educational programs, not only to meet demand, but also to create it. In this they succeeded. During its first decade, the AEC sold close to 64,000 shipments of original radioactive materials to research laboratories, companies, and clinics.[5] Needless to say, the scale of the radioisotope production at Oak Ridge was small compared with the AEC's manufacture of nuclear weapons. By the early 1950s the government's capital investment in atomic energy plant facilities and equipment totaled to "more than the capital investments of US Steel and General Motors combined."[6] As the chief of the AEC's Isotopes Division quipped, "Atomic energy is truly a big business."[7]

The centrality of the federal government to the business of radioisotopes raised difficulties regarding two of the AEC's statutory goals, which were themselves in tension. On the one hand, the agency was responsible for "strengthening free competition among private enterprise" in the development of atomic energy; on the other, it was to "insure adequate health protection to the public as well as the users of atomic energy (radioisotopes)."[8] The AEC fostered commercial involvement in the sales of radioactive materials to the extent that it could, given the restrictions of the 1946 Atomic Energy Act and the fact that, by virtue of infrastructure alone in the early years, the government had no rivals in producing inexpensive radioisotopes. Rather than competing directly with Oak Ridge, retail vendors began offering radiolabeled compounds and radiopharmaceuticals in conjunction with the AEC's role as wholesaler. On the regulatory side, the agency's safety controls over civilian uses of radioisotopes lagged behind its promotional efforts. Initially, the transfer of radioactive materials from custody of the federal government to research institutions

5. AEC, *Eight-Year Isotope Summary* (1955), p. 2.

6. Paul C. Aebersold, "Current Uses of Radioactive Isotopes in Industry," 4 Dec 1952, talk given at a meeting of the American Management Association in Cleveland, Aebersold papers, box 6, folder 6-53, p. 2.

7. Ibid.

8. "Revised McMahon Bill" (1946), p. 2; Conference on Commercial Distribution of Isotope-Labeled Compounds, Minutes of the Meeting, 30 Oct 1947, Appendix "E" to AEC 108, NARA College Park, RG326, E67A, box 47, folder 3 Isotopes: Labeled Compounds, p. 27.

and hospitals relied on trust of the recipients to use them safely—and to assume liability if they did not. The AEC promulgated but did not necessarily enforce guidelines for the safe handling and disposal of radioisotopes, particularly in the early years of the program.

The Atomic Energy Act of 1954 relaxed the strictures around owning fissionable materials and patenting in order to promote the nuclear power industry. Over the next decade, this had an impact on the sales of radioisotopes in two key ways. First, the growing number of civilian reactors and increasing demand for radioisotopes finally led industry to challenge the government's dominant position as supplier. Companies petitioned the AEC to let the private sector take over the production of reactor-generated radioisotopes. The X-10 reactor was shut down in 1963, although the federal government (which had other more up-to-date reactors) withdrew from manufacturing isotopes more slowly and unevenly than this closure would suggest. Second, the AEC developed a more stringent regulatory apparatus, aimed at the emerging nuclear power industry but also affecting users of radioisotopes. Even so, by the 1960s, questions arose about the adequacy of government regulation, particularly as growing numbers of civilian reactors were constructed for utility companies. The conflict of interest between the agency's role as both promoter and regulator of atomic energy—an issue that led in 1975 to the dissolution of the AEC and the formation of an energy agency (subsequently renamed the Department of Energy) and a regulatory agency (the Nuclear Regulatory Commission)—was evident early on in the sales and safety oversight of radioisotopes from Oak Ridge.[9]

The AEC and Free Enterprise

The national security provisions of the 1946 Atomic Energy Act put strict limits on the involvement of industry. Ownership of all fissionable materials, reactors, and manufacturing plants for atomic weapon components belonged to the United States, to be administered by the Commission. In addition, the law prohibited individuals and corporations from taking out new patents on nuclear inventions or technologies. Preexisting patents were subject to compulsory purchase by the Commission.[10] These features

9. Mazuzan, "Conflict of Interest" (1981).

10. "Revised McMahon Bill" (1946), p. 5; Turchetti, "Slow Neutrons" (2006), p. 14; idem, "Invisible Businessman" (2006); Wellerstein, "Patenting the Bomb" (2008).

of the legislation had been attacked on the floor of Congress as socialist, but they survived nonetheless.[11] This produced the anomalous situation of a new industry developing in the near-complete absence of commercial patents.[12]

The AEC acknowledged that its own operations were not only distinct from, but even antithetical to, commercial ventures: "Inasmuch as private custody and ownership of fissionable material is prohibited by the Act, the essential core of the atomic energy enterprise is removed altogether from the sphere of possible private enterprise development."[13] This created a political dilemma for the AEC. Conservative Congressmen could criticize the agency for both for its allegedly lax national security measures and its antibusiness outlook. The only companies directly involved in atomic energy were government contractors, who managed AEC laboratories and plants. Commissioners viewed this as one reason that its contractor system was so valuable, but the sponsorship of industry through these contracts remained far from the model of competitive capitalism.[14]

David E. Lilienthal, first chair of the Commission, was already viewed with deep suspicion by the business community when appointed.[15] As chairman of the Tennessee Valley Authority from 1933 to 1946, he had chosen to have power from the massive dams distributed through rural power cooperatives, rather than by private utility companies. Lilienthal's faith in the federal government to develop national resources clearly shaped his vision of the AEC as an atomic TVA. In a 1947 article he published in *Collier's* magazine, "The Atomic Adventure," he contended that the country possessing the greatest production of electrical and mechanical energy per person would enjoy the greatest military security. Developing atomic energy for domestic power was a way to "enlist the broad

11. The exclusion of much atomic energy technology from protection under the Patent Act was an amendment to McMahon's original bill. Hewlett and Anderson, *New World* (1962), pp. 495–98. As Turchetti notes ("Slow Neutrons" [2006], p. 15n44), the provisions were attacked by various interest groups after passing the Senate, including the American Bar Association, the Association of Manufacturers, and the National Patent Council. See also Miller, "Law Is Passed" (1948), p. 816.

12. Turchetti, "Contentious Business" (2009), p. 192.

13. AEC 297, AEC Policy of Operating through Contractors, 14 Feb 1950, NARA College Park, RG326, E67A, box 9, folder 1 AEC Relationship with Contractors, vol. 2, p. 5.

14. AEC, *Second Semiannual Report* (1947), pp. 5–6. For a list of AEC's majors contractors, Hewlett and Holl, *Atoms for Peace and War* (1989), see pp. 10–11.

15. Hughes, "Tennessee Valley" (1989); Wellerstein, *Knowledge and the Bomb* (2010), ch. 7.

participation of the American people" in this new resource, especially the private sector.[16] Lilienthal focused on the key role of industry in this enterprise in part to mollify critics of his prior activities as a New Dealer. But he also perceived the difficulties of advancing the interests of a federal agency without the support of an economic interest group. He fully intended to bring the emerging nuclear industry into alliance with the AEC.[17] (See figure 6.1.)

Lilienthal's hopes for rapid nuclear power development proved too optimistic. The AEC's General Advisory Committee, headed by J. Robert Oppenheimer, drafted a memorandum in July 1947 explaining that development of economical breeder reactors would take years.[18] (A breeder reactor is a nuclear reactor that can generate more fissile material than it uses.) Because of the immediate priority of using available fissionable materials for atomic weapons, without breeder reactors it would be decades before sufficient nuclear fuel could be accumulated for electrical power. As the official history of the AEC puts it, "The draft struck the Commissioners like a sledge hammer."[19] The Commissioners asked James Conant and Oppenheimer of the committee to revise the memorandum somewhat, adding a paragraph to mention the benefits of radioisotopes to science, medicine, and industry.[20] The scientists did not change the sober tenor of their appraisal, which was published (without a paragraph on radioisotopes) in July 1948.[21]

16. Lilienthal, "Atomic Adventure" (1947), p. 12. I find that "atomic TVA" describes well Lilienthal's forecast of nuclear energy development in this piece. For a similar usage, see Weart, *Nuclear Fear* (1988), p. 159. That said, the term would have been one of opprobrium among Lilienthal's Congressional critics, who opposed his appointment as chair of the Commission precisely on his commitment to public development of energy at the TVA. On this issue, see Lowen, "Entering the Atomic Power Race" (1987), p. 466.

17. Political scientists refer to the mutually reinforcing alliances of congressional committees, economic interest groups, and administrative agencies as "iron triangles." Having lacked the support of an economic interest group at the TVA, Lilienthal set out to develop it for the AEC, though the conflicts among the agency, the JCAE, and the nuclear industry prevented formation of an effective iron triangle. Balogh, *Chain Reaction* (1991), p. 16 and chapter 3.

18. The second draft, dated 29 Jul 1947, is attached to AEC Info Memo 264, NARA College Park, RG 326, E67A, box 56, folder 10 Development of Atomic Power–General.

19. Hewlett and Duncan, *Atomic Shield* (1969), p. 100.

20. Ibid.

21. AEC, *Recent Scientific and Technical Developments* (1948), pp. 43–46. The report concludes that "we do not see how it would be possible under the most favorable circumstances to have any considerable portion of the present power supply of the world replaced

FIGURE 6.1. Images illustrating articles by David Lilienthal, "The Atomic Adventure," and Lester Veile, "Inertia–U.S.A." in a special "Man and the Atom" feature in *Collier's* 119 (3 May 1947): 12–13. On the left is a Japanese landscape after an atomic blast; on the right are power plants.

---

by nuclear fuel before the expiration of 20 years." On the General Advisory Committee's several revisions of the statement, see Balogh, *Chain Reaction* (1991), p. 83. Industry calculations showed the cost of nuclear energy in the United States to be about ten times higher than coal- and hydroelectrically generated power; see Hood Worthington to David E. Lilienthal, 2 Feb 1948, in AEC Info Memo 264, NARA College Park, RG 326, E67A, box 56, folder 10 Development of Atomic Power–General.

Despite the diminished expectations for nuclear power, the Commission appointed an Industrial Advisory Committee to assess how the agency could increase commercial involvement in atomic energy. In an address delivered in Detroit on October 6, 1947, Lilienthal asserted that the AEC intended "to move away from the present Government monopoly" and "find opportunities for industrial participation; that is, opportunities for profit."[22] The group, chaired by James W. Parker, President of the Detroit Edison Company, included executives from electric power companies and from industrial research corporations.[23] Over the course of six months or so, the Industrial Advisory Group met with members of the Commission, staff from several of the AEC's principal contractors, and individuals who had been employed on work with the Manhattan Engineer District. The group also visited the Commission's installations at Oak Ridge, Argonne, Hanford, and Schenectady.[24]

The Industrial Advisory Group found the main obstacle to commercialization of atomic energy to be national security regulations; firms could not even obtain sufficient information to ascertain what commercial opportunities existed.[25] The AEC's security measures had made it hard for even this group to assess the situation: "Difficulties in connection with clearances, the complicated mechanisms of arranging for access to people and installations, the elaborate procedures for the safekeeping of notes and documents, as well as other secrecy restrictions, together constitute a formidable impediment to any attempt to study and understand the enterprise."[26] Radioactive and stable isotopes seemed the "only important

22. As quoted in Conference on Commercial Distribution of Isotope-Labeled Compounds, Minutes of the Meeting, 30 Oct 1947, Appendix "E" to AEC 108, NARA College Park, RG326, E67A, box 47, folder 3, Isotopes: Labeled Compounds, p. 26.

23. Mazuzan and Walker, *Controlling the Atom* (1985), p. 17. The other members were Bruce K. Brown, Gustav Egloff, Paul D. Foote, Isaac Harter Sr., Jerome C. Hunsaker, Gabriel O. Wessenauer, and Robert E. Wilson. Two members of the original group, Oliver E. Buckley and Donald F. Carpenter, resigned before its activities were concluded to accept other government appointments. Walter Cisler, who had previously been a consultant for the Commission, worked closely with the Industrial Advisory Group.

24. Preliminary Draft of Report to the United States Atomic Energy Commission by the Industrial Advisory Group, 29 May 1948, NARA College Park, RG 326, E67A, box 25, folder 4 Industrial Advisory Group, p. 2.

25. Preliminary Draft of Report, 29 May 1948, p. 6; Report to the United States Atomic Energy Commission by the Industrial Advisory Group, 15 Dec 1948, attached with letter from James W. Parker to David E. Lilienthal, 22 Dec 1948, NARA College Park, RG326, E67A, box 25, folder 4 Industrial Advisory Group, pp. 10–11.

26. Report to the United States Atomic Energy Commission, p. 3.

open field in connection with atomic energy."[27] Even here, the committee mostly focused on lagging industrial consumption of nuclear materials, not on production or new commodities. More companies would seek to apply these newly available tools "if simple instructions for the safe use of radioactive tracers were furnished on request."[28] The contractor system itself was highly problematic as a spur to free enterprise, as the group noted. The federal government, by selecting the companies and assigning them specific tasks, failed to foster competition between firms and so bled the entrepreneurialism out of its industrial partners.[29] Moreover, companies viewed the government's patent policy as "prejudicial to industrial participation in atomic energy."[30] The Commission had done little to dispel such perceptions.

The Report of the Industrial Advisory Group was published in the *Bulletin of Atomic Scientists*, along with a reply by Lilienthal.[31] No doubt its criticisms of the extensive secrecy barriers around nuclear science and technology cheered the journal's editors and readership, those scientists who had lobbied for the demilitarization of atomic energy at the end of the war.[32] That same year the *Bulletin of Atomic Scientists* carried other articles that highlighted the strictures the Atomic Energy Act placed on commercial activity, particularly David Lilienthal's "Private Industry and the Public Atom" and James Newman and Byron Miller's widely cited piece, "The Socialist Island."[33] In addition to appointing further industry groups to study the problem, the AEC launched programs to educate engineers in nuclear technology and to offer government financing for companies

27. Ibid., p. 6.

28. Preliminary Draft of Report, p. 5.

29. Report to the United States Atomic Energy Commission, p. 8. The legal and economic features of the contractor system were examined more carefully in the 1950s; see Palfrey, "Atomic Energy" (1956); Tybout, *Government Contracting in Atomic Energy* (1956).

30. Report to the United States Atomic Energy Commission, p. 15.

31. "Report of the AEC Industrial Advisory Group" (1949).

32. See Smith, *Peril and a Hope* (1965).

33. Lilienthal, "Private Industry and the Public Atom" (1949); Newman and Miller, "Socialist Island" (1949). The latter piece was a reprinting of a chapter of Newman and Miller's 1948 book, *The Control of Atomic Energy*. In its 1949 article form, the piece was followed by an article criticizing its claims: Lerner, "Control of Atomic Energy" (1949). Lilienthal was more vocal on this issue after he left the Commission, declaring in 1950, "No Soviet industrial monopoly is more completely owned by the state than is the industrial atom in free-enterprise America." Lilienthal, "Free the Atom" (1950), p. 13.

to participate in reactor design.[34] Yet the legal provisions that discouraged commercial involvement remained in place. In the view of George T. Mazuzan and J. Samuel Walker, official historians of the Nuclear Regulatory Commission, perhaps the most important contribution of the Industrial Advisory Group was simply the level of exposure to atomic energy developments gained by these influential leaders of corporate America.[35]

Fostering a Nuclear Industry with Radioisotopes

As the Industrial Advisory Group noted about the atomic energy enterprise, only radioisotopes seemed ready for commercial participation. More specifically, the synthesis and sale of radiolabeled compounds provided an entry point for industry.[36] For many experiments, the isotopic atoms themselves were not the reagents scientists used—they often wanted a particular compound labeled with a radioisotopic atom. (That radioactive label was frequently carbon-14, due to its very long half-life and the ubiquity of carbon in organic compounds.) Research laboratories adept at tracer work generally included organic chemists on staff who could synthesize the precise radiolabeled compounds of interest, but most laboratories lacked this chemical expertise. In the judgment of the AEC, "The present situation, in which individual workers must usually synthesize for themselves single lots of the desired labeled compounds, or else engage someone to do it for them, is not a satisfactory or efficient manner of meeting the overall demands."[37] Already there were certain standard compounds that many researchers appeared eager to use if they were available in labeled form.[38] The wider utilization of radioisotopes in biology and

34. Mazuzan and Walker, *Controlling the Atom* (1985), p. 19; Balogh, *Chain Reaction* (1991), p. 97. It was not only the issues of security clearance that made educational initiatives difficult to implement; the Commission's own contractors resisted bringing in the employees of other companies to the facilities they ran. Hewlett and Duncan, *Atomic Shield* (1969), p. 436.

35. Mazuzan and Walker, *Controlling the Atom* (1985), p. 17.

36. Lenoir and Hays, "Manhattan Project for Biomedicine" (2000).

37. Isotopes Division Circular E-21, Special Considerations in the Synthesis and Distribution of C14-Labeled Compounds, 15 Apr 1948, Appendix "G" to AEC 108, NARA College Park, RG326, E67A, box 47, folder 3 Isotopes: Labeled Compounds, p. 35.

38. AEC 108/1, NARA College Park, RG 326, E67A, box 47, folder 3 Isotopes: Labeled Compounds, p. 4.

medicine seemed to hinge on establishing retail suppliers of radiolabeled compounds and radiopharmaceuticals.

On October 30, 1947, the AEC held a Conference on Commercial Distribution of Isotope-Labeled Compounds at Oak Ridge. The roster of companies who sent representatives to the meeting included Tracerlab, Sun Oil, Dow Chemical, Houdry Process, Monsanto Chemical, Stuart Oxygen, Eastman Kodak, American Cyanamid, and Abbott Laboratories. Tracerlab, which sent two representatives, had just been founded as a company dedicated to nuclear technology. A 1947 feature in *Fortune* magazine described it as "the first real business to be built entirely out of byproducts of the atom bomb."[39] The company focused on synthesis of radiochemicals, radiation-counting instruments, and development of industrial control methods using isotopes. (Its specialization in radiation detection equipment would lead the company to design instruments for airborne surveys of fission products, used to detect the first Soviet atomic weapons test in 1949.)[40] Representatives from Tracerlab had already been corresponding with the AEC about whether it could process Oak Ridge radioisotopes and resell them to researchers. The official who responded said that while processing was fine, resale was impermissible.[41] Otherwise, the attendees at the AEC's meeting were dominated by big players in chemicals, with only two pharmaceutical firms represented.

The aim of the conference was to explain the need for the production of radiolabeled compounds and "to encourage interest of firms in a production and distribution program—both immediate and long range."[42] The agency hoped to see progress within six to eighteen months. One representative from industry asked the AEC whether firms would have "patent rights on processes and new production methods."[43] As the agency itself noted, research support from the US government often entailed the reservation of some patent rights.[44] The Atomic Energy Act's prohibition of

39. "Business in Isotopes" (1947), p. 121.

40. Gordin, *Red Cloud at Dawn* (2009), chapters 5 and 6.

41. See Apr 1947 correspondence between F. C. Henriques, Jr. of Tracerlab and T. Raymond Jones of Oak Ridge, NARA Atlanta, RG 326, MED CEW Gen Res Corr, Acc 67B0803, box 175, folder AEC 441.2 (R–Tracerlab, Inc.).

42. Conference on Commercial Distribution of Isotope-Labeled Compounds, Minutes of the Meeting, 30 Oct 1947, Appendix "E" to AEC 108, NARA College Park, RG326, E67A, box 47, folder 3 Isotopes: Labeled Compounds, p. 25.

43. Ibid., p. 28.

44. Eisenberg, "Public Research" (1996).

private patents on nuclear technologies was a further sore spot with these private interests.[45] Another industrial representative at the meeting raised objections to the clause in the agreement form for purchasing AEC radioisotopes that required the consumer to report on the use of materials. He argued that "there is a police power inherent in the requiring of reports."[46] In general, representatives from industry felt that such reports might betray potential patent disclosures.[47] Questions about patents and concerns about reporting were "closely related in the minds of industry," reflecting sensitivity over the degree of government control and surveillance in the field of atomic energy.[48] The possibility of government price controls also rankled industrial representatives. AEC personnel suggested that companies might synthesize the compounds and then return them to the agency for sales and distribution, but this option was unacceptable to industry.

Paul Aebersold, head of the Isotopes Division, presented the industrial representatives with a list of the twenty-five most desired radiolabeled compounds, most of which were intermediates or general products of biosynthesis.[49] He mentioned the strong interest in labeled sugars (such as fructose and glucose), steroid hormones, amino acids, and some fatty acids. Tracerlab had already presented the AEC with a proposal for producing eleven of the compounds on the agency's list, and had also developed synthesis schemes for some the AEC had not listed, including glycine, acetyl chlorine, and testosterone. By contrast, the representatives of Cyanamid and Dow were uninterested in preparing radiolabeled compounds on a commercial basis, even if they were doing it in-house. Stuart Oxygen Company and Eastman Kodak were more interested in stable isotope compounds. The AEC did admit that the scale of production for ^{14}C-labeled compounds "would not constitute 'big business,' especially if the total sales volume were shared by several firms, as will no doubt be the case."[50] Under these conditions, the question facing the agency was how

45. As it turned out, the AEC did not extend patent restrictions to most biological and medical applications as the agency did to reactor development. Minutes, Atomic Energy Commission Meeting No. 882, 26 Jun 1953, in AEC 615/16, Patent Policy in Connection with Industrial Development of Atomic Power, NARA College Park, RG 326, E67B, box 35, folder 4 Legislation on Industrial Participation Program vol. 2, p. 3.

46. Conference on Commercial Distribution, 30 Oct 1947, p. 28.

47. AEC 108/1, NARA College Park, RG 326, E67A, box 47, folder 3 Isotopes: Labeled Compounds, pp. 8–9.

48. Conference on Commercial Distribution, 30 Oct 1947, pp. 28–29.

49. Ibid., p. 30.

50. Isotopes Division Circular E-21, 15 Apr 1948, p. 37.

to make development of these products attractive to industry in the first place.

In reality, the companies were not as up-to-speed on producing radiolabeled compounds as the AEC's own laboratories. In July 1947, Melvin Calvin of the Radiation Laboratory at the University of California, Berkeley, had presented a proposal to the AEC that staff in his bio-organic group could synthesize a number of ^{14}C-labeled compounds for distribution.[51] Researchers at Los Alamos Scientific Laboratory were isolating radiolabeled compounds from bacteria that were fed carbon-14, though this method did not compete with Calvin's approach of chemical synthesis.[52] Members of the Isotopes Division felt the Commission had "certain obligations in meeting the needs" for ^{14}C-labeled compounds, and progress with involving industry was slow.[53] They viewed Calvin's proposal as a straightforward stopgap measure. Companies were less certain. By October, concern had been expressed by Tracerlab and by Monsanto that the program might have a negative effect on "private enterprise and free competition in undertaking the commercial provision of isotope labeled compounds."[54] Leaders at Oak Ridge insisted that the capabilities in the private sector were not yet sufficient to meet research demand, and that the Commission's involvement in this supply would be temporary.[55]

On June 9, 1948, the Commission approved a proposal that authorized "production at Commission installations, for off-project distribution, of chemical compounds labeled with radioactive isotopes."[56] Its provisions aimed to ensure that government activity did not discourage or preclude commercial involvement. The Commission would sell only radiolabeled compounds that companies were not prepared to begin making, and

51. Distribution of C 14 Labeled Compounds by the Atomic Energy Commission, 24 Jul 1947, Tab to Annex to Appendix "D" of AEC 108, NARA College Park, RG326, E67A, box 47, folder 3 Isotopes: Labeled Compounds.

52. AEC, *Atomic Energy and the Life Sciences* (1949), p. 79.

53. The sense within the Isotopes Division (most likely emanating from Paul Aebersold) of moral obligation in providing radiolabeled materials is striking; this passage continues, "There is a real responsibility to the over-all national welfare and especially to all scientific and technical personnel who are using, or would use, radioactive isotopes." Isotopes Division Circular E-21, 15 Apr 1948, p. 37.

54. J. C. Franklin and Paul C. Aebersold, "Proposal for Distribution of C 14-Labeled Compounds Synthesized Under Berkeley Area Program," 9 Oct 1947, NARA Atlanta, RG 326, MED CEW Gen Res Corr, Acc 67B0803, box 157, folder 441.2 (R–Instr.), p. 1.

55. Ibid., p. 4.

56. AEC 108, Isotope Labeled Compounds, 2 Jun 1948, NARA College Park, RG 326, E67A, box 47, folder 3 Isotopes: Labeled Compounds.

"would withdraw from production of any compound when a commercial firm demonstrates it can make satisfactory synthesis."[57] Since the previous fall, the Isotopes Division had made some progress in working with industrial partners, having concluded arrangements with Tracerlab and Abbott Laboratories. The AEC funded the development costs incurred by these companies with the expectation that they would not be passed along to consumers.

The background report for this decision included considerations beyond industrial reluctance. At some level, the public service philosophy that underlay radioisotope production did not easily mesh with a free market framework. For example, the pricing of radioisotopes involved a substantial government subsidy. Should companies using these isotopes as raw materials for commercial products receive this subsidy in full?[58] Also, the involvement of commercial purveyors in the sales of radiolabeled compounds meant that the AEC had to develop an apparatus for legalizing and regulating a secondary market in isotopes.[59] The radioisotope distribution plan had been established with explicit rules against secondary distribution, so that the AEC's Isotopes Branch could make sure that every institution receiving radioisotopes had submitted the proper certificates and received authorization on the basis of adequate facilities and safeguards for handling of radioactive materials. Could the agency trust companies to make sure that every purchaser had been cleared by the AEC? More to the point, how could the AEC exercise effective oversight over the companies at the same time that it was trying to elicit their participation as suppliers of radiolabeled compounds in the first place? The agency was ready to make certain concessions to industry. In particular, the AEC would not reserve patent rights associated with the manufacturing and use of isotope-labeled compounds.[60]

On March 28, 1949, the AEC finally advertised four compounds for sale from the Berkeley Rad Lab, all labeled with carbon-14 on the carboxyl group: sodium butyrate, sodium valerate, sodium caproate, and so-

57. Ibid., p. 2.

58. Ibid., p. 11.

59. Acceptance of Terms and Conditions for Order and Receipt of Byproduct Materials, Appendix "F" to AEC 108, NARA College Park, RG326, E67A, box 47, folder 3 Isotopes: Labeled Compounds.

60. AEC 108, Isotope Labeled Compounds, NARA College Park, RG 326, E67A, box 47, folder 3 Isotopes: Labeled Compounds, p. 13; Lilienthal to Hickenlooper, 24 Jun 1948, NARA College Park, RG326, E67A, box 47, folder 3 Isotopes: Labeled Compounds.

dium heptanoate. The Commission's circular on the availability of these compounds stated that the price of $150/millicurie had been "based on a formula designed to give a reasonable cost of the production of the material." The price the agency charged for the ^{14}C-labeled chemicals was admittedly lower than what a commercial outfit might charge, but the Commission did not believe this would "discourage commercial production."[61] The Berkeley scientists requested credit, if not profit, by having the compounds "bear the name of the laboratory of their origin, in our case 'Bio-Organic Division, Radiation Laboratory, University of California, Berkeley,' as one of the laboratories of the Atomic Energy Commission."[62]

The AEC soon involved companies in supplying radiolabeled reagents for research, contracting with six private firms for the production of 65 labeled compounds later that year.[63] Despite this expansion, the commercial market did not take off as quickly as the Commission expected. As Shields Warren, the Director of the Division of Biology and Medicine, noted to Kenneth Pitzer, the Division of Research, only compounds that "can be marketed in substantial quantities" seemed economically feasible for vendors to prepare.[64] Many investigators had expressed to Warren their desire to obtain small amounts of a wider array of radiolabeled compounds. The Division of Biology and Medicine was willing to make $50,000 available in 1950 and the same amount in 1951 for the Isotopes Division to fund the

61. Isotopes Division Circular E-39, 28 Mar 1949, NARA College Park, RG 326, E67A, box 47, folder 3 Isotopes: Labeled Compounds. However, the most important patent for manufacturing radioisotopes, that filed by Enrico Fermi and colleagues for slowing down neutrons, was acquired by the AEC. The compensation was a fraction of its worth. Turchetti, "Slow Neutrons" (2006).

62. Distribution of C 14 Labeled Compounds by the Atomic Energy Commission, 24 Jul 1947, Tab to Annex to Appendix "D" of AEC 108, NARA College Park, RG326, E67A, box 47, folder 3 Isotopes: Labeled Compounds, p. 23. An appended memo to the Manager, Chicago Directed Operations, authorized funding up to $50,000 as charged by Calvin to synthesize compounds for the Isotopes Division of Oak Ridge National Laboratory. On the development of a formula for pricing, see Memorandum from F. H. Belcher, Oak Ridge Laboratory Division, and J. C. Franklin, Manager, 19 Jul 1948, NARA Atlanta, RG 326, OROO Files Relating to K-25, X-10, Y-12, Acc 67A1309, box 65, folder Isotopes Program 3 Distribution.

63. Press Release "Progress in Radioisotopes Program," 28 Nov 1949, NARA College Park, RG 326, E67A, box 45, folder 3 Distribution of Radioisotopes–Domestic.

64. Memorandum from Shields Warren, Director, Division of Biology and Medicine, to Kenneth S. Pitzer, Director, Division of Research, 14 Mar 1950, NARA Atlanta, RG 326, New York Operations, Acc 68B0588, box 29, folder Isotopes Program 1—General Policy.

synthesis and distribution of requested compounds. The hope was that such subsidies would not only "accelerate scientific progress" but also develop a commercial market by stoking demand among researchers.[65]

The medical market presented its own challenges and opportunities. The AEC did not guarantee the safety of the radioisotopes shipped from Oak Ridge for direct human use. Abbott Laboratories, a pharmaceutical company that already focused on compounds used clinically, contracted with the Commission to produce certain radiolabeled drugs for medical researchers.[66] When Cornelius Tobias of the University of California requested radioactive gold sodium thiosulfate in a form suitable for medical experiments, the Isotopes Branch contacted Abbott about producing this material.[67] Others (including Paul Hahn) became interested in using colloidal radiogold for experimental cancer therapy, and Abbott entered into a contract to prepare it from gold irradiated at Argonne National Laboratory, on account of its proximity to the company's Chicago facilities.[68]

By this time a large demand existed for clinically ready radiophosphorus and radioiodine. Donalee L. Tabern of Abbott proposed that the company set up a facility near the X-10 reactor to purify and prepare government-produced radiophosphorus in sterile standardized solutions for hospitals and laboratories to purchase.[69] (Tracerlab was interested in providing a similar service, at least on a case-by-case basis, for iodine-131.)[70] Despite a concern that granting one company such an exclusive

65. Ibid.

66. These included, among others, ^{14}C-labeled barbiturates and thiobarbiturates as well as penicillin labeled with radioiodine and radiosulfur. The contract for July 1, 1950, to June 30, 1951, was $9,900, and was renewed at about the same rate for the two years following. See documents in contract #AT-(40-1)-290 with Abbott Laboratories, NARA Atlanta, RG 326, series 16, DOE Contracts–Retired.

67. Paul C. Aebersold to D. L. Tabern, 8 Sep 1947, NARA Atlanta, RG-326, MED CEW Gen Res Corr, Acc 67B0803, box 144, folder AEC 441.2 (R–Abbott Laboratories). For more on Paul Hahn's research trajectory, see chapter 8.

68. In doing so, Abbott was "donating Tabern's service," though it was clear that this arrangement would not last forever. Memorandum from Paul C. Aebersold to Shields Warren, 15 Apr 1948, OpenNet Acc NV0720838. On the radiogold contract, see documents in NARA Atlanta, RG-326, MED CEW Gen Res Corr, Acc 67B0803, box 144, folder AEC 441.2 (R–Abbott Laboratories).

69. Letter from D. L. Tabern to Fenton Schaffer, copied to Paul C. Aebersold, 23 Jun 1948, in NARA Atlanta, RG-326, MED CEW Gen Res Corr, Acc 67B0803, box 144, folder AEC 441.2 (R–Abbott Laboratories).

70. See William E. Barbour, Jr. to Paul C. Aebersold, 6 Feb 1947, NARA Atlanta, RG-326, MED CEW Gen Res Corr, Acc 67B0803, box 144, folder AEC 441.2 (R–Tracerlab).

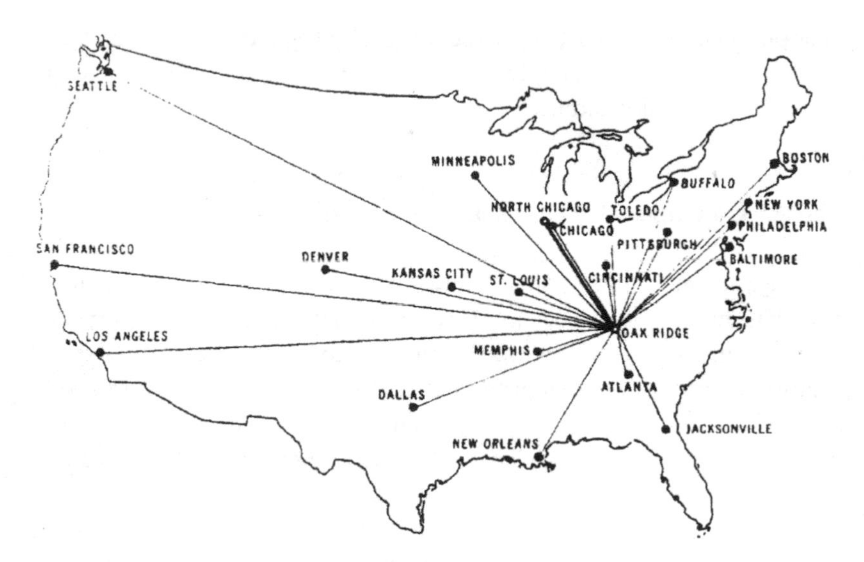

FIGURE 6.2. "The Abbott system of branches interconnected with Oak Ridge and North Chicago by teletype makes possible 'same day' shipment of needed supplies." Image and caption from *Nuclear Medicine for the Modern Physician and His Hospital*, pamphlet published by Abbott Laboratories, 1 Jan 1956, DOE Opennet document NV0723949, p. 6. Reproduced by permission of Abbott Laboratories.

right might hinder "free competition" in the field, the AEC acceded to such an arrangement.[71] In 1951 Abbott built a radiopharmaceutical production plant at Oak Ridge National Laboratory to prepare clinically ready isotopes and labeled compounds on site, shipping those with short half-lives on the same day as the radioisotopes were supplied.[72] By 1953, the firm had introduced "Radiocaps," capsules containing a precise radioactive dose of iodine-131 for the diagnosis and treatment of thyroid disorders. By 1954, Abbott Laboratories sent out more shipments each month from Oak Ridge than the government.[73] Within two more years, Abbott was making more than 30,000 shipments a year, predominantly iodine-131, phosphorus-32, and gold-198. The company's early pursuit of the

71. Paul C. Aebersold to D. L. Tabern of Abbott, 9 Apr 1948, NARA Atlanta, RG-326, MED CEW Gen Res Corr, Acc 67B0803, box 144, folder AEC 441.2 (R–Abbott Laboratories).

72. *Nuclear Medicine for the Modern Physician and His Hospital*, pamphlet published by Abbott Laboratories, 1 Jan 1956, OpenNet Acc NV0723949, p. 4 and p. 8; Johnson and Schaffer, *Oak Ridge National Laboratory* (1994), p. 35.

73. Capsule Summary of Isotopes Distribution Program, Sep 1954, in Aebersold papers, box 1, folder 1-16 Gen Corr 1954.

radiopharmaceutical market proved astute. By 1956, the AEC estimated that more than 500,000 people a year were receiving either treatment or diagnosis with radioisotopes.[74] (See figure 6.2.)

The AEC's Unhurried Exit from the Radioisotope Supply Chain

The 1954 Atomic Energy Act allowed companies to own reactors, patent atomic technologies, and license fissionable materials, nominally ending the government's monopoly on the atom.[75] It was not the economic viability of nuclear power that prompted the legislation, but political anxieties that the United States was falling behind other countries.[76] The AEC funded the first reactor to produce electricity commercially for the grid, built at Shippingport, Pennsylvania, by Westinghouse Electric Corporation.[77] The reactor started up in 1957, but the electricity it generated was several-fold costlier than that of coal-powered plants.[78] The Commission also launched a five-year, $200 million experimental reactor program, relying on contracts with industry, rather than competition, as a means to achieve technological advances.[79]

In principle, this new infrastructure meant that companies could begin to acquire radioisotopes from civilian reactors rather than from the AEC. In reality, this was not yet economical, despite the existing market for radioisotopes and radiolabeled compounds, with sales of more than $500,000 in 1953.[80] Tracerlab and Bendix Aviation Corporation of Detroit conducted a study and concluded that "complete commercial handling of isotope production and distribution" was not yet feasible, and would not be for at least five years.[81] Companies in the radioisotope retail market

74. *Nuclear Medicine for the Modern Physician and His Hospital*, pamphlet published by Abbott Laboratories, 1 Jan 1956, OpenNet Acc NV0723949, p. 4.

75. Palfrey, "Atomic Energy" (1956).

76. Mazuzan and Walker, *Controlling the Atom* (1985); Lowen, "Entering the Atomic Power Race" (1987); Walter L. Cisler and Mark E. Putnam to Gordon Dean, 16 Apr 1953, attached to AEC 615, Patent Policy in Connection with Industrial Development of Atomic Power, NARA College Park, RG 326, E67B, box 35, folder Legislation on Industrial Participation Program vol. 2.

77. Mazuzan and Walker, *Controlling the Atom* (1985), p. 21.

78. Lowen, "Entering the Atomic Power Race" (1987), p. 477.

79. Palfrey, "Atomic Energy" (1956), p. 372.

80. AEC, *Fifteenth Semiannual Report* (1954), p. 40.

81. Ibid., p. 40.

seemed content with the AEC's role as a wholesaler, especially as government involvement saved them from needing to make large capital investments. The Commission, in turn, bowed out of competing directly with the retail industry. In 1955, the ORNL Chemistry Division held a "going-out-of-business" sale of those ^{14}C-labeled compounds still in its inventory.[82] Along similar lines, a program the agency launched in 1949 to distribute cyclotron-produced radioisotopes was discontinued in 1955 because "private industry appears to be prepared to assume this function."[83] This general principle, that the AEC would withdraw once the private sector was viable, became a defining feature of the agency's policy.

The research market for radiolabeled compounds grew markedly over the 1950s, in part reflecting the large increases in US government funding for biomedical research.[84] In 1956, two Tracerlab employees, Seymour Rothschild and Ed Shapiro, left to form their own firm, New England Nuclear, based in Boston.[85] New England Nuclear bought radioisotopes in bulk from Oak Ridge (and, in special cases, from cyclotrons), which they incorporated into compounds that scientists could purchase. The initial group of employees took pride in their close connection to the scientific community, and set up their production line so that they could often ship out a radiolabeled compound within 24–48 hours of receiving the order. They focused on radiochemicals containing carbon-14 and hydrogen-3, which were becoming preferred tracers in life science research.

New England Nuclear, like other commercial suppliers, could sell their radiolabeled compounds only to licensed scientists, institutions, and hospitals, and had to report on each shipment to the federal government.[86] By 1962 their product line included 400 labeled compounds, of which 300 contained carbon-14, and their sales totaled $1 million.[87] Through the

82. Rupp and Beauchamp, "Early Days" (1966), p. 38.

83. Press Release, "AEC to Discontinue Distribution Program of Cyclotron-Produced Radioisotopes," 30 August 1955, copy in Hickenlooper papers, Senate Committee Files, folder JCAE–Isotopes 1951–1964.

84. Strickland, *Politics, Science, & Dread Disease* (1972).

85. Interviews with Charles Killian, 6 Mar 2003; Paul McNulty, 19 Mar 2003; Robert Ludovico, 19 Mar 2003; Rheinberger, "Putting Isotopes to Work" (2001). There were many other companies in the late 1950s who sold radioisotopes and isotopically tagged compounds. A 1958 issue of *Nucleonics*, for instance, lists 23 companies selling isotope-labeled compounds, 36 selling radioactive isotopes, and 16 selling stable isotopes. The major competitors of New England Nuclear for the research market were Nuclear Chicago and Amersham.

86. Interview with Robert Ludovico, initial accountant and then Treasurer for New England Nuclear, 29 Mar 2002, Wellesley, Massachusetts.

87. Tivnan, "Firm's Annual Report" (1962).

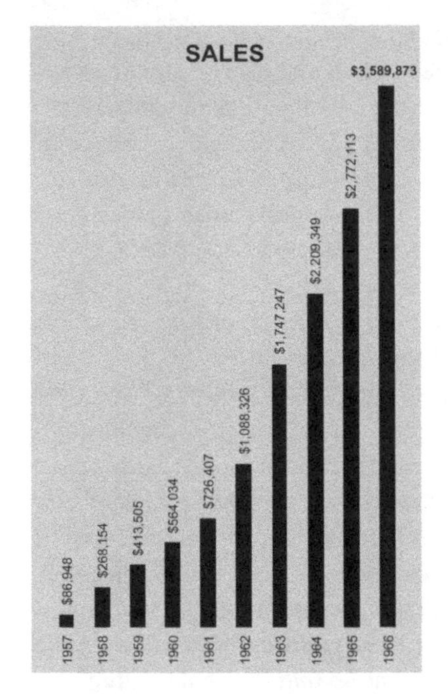

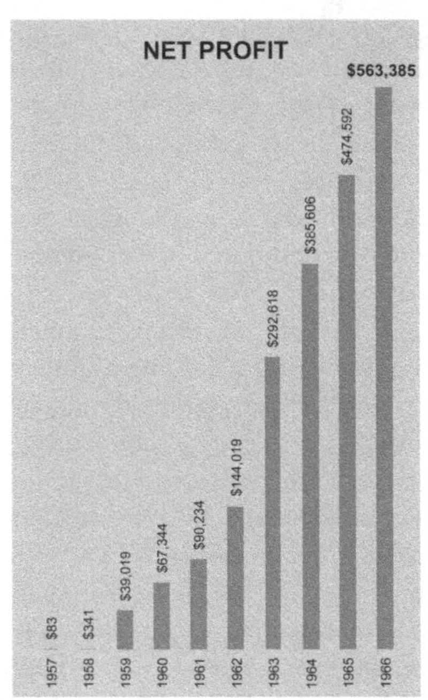

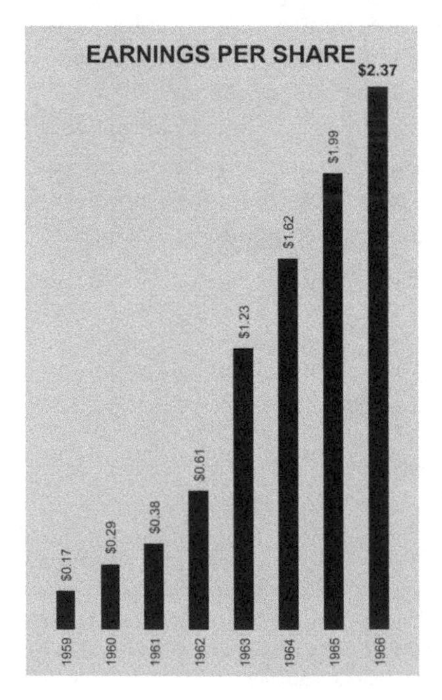

FIGURE 6.3. Table of sales, net profit, and earnings per share for first decade of New England Nuclear. Adapted from New England Nuclear Annual Report 1964. Courtesy of Paul McNulty.

early 1960s, sales grew by an average of more than 50% per year, and New England Nuclear was supplying a majority of the domestic market for radiolabeled compounds.[88] (See figure 6.3.)

The other major vendors of radioisotope-based reagents, drugs, and diagnostic tests were Abbott Laboratories, Amersham, Bio-Rad, California Corporation of Biochemical Research, Merck, Nucleonic Corporation, Tracerlab, and Volk Radiochemical.[89] By 1959, US-based commercial suppliers sent out about 100,000 shipments of radioactive materials per year. This continued to be a secondary market: many of the 12,000 or so radioisotope shipments from Oak Ridge that year were destined for these retail suppliers.[90] The International Directory of Radioisotopes, published in 1959 by the International Atomic Energy Agency, listed Oak Ridge National Laboratory as the "principal US source of radioisotopes" among 44 purveyors worldwide.[91] The retail market meant that purchasers did not necessarily perceive the role of the government as the chief radioisotope wholesaler, even though it remained central to the supply chain.

By 1965, there were estimated to be 140 civilian reactors in operation in the United States, including power stations, experimental reactors, testing facilities, research reactors, and teaching reactors.[92] (This number included 49 wholly owned by the AEC, which also co-owned or helped

88. Interview with Paul McNulty, 29 Mar 2002, Newton, Massachusetts; average growth calculated from Annual Report, year ending 28 Feb 1963, New England Nuclear Corporation, copy courtesy of Paul McNulty.

89. International Atomic Energy Agency, *International Directory of Radioisotopes, Vol. II* (1959), pp. i–ii. Ten years later, there were nearly twenty companies in this field. See "U.S. Radioisotope Industry–1966," (1967), p. 209.

90. AEC, *Radioisotopes in Science and Industry* (1960), p. 2. The smaller market of cyclotron-produced radioisotopes reflected a partnership of public sector and private sector like that for the larger supply of reactor-produced radioisotopes. The AEC withdrew from the processing and distribution of cyclotron-produced materials in 1955, although it remained a major supplier of the radioisotopes themselves. The principal commercial suppliers of the processed cyclotron-generated materials in the mid-1960s were Abbott, Cambridge Nuclear Corporation, New England Nuclear, Nuclear Science and Engineering Corporation, US Nuclear Corp., and Nuclear Consultants Corp. (a division of Mallinckrodt). "U.S. Radioisotope Industry–1966" (1967).

91. This list includes thirteen government agencies or installations, as well as major commercial suppliers of radiolabeled compounds, radiopharmaceuticals, and various medical and industrial radiation sources in several nations. International Atomic Energy Agency, *International Directory of Radioisotopes, Vol. I* (1959), p. vii.

92. Eisenbud, *Environmental Radioactivity* (1963), Appendix 8-1. Strikingly, one private reactor was built in 1956 with funds from American Tobacco–Medical College of Virginia, in

finance many other reactors located outside its facilities.)[93] The AEC's own radioisotope production pipeline diversified. The X-10 reactor in Oak Ridge, increasingly obsolete, was shut down in 1963, twenty years to the day after it became operational.[94] The agency had already begun using other facilities at Oak Ridge, Brookhaven, and Argonne to manufacture radioisotopes.[95] The Isotopes Division developed contracts with Hanford to purify fission products and with thc Savannah River Plant (the Commission's other plutonium-production facility) to prepare cobalt-60, reflecting increased demand for large-scale radiation sources.[96]

Some firms—though generally not the radiochemical and radiopharmaceutical retailers—became restive under the government's domination of the isotope wholesale market. The Commission's stated policy was to discontinue producing radioisotopes where commercial sources could reasonably provide the supply. On October 1, 1963, the AEC stopped producing iodine-125 and iodine-131, the latter being the isotope used in greatest quantity.[97] At that point there existed several companies selling radioiodine, including Abbott Laboratories from their Oak Ridge plant; General Electric Company in Pleasanton, California; Iso/Serv of Cambridge, Massachusetts; Nuclear Consultants Corporation of St. Louis; Nuclear Science and Engineering Corporation of Pittsburgh; E. R. Squibb of New Brunswick, New Jersey; Union Carbide of Tuxedo, New York; and Volk Radiochemical Company of Skokie, Illinois. The following year, the

---

part to provide isotopes for the study of polonium-210 in cigarettes. Proctor, *Golden Holocaust* (2011), p. 184; Rego, "Polonium Brief" (2009).

93. The AEC's role should not be underestimated. As Balogh notes, "Every power reactor built before 1963 received some kind of direct or indirect federal assistance." *Chain Reaction* (1991), p. 118.

94. There are media clippings covering the reactor's "retirement party" in the Aebersold papers, box 20, folder 20-3 Clippings about ORNL, Sep–Dec 1963.

95. Argonne supplied isotopes not normally available from Oak Ridge as well as irradiation services. Brookhaven supplied fluorine-18, iodine-132, iodine-133, magnesium-28, and standard reactor units, as well as cyclotron beam bombardment, gamma ray irradiation, and hot laboratory services. International Atomic Energy Agency, *International Directory of Radioisotopes, Vol. 1* (1959), pp. iii–iv. The involvement of AEC facilities beyond Oak Ridge National Laboratory had started in the mid-1950s; Press Release, 27 Jan 1955, NARA College Park, RG 326, E67B, box 58, folder 6 Organization & Management 2 Division of Licensing and Regulation.

96. Rohrmann, "Hanford Isotopes Plant" (1964–65); "Growing Demand for Cobalt-60," (1965).

97. "AEC Withdraws from Routine Production of Radioiodine" (1963–64).

AEC withdrew from producing six other radioisotopes.[98] In March 1965, the Commission established a procedure whereby firms could petition the government to withdraw from production and sale of specific radioisotopes.[99] Soon thereafter Nuclear Science and Engineering Corporation filed such a petition for nineteen more radioisotopes. These included several that were important in biology and medicine: phosphorus-32, sulfur-35, sodium-24, and gold-198. On January 25, 1966, the AEC announced that it would withdraw in March from the routine production and sale of these radioisotopes, ceding the market to this company.[100] Production of radioisotopes from Oak Ridge that month, as reported by Union Carbide (the contractor), still amounted to 577 shipments of radioisotopes, over 163,000 curies of radioactivity.[101] For those isotopes that the agency continued to sell, they adjusted prices in 1965 so as to recover full costs of production and distribution.[102]

This shift in supply was stimulated by substantial growth in the radioisotope business in the mid-1960s. In 1967, five private reactors reported sales of $5.6 million for radioisotopes. Retail numbers were also up: sales of radiochemicals (predominantly radiolabeled compounds) by nineteen companies were $8 million, marking a growth of 15% since 1966; sales of radiopharmaceuticals by eight companies amounted to $12 million, showing 20% growth. Sales of encapsulated radiation sources by 18 companies reached $3 million, an increase of 20% since the prior year.[103] Rather than subsidize its radioisotopes, the Commission began charging as much as possible. As an article in *Business Week* noted, "AEC now arbitrarily sets prices once a year for its bulk sales of isotopes; with no free market in the materials, it tends to err on the high side, and added costs push prices up

98. "AEC Ends Routine ^{85}Sr Production" (1965).

99. "Formal Procedure Adopted for Withdrawal" (1965).

100. See documents in NARA Atlanta, RG 326, OROO Office of Public Information, Acc 73A0898, box 202, folder Isotopes: AEC Withdrawal from Routine Production and Sale of 19 Radioisotopes.

101. Ibid.

102. "AEC Announces Change in Prices" (1965).

103. These figures do not indicate whether the radionuclides used to prepare radiochemicals, radiopharmaceuticals, and encapsulated radiation sources were from government or private reactors. Martin Moon to Mr. Downey, Sale of Isotopes by Private Reactor Operators during 1967, 18 Jan 1968, NARA Atlanta, RG 326, OROO Office of Public Information, Acc 73A0898, box 224, folder Isotopes Program–8 Reports & Data.

still more before the final product reaches the user."[104] These efforts aimed in part at making sure prices were robust enough to encourage a transition to commercial wholesalers, but they squeezed the retail industry.

The wholesaling of reactor-produced radioisotopes for radiopharmaceuticals shifted to the private sector earliest, due to production by Union Carbide Corporation, General Electric Company, and Nuclear Science and Engineering Corporation. General Electric had entered the radioisotope production market in 1959, but it took four years to top $400,000 in sales.[105] Oak Ridge National Laboratory supplied radiopharmaceutical makers only with radioisotopes not commercially available, though the Canadian atomic energy installation at Chalk River operated under no such restrictions and remained a major nondomestic supplier, competing with American firms.[106] This shift can be seen vividly in the graphs of the government's market share of iodine-131 and phosphorus-32, mainstays of nuclear medicine. (See figure 6.4.) In terms of curies, the amount of iodine-131 produced at Oak Ridge peaked in 1958, as did and the amount of phosphorus-32 produced. Even so, in 1966 the AEC remained the principal domestic producer and distributor of reactor-produced radioisotopes.[107] In large part, this derived from the government's central role in providing large-scale radiation sources. The AEC also remained the major supplier of carbon-14, an isotope for which the sales increased due to the growing research demand for radiolabeled compounds.

The AEC emphasized the level and vitality of industry involvement, and their role in enabling, rather than prohibiting, free enterprise. By the same token, the entire American nuclear industry remained heavily dependent on the US federal government. The Commission supported private reactor construction into the early 1960s by providing technical information and reactor fuel free of charge for seven years. It also conducted research and development on radioactive waste disposal. The Price-Anderson Act of 1957 limited the nuclear power industry liability in the event of an accident, and required the government to underwrite the insurance (though utilities paid the premiums). The AEC procured and processed—and until 1964, owned—all nuclear materials. Needless

104. "Radioisotope Business . . . Is Booming" (1963), copy in Aebersold papers, box 20, folder 20-1 General Newspaper Clippings, 1963–1970.

105. Ibid.

106. Virgona, "Radiopharmaceutical Production at Squibb" (1967), p. 222.

107. "U.S. Radioisotope Industry–1966" (1967), p. 207.

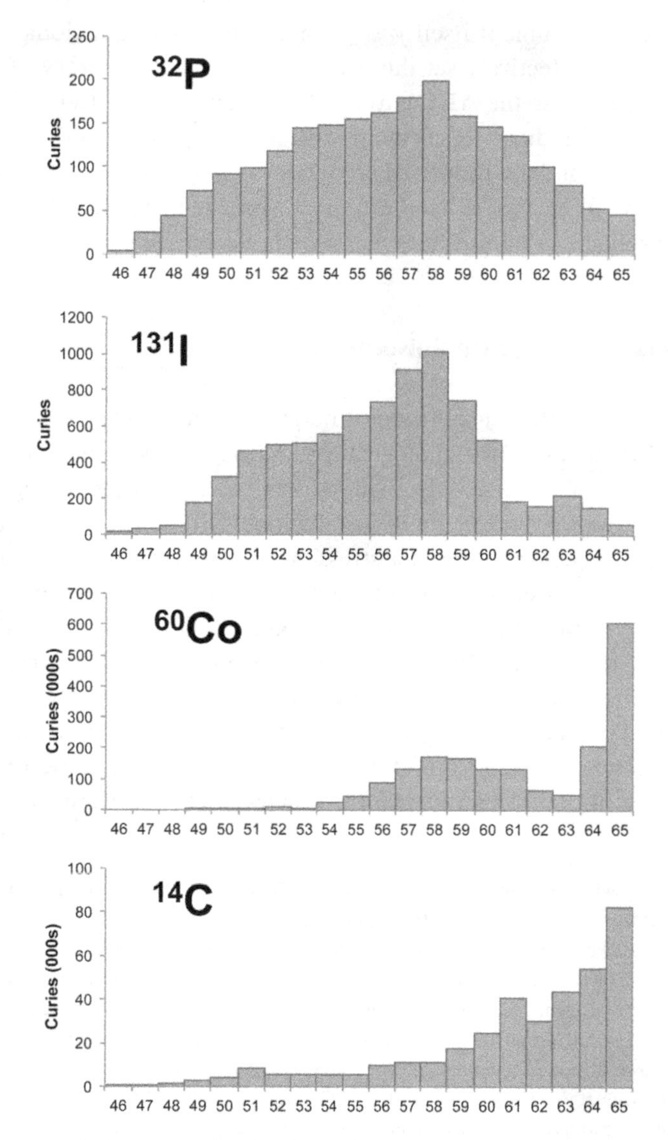

FIGURE 6.4. Graph showing yearly sales of phosphorus-32, iodine-131, cobalt-60, and carbon-14 in curies (top two) or kilocuries (bottom two). Adapted from Oak Ridge National Laboratory, 1946–1965. "Twenty Years of Radioisotopes," *Isotopes and Radiation Technology* 4:1 (1966): 66–67.

to say, the government itself was a major consumer of fissionable materials, and it effectively set the prices. As one contemporary observer put it, "Every time the AEC burps, the industry excuses itself."[108] In the wholesaling of radioisotopes, the transition from public to private sector remained incomplete. Industrial production—having been coaxed by the government in the first place—did gain momentum, but Oak Ridge continued producing radioisotopes that were not commercially viable.

Sales and Safety for Radioisotope Users

The passage of the 1954 Atomic Energy Act effectively expanded the scope of regulation for radiological protection to a new set of licensees in the private sector. The AEC had to focus more attention on regulating radiation exposure outside of its installations.[109] In part due to the fallout debates, the agency put greater stress on protecting the public from undue exposure as well as on regulating licensees of radioactive material.[110] This involved not only oversight of radiological protection in worksites, but also safe disposal of radioactive waste from civilian facilities.[111] The AEC established a new Division of Inspection to deal with licensees that were building and operating civilian reactors, and agency bureaucrats in Washington recommended that it take over inspecting licensees of radioisotopes as well.[112] This represented a marked change from the degree

108. "Growing Market" (1963), p. 175; Walker, *Containing the Atom* (1992), p. 35.

109. Mazuzan and Walker, *Controlling the Atom* (1985), pp. 55–56.

110. Dunlavey, "Federal Licensing and Atomic Energy" (1958).

111. Radioactive waste disposal was even more of a concern in hospitals than in laboratories, because therapeutic doses used so much more radioactivity than the tracer uses common to research. Hospitals using radioactive iodine and radioactive phosphorus generally disposed of waste down the drain. (Undoubtedly laboratory scientists did the same.) Growing concern about the hazards this might pose led the AEC to fund a study at Mount Sinai Hospital on the release of radioactive waste in public sewers. The study involved dismantling plumbing fixtures through which radioactive waste had passed; the level of contamination was deemed insufficient to make them hazardous to any plumbers that might work on the fixture. As for the sewage disposal workers, the agency determined that "the volume and rate of flow under normal circumstances dilutes the active material to a safe level." AEC, *Tenth Semiannual Report* (1951), p. 42.

112. T. H. Johnson to S. R. Sapirie, Re: Licensing and Inspection Functions of the Isotopes Division Under the Atomic Energy Act of 1954, Aebersold papers, box 2, folder 2-1 Gen Corr Jan–June 1955, pp. 7–8.

of oversight the Isotopes Division previously exercised over radioisotope purchasers.

From the outset, the AEC recognized that radioisotope supply posed a potential liability issue. Did the agency bear responsibility for those exposed to dangerous levels of radiation from radioisotopes purchased from Oak Ridge? As General Manager Carroll Wilson told the Medical Review Board at their June 1947 meeting, the Atomic Energy Act's mandates for protecting the public were difficult to meet when it came to isotope distribution:

> The Commission also has the responsibility to see that (and this is spelled out under the statute)—to see that people don't get hurt as a result of what the Commission does. For example, in the production and sale of isotopes, the Commission has a responsibility to see that to the best of its efforts, or to the best of its ability, people who work with and use these isotopes do not endanger themselves or their surroundings. The extent to which the Commission can discharge this is bound to be related to practical realities. It cannot police twenty-four hours a day everybody who uses isotopes. It shouldn't. On the other hand, it is necessary to take certain measures.[113]

Initially, the agency sought to address the liability problem through a "contract" method of distribution. Prior to submitting their purchase orders, recipients of radioisotopes from Oak Ridge had to sign an agreement that neither Monsanto (the contractor for the facility in 1946) nor the federal government would be responsible for "injury to persons or other living material or for any damage to property in the handling or application of the material."[114] The receiving institution assumed these responsibilities, and individuals who would "supervise the proper control of radiation and contamination on the part of the Institution" had to be specified in the application for purchase. Recipients also had to agree to dispose of radioactive waste in such a way that irradiation of humans or nonexperimental animals would be "below-tolerance," and with minimum contamination

113. Transcript of the Discussion at the First Meeting of the Medical Review Board, AEC, Washington, DC, 16 Jun 1947, OpenNet Acc NV0709599, pp. 18–19.

114. The signed copies of the AEC's form, Agreement and Conditions for Order and Receipt of Radioactive Materials, as well as the Certificates issued, are archived in NARA Atlanta, RG 326, Files Relating to K-25, X-10, Y-12, Acc 67A1309. The initial contracts authorized purchasers to order radioactive materials for twelve months before reapplying.

to facilities. Once this "Application for Radioisotope Procurement" was filed, the Isotopes Division issued the applicant an "Authorization for Radioisotope Procurement," providing the necessary evidence to the supplier that shipments could be sent out.[115]

The AEC issued publications on safe handling and shielding methods for radioisotopes and radiation sources, as well as methods for radioactive waste disposal.[116] In January 1947, the first month the agency was in operation, the Isotopes Branch published two circulars for radioisotope users: General Rules and Procedures Concerning Radioactive Hazards, and Health-Protection in Handling Radioisotopes. The General Rules and Procedures were excerpted from the regulations that governed work with radioactive materials at the AEC's Clinton Laboratories (soon to be renamed Oak Ridge National Laboratory), where the X-10 reactor was located. This pamphlet, which totaled 12 mimeographed pages, specified maximum permissible exposures for each kind of radiation, while emphasizing the importance of striving for the "lowest possible daily total exposure in every operation."[117] Mandated protective measures included monitoring of personnel for radiation exposure (via dosimeters), the use of protective clothing, gloves, and devices, and a ban on eating or smoking in "hot" laboratories. There were instructions on decontamination of laboratory spaces and of hands, transportation of radioactive materials, and solid and liquid radioactive waste disposal. Many of the regulations pertained to plutonium use, reflecting the military orientation of work at Clinton Laboratories. In effect, then, users of radioisotopes purchased from Oak Ridge were supposed to comply with the AEC's rules for its own employees.

115. By 1950 the "Application for Radioisotope Procurement" was Form AEC-313, and the "Authorization for Radioisotope Procurement" was Form AEC-374. AEC 398, Atomic Energy Commission, Regulations for the Distribution of Radioisotopes, 22 Jan 1951, NARA College Park, RG 326, E67A, box 45, folder 1 Regulations for the Distribution of Radioisotopes, p. 2.

116. Order Blank for Isotopes Division Circulars on the Techniques and Uses of Radioisotopes, Isotopes Division Circular E-15, 29 January 1948, Appendix A to "Study of Wider Use of Isotopes," Info Memo 48-92, 29 Jul 1948, NARA College Park, RG 326, E67A, box 45, folder 6 Study of Wider Use of Isotopes. A few years later the Isotopes Division began publishing information of this sort in a serial, *Isotopics*. E.g., Ward, "Design of Laboratories" (1951).

117. Isotopes Branch Circular No. B-1, General Rules and Procedures Concerning Radioactive Hazards (Excerpts from Clinton Laboratories Regulations), 8 Jan 1947, Evans papers, box 1, folder 1 Isotopes–Clinton Lab, p. 1.

The other pamphlet, Health-Protection in Handling Radioisotopes, was geared specifically to radiological safety in research laboratories. Written by Arthur K. Solomon, a biophysicist at Harvard Medical School, the six-page mimeograph was meant to serve "as an indication of regulations that may be suitable for adoption by other small-scale users of radioisotopes."[118] The Isotopes Branch set up an Advisory Field Service whose personnel visited isotope users to "give on-the-spot advice concerning radiation protection, monitoring and measuring equipment, standards of radioactivity, laboratory design, remote handling equipment, and radioactive waste handling."[119] In fact, the Advisory Field Service had authority to stop unsafe uses of radioisotopes, though their focus was on "education and information."[120] But the agency relied on scientists themselves to provide oversight. As Solomon's pamphlet specified, "In each laboratory one responsible member should be delegated to supervise all local protection measures."[121]

The AEC looked to the National Committee on Radiation Protection (NCRP) to set permissible exposure levels. This group was founded in 1929 as the Advisory Committee on X-Ray and Radium Protection, a year after the International X-Ray and Radium Protection Committee formed at the Second International Congress of Radiology. From the outset, the American group maintained a close connection with the US National Bureau of Standards, in part because its chair, physicist Lauriston S. Taylor, worked on x-ray standards at the agency. The National Bureau of Standards published the NCRP's reports, though without endorsing them.[122] As mentioned in chapter 5, the group's original recommendations specified a "tolerance" dose of radiation, first for external radiation in 1934 (0.1 roentgen per day whole-body exposure), then for internally deposited radiation, or "internal emitters," in 1941 (not to exceed 0.1 microcuries of radium-226 per day). These provided the basis for the radiological

118. Isotopes Branch Circular No. B-2, Health-Protection in Handling Radioisotopes by Arthur K. Solomon, 20 Jan 1947, Evans papers, box 1, folder 1 Isotopes–Clinton Lab, p. 3.

119. AEC, *Eighth Semiannual Report* (1950), p. 75. See also Paul Aebersold, "Isotopes and Their Application to Peace Use of Atomic Energy," 29 Dec 1947, presented at the 114th meeting of the American Association for the Advancement of Science, Appendix B to "Study of Wider Use of Isotopes," Info Memo 48-92, 29 July 1948, NARA College Park, RG 326, E67A, box 45, folder 6 Study of Wider Use of Isotopes, p. 9.

120. Bizzell, "Early History" (1966), p. 31.

121. Isotopes Branch Circular No. B-2, p. 1.

122. Mazuzan and Walker, *Controlling the Atom* (1985), p. 34.

safety regulations used by the Manhattan Engineer District during World War II.[123]

There was substantial overlap between members of the NCRP's subcommittees and those of the AEC's advisory bodies, particularly its Advisory Committee on Biology and Medicine and Advisory Committee on Isotope Distribution. While this facilitated the flow of information between these organizations, there were also tensions. At one point the AEC pressed the NCRP for advance information on the permissible doses they would set for radiation workers. As Taylor puts it, "The AEC was given some 'off the record' information as to the new levels that the NCRP would probably recommend."[124] Yet the leadership of the NCRP sought to maintain independence and avoid being influenced by government agencies or industries with an interest in their assessments.[125] The NCRP did accept financial support from the AEC to reimburse its members to attend committee meetings, a measure that reflected its voluntary character.[126]

In 1949 the National Bureau of Standards published an NCRP handbook on "Safe Handling of Radioactive Isotopes." The handbook had been prepared by the NCRP's Subcommittee on the Handling of Radioactive Isotopes and Fission Products, whose eight members included Aebersold and Joseph Hamilton. Before detailing the hazards of radioisotopes and how to protect against them, the guide listed the principal radioisotopes in use, indicating the half-life, type of emission, and typical uses of each. A table groups the most widely used radioisotopes in terms of their danger, and represents the hazard in terms of an "activity" scale for increasing curie levels. Notably, the isotopes most commonly employed in laboratories and hospitals—tritium, carbon-14, phosphorus-32, iodine-131—were all ranked as moderately dangerous. As the NCRP explained, there are four hazards associated with handling radioisotopes, in order of importance: deposition of radioisotopes in the body after ingestion, inhalation, or absorption; exposure of the whole body to gamma radiation; exposure of the body to beta radiation; and exposure of the hands to gamma or beta radiation. The handbook stressed good housekeeping and work habits as

123. Walker, *Permissible Dose* (2000), pp. 8–9; Hacker, *Dragon's Tail* (1987), p. 25.

124. Taylor, *Organization for Radiation Protection* (1979), p. 7-016.

125. Mazuzan and Walker, *Controlling the Atom* (1985), p. 37.

126. Taylor, *Organization for Radiation Protection* (1979), p. 7-032. As Elizabeth Rolph puts it, "Although the NCRP was reputable, it was a private organization with no government charter and virtually no access to research funds." Rolph, *Nuclear Power and the Public Safety* (1979), p. 45.

well as personal cleanliness in preventing contamination. Personnel who failed to develop radiological protection skills ("regarding food handling, checks of personnel activity, waste disposal, etc.") were "advised to transfer to other occupations."[127]

The NCRP recommended weekly monitoring of exposure of the (preferably neat and conscientious) personnel, to ensure that exposure to radiation remained below specified levels.[128] Dosimetry (or film) badges enabled the tracking of external radiation exposure by workers.[129] The NCRP also advised personnel handling radioactive solutions or radiation sources to wear finger rings detecting exposure. In an effort to promote compliance with these safety guidelines, the Isotopes Division made film badges available to radioisotope purchasers, who could return them to Oak Ridge for processing.[130] Film badges could also be obtained commercially—Tracerlab sold them, along with the weekly processing, and argued that their service was superior to that provided by the AEC.[131]

The complexity of providing radiation protection in the atomic energy program, as compared with earlier medical and industrial uses of radium, derived from the heterogeneous hazards posed by various fission products, artificial radioisotopes, and radioactive sources. Alpha particles, beta particles, and gamma rays, given off at various energies by different radioactive disintegrations, penetrated human skin to varying degrees, and so required different levels of shielding. (Alpha particles were not penetrating; gamma particles were highly penetrating; and beta particles required a different kind of dosimeter than the other two.) The biological utility of certain radioelements, such as carbon-14, also contributed to their hazard

127. National Committee on Radiation Protection, *Safe Handling of Radioactive Isotopes* (1949), p. 10.

128. These levels were 300 millirem per week for gamma radiation and 500 millirep per week for beta radiation. The NCRP handbook states: "According to present knowledge, this general exposure to gamma radiation is believed to be safe as far as any bodily injury is concerned, when there is no other type of radiation exposure. The importance of possible genetic change effective in later generations has not been established." Ibid., p. 6.

129. At the time the handbook was published, the NCRP advised use of both pocket ion chambers and film badges to register gamma and beta radiation exposure. But it appears that film badge services were available that could detect both types of radiation. See "Intercomparison of Film Badge Interpretations" (1955).

130. Order Blank for Isotopes Division Circulars on the Techniques and Uses of Radioisotopes, Isotopes Division Circular E-15, 29 Jan 1948, Appendix A to "Study of Wider Use of Isotopes," Info Memo 48-92, 29 July 1948, NARA College Park, RG 326, E67A, box 45, folder 6 Study of Wider Use of Isotopes.

131. "Business in Isotopes" (1947), p. 158.

if ingested—as the report put it, radiocarbon "would be built into virtually any part of the body structure like stable carbon, and unless excreted it would emit soft beta radiation during the lifetime of the victim."[132] By contrast, ingested radioiodine would concentrate only in the thyroid gland, although its strong gamma and beta radiation could damage that organ, and so affect the entire body.

According to the agency, there was only one significant radiation injury from a radioisotope sold by Oak Ridge, a case in which a graduate student was accidentally exposed to a 200-curie source of cobalt-60. He developed acute radiation sickness after receiving a total body exposure of 250–300 rad. (This was, in 1950, about twenty times the NCRP's permissible dose for an entire year.) Nonetheless, he was reported as having made a "satisfactory recovery without specific therapy."[133] If the bar for occupational hazard was acute radiation sickness, then it was not difficult for the AEC to claim that their radioisotope program was completely safe (though NCRP levels had been set far below visible injury levels in acknowledgment that lower doses could cause harm). That said, most researchers used radioisotopes as tracers in the microcurie range, putting their exposures well below the NCRP maximum permissible limits. The development of highly intense cobalt-60 and cesium-137 sources, such as the one the graduate student was exposed to, posed a much higher danger.

The Advisory Field Service developed scale-models of laboratories and the equipment needed for using radioisotopes, including a two-room version suitable for beta-emitting radioisotopes and a three-room laboratory designed to handle both beta-emitting and gamma-emitting radioisotopes. (See figure 6.5.) The cost of outfitting a larger radioisotope laboratory capable of handling up to 500 millicuries of gamma-emitters was $20,000.[134] Obviously, this was a significant expense. The Manager of the Oak Ridge Directed Operations suggested that the agency make funds available to nonprofit organizations to pay for the setting up of "semi-hot" isotope laboratories in existing facilities, at a cost of $10,000-$20,000 per laboratory. However, this does not appear to have been enacted.[135]

132. AEC, *Eighth Semiannual Report* (1950), p. 13.

133. Bizzell, "Early History" (1966), p. 31.

134. AEC, *Eighth Semiannual Report* (1950), p. 76.

135. "Study of Wider Use of Isotopes," Info Memo 48-92, 29 Jul 1948, NARA College Park, RG 326, E67A, box 45, folder 6 Study of Wider Use of Isotopes, p. 6. The cost figure is from p. 26.

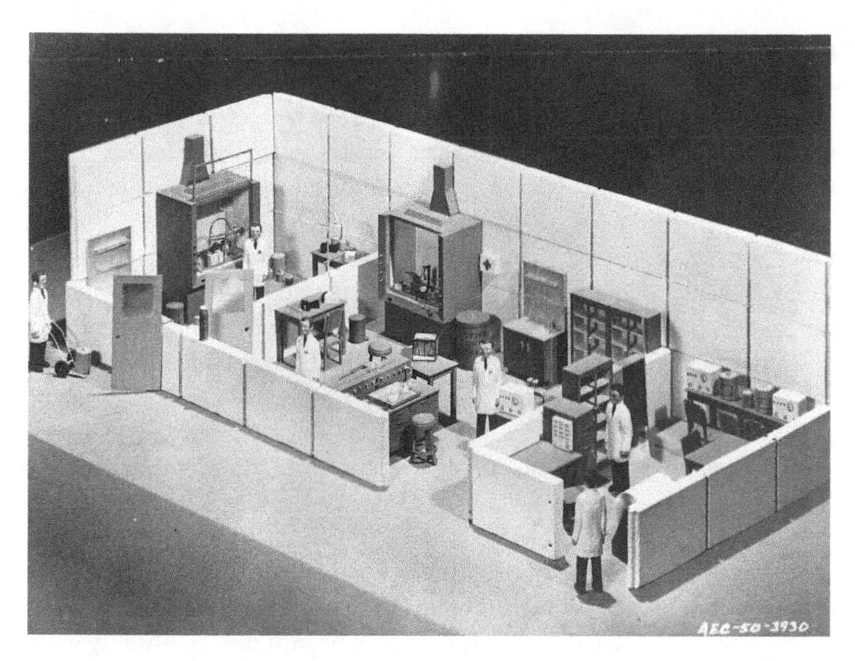

FIGURE 6.5. AEC's model of a three-room laboratory designed to handle both beta- and gamma-emitting radioisotopes, with a high-level room on the left, a low-level room in the center, and a radiation counting room on the right. This arrangement prevents random radiations from the high-level laboratory from registering false "counts" on the counting room instruments and minimizes the possibility of spreading contamination. The large white instrument on the right side of the center room is used by each worker to examine his clothing and hands for contaminating radioactivity before entering the counting room. The hoods include trays to catch any spills and lead bricks to use as shields against gamma radiation. National Archives, RG 326-G, box 2, folder 2, AEC-50-3930.

A little-known but important piece of Congressional legislation provided some impetus for strengthening the AEC's regulatory apparatus. The Administrative Procedure Act, passed in 1946, specified how federal agencies should make and adjudicate their rules, and created a process whereby federal courts could review agency decisions and regulations.[136] In response to this law, the agency's Isotopes Division was asked in 1950 to formulate their procurement instructions for radioisotopes. The resulting guidelines shifted the control of isotope distribution from the contract method to government regulation. The AEC viewed the regulatory

136. Shapiro, "APA" (1986).

framework as providing several distinct advantages, from reducing the number of forms required to encouraging more industrial users of radioisotopes, who might otherwise be deterred from the contract's liability requirements.[137] At the same time, the new rules gave greater leeway to institutions that had already been using radioisotopes, allowing those with experience to obtain broad institutional licenses for their usage.[138] Radioisotopes below a specified amount of radioactivity (in the range of microcuries, though amounts varied depending on the isotope) were considered generally licensed and exempted from the application requirements.[139]

By the time the draft regulations reached the Commissioners for consideration at their January 25, 1951, meeting, they had already been approved by the Division of Research, the Isotopes Division (which was organizationally a part of the Division of Research), the Division of Biology and Medicine, and the Division of Reactor Development. The Commissioners approved them as well. The Regulations for the Distribution of Radioisotopes, Title 10 of the Code of Federal Regulations Part 30, became effective upon their publication in the Federal Register on April 13, 1951.[140] Part 30 was amended early in 1956 to bring it into conformity with the Atomic Energy Act of 1954, and to codify a simpler and less restrictive procedure for radioisotope export.[141]

One aspect of radioisotope use was, however, omitted from these regulations. The Advisory Committee on Isotope Distribution reviewed a draft section on safety standards in March 1950, but could not come to consensus on permissible radiation exposures. That entire section of regulation was referred to a subcommittee for further consideration; a report was not expected for at least one year.[142] By 1953, the Isotopes Division's licensing terms included agreement to "comply with all applicable Govern-

137. AEC 398, Atomic Energy Commission, Regulations for the Distribution of Radioisotopes, 22 Jan 1951, NARA College Park, RG 326, E67A, box 45, folder 1 Regulations for the Distribution of Radioisotopes, p. 9.

138. Bizzell, "Early History" (1966), p. 32.

139. See Schedule B of AEC 398.

140. The regulations were initially published in the Federal Register on February 6, 1951, according to AEC 398, Regulations for the Distribution of Radioisotopes, but they became effective on April 13, according to AEC, *Tenth Semiannual Report* (1951), p. 35.

141. AEC, *Radioisotopes in Science and Industry* (1960), p. 103.

142. AEC 398, Atomic Energy Commission, Regulations for the Distribution of Radioisotopes, 22 Jan 1951, NARA College Park, RG 326, E67A, box 45, folder 1 Regulations for the Distribution of Radioisotopes, p. 5.

ment safety regulations and instructions of the AEC."[143] But the formally codified regulations were still in development.

Not until July 1955 were the AEC's proposed radiation protection standards published for public comment. In February 1957 the final version became law as Title 10 Code of Federal Regulations Part 20. For the research and medical uses of radioisotopes, the Part 20 regulations stipulated requirements for radiological protection of workers and disposal of radioactive waste. Any employee likely to be exposed to more than 25% of the permissible dose was required to wear a monitoring device (usually a dosimetry badge). Records of every employee had to be maintained, and the licensee (usually an institution) was also required to conduct surveys of radiation to detect contamination in the facility.[144]

In conjunction with these new regulations, the Commission considered which part of the agency should enforce them. The Director of the Isotopes Division, Paul Aebersold, argued strenuously against transferring the licensing, regulation, or inspection of isotope users to another AEC division. What had begun as the Advisory Field Service in 1947 had grown by the mid-1950s into a Radiological Safety Branch (RSB). Its staff of eight visited 400–500 institutions a year to inspect licensees for compliance with regulations and instructions, and to evaluate facilities for their adequacy from a radiological health point of view. Violations of safety regulations could result in criminal prosecution, but the RSB usually dealt with "unsatisfactory conditions" through issuing recommendations. As Aebersold bragged, "The RSB even now resolves better than 95% of its visits and problems through education. Nearly all users are willing and anxious to conform both to safe practices and AEC regulations." He argued that cases should be referred to the Inspection Division only "where an obvious violation of law is apparent and criminal prosecution is contemplated."[145] Otherwise, he insisted routine regulatory oversight should remain within his unit.

143. Aebersold, "Philosophy and Policies" (1953), p. 4. By 1950 the form applicants filled out was referred to as AEC-313, "Application for Radioisotope Procurement." This form and the general procedures are described in Bulletin GM-161, Procedure for Securing Isotope Materials and Irradiation Services, effective date 1 Jul 1950. Copy in AEC Records, NARA College Park, RG 326, E67A, box 45, folder 1 Regulations for the Distribution of Radioisotopes.

144. AEC, *Radioisotopes in Science and Industry* (1960), p. 104.

145. Paul C. Aebersold to N. H. Woodruff, 6 Jan 1955, Re: Functions of the Radiological Safety Branch, Isotopes Division, Aebersold papers, box 2, folder 2-1 Gen Corr Jan–June 1955, pp. 4–5.

Aebersold's memoranda illustrate the priority he placed on promoting radioisotopes, and the concern that stricter oversight might inhibit their wider use. As he wrote Nathan Woodruff on October 15, 1955,

> A good balance has been maintained between regulatory controls and activities to encourage and facilitate the wider use of radioisotopes. . . . Although less than 10% of the Division workload consists of education or promotion, such activities have been highly effective in encouraging and assisting in the rapid growth of radioisotope uses.[146]

Undoubtedly the primacy Aebersold placed on increasing radioisotope consumption shaped how the AEC dealt with problematic licensees. Some applications to the Isotopes Division were turned down because of inadequate radiological safety. As one memo illustrated, "In one instance an applicant proposed to receive a weekly radiation dose which, over a period of several weeks, would doubtless have resulted in extremely serious radiation injury and possibly death."[147] Members of the licensing group in the Isotopes Division worked with such applicants to develop a safer research proposal. As the *Eighth Semiannual Report* to Congress described the agency's approach to oversight, "The Commission has a positive program of encouraging and developing health safety standards in the use of the radioisotopes available for distribution."[148]

This mindset also guided Aebersold's view on pricing of radioisotopes, which he felt should be set low enough to promote full utilization, even if government subsidies were needed. Only after sales of a particular isotope were established should prices be set to recover costs.[149] Along similar lines, he argued that the best way to promote private sector involvement in radioisotope production was for the government to foster and stabilize demand to a level at which profits were assured. As he remarked in 1957, "Private enterprise participation in radioisotope production appears imminent. May need further development of radioisotope market to support

146. Paul C. Aebersold to N. H. Woodruff, 13 Oct 1955, Re: Manager's Meeting, Aebersold papers, box 2, folder 2-3 Gen Corr Aug–Dec 1955.

147. S. R. Sapirie to T. H. Johnson, 15 Nov 1954, Re: Licensing and Inspecting Functions of the Isotopes Division Under the Atomic Energy Act of 1954, Aebersold Papers, box 2, folder 2-3 General Correspondence Aug–Dec 1955, p. 6.

148. AEC, *Eighth Semiannual Report* (1950), p. 75.

149. Background material for JCAE Hearings [March 1956], Aebersold papers, box 2, folder 2-4 Gen Corr Jan–Mar 1956, pp. 4–5.

this private activity."[150] Radioisotopes were too important to be left solely to free enterprise.

Aebersold's aspiration to retain radioisotope licensing, regulation, and inspection together—where all could be managed to maximize sales—did not survive the political and organizational changes besetting the agency. In 1957, the Commission decided to separate the regulatory and promotional functions that had been assigned to the Division of Civilian Application.[151] A Division of Licensing and Regulation came into existence, under Harold Price (formerly director of Civilian Applications).[152] Most of the personnel from the Isotopes Extension went into this division to manage licensing for radioisotopes.[153] The division laid out more specific criteria by which applicants for by-product material licenses would be evaluated; these included the training of personnel in the principles and practices of radiological protection.[154]

Aebersold, dubbed "Mr. Isotope," had already moved from Oak Ridge in 1957 to the agency's headquarters in Washington, DC, to encourage greater industrial utilization of radioisotopes, an activity he continued after the reorganization.[155] The AEC established an Isotopes and Radiation Development Advisory Committee, with members from chemistry and industry, with subcommittees on Commercial Activities, Process Radiation,

150. Paul C. Aebersold, Introductory Remarks, NICB Round Table Conference, How to Develop Better Products Through Radioisotopes, 14 Mar 1957, Aebersold papers, box 2, folder 2-10 Gen Corr Mar 1957.

151. On the background to this decision, in which the JCAE played a key role, see Mazuzan and Walker, *Controlling the Atom* (1985), ch. 6 and 7; Rolph, *Nuclear Power and the Public Safety* (1979), pp. 38–43.

152. The reorganizations continued over the years, resulting in a Division of Regulation in 1961, still headed by Harold L. Price, which reacquired the function of inspection in 1964. See AEC Press Release No. G-64, AEC Makes Organizational Changes in Its Regulatory Program, 28 Mar 1964, Records of the Nuclear Regulatory Commission papers, NARA College Park, RG 431, Entry 16, box 11, folder Organization & Management 2, Safety Standards, Division of, 3/28/64–10/12/66.

153. Paul C. Aebersold to R. Maxil Ballinger, 20 Jan 1958, Aebersold papers, box 3, folder 3-1 Gen Corr Jan–Feb 1958.

154. AEC-R 8/5, Policies, Regulations and Procedures Governing the Licensing of Atomic Energy Materials and Facilities, Records of the Nuclear Regulatory Commission papers, NARA College Park, RG 431, Entry 16, box 9, folder PFC 1–1 Radiation Protection, vol. 2. It is possible the six criteria listed on p. 5 of this document were not new, but I have not seen them listed in this way in earlier documents from the Isotopes Division.

155. Barbara Land, "Dr. Paul C. Aebersold: Mister Isotope," *Science World*, 4 May 1960, copy in Aebersold Papers, box 1, folder 1-1 Biographical Materials.

Radioisotope Power (largely thermal applications), and Radiation Systems.[156] The minutes of this group give almost no attention to tracer uses of radioisotopes in science or medical therapy; it was no longer a frontier.[157] In part reflecting the growing production of cobalt-60 and cesium-137 as high-curie radiation sources, the agency was trying to encourage use of irradiation as a means of preserving food and sterilizing medical equipment.[158] The AEC's support of industrial uses of radioisotopes drew political criticism. A *Washington Post* editorial raised questions about "whether private companies are led to lean too heavily upon the Government."[159] For its part, the Office of Isotopes Development itself adopted the language of business, referring to the "AEC radioisotope production, marketing, and sales program."[160]

Conclusions

In 1951, the Radioisotopes Laboratory at Case Institute of Technology in Cleveland (with cooperation and funding from the AEC) convened a conference on "Radioisotopes in Industry." There the AEC Commissioner T. Keith Glennan described the unusual role of the federal government in radioisotope production as intruding into the "private domain of science":

> In a sense, the presence of government in the isotope business is artificial; it is not there because isotopes are secret or need to be controlled for security reasons, but rather because it is virtually the only source of supply. The reason

156. The Advisory Committee on Isotopes and Radiation Development was founded in 1958 when the Advisory Committee on Isotope Production and Distribution was terminated, as the Commission's regulatory and licensing responsibilities were separated from its operational activities. "U.S. Radioisotope Industry–1966" (1967), p. 213.

157. Nuclear Regulatory Commission papers, NARA College Park, RG 431, Entry 16 Regulatory Program Gen Corr Files, 1956–1972, box 12698, folders on Organization and Management 7 Isotopes & Radiation Dev., Adv. Cmtee. [1964–1968].

158. Buchanan, "Atomic Meal" (2005); Zachmann, "Atoms for Peace" (2011).

159. "Nuclear Enterprise" (1958), p. A14. Paul C. Aebersold published a rebuttal on 3 Nov 1958, p. A12.

160. Previous to this reorganization, the agency's activities promoting the use of isotopes in industry had been handled by the Isotopes Development Staff in the Office of Industrial Development. This group was merged with the Isotopes Division. Announcement No. PSMO-120, 18 Mar 1959, Organization and Principal Staff of the Office of Isotopes Development, Aebersold papers, box 1, folder 1-1 Biographical Materials.

> for this is quite unrelated to isotope utilization—the fact that the vast majority of isotopes are produced in nuclear reactors which contain fissionable materials that can also be used in weapons.

He went on to assert that "it is the ultimate objective of the Atomic Energy Commission to work itself out of the isotope business."[161] This statement became a refrain, frequently repeated over the next dozen years, although the US government stayed involved in supplying radioisotopes—and still is to this day. Radiochemical and radiopharmaceutical companies had no reason, even after the 1954 Atomic Energy Act, to build their own reactors. The government's reactor at Oak Ridge had been constructed and paid for as part of the Manhattan Project (and the AEC's other reactors similarly paid for by public funding), and these costs were never passed down to consumers, including secondary distributors. In addition, the production of certain starting materials—such as tritium—were by-products of continued nuclear weapons production and this was reflected in their low cost.

This complex mingling of government and corporate involvement in developing atomic energy went against the clear divide between private enterprise and public sector that Americans imagined to be crucial to their capitalist economy. It must be stressed that atomic energy was far from the only realm in which such ideals did not match with reality—the transportation and communication industries were characterized by state promotion as well as (even through) regulation.[162] Yet the incompatible responsibilities assigned to the AEC with regard to the private sector were especially striking. John Gorham Palfrey, a lawyer who worked for the Commission in its early years, said that in passing the 1954 Atomic Energy Act, "Congress tried to have its cake and eat it too." The law attempted to "secure the greatest possible public benefit from government monopoly, competitive government operation, government regulation, private competition, and government promotion of private activity." He questioned whether "competition, promotion, and regulation . . . will operate effectively side by side," anticipating the inherent conflicts of interest that led to the AEC's demise.[163]

161. Glennan, "Radioisotopes: A New Industry" (1953), pp. 7–8.
162. Horwitz, *Irony of Regulatory Reform* (1989).
163. Palfrey, "Atomic Energy" (1956), p. 390.

For the radioisotope program, the peculiar path along which nuclear technology developed was especially consequential. Purchasers of radioisotopes, either directly from the AEC or indirectly from retail companies, benefited from the enormous public investment into nuclear reactors—and the political circumstances that necessitated evidence of civilian payoffs from a technology developed for nuclear weapons. The combination of government subsidies and promotional activities undoubtedly accelerated the consumption of isotopes in science, medicine, and industry. Just as significantly, the AEC jump-started business in radiolabeled compounds and radiopharmaceuticals, even as the agency championed the role of free enterprise in the emerging nuclear industry.[164]

In the mid-1950s, at the same time that the US government tried to increase private sector involvement in atomic energy, the AEC increased their subsidies on radioisotopes used in research. The 20% discount the agency previously offered on radioisotope shipments used in cancer research, therapy, and diagnosis was extended on July 1, 1955, to domestic shipments for all areas of biomedical and agricultural research, medical therapy, and diagnosis.[165] Other countries with atomic energy installations had policies with similar effects. As Néstor Herran and Xavier Roqué surmise: "Huge investments by the atomic powers in nuclear programmes created a 'false' economy of isotopes which effectively subsidized radioactive materials and made them available at delivery, rather than production, cost."[166] The disruption of commercial patent practices by atomic energy legislation further deflated prices in the economy of isotopes.[167]

It was not only prices that were kept low, but also the priority placed on enforcing radiological protection for consumers of radioisotopes. As Elizabeth Rolph notes, during the early years, "both the Commission and the public seemed comfortable with a rather murky and varying definition of 'safe.'"[168] To the degree that safety meant the lack of acute radiation injury, the AEC could boast of a near-perfect record—as they did in their 1950 *Semiannual Report*. Yet accumulating evidence through the 1950s and 1960s that low-dose radiation exposure could increase risk of leukemia and other cancers put the meaning of safety in a different light, as the

164. Herran and Roqué, "Tracers of Modern Technoscience," p. 129.
165. AEC, *Eighteenth Semiannual Report* (1955), pp. 92–93.
166. Herran and Roqué, "Tracers of Modern Technoscience" (2009), p. 129.
167. Ibid.; Turchetti, "Contentious Business" (2009).
168. Rolph, *Nuclear Power and the Public Safety* (1979), p. 49.

NCRP had already recognized. In general, the Commission followed the lead of the NCRP in radiological protection, in part to avoid setting its own standards. For example, when this advisory group lowered its permissible dose for cumulative radiation exposure in 1958, the AEC updated its regulations for radiation protection accordingly. When the Joint Committee on Atomic Energy conducted hearings on Radiation Safety and Regulation in June 1961, Walter Zinn, a long-time AEC consultant, asserted that the Commission's goal was to "reduce the probability of serious hazard to the public to a low enough value so that the risk [was] comparable to other risks which are found acceptable in our society."[169] This statement reflected the acknowledgement that "safety," a subjective perception, needed to be replaced with the more quantifiable risk.[170] But who would define what risk was acceptable?

Most of the public debate over the safety of atomic energy development revolved around exposure to radiation from atomic weapons fallout and from civilian power reactors. By contrast, radiation exposure involved with using isotopes was largely occupational rather than environmental (though disposal of radioactive waste was also a concern). Yet for a dozen years in the AEC's Isotopes Division, the regulation of users of radioisotope took a backseat to the promotion of their use. This mattered more for hospitals than research laboratories, because the doses used in therapy tended to be much higher than the tracer amounts most scientists used. Moreover, hospital personnel were less likely to be well-versed in the principles behind radiological protection. Paul Aebersold's concern in the late 1950s was that public misunderstanding of radiation hazards—and increased regulation—would slow the sales of radioisotopes.[171] However, by the time that new concerns about low-dose exposure to radioactivity translated into more stringent regulation of users, reliance on radioisotopes had become entrenched in laboratory research and clinical practice. Government regulation then became an enduring—and expanding—aspect of the routine use of radioisotopes in science and medicine.

169. As quoted in Rolph, *Nuclear Power and the Public Safety* (1979), p. 50.

170. Beck, *Risk Society* (1992).

171. As summarized in remarks for a "Compliance Inspection Meeting": "Balance [is] needed between licensing, inspection and promotion—all must keep pace for proper expansion in safe usage. In final analysis everyone in AEC (even lawyers) interested in seeing healthy rapid growth of all civilian uses of AE." 22 Jul 1958, Aebersold papers, box 8, folder 8-35, p. 1.

CHAPTER SEVEN

Pathways

The fact that particular atoms and molecules can be identified by Geiger or scintillation counters has made it possible to add time to the known dimensions of biochemical processes in plants, animals, and man, and it has added a degree of specific understanding of biological processes that would be possible in no other way. — Shields Warren, 1956[1]

As the epigraph suggests, isotopic tracers brought out the temporality of biological systems in new ways. First, the use of radioactive tracers to study biochemical pathways made it possible for researchers to track the appearance (and disappearance) of metabolites over time, as tagged compounds that could be followed as they underwent the sequence of reactions generating a particular biological molecule. These changes through time were represented in schematic diagrams as changes over *space*, most notably in the map of a metabolic pathway but also in the cycling of an element through a body or a landscape. Second, experiments with both stable and radioactive isotopes showed that the turnover of molecules in the mammalian body was much more rapid than anyone had imagined. Isotopes revealed that at the molecular level, life was sustained in the midst of flux, not fixity.

Third, radioisotopes have a built-in time dimension, namely their half-lives. Tracer research could exploit this, so long as the radioisotope selected had sufficient radioactive decay to enable detection of an element as it was chemically transformed, but not so high a rate of decay that it affected the process being studied. In other words, one wanted the half-life and the biological transformation being studied to have commensurate activities and speeds. Han-Jörg Rheinberger describes eloquently the "paradox of the production of traces as such" in a radiolabeling experiment:

1. Davis, Warren, and Cisler, "Some Peaceful Uses" (1956), p. 294.

"A radiogram makes something visible in situ that no longer exists: at the very moment that the trace is produced the marker by decaying irrevocably abolishes itself."[2] However, not all the radioactivity that enters into systems of study is obliterated during an experiment. For radioisotopes with long half-lives, the perpetuation of radioactivity, particularly in the bodies of patients or human subjects (taken up in chapters 8 and 9), and in the environment (covered in chapter 10), brought a new awareness of the lasting consequences of tracer research—and of radioactive contamination generally.

This chapter considers how the initial biological experiments with isotopes put tracer methodology in motion, then focuses on two case studies that exemplify how radioisotopes could illuminate molecular transformations over time. One concerns the most familiar metabolic process in plants, photosynthesis, whereby plants turn carbon dioxide and sunlight into carbohydrates and oxygen. Researchers at the Radiation Laboratory in Berkeley during the 1940s and 1950s used radiocarbon to determine the precise chemical reactions that take place in photosynthesis, beginning with carbon dioxide and generating fructose and sucrose. This celebrated example of the utility of carbon-14 led to a much broader application of this isotope in studies of intermediary metabolism.[3] The elucidation of the steps of photosynthesis was also hailed for its importance to agricultural research. Some commentators predicted that the knowledge gained would enable scientists to harness the chemical process for the artificial production of foodstuffs.[4]

The other case study focuses on researchers of bacterial viruses, who extended the biochemical practice of radiolabeling to problems of molecular structure and heredity. For example, isotopic labels could be used to follow the transfer of material in genetic replication (as compared with the passage of atoms down a metabolic pathway). The 1952 Hershey-Chase experiment took just this tack, employing phosphorus-32 to label the DNA and sulfur-35 to label the protein of a bacterial virus, or bacteriophage.

2. Rheinberger, *Epistemology of the Concrete* (2010), p. 230. For related insights on temporality, Landecker, "Living Differently" (2010).

3. Ashmore, Karnovsky, and Hastings, "Intermediary Metabolism" (1958).

4. E.g., "Photosynthesis and Biosynthesis," in AEC, *Some Applications of Atomic Energy* (1952), pp. 126–33. The idea that understanding photosynthesis would help address nutritional problems and increase food production went back to the nineteenth century: Nickelsen, *Of Light and Darkness* (2009).

Subsequent infection with each type of labeled phage showed that only the phosphorus-32-labeled nucleic acid component of the virus entered the bacterial cell to a significant extent.[5] Up until that point, most biologists assumed that protein, perhaps in conjunction with nucleic acid (as a "nucleoprotein"), was the hereditary material of living organisms, including viruses.[6] Alfred Hershey and Martha Chase's surprising outcome challenged this presumption—their radiolabels traced infectiousness and heredity to the bacteriophage's DNA, reorienting biologists to the importance of nucleic acids.[7] A line of similar experimentation connects this work to the other classic label-transfer experiment in molecular biology. The Meselson-Stahl experiment demonstrated the semiconservative replication of DNA using nitrogen-15, a stable isotope.[8]

The early emphasis on radioisotopes as magic bullets against cancer missed the eventual importance of radiotracers to biology and even medicine. From the outset of the radioisotope program, the AEC differentiated tracer applications of radioisotopes from their use as sources of radiation.[9] The distinction played out at several levels. First, radioisotopes were employed as tracers principally in research contexts. By comparison, most clinical uses were aimed at replacing radiation sources such as x-rays and radium with specific radioisotopes, usually in hopes of better localizing the radiation exposure. The application of radioisotopes in medical diagnostics was an exception to this generality, employing them as tracers. Still, the commonplace distinction between radioisotopes as tracers and

5. Hershey and Chase, "Independent Functions" (1952).

6. Olby, *Path to the Double Helix* (1974); Kay, "Protein Paradigm" in *Molecular Vision of Life* (1993), pp. 104–20; Creager, *Life of a Virus* (2002), ch. 6.

7. Hershey and Chase's experiment was not the first to suggest that genes were composed of DNA (that of Avery, MacLeod, and McCarty came eight years earlier), but it is the generally the one credited with persuading biologists. For various perspectives including citations to earlier work, see Stent, *Molecular Genetics* (1971), p. 315; Judson, *Eighth Day of Creation* (1979), pp. 130–31; Echols, *Operators and Promoters* (2001), pp. 12–13. On how the Hershey-Chase experiment is simplified in popular and pedagogical representation, see Wyatt, "How History Has Blended" (1974).

8. Semiconservative replication, as discussed below, means that each "parent" strand of DNA is the template for a daughter strand, yielding replicated molecules of DNA that are half original material and half newly synthesized material. The AEC's isotope distribution program included heavy isotopes, such as the nitrogen-15 used by Meselson and Stahl, but they were not as much in demand as radioisotopes by biomedical researchers. Meselson and Stahl, "Replication of DNA" (1958).

9. E.g., AEC, *Fourth Semiannual Report* (1948), p. 5.

as radiation sources conveyed a related differentiation between research and medical use. Second, these two classes of application required vastly different amounts of material—in general, the therapeutic dose of an isotope such as phosphorus-32 required more than a thousand-fold greater radioactivity than its use as a tracer.[10] Besides medical therapy, most of the applications of radioisotopes as sources were industrial, such as food irradiation.[11] Some biophysicists and geneticists were interested in using reactor-generated radiation sources, including radioisotopes, to study the biological effects of radiation. Generally, however, radioisotopes failed to supplant older radiation sources, especially ultraviolet radiation and x-rays, that scientists used to induce mutations and cytological alterations.

This categorical distinction between tracers and radiation sources that characterized early postwar work with radioisotopes created blind spots. Researchers using radioisotopes as tracers did not usually reckon with the biological effects of the radiation they put into their systems.[12] In fact, it was the observation that low-level amounts of radiation did *not* disturb fundamental living processes that legitimated the use of radioisotopic tracers as probes.[13] Yet even tracer uses of radioisotopes sometimes exceeded this limit. In the research that led up to the Hershey-Chase experiment, for instance, biologists at Washington University found the radiation effects of their isotopic label could not be ignored. Once some bacterial virus was labeled for a gene transfer experiment, the specific activity of the incorporated phosphorus-32 was high enough that its radioactive decay degraded the phage DNA. This observation led researchers to begin to investigate the intracellular genetic effects of incorporated radioisotopes under the rubric of "suicide experiments" (because the radioactive decay tended to be lethal). In this way phosphorus-32 came to serve as the molecular radiation source for a new kind of radiobiology experiment in which the distribution of the radioisotope in phage particles over time

10. On the thousand-fold difference between these kinds of applications, see Human Radiation Studies: Remembering the Early Years, Oral History of Biochemist Waldo E. Cohn, Ph.D. Conducted January 18, 1995 through the Department of Energy by Thomas Fisher, Jr. and Michael Yuffee, published at http://www.hss.energy.gov/HealthSafety/ohre/roadmap/histories/index.html.

11. AEC, *Twenty-Second Semiannual Report* (1957), p. 34; Buchanan, "Atomic Meal" (2005); Zachmann, "Atoms for Peace" (2011); idem, "Atoms for Food" (2013).

12. This had particularly troublesome consequences in human experiments; see chapter 8.

13. Kamen, *Radioactive Tracers in Biology* (1951), p. 122.

could be registered by the survival curves. In these studies of virus reproduction, phosphorus-32 had an unusual, perhaps even unique, use as both a tracer and a radiation source.

Early Isotopic Tracers

The use of isotopic labels in life science research preceded the atomic age by more than two decades. George de Hevesy performed the first biological experiment with radioisotopes in 1922 when he used thorium B, an old name for the isotope lead-212, to follow its uptake in plant tissues.[14] An array of similar studies followed this example, some conducted by Hevesy but many by other scientists, monitoring the incorporation and movement of heavy radioactive elements—such as bismuth, thorium, and polonium—into animal and plant tissues.[15] Researchers could examine which tissues or organs they localized to, and how rapidly they were excreted. Humans were not exempt. In 1924, Herrmann Blumgart and coworkers at Harvard Medical School injected radium C (bismuth-214) into a clinical subject's arm and determined how long it took for the radioactivity to reach the other arm, as detected in a Wilson cloud chamber.[16]

There were other practical incentives for studying the effects of such materials used by industry, especially radium. Toxicological studies of the distribution of radium in animals and humans dated back to the early twentieth century.[17] As the health hazards of radium became vivid in the tragic suffering and deaths of watch dial painters in the 1920s, knowledge about the localization and biological effects of radioactive substances took on an urgent medical relevance.[18] Yet neither radium nor most of the heavy radioactive elements used in early tracer experiments are usually found in living organisms, so these studies did not shed direct light on physiological processes. In order to use radioisotopes to follow the dynamics of life processes, especially metabolism, scientists needed isotopes of lighter

14. Hevesy, "Absorption and Translocation" (1923); idem, "Historical Progress" (1957).

15. Broda, *Radioactive Isotopes* (1960), p. 1; Fink, *Biological Studies* (1950).

16. See Early, "Use of Diagnostic Radionuclides" (1995), p. 650; Blumgart and Yens, "Study on the Velocity of Blood Flow, I." (1927); Blumgart and Weiss, "Studies on the Velocity of Blood Flow, II." (1927).

17. E.g., Seil, Viol, and Gordon, "Elimination of Soluble Radium Salts" (1915).

18. Clark, *Radium Girls* (1997); Hacker, *Dragon's Tail* (1987), ch. 1.

elements that were the main constituents of living matter—namely carbon, hydrogen, oxygen, nitrogen, phosphorus, and sulfur. These were first available with stable isotopes.

Harold Urey identified deuterium (a heavy isotope of hydrogen) in 1932 and used it to prepare "heavy" water (^2H_2O). He eagerly sought biological applications of this new isotopic material, and as sufficient quantities became available, researchers investigated the effects of heavy water as a medium on various biological processes—respiration of fish, division of eggs, growth of fungi, and germination of plant seeds.[19] As a stable isotope of hydrogen, deuterium does not decay (as tritium, a radioactive isotope of hydrogen, would do), but its greater mass, due to the existence of an extra neutron, allows its presence to be detected in a mass spectrometer. Deuterium could also be used as a constituent of biological molecules to follow living processes from within, not just altering their properties from without.

As it happened, Urey's interest in stable isotopes coincided with keen interest among biochemists in work on intermediary metabolism.[20] These researchers were investigating the myriad chemical transformations involved in synthesizing and degrading the key molecules of living organisms—proteins, fats, carbohydrates, and nucleic acids. In effect, they were opening up the organismal black box of nineteenth-century intake-output physiology.[21] Whereas nineteenth-century chemists had established the identity and structure of many biological compounds (e.g., sugars, amino acids, fatty acids, dicarboxylic acids, and keto acids), their successors focused their efforts on understanding the chains of reactions connecting them in vivo, each step apparently controlled by a specific enzyme.[22] Otto Meyerhof and W. Kiessling published the reaction steps in anaerobic carbohydrate metabolism (glycolysis), soon termed the Embden-Meyerhof pathway, in 1936. The following year, Hans Krebs, building on his earlier elucidation of the urea cycle, announced the steps of the citric acid cycle ("Krebs cycle").[23] His cycles became paradigmatic for biochemists,

19. Kohler, "Rudolph Schoenheimer" (1977), p. 269; Hevesy and Hofer, "Diplogen and Fish" (1934).

20. Holmes, "Manometers, Tissue Slices, and Intermediary Metabolism" (1992), p. 151.

21. Holmes, "Intake-Output Method" (1987).

22. Kohler, "Enzyme Theory" (1973).

23. Krebs, "Cyclic Processes in Living Matter" (1947); Nickelsen and Graßhoff, "Concepts from the Bench" (2009).

functioning as exemplars to researchers working out the intermediate steps of the synthesis and degradation of a wide array of metabolites.[24]

Isotopes were even better suited than chemical manometers for detecting metabolic intermediates, which were produced in small quantities and quickly transformed.[25] Urey's deuterium could be put to use in determining the sequences of metabolic reactions, and a biochemist with the skills to advance this line of research arrived at Columbia in 1934. Rudolph Schoenheimer had worked previously on the biochemistry of fatty acids and their role in atherosclerosis. He rapidly saw how deuterated compounds could be used to investigate steroid metabolism.[26] The Rockefeller Foundation generously funded this project as part of their program for applying tools and methods from the physical sciences to biology.[27]

Schoenheimer and his colleague David Rittenberg fed deuterated linseed oil to mice, and were surprised by how much of the ingested fatty acids was taken up into the body's fat deposits, even when the mice were losing weight. Thus fats, far from being a metabolically inert currency of energy, were in a state of rapid metabolic flux.[28] Schoenheimer went on to show that cholesterol was synthesized in the tissues of mammals and not taken up only from dietary sources. These experiments illustrated the power of isotopic labels for elucidating biochemical pathways, often overturning views based on simple feeding experiments.[29]

Urey went on to concentrate other naturally occurring isotopes, such as oxygen-18, nitrogen-15, and carbon-13. These could be substituted for ordinary atoms in a variety of biological molecules to track compounds through metabolic transformations.[30] In a set of elegantly conceived experiments on protein metabolism, Schoenheimer and Konrad Bloch used nitrogen-15 to show how creatine was synthesized from arginine and gly-

24. Holmes, *Between Biology and Medicine* (1992), p. 77.

25. Holmes, "Manometers, Tissue Slices, and Intermediary Metabolism" (1992), p. 152.

26. Kohler, "Rudolph Schoenheimer" (1977), p. 274.

27. Kohler, *Partners in Science* (1991); Abir-Am, "Discourse of Physical Power" (1982); Kay, *Molecular Vision of Life* (1993).

28. Kohler, "Rudolph Schoenheimer" (1977), pp. 277–79. Schoenheimer's findings reinforced Frederick Gowland Hopkins's vision of biochemistry as the study of the chemical dynamics of life. See Kamminga and Weatherall, "Making of a Biochemist, I." (1996).

29. Kohler, "Rudolph Schoenheimer" (1977).

30. Cohn, "Some Early Tracer Experiments" (1995); idem, "Atomic and Nuclear Probes" (1992); Schoenheimer and Rittenberg, "Application of Isotopes" (1938).

cine.[31] Schoenheimer also found that the rate of protein turnover in the tissues was much higher than anyone had imagined. He surmised that vital compounds were being continuously broken down and regenerated from a metabolic pool, part of what he termed the "dynamic state of body constituents."[32] However, just as this work was taking off, radioactive isotopes began competing with stable isotopes as tracers.[33] After the war, the inauguration of the US government's distribution program tilted the balance decisively in favor of radioactive tracers.

Radiocarbon in Photosynthesis Research

Radioactive carbon, first available through the isotope carbon-11, offered the promise of tagging nearly any biological molecule of interest. In Berkeley, Sam Ruben collaborated with physiologist Israel L. Chaikoff to use carbon-11 to study carbohydrate metabolism in rats. They aimed to use biology rather than chemistry to prepare radiolabeled sugar, by employing plants to convert ^{11}C-labeled carbon dioxide, via photosynthesis, into ^{11}C-labeled glucose. The radiolabeled glucose could then be fed to the rats and the metabolic fate of the carbon-11 label tracked. This complex scheme proved unworkable, but Ruben realized that carbon fixation in photosynthesis itself was just as attractive a research problem for tackling with radiocarbon. Martin Kamen, the radiochemist supplying the starting material, joined Ruben on this effort. They recruited plant biochemist William Zev Hassid to collaborate on the experiments; their work led to the first published paper on the use of radiocarbon as a biological tracer.[34]

Using ^{11}C-labeled carbon dioxide and barley leaves, Ruben, Kamen, and Hassid found that the first product of green plant photosynthesis was the transfer of carbon from CO_2 to a carboxyl group on an unknown

31. For a fuller discussion and citations, see Fruton, *Molecules and Life* (1972), pp. 461–63.

32. Schoenheimer, *Dynamic State* (1942). As Kohler points out, Schoenheimer's work built on Hevesy's findings of "molecular rejuvenation" (as he termed it) of creatine phosphate in muscle tissue and S. C. Madden and George H. Whipple's work on the rapid turnover of plasma proteins. Kohler, "Rudolph Schoenheimer" (1977), pp. 289–90.

33. Kohler, "Rudolph Schoenheimer" (1977), pp. 294–95.

34. Ruben, Hassid, and Kamen, "Radioactive Carbon" (1939); Kamen, *Radiant Science* (1985), pp. 81–86. For a more detailed—and excellent—account, see Nickelsen, *Of Light and Darkness* (2009), ch. 5.

molecule. Surprisingly, the carbon dioxide fixation step occurred even in the absence of light.[35] They set out to conduct the same experiment using the single-cell green alga *Chlorella pyrenoidosa*. Otto Warburg had already established the utility of this species for photosynthesis research; the algae were more convenient than intact plants or leaves, as they could be grown like microbes in aqueous suspensions. In addition to being well suited for quantitative studies, *Chlorella* were more metabolically active than higher plants. As Ruben, Kamen, and Hassid noted, "100 cu. mm. of *Chlorella* cells suspended in 10 cc. of water reduces more $C^{11}O_2$ than a quantity of barley sufficient to fill a 10-liter desiccator."[36] Warburg used *Chlorella* in conjunction with his manometric apparatus, which detected gas exchange in the microliter range.[37] Eventually he was able to use this apparatus to determine the minimum number of photons of light needed to produce a molecule of oxygen. That number, he and his coworkers determined, was four.[38]

By contrast, Kamen, Hassid, and Ruben were interested in using *Chlorella* to determine the chemical intermediates on the photosynthetic pathway from carbon dioxide to sugar. After exposing the *Chlorella* cultures to ^{11}C-labeled carbon dioxide, they isolated several radioactive chemical intermediates and analyzed them as completely as they could using chemical analysis and ultracentrifugation. Given the very short half-life of the label (21 minutes), results were meager.[39] But the negative results held significance. Most researchers thought the initial step of photosynthesis involved the formation of a compound between carbon dioxide and chlorophyll followed by a photochemical reduction step to yield formaldehyde.[40] The formaldehyde could then polymerize into sugar, releasing oxygen as gas.

35. Ruben and Kamen, "Radioactive Carbon in the Study of Respiration" (1940); Kamen, "Cupful of Luck" (1986), p. 6; Zallen, "'Light' Organism for the Job" (1993), p. 71.

36. Ruben, Kamen, and Hassid, "Photosynthesis with Radioactive Carbon, II." (1940), p. 3443.

37. Nickelsen, "Construction of a Scientific Model" (2009).

38. This number fit Warburg's "romantic notion of how photosynthesis was perfect," but it was incorrect. The current number is 9–10 per O_2 produced. Robert Blankenship, pers. comm., 4 Jun 2012; Nickelsen and Govindjee, *Maximum Quantum Yield Controversy* (2011).

39. It did not help that the ultracentrifuge they first used was located at Stanford. Kamen, "Early History" (1963), p. 588. Kamen tried to obtain some carbon-13 for the work with Ruben; see Martin Kamen to Ed [McMillan], 12 Jun 1940, EOL papers, series 1, reel 14, folder 10:10 Kamen, Martin D.

40. Willstätter and Stoll, *Untersuchungen über die Assimilation* (1918); Myers, "Conceptual Developments in Photosynthesis" (1974).

Formaldehyde, however, was conspicuously lacking among the ^{11}C-labeled compounds Ruben, Kamen, and Hassid isolated.[41] In *Chlorella* as in their earlier experiment with barley, labeled carbon dioxide was converted into a carboxyl compound.

In the meantime, Harold Urey was vigorously promoting his stable isotopes as tracers for biological work.[42] In September 1939 E. O. Lawrence, in a spirit of competition, authorized Kamen to use both the 37-inch and 60-inch cyclotrons to search systematically for longer-lived radioactive isotopes of hydrogen, carbon, nitrogen, and oxygen.[43] This effort led Kamen and Ruben in February 1940 to isolate the radioisotope carbon-14, which offered an excellent alternative to carbon-11 with its estimated half-life of 4,000 years. (This was later updated to 5,730 ± 40 years.)[44] Urey wrote Lawrence that fall, admitting that carbon-14 was "very definite competition" for carbon-13. He also inquired how cheaply carbon-14 might be made, as he was advising the Eastman Kodak Company on their plans to produce carbon-13 commercially. Lawrence informed him that cyclotrons would not be able to produce enough carbon-14 for scientific demand.[45] In the end, the wartime mobilization pushed off the immediate prospects of commercialization of either. It also postponed Kamen and Ruben's further work on photosynthesis.[46] Kamen became involved in the Rad Lab's work on uranium separation, and Ruben undertook war work with poisonous gases.[47] Tragedy intervened even more decisively than war. An accident with phosgene took Ruben's life in late September 1943. The following

41. Ruben, Kamen, and Hassid, "Photosynthesis with Radioactive Carbon, II." (1940). On prevailing ideas, see Ruben and Kamen, "Photosynthesis with Radioactive Carbon, IV" (1940), p. 3453.

42. Ruben and Kamen made use of a stable isotope themselves, in their case oxygen-18, in a tracer experiment on photosynthesis. They determined that the oxygen produced in photosynthesis originates from the water rather than from the carbon dioxide. Ruben, Randall, Kamen, and Hyde, "Heavy Oxygen" (1941).

43. Urey's were hydrogen-2 (deuterium), carbon-13, nitrogen-15, and oxygen-18. Kamen, "Cupful of Luck" (1986), p. 9.

44. Godwin, "Half-Life of Radiocarbon" (1962).

45. Harold Urey to E. O. Lawrence, 7 Oct 1940 (quote); Urey to Lawrence, 27 Sep 1940; Lawrence to Urey, 3 Oct 1940, EOL papers, series 1, reel 14, folder 10:10 Kamen, Martin D.

46. Lawrence's Rad Lab was involved in war work many months before Pearl Harbor. Kamen, "Cupful of Luck" (1986), p. 10.

47. Chemical weapons, although not used by the United States in the major battles of World War II, were objects of research and strategic consideration. Moon, "Project SPHINX" (1989); Price, *Chemical Weapons Taboo* (1997).

summer Kamen was fired from the Rad Lab over security issues.[48] He took a job in the shipyards north of Berkeley to make ends meet.

Kamen managed to continue research quietly on the Berkeley campus thanks to biochemist Horace A. ("Nook") Barker, who obtained a small amount of ^{14}C-labeled carbonate from a former graduate student of Ruben's.[49] Kamen would come to Barker's laboratory at night after his shift at the shipyards was over. Because they possessed only microcuries of the carbon-14 and could not obtain more, they sought a problem and organism that would enable them to recover the label and reuse it. They selected *Clostridium thermoaceticum*, an anaerobic microorganism that thrived at high temperatures and metabolized glucose into acetic acid.[50] Chemical analysis of radioactive acetic acid, generated by *C. thermoaceticum* cultures grown in ^{14}C-carbonate, showed the label to be equally represented in the molecule's methyl and carboxyl groups.[51] This supported Barker's idea that glucose fermented into two molecules of acetic acid plus two molecules of carbon dioxide, which could join to form a third molecular of acetic acid.[52] It also showed that with radioactive carbon, researchers could tag not only specific compounds but also specific atoms within compounds. It was thus as sensitive a marker as Urey's carbon-13. Kamen's situation improved beyond these gratifying results. By the time their experiment was published in late 1945, Washington University had hired him to supervise the operation of their cyclotron and develop its uses in medical and biological research.[53]

After the war, those who were aware that carbon-14 would soon be available from nuclear reactors viewed photosynthesis as a puzzle just waiting to be solved. Just two months after the dropping of the atomic bomb over Nagasaki, William Laurence (the Manhattan Project's semi-

48. Kamen, *Radiant Science* (1985), pp. 165–67. For more on the events leading to Kamen's dismissal, see chapter 2.

49. Although my sources do not confirm this, the former graduate student of Ruben's was almost certainly Andrew Benson. (See below.)

50. Kamen, *Radiant Science* (1985), pp. 171–72.

51. Barker and Kamen, "Carbon Dioxide Utilization" (1945). Barker and Kamen soon published two more papers studying fermentation in two other anaerobic fermenters: Barker, Kamen, and Haas, "Carbon Dioxide Utilization" (1945); Barker, Kamen, and Bornstein, "Synthesis of Butyric and Caproic Acids" (1945).

52. Harland Wood and his colleagues took up work on the metabolic pathways of fermentation in these microbes; see Ljungdahl and Wood, "Total Synthesis of Acetate" (1969).

53. Kamen, *Radiant Science* (1985), p. 175.

official reporter) featured photosynthesis in a *New York Times* story about the peacetime applications of atomic energy. As he put it, "With new types of 'tagged atoms' now made available, a new approach can be made toward solving one of the major mysteries of nature, the process whereby plants are able, by the use of the green coloring substance named chlorophyll, to harness the energy of the sun."[54] The use of radiocarbon to investigate photosynthesis had begun in the Rad Lab, and E. O. Lawrence was keen to see it completed there. Late in 1945, he persuaded Berkeley chemist Melvin Calvin to begin studying photosynthesis with the supply of carbon-14 available at Berkeley. The two of them had become acquainted in the Manhattan Project through working on uranium-plutonium fission product extraction.[55]

Calvin invited Ruben's former collaborator, Andrew Benson, to lead the effort in photosynthesis research.[56] Benson, in fact, already possessed the entire supply of barium ^{14}C-labeled-carbonate; Ruben had given it to him before he died.[57] Calvin and Benson soon filed four papers, representing ongoing collaborative work with Berkeley plant biochemists Hassid and Barker, as Manhattan District declassified reports.[58] Because the work was being conducted in conjunction with the Rad Lab, it was soon supported through an Atomic Energy Commission grant to Lawrence.[59] In fact, the research on photosynthesis with carbon-14 quite literally took over the old Rad Lab. As the 37-inch cyclotron was superseded by the 60-inch one and donated to the UCLA Department of Physics, the wooden building that had housed it was released for Calvin's work.[60]

54. Laurence, "Atomic Key" (1945), p. 6. See also Laurence, "Is Atomic Energy the Key to Our Dreams?" (1946), p. 41. On Laurence, one of the most important boosters of atomic energy in the media, see Gordin, *Five Days in August* (2007), p. 109–11.

55. Calvin recalled Lawrence telling him they should do something "useful" after their involvement in the Manhattan Project. Calvin, *Following the Trail of Light* (1992), p. 51.

56. Ibid., p. 53, and Seaborg and Benson, "Melvin Calvin" (1998), p. 9.

57. Benson, "Following the Path of Carbon" (2002), p. 34.

58. In the AEC bibliographies these are reported (undated) as A. Benson and M. Calvin, "Dark Reductions of Photosynthesis," MDDC 1027; S. Aronoff, A. Benson, W. Z. Hassid, and M. Calvin, "Distribution of C^{14} in Photosynthetic Barley Seedlings," MDDC 965, published under nearly the same name in 1947; S. Aronoff, H. A. Barker, and M. Calvin, "Distribution of Labeled Carbon in Sugar from Barley," MDDC 966; and S. Aronoff and M. Calvin, "Phosphorus Turnover and Photosynthesis," MDDC 1589.

59. It was AEC Contract #W-7405-Eng-48. Aronoff, Benson, Hassid, and Calvin, "Distribution of C^{14}" (1947), note 1; Seidel, "Accelerating Science" (1983).

60. Seaborg and Benson, "Melvin Calvin" (1998), p. 9.

After some further work with barley seedlings, the Berkeley effort returned to convenient *Chlorella*, which became their mainstay model organism.[61] Ruben and Kamen had found that the initial reduction of carbon dioxide in photosynthesis took place even in the absence of light, showing that the metabolic pathway leading to carbohydrates could be distinguished from the photochemical step. Following this lead, Benson initially sought to identify the intermediate possessing the carboxyl group from labeled carbon dioxide. It took three years to crystallize the intermediate, succinic acid, and by then Benson realized it was not actually the first intermediate product of carbon dioxide fixation.[62]

Not surprisingly, Calvin became an early purchaser of Oak Ridge carbon-14. Its higher specific activity than carbon-14 from the Berkeley cyclotrons made it feasible for Calvin's group to isolate the earliest, fleeting products of carbon dioxide fixation.[63] Benson devised an ingenious piece of glassware, nicknamed the "lollipop," holding the culture suspension into which ^{14}C-labeled CO_2 could be injected.[64] A glass tube went into the top of the lollipop, enabling air to be bubbled in while the culture was exposed to light. This would allow photosynthesis to occur at an active rate. Then the bubbler would be removed, the remaining air would be flushed out with nitrogen, and immediately a solution of bicarbonate labeled with carbon-14 would be added. The flask would then be sealed and shaken in the light, as the radiolabeled carbon was taken up. At the end of a predetermined period of time, from seconds to minutes, the researcher would then drain the suspension into boiling ethanol to kill the cells.[65] (See figure 7.1.)

The next step was analysis of the radioactive contents of the *Chlorella*. Assuming all of the assimilated carbon dioxide entered the photosynthetic pathway, every metabolic intermediate of the reduction from CO_2 to sugar should be labeled. Limited progress was made in the first few years identifying the labeled intermediates.[66] One complexity was that some of the

61. Aronoff, Barker, and Calvin, "Distribution of Labeled Carbon" (1947); Aronoff, Benson, Hassid, and Calvin, "Distribution of C^{14}" (1947). Initially, Calvin and Benson worked with another algae, *Scenedesmus* D_3.

62. Benson, "Following the Path of Carbon" (2002), 34.

63. Seaborg and Benson, "Melvin Calvin" (1998), on p. 10.

64. Fuller, "Forty Years" (1999), pp. 8–9.

65. Bassham, "Mapping the Carbon Reduction Cycle" (2003), p. 40.

66. Ibid., p. 40; Nickelsen, *Of Light and Darkness* (2009), ch. 5; idem, "Path of Carbon" (2012).

FIGURE 7.1. Picture of "lollipop" apparatus used to grow *Chlorella pyrenoidosa* with ^{14}C-labeled carbon dioxide. Credit: University of California, Lawrence Berkeley National Laboratory.

fixed $^{14}CO_2$ fed into other pathways, so that the researchers could not assume that every labeled compound was a photosynthetic intermediate. Because algae cells that had been "pre-illuminated" for at least ten minutes prior to exposure to carbon dioxide subsequently fixed larger amounts of the gas, Berkeley researchers relied on this technique to maximize the assimilation of labeled carbon into the products of photosynthesis. But the assumption that this would eliminate confounding metabolic reactions proved simplistic.[67]

Another challenge was simply to identify the labeled compounds at all. Initial studies relied on traditional techniques of chemical extraction and analysis, but in 1948 collaborator William A. Stepka of the Department of Plant Nutrition introduced a newer separation method, paper chromatography, which could distinguish the groups of chemically similar radiolabeled compounds.[68] Researchers separated the algae juices using

67. Nickelsen, *Of Light and Darkness* (2009), p. 278; idem, "Path of Carbon" (2012).

68. Benson et al., "C^{14} in Photosynthesis" (1949); Stepka, Benson, and Calvin, "Path of Carbon in Photosynthesis, II" (1948); Nickelsen, "Path of Carbon" (2012). On the role of Stepka, see Benson, "Paving the Path" (2002), p. 11; Calvin, "Intermediates in the

two different eluting fluids in sequence, along two perpendicular sides of the paper. Different chemical compounds migrated in the two-dimensional space as discrete spots. Exposing the paper chromatogram to medical x-ray film enabled the researcher to pinpoint the radioactive compounds, which could then be cut out from the paper for chemical analysis.[69] (See figure 7.2.)

By comparing autoradiograms from short exposures of carbon dioxide to those of longer exposures, a researcher could follow the appearance of radioactivity in new compounds. The appearance of these new spots over time revealed the chemical transformations involving the labeled carbon as it proceeded down various metabolic pathways, including—and especially—that for photosynthesis. As Benson later reflected, being able to analyze everything from the algal extract on one chromatogram was essential to their success in identifying every photosynthetic intermediate.[70]

The group's chromatographic method was not only more effective than traditional extraction techniques but visually impressive. On August 26, 1948, the physicist Freeman Dyson attended a lecture by Calvin on these results, and described in admiring terms the way the resulting pictures "show, in the most direct possible way, the progress of the delicate and transitory reactions through which the radio-carbon passes as it is assimilated." For Dyson, Calvin's research provided dramatic evidence of the advances brought by atomic science:

> The long-sighted people said, when nuclear energy first came on the scene, that the application to biological research would be more important than the application to power. But I doubt if anyone expected that things would actually get going as fast as they have. This blotting-paper-plus-radio-activity technique is completely revolutionary because it means that *any* substance can be fed to a cell and its transformations followed second by second in detail, even in

Photosynthetic Cycle" (1989), p. 404. Paper chromatography had been developed a few years earlier in England: Martin and Synge, "Analytical Chemistry" (1945).

69. Borrowing another new technique, ion-exchange resin, Calvin and Benson identified the first product of carbon dioxide assimilation to be phosphoglyceric acid. Benson performed an elegant double-labeling experiment, using both carbon-14 and phosphorus-32, to demonstrate that the compound to which carbon dioxide was added to make two molecules of phosphoglyceric acid was ribulose-1,5,-diphosphate. This same compound is now referred to as ribulose-1,5-bisphosphate. Benson, "Identification of Ribulose" (1951).

70. Benson, "Paving the Path" (2002), p. 12.

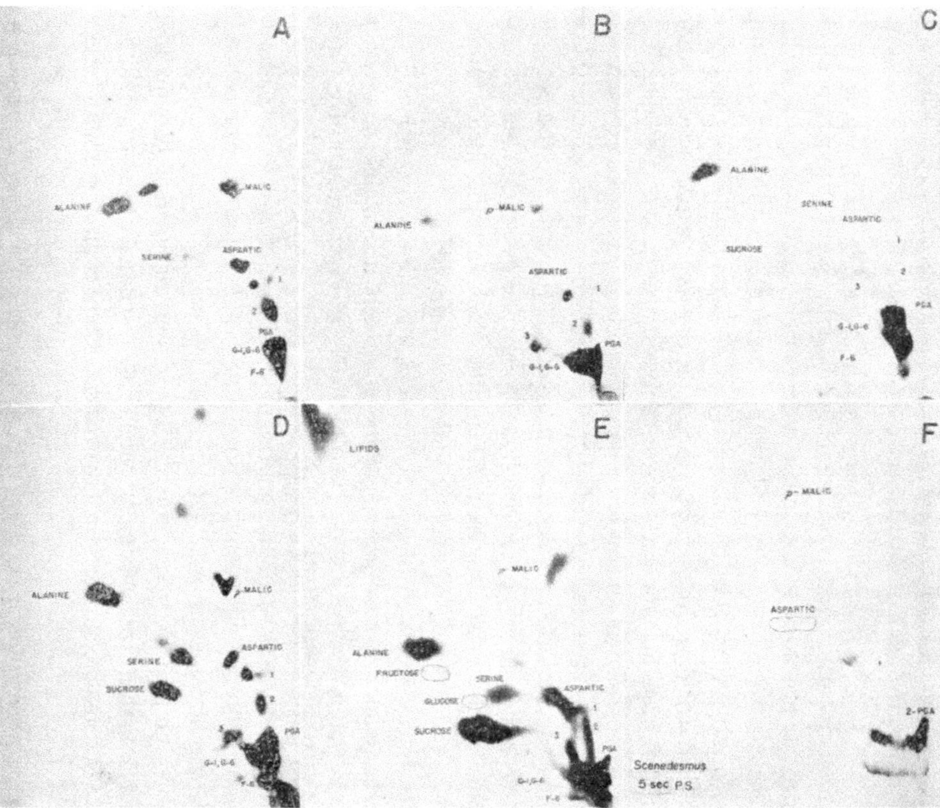

FIGURE 7.2. Chromatograms of extracts from algae indicating uptake of radiocarbon during photosynthesis at several time-points. The sequence B-E shows the appearance of new compounds as the radiolabeled carbon proceeds down metabolic pathways, particularly that for photosynthesis. Dotted circles indicate the positions of fructose and glucose, which are not radioactive, whereas sucrose is. (A) 15-sec. dark fixation by *Chlorella* which had been pre-illuminated for 15 min. This demonstrates that light absorption and carbon dioxide fixation can occur separately. (B) 5-sec. photosynthetic fixation by *Chlorella*. (C) 30-sec. photosynthetic fixation by *Chlorella*. (D) 90-sec. photosynthetic fixation by *Chlorella*. (E) 5-min. photosynthetic fixation by *Chlorella.* (F) 5-sec. photosynthetic fixation by *Scenedesmus*, another species of algae. Reproduced from Melvin Calvin and Andrew A. Benson, "The Path of Carbon in Photosynthesis, IV: The Identity and Sequence of the Intermediates in Sucrose Synthesis," *Science* 109 (1949): 140–42, on p. 141. Reprinted with permission from American Association for the Advancement of Science.

> quantities too small to be seen or weighed, and with substances too unstable to stand old-fashioned stewing and chemical extraction.[71]

Given his background, Dyson's exuberance about how this tool from physics could transform biology might not be surprising. But chemists and life scientists were no less excited by the possibilities.[72] As Glenn Seaborg reported in 1947, "Organic chemists, biochemists, physiologists, and men of medicine have dreamed for years of the day when a radioactive isotope of carbon suitable for tracer investigations should become available."[73]

By 1958 Calvin and his remaining coworkers had elucidated each step in the pathway and composed what became an eponymous schematic diagram of interlocking cycles. (See figure 7.3.) This feat involved not only tracing the pathway of radiocarbon through photosynthesis, but also distinguishing this pathway from others into which the radiolabeled carbon intermediates fed.[74] Most of the pathway was solved by 1954, when Calvin summarily dismissed Benson, for reasons that still remain unclear.[75] The consequences of this severed collaboration were undeniable, however; in 1961 Calvin was the sole recipient of the Nobel Prize in Chemistry, which many from his laboratory feel Benson should have shared.[76]

From the late 1940s, the determination of the photosynthetic pathway stood as a stunning example of how radioisotopes could unlock biochemical puzzles, and the AEC referred to it frequently.[77] Accompanying the scientific tributes were frequent expectations of practical benefits that would materialize from understanding photosynthesis. Seaborg predicted that a complete elucidation of the chemical steps of photosynthesis "might give men the ability to synthesize food and fuel at will, using this principle."[78] According to his logic, humans could use biochemical knowledge to

71. Freeman Dyson, letter to family from Berkeley, 26 Aug 1948, Dyson personal papers.

72. One review begins: "The use of isotopes in the study of intermediary metabolism has become such a powerful tool and is so widely used that it would be impossible in one chapter to do justice to the numerous advances that have been made in this subject in recent years." Ashmore, Karnovsky, and Hastings, "Intermediary Metabolism" (1958).

73. Seaborg, "Artificial Radioactive Tracers" (1947), p. 351.

74. For an elegant analysis of this process and the complexities involved, see Nickelsen, "Path of Carbon" (2012).

75. Benson, "Last Days" (2010).

76. Fuller, "Forty Years" (1999), esp. p. 10.

77. E.g., AEC, *Fourth Semiannual Report* (1948), p. 5.

78. Seaborg, "Artificial Radioactive Tracers" (1947), p. 352.

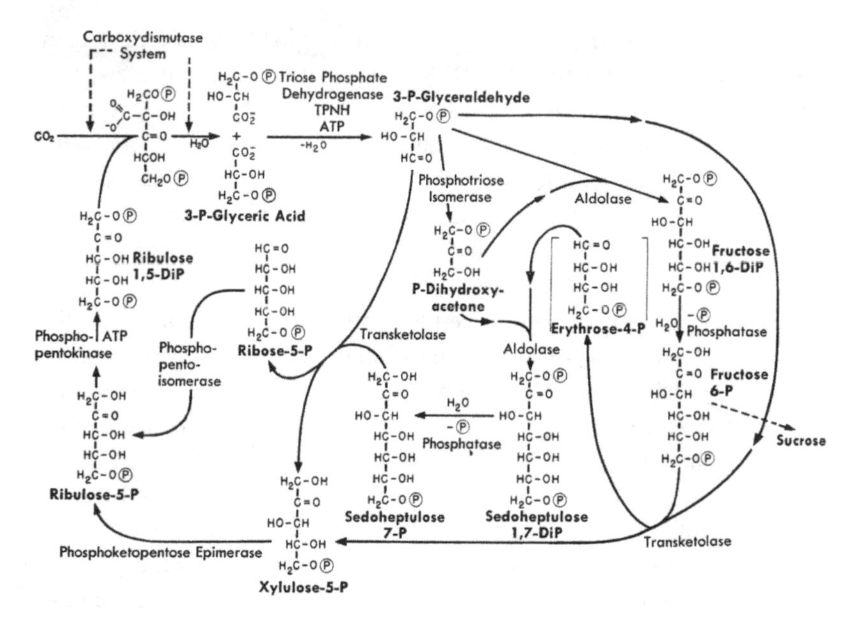

FIGURE 7.3. Schematic diagram of the photosynthetic carbon cycle ("Calvin-Benson cycle"). Reproduced from Melvin Calvin, "Photosynthesis," in *Radiation Biology and Medicine: Selected Reviews in the Life Sciences*, ed. Walter D. Claus (Reading, MA: Addison-Wesley, 1958), pp. 826–48, on p. 836. © 1958. Reprinted by permission of Pearson Education Inc., Upper Saddle River, NJ.

harness the sun's energy directly. Popular accounts similarly represented Calvin and Benson's work as ushering in a new era of agricultural productivity. The *Christian Century* hailed the "atomic scientists" for solving the age-old problem of hunger: "recent experiments promise a vast increase in the world's food supply within the next year or so!"[79] However, solving the puzzle of photosynthesis did more to illustrate the power of carbon-14 in research than to revolutionize agriculture.

As a nearly universal label for biological molecules, carbon-14 was especially crucial to postwar biochemistry. Carbon-14 had an uncommonly long shelf-life, and it could be detected readily. Given material of sufficient specific activity, carbon-14 could be diluted one-billion-fold and still be detected through its radioactive decay.[80] The AEC regularly cited its sales of carbon-14 to illustrate the immense savings made possible by using a

79. "Atomic Research May End World's Hunger" (1948), p. 749.

80. See Kamen, "Early History" (1963); Libby, "Radiocarbon Story" (1948).

nuclear reactor to produce isotopes. In one report to Congress, the agency estimated that whereas the Oak Ridge pile could manufacture 200 millicuries of carbon-14 in a few weeks, at a cost of about $10,000, "it would take one thousand cyclotrons to equal this output, and the operating cost would be well over a hundred million dollars."[81] As one historian of biochemistry has put it, the growing reliance on isotopic tracers, above all carbon-14, resulted in "so elaborate a proliferation of metabolic pathways as to boggle the minds of students of biochemistry."[82]

Over time, two trends reinforced the utilization of carbon-14 in biochemistry. The first was the growing availability of radiolabeled compounds for purchase. Calvin's group synthesized ^{14}C-labeled compounds for the AEC Isotopes Division to sell before companies took over this role.[83] This meant that even biochemists who lacked the synthetic chemistry expertise to label the compounds they wanted could take advantage of the new tool. The second had to do with the changing array of detection instruments available to researchers. Carbon-14 decay was difficult to detect using Geiger counter technology; its beta emissions were of such low energy they did not penetrate the walls of counting tubes. (This problem also affected tritium, which could not be detected at all in such tubes.) An alternative radiation detection technology, scintillation counting, was being developed commercially in the postwar decade.[84] Scintillation counters were well suited for measuring carbon-14 and tritium, which opened up these labels for broader use.

These two trends were mutually reinforcing.[85] The wide array of radiolabeled compounds made them attractive to use in enzyme assays and other biochemical tests. The counting instrument, particularly Packard's Tri-Carb liquid scintillation counter that became available in 1953, enabled researchers to measure these labeled compounds easily (or, in the case of many enzyme assays, the resulting labeled product). In fact, scintillation counters were furnished with automatic sample counters that would let researchers load in hundreds of tubes, often to be counted overnight.

81. AEC, *Fourth Semiannual Report* (1948), p. 10. Scientists took advantage of the affordability of reactor-produced carbon-14, which was the third most commonly purchased isotope from Oak Ridge after phosphorus-32 and iodine-131, both of which were used in therapy. This was true for both domestic and foreign sales. AEC, *Isotopes* (1949), pp. 53–54.

82. Fruton, *Molecules and Life* (1972), p. 446.

83. See chapter 6.

84. Rheinberger, "Putting Isotopes to Work" (2001).

85. Ibid., p. 163.

Phosphorus-32 in Gene Transfer Experiments

Like the photosynthesis work described above, the use of phosphorus-32 in studies of phage reproduction involved researcher Martin Kamen, after he moved to Washington University in St. Louis. According to Kamen, soon after arriving in 1945 he was approached by the new chair of the Chemistry Department, Joseph W. Kennedy, asking if he was interested in understanding how viruses multiply.[86] Kamen replied drily that "a free trip, lodging, and a state dinner in Stockholm awaited the successful solver of this problem."[87] Kennedy proposed that Kamen conduct an experiment in which he would label tobacco mosaic virus with phosphorus-32, infect plants with the labeled virus, and see if the label was carried to the progeny of the infecting virus. If not, Kennedy inferred, that would show the parent was just a template for virus replication. If the label was evenly distributed in the progeny, by contrast, that would suggest the parent virus was fragmented into pieces that were subsequently incorporated into newly synthesized viruses. Kamen suggested that there was a better system than tobacco mosaic virus for this problem, at least in terms of quantifying infective units. That was bacteriophage, and Washington University possessed one of the bacterial virus's experts, Alfred D. Hershey, an Associate Professor of Bacteriology and Immunology.

Howard Gest, Kamen's graduate student, became involved in this joint effort. Gest happened to possess a background in phage research, because this subject interested him as a college student at UCLA in 1940. He worked as a research assistant to Max Delbrück and Salvador Luria during the summers at Cold Spring Harbor in 1941 and 1942 and began doctoral research with Delbrück at Vanderbilt in the fall of 1942. The war effort interrupted Gest's studies. After the war, he resumed graduate work at Washington University with Kamen, who was continuing research on photosynthesis.[88] Gest studied the uptake of radiolabeled inorganic

86. Kennedy served on the MED's and then AEC's Advisory Committee on Isotope Distribution, and so was well aware of the potential uses of radioisotopes. "Committee on Isotope Distribution: Report by the Manager of the Office of Oak Ridge Directed Operations in Collaboration with the Directors of the Division of Research and the Division of Biology and Medicine," Dec 1947, AEC General Secretary Records, RG 326, E67A, box 25, folder 7 Isotope Distribution, Committee on.

87. Kamen, "Cupful of Luck" (1986), p. 13.

88. Gest, "Photosynthesis and Phage" (2002), p. 333; Ruben, "Photosynthesis and Phosphorylation" (1943). Gest had access to phosphorus-32 because Kamen was responsible for

phosphate (P_i) in three species of phosphosynthetic bacteria and algae, and found that all three organisms took up more P_i when illuminated.[89] Gest conjectured that this inorganic phosphate was converted to low-molecular-weight organic phosphoryl compounds, which in turn were precursors of "energy-rich" phosphoryl compounds such as ATP. Unfortunately for Gest, experiments by Calvin did not confirm their findings, although Calvin's results were later disproven.

Shifting his attention from photosynthesis to phage, Gest was entering another competitive field for radiotracer work. At the University of Chicago, Frank W. Putnam and Lloyd M. Kozloff labeled T6 bacteriophage by growing them in the presence of phosphorus-32. When they subsequently infected unlabeled *Escherichia coli* with this "hot" phage, they could track the movement of the phosphorus. What they found was that nearly 70% of the phosphorus in progeny phage came from the medium, presumably through a bacterial pathway.[90] However, their experiments did not settle what happened to the atoms of an individual virus during replication. Paraphrasing Ole Maaløe and James D. Watson on this issue, the biochemical problem of reproduction could be seen in the fact that *atoms* do not reproduce but *genes* do. Where do the atoms that form new genes come from?[91] For a generation of biologists, the problem of virus reproduction seemed to hold the key to understanding the nature of the gene, and radioisotopes offered a tantalizing molecular flashlight for examining the process.

Building on his familiarity with phosphorylated compounds and their metabolism, Gest—in collaboration with Hershey, Kennedy, and Kamen—designed an experiment to "trace the fate of radioactive phosphorus in a single phage particle during its multiplication in a single *E. coli* cell."[92] Would the radiophosphorus be transferred from parental phage to its progeny or would it remain in the original phage particle? The re-

---

preparing batches of this radioisotope in the Washington University cyclotron for "use by Institute of Radiology clinicians in treating certain blood diseases." Gest, "Photosynthesis and Phage" (2002), p. 334.

89. Gest began using cyclotron-produced phosphorus-32 as a tracer to test Sam Ruben's hypothesis that light was converted to chemical energy in photosynthesis by means of "high energy phosphate compounds." Gest, "Photosynthesis and Phage" (2002), pp. 333–34.

90. Putnam and Kozloff, "Origin of Virus Phosphorus" (1948); Kozloff and Putnam, "Biochemical Studies" (1950); Putnam and Kozloff, "Biochemical Studies" (1950).

91. Maaløe and Watson, "Transfer of Radioactive Phosphorus" (1951), p. 507.

92. Gest, "Photosynthesis and Phage" (2002), p. 335.

searchers obtained phosphorus-32 of high specific radioactivity from Oak Ridge to label a strain of T2 phage that Hershey worked with. The ^{32}P-labeled phage was highly labeled, and it turned out that Hershey's teaching duties postponed the infection experiment by a few weeks. This delay proved significant. Re-assaying the radioactivity and phage titer before beginning the experiment, Gest and Hershey were dismayed to find that the phage titer had decreased significantly. Another test a few weeks later showed a further decline in phage titer. Gest recollects: "Finally, it dawned on us that a certain number of ^{32}P disintegrations *within* a phage particle leads to biological inactivation. We had accidentally discovered the phenomenon of phage 'suicide' caused by ^{32}P β-decay."[93]

The goal of the collaboration then shifted from following the dynamics of phosphorus transfer during infection to studying the survival rate of ^{32}P-labeled phage. As Hershey, Kamen, Gest, and Kennedy (hereafter "Hershey et al.") reported in their paper, their assays gave two kinds of information: the rate of inactivation indicated the specific activity of the radiophosphorus in phage, and the survivor curve shed light on how the radioactivity was distributed within the phage population.[94] The survival of T-phages after exposure to ultraviolet radiation had been studied intensively throughout the 1940s, as part of a broader literature on virus inactivation with x-rays, γ rays, α rays, electrons, neutrons, and deuterons. Not all these radiation agents acted directly on the virus. Studies of x-ray inactivation of a variety of viruses (first papilloma virus, then phage, then plant viruses) showed the effects of this agent to be indirect, resulting from the products of ionizing radiation.[95] Indirect effects were distinguished by direct effects on the basis of whether alterations of the medium could be made to protect the phage—as Salvador Luria put it, "a *direct effect* of ionizing radiations is defined as a 'nonprotectable' effect."[96] Researchers ascertained that one radiation "hit" inactivated a virus particle, and this provided the basis for much of the research associated with target theory.[97]

93. Ibid., p. 335. Emphasis in original.

94. Hershey et al., "Mortality of Bacteriophage" (1951), p. 305. Kennedy was a coauthor of the publication.

95. See Luria, "Radiation and Viruses" (1955). As Luria makes clear in a footnote, the review was written in 1951 and not updated before publication.

96. Ibid., p. 337. Emphasis in original.

97. Summers, "Concept Migration" (1995).

Hershey et al. argued that their results could be best understood as compatible with the assumption that the inactivation of a phage particle was "the consequence of the disintegration of a single atom (not necessarily the first) of its assimilated P^{32}." From their extensive study of the radiosensitivity of ^{32}P-labeled T4 phages, the authors determined that "a phage particle dies . . . after an average of about 11.6 disintegrations."[98] The low efficiency of inactivation suggested that the phage were killed as a direct result of the nuclear reaction. But this did not resolve the exact cause of death, which could be attributable to any one of several effects of phosphorus-32 decay: the release of energy at decay, the absorption by the nucleus of the 30 electron volts released, other energy dissipation associated with the rearranged electrons, or the fact that a sulfur atom is left in the place of the phosphorus.[99]

In their 1951 paper, Hershey et al. inferred from the localization in phage particles of radiophosphorus that "the vital structures contain nucleic acid."[100] Hershey embarked upon further investigation of bacteriophage's "vital structures" using radioisotopic tracers in 1950, when he took up a new post in the Department of Genetics at Cold Spring Harbor. Hershey wrote in his 1950–1951 Carnegie Institution Research Report, "If . . . labeled atoms were transferred in the form of special hereditary material, the progeny of a first cycle of growth from radioactive seed would contain radioactive atoms principally in this special material. During a second cycle of growth, therefore, radioactivity should be more efficiently conserved."[101]

In fact, a similar experiment was already being done in Copenhagen. In 1951, Ole Maaløe and James Watson published the results of an experiment designed to follow ^{32}P-labeled phage through two generations. As mentioned, Putnam and Kozloff had already demonstrated that 30% of the isotopic label was transferred from parent to progeny phage, but the localization and distribution of this label in the progeny was not clear. They suggested that the phage might be composed of both genetic and nongenetic components, each of which was labeled with phosphorus-32. The portion that was not transferred to the next generation would then be the nongenetic portion of the virus. At the 1950 summer phage meeting

98. Hershey et al., "Mortality of Bacteriophage" (1951), pp. 308 and 315.

99. Ibid., p. 316.

100. Ibid., p. 317.

101. Hershey et al., "Growth and Inheritance" (1951), p. 175.

at Cold Spring Harbor, Seymour Cohen noted that this hypothesis could be tested by taking the labeled phage through a subsequent generation; Maaløe and Watson referred to this idea as the "second generation experiment," and it formed the inspiration for their experiment.[102]

What Maaløe and Watson found was that 30% of the isotopic label was transferred from the progeny of the first infection to the second generation. This meant that the phosphorus was similarly localized in both the parents and the progeny. Since the original radioactive phage particles were presumably uniformly labeled, their progeny must also be uniformly labeled. This ruled out Kozloff and Putnam's suggestion that some of the phosphorus-32 might be labeling a genetic portion of the virus, and some a nongenetic part, such that the 30% represented the label on the genetic portion that was transferred. But, as Maaløe and Watson qualified, their experiment addressed only the fate of the phosphorus atoms. "A different answer might be obtained with a label like sulphur, that would label specifically the protein moiety of the phage."[103]

This is exactly the experiment Hershey and Martha Chase performed, resulting in their renowned paper, "Independent Functions of Viral Protein and Nucleic Acid in Growth of Bacteriophage."[104] Chase, Hershey's able technician, undertook the labeling experiments.[105] Hershey's own preliminary labeling experiments had not suggested a special role for nucleic acid.[106] But the reinvestigation of the problem with Chase also brought a new technique into play: the use of a kitchen blender. Blending

102. Maaløe and Watson, "Transfer of Radioactive Phosphorus" (1951), p. 508.

103. Ibid., p. 508.

104. Hershey and Chase, "Independent Functions" (1952).

105. Chase is one of those figures in the history of molecular biology who is virtually unknown except for the eponymous experiment—otherwise, she is one of the field's "invisible technicians," to use Steven Shapin's phrase in "The Invisible Technician" (1989). Hershey did value her contributions highly, telling Bruce Wallace that only Chase "had the concentration needed to carry out the protocol that led to the Hershey-Chase experiment." Stahl, *We Can Sleep Later* (2000), p. 99.

106. Hershey's early experiments using sulfur-35 to label parental phage protein showed that about a third of the label from either ^{35}S-labeled parental protein ended up in phage progeny, about the same amount as ^{32}P-labeled parental DNA that was transferred. In addition, this 35% transfer rate was seen in both the first and second cycles of growth. According to Hershey, "This means that neither phosphorus nor sulfur is transferred from parent to progeny in the form of special hereditary parts of the phage particles." The experiment with Chase overturned this interpretation. Quote from Hershey, Roesel, Chase, and Forman, "Growth and Inheritance" (1951), p. 198.

the infected cultures disrupted the attachment of phage to the outside of bacterial cells, so that intracellular virus particles and extracellular virus particles could be physically separated. This proved decisive, revealing a dramatic difference between the transfer of labeled phage protein and that of phage nucleic acid. Eighty percent of the ^{35}S-labeled phage protein remained outside the bacterial cells (and so was agitated off by the blender and recovered in the supernatant), as compared with only 30% of the ^{32}P-labeled DNA. The remaining 70% of labeled viral DNA was within the cells, having entered the bacteria shortly after phage adsorption.[107] In other words, very little if any of the ^{35}S-labeled phage protein entered the bacterial cell, unlike the ^{32}P-labeled DNA, which seemed the active agent of infection—and phage heredity. (The failure to recover all of the sulfur-35 in the supernatant could be attributed to phage that remained attached to the cells despite the operation of the blender.) (See figure 7.4.)

Based on the amount of each label that entered the bacterial cell in their experiments with phosphorus-32 and sulfur-35, Hershey and Chase suggested that the viral DNA was the active agent of phage reproduction.[108] Hershey himself was surprised by this outcome.[109] Its implications were that the bacteriophage should no longer be regarded as an indivisible unit—and should not be called a nucleoprotein, as had been conventional among virus researchers for more than a decade.[110] The protein and nucleic acid had distinct biological roles. Hershey referred to the protein as the "membrane" that surrounds, carries, and delivers the phage's genetic material, which was solely nucleic acid.

Hershey's ideas along this line were inspired by Thomas Anderson's electron micrographs of phage particles attached by their tails to bacterial cells, as well as by Roger Herriott's finding that osmotic shock could release phage DNA into solution leaving "ghosts," or protein shells, behind.[111] Hershey inferred that the phage protein was not hereditary material, but its vehicle, adsorbing to the bacteria and functioning "as an

107. On the experiment with Chase as reinvestigating the issue, see Hershey, "Intracellular Phases" (1953), p. 102.

108. As Hershey and Chase understatedly put it, "We infer that sulfur-containing protein has no function in phage multiplication, and that DNA has some function." Hershey and Chase, "Independent Functions" (1952), p. 54.

109. On Hershey's expectations, see Szybalski, "In Memoriam" (2000), p. 19. On reasons for lack of interest in phage DNA, see Hershey, "Injection of DNA into Cells" (1966).

110. Hershey, "Intracellular Phases" (1953), p. 99.

111. Anderson, "Techniques for the Preservation" (1951); Herriott, "Nucleic-Acid-Free T2 Virus 'Ghosts'" (1951).

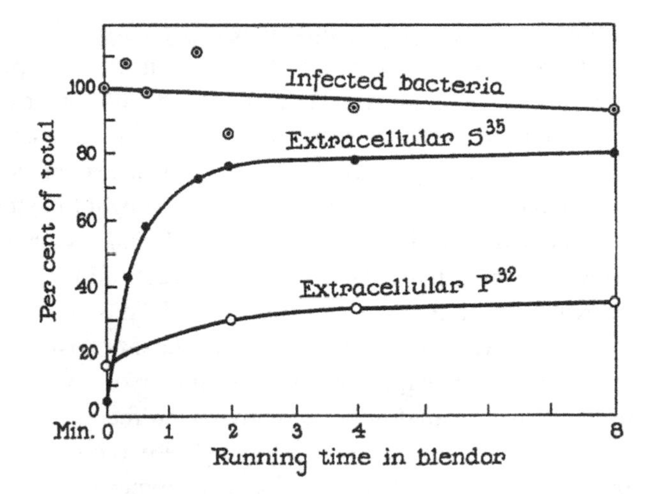

FIGURE 7.4. Schematic diagram showing removal of sulfur-35 and phosphorus-32 from bacteria infected with radioactive phage, and survival of the infected bacteria, during agitation in a Waring blender. © Rockefeller University Press. Originally published in A. D. Hershey and M. Chase, "Independent Functions of Viral Protein and Nucleic Acid in Growth of Bacteriophage," *Journal of General Physiology* 36 (1952): 39–56, p. 47.

instrument for the injection of the phage DNA into the cell."[112] Less could be said about the fate of the phage DNA. "Parental DNA components are, and parental membrane [protein] components are not, materially conserved during reproduction. Whether this result has any fundamental significance is not yet clear."[113]

Irradiating DNA with Phosphorus-32

In the mid-1950s, Hershey continued to pursue the question of material transfer from parent to progeny through use of phosphorus-32 as a tracer.[114] Others in the phage group turned to exploiting the isotope's radiobiological potential.[115] This style of experiment seems to have been

112. Hershey and Chase, "Independent Functions" (1952), p. 56.

113. Hershey, "Intracellular Phases" (1953), pp. 110–11.

114. See Hershey, "Conservation of Nucleic Acids" (1954); Hershey and Burgi, "Genetic Significance" (1956).

115. The correspondence cited here attests to the vitality of an in-group of phage researchers closely affiliated with Delbrück, Luria, and Hershey, the three who are conventionally credited with paternity rights for the "phage group." At the same time, these radiolabeling

especially attractive to those phage workers who came from the physical sciences (e.g., Gunther Stent, Cyrus Levinthal, Seymour Benzer), perhaps because it continued the line of research associated with target theory, itself an application of physics to genetics.[116] In particular, researchers emulated the state-of-the-art phage experiments with ultraviolet radiation using incorporated phosphorus-32. In doing so, they sought to determine whether key genetic discoveries with UV-irradiated phage, such as multiplicity reactivation, cross-reactivation, and photoreactivation, could also be observed with radiation from phosphorus-32.[117] These experiments were aimed, in other words, at using biophysical tools to answer fundamental questions about the nature of the gene. After World War II, most of these investigations related either directly or indirectly to concerns about the genetic hazards of radiation, a topic of research supported by national governments as part of efforts to develop atomic energy.[118]

In 1952, geneticist Guido Pontecorvo observed that one could define the gene as a unit of recombination, a unit of mutation, or a unit of physiological activity.[119] Each was valid but in certain cases discrepancies arose, and the inconsistencies were most pronounced at the level in which genetics and biochemistry intersected. Many of the biophysically minded phage

experiments expose a wider circle of participants, including those either not part of the clique or more marginal, such as Seymour Cohen and Lloyd Kozloff.

116. Sloan and Fogel, *Creating a Physical Biology* (2011).

117. *Multiplicity reactivation* refers to Salvador Luria's observation that two or more UV-inactivated phage particles, if they infect the same bacterium, can cooperate or combine to produce viable progeny. *Cross-reactivation* is also called marker rescue; it occurs in mixed infection when a genetic marker of an inactive irradiated bacteriophage appears in the progeny when crossed with active phage. Renato Dulbecco discovered *photoreactivation* in 1950 when he observed that UV-inactivated phage could be reactivated through illumination by a visible light source. Luria, "Reactivation of Irradiated Bacteriophage" (1947); Dulbecco, "Experiments on Photoreactivation" (1950); Luria, "Reactivation of Ultraviolet-Irradiated Bacteriophage" (1952); and, for general explication, Stent, *Molecular Biology* (1963), especially pp. 282–91.

118. For examples see Beatty, "Genetics in the Atomic Age" (1991); Lindee, *Suffering Made Real* (1994); de Chadarevian, "Mice and the Reactor" (2006); Rader, "Hollaender's Postwar Vision" (2006); de Chadarevian, "Mutations in the Nuclear Age" (2010). On the relationship between biophysics and molecular biology: Rasmussen, "Mid-Century Biophysics Bubble" (1997); idem, *Picture Control* (1997); de Chadarevian, *Designs for Life* (2002); Strasser, *La fabrique d'une nouvelle science* (2006).

119. Pontecorvo, "Genetic Formulation of Gene Structure" (1952). His brother, Bruno Pontecorvo, was a nuclear physicist and suspected spy who defected to the Soviet Union in 1950. Turchetti, *Pontecorvo Affair* (2012).

researchers sought to apprehend genes as physical entities, precisely in this realm of ambiguity.[120] One experiment that figured prominently in this arena of phage radiobiology was that of Luria and Raymond Latarjet, in which bacteria already infected with phage were exposed to various doses of radiation, to assess the radiosensitivity of phage that was already in the process of reproduction.[121] They found that the sensitivity of T2 phage to ultraviolet radiation decreased significantly in the early infection period, then later became multiple-target in character, and finally showed an increase again in ultraviolet sensitivity. Seymour Benzer repeated this experiment with T7 phage, since unlike T2, it did not show genetic recombination between inactivated phage particles (so-called multiplicity reactivation). The results were more straightforward than Luria and Latarjet's: there appeared "simply an increase with time of the average number of targets per cell, each target being similar radiologically to a T7 particle."[122] Yet even when results accorded with target theory, as in this case, it proved difficult to pinpoint the character of the gene using the tools of radiation biology.

Gunther Stent first began to experiment with ^{32}P-labeled phage while on a postdoctoral fellowship in Copenhagen with Herman Kalckar. After leaving Nazi Germany in 1940, Stent completed a PhD in physical chemistry at the University of Illinois before becoming interested in phage research. His project in Copenhagen was aimed at investigating "by means of radioactive tracers the kinetics of the processes by which the virus-infected host cell assimilates phosphorus and incorporates it into bacteriophage material."[123] Stent collaborated with Maaløe, following up Watson and Maaløe's work on second-generation phage label transfer experiments. They used phosphorus-32 to show that there existed phosphorus-containing "phage-like structures" before the release of infectious viruses,

120. Seymour Benzer exemplified this trend and cited Pontecorvo's paper; see Holmes, *Reconceiving the Gene* (2006).

121. Luria and Latarjet, "Ultraviolet Irradiation of Bacteriophage" (1947). To use the language of phage biology, the Luria-Latarjet experiment examined the radiosensitivity of *vegetative* phage "at various stages of the latent period." Stent, *Molecular Biology* (1963), p. 300. The period after phage infects bacteria, when infective particles cannot be recovered, is called the "dark" or eclipse period.

122. Benzer, "Resistance to Ultraviolet Light" (1952), p. 69.

123. Gunther Siegmund Stent, "Fellowship Summary, August 1950–July 1951, Radioactive Phosphorus Tracer Studies on the Reproduction of T4 Bacteriophage," submitted with letter to Charles E. Richards, National Research Council, 16 Aug 1951, Stent papers, box 1, folder American Cancer Society.

which were identified through sedimentation in the centrifuge, adsorption onto sensitive bacteria, and precipitation with antiphage serum.[124]

In one respect, this was a new approach to an old problem: Delbrück and Luria had attempted in their initial collaborative experiments to use superinfection with more than one kind of bacteriophage to capture and analyze viral replication intermediates. They reasoned that if they could infect a suitable host with two different phages, one might lyse the bacteria while the other was in the process of reproducing, revealing an "intermediate stage of virus growth" usually hidden within the cell.[125] Instead, as Hershey put it a few years later, "this experiment led into a number of still half-explored byways, and eventually to the discovery of genetic recombination of viruses."[126] Their joint work along this line also revealed the phenomenon of interference—that infection with one bacteriophage could prevent altogether infection by the second.[127] However, the experiments did not make visible the mode of reproduction of bacteriophage.

Radiolabeling seemed to offer another, more promising way to visualize the intermediate stages of virus replication. In the fall of 1952, Stent continued this use of radiolabels to investigate intracellular phage development at Berkeley, where he joined Wendell Stanley's Virus Laboratory. As Stent put it in a summary for the Microbial Genetics Bulletin: "I am engaged in a study of the replication of the nucleic acids of bacteriophages by means of radioactive tracers, as well as in searching for effects of the transmutation of radiophosphorus on the genetic character of bacteriophages into which it has been incorporated."[128]

124. Maaløe and Stent, "Radioactive Phosphorus Tracer Studies" (1952).

125. Delbrück and Luria, "Interference between Bacterial Viruses" (1942), p. 111.

126. Hershey, "Reproduction of Bacteriophage" (1952), p. 125.

127. Specifically, infection with one bacteriophage (γ, later called T2) prevented the bacteria from producing another (α, later called T1), a phenomenon termed "interference." Subsequent studies enabled them to differentiate "mutual exclusion" (in which a single bacterium would only produce one type of virus at a time) from the "depressor effect," which referred to the observation that a bacterium infected with more than one phage produced less of the prevailing phage than it would otherwise. While interesting, these findings did not advance the understanding of virus reproduction, to the evident frustration of Delbrück. As he asserted in 1946: "Remember that what we are out to study is the multiplication process proper, we want to get to the bottom of what goes on when more virus particles are produced upon the introduction of one virus particle into a bacterial cell. All our work has circled around this central problem." Delbrück, "Experiments with Bacterial Viruses" (1946), p. 162.

128. Gunther S. Stent to Evelyn Witkin, 2 Dec 1952, Stent papers, box 16, folder Witkin, Evelyn.

Cyrus Levinthal, a physicist-turned-phage researcher, set up a similar research program at the University of Michigan, to follow up observations using ultraviolet radiation of multiplicity reactivation. His correspondence with Stent reveals just how hard it was to get these kinds of experiments with radioactive phosphorus working. This was in part due to the challenges of getting carrier-free phosphorus-32 with high enough specific activity in sufficient quantities. (Stent ended up importing the radioisotope from the British atomic energy installation in Harwell, England, though he continued to have problems with both the quality of material and the speed of delivery.)[129] As Levinthal cautioned Stent, "If our experiences are any indication, your problems with the suicide experiments will not be entirely over when you get the high specific activity P^{32}."[130]

Stent did manage to get the bacteriophage suicide studies working early in 1953. He viewed these experiments with "super P^{32}-hot T2 and T3" as "something like a cross between the Hershey, Kamen, Kennedy and Gest and the Luria-Latarjet experiments."[131] Stent was combining various phage mutants, labeled or not with phosphorus-32, and analyzing the mixed infections over time, freezing aliquots in liquid nitrogen at various time points. The so-called eclipse period of bacteriophage infection for T2 was a brief thirteen minutes, whereas the half-life of phosphorus-32 was fourteen days.[132] Thus one had to slow down the replication process—by freezing the infected cells in liquid nitrogen—to allow the radioactive decay to occur, a process unaffected by temperature. The aim was to assess how phage mortality due to radioactive decay varied over the course of the viral reproduction process, as a way to ascertain when in this cycle the infecting phage is genetically vulnerable.

129. See letters from 1952–1955 in Stent papers, box 1, folder Atomic Energy Research Establishment; box 7, folder Isotopes; and box 11, folder Oak Ridge. Stent also explored procuring the radiophosphorus from the Canadian atomic energy installation at Chalk River. Only in 1951 did the US AEC complete arrangements to allow researchers in the United States to import radioisotopes from the United Kingdom. See AEC 231/16 in NARA-College Park, RG 326, E67A, box 47, folder 1 Foreign Distribution of Radioisotopes, vol. 3. On the British radionuclide supply, see Kraft, "Between Medicine and Industry" (2006).

130. Cyrus Levinthal to Gunther S. Stent, Nov. 17, 1952, Stent papers, box 9, folder Levinthal, Cyrus.

131. Gunther Stent to A. H. Doermann, 19 Feb 1953, Stent papers, box 4, folder Doermann, A. H.

132. On timing, see Stent, "Decay of Incorporated Radioactive Phosphorus" (1955), p. 855.

This approach drew on the earlier radiolabel transfer experiments (as developed by Hershey, by Kozloff and Putnam, and by Watson and Maaløe), with a twist: Here the experimental design was aimed at assessing the genetic effects of the phosphorus-32 decay rather than simply following the movement of the radiolabel from infecting virus to progeny. In his paper at the 1953 Cold Spring Harbor Symposium on "Viruses," Stent framed the line of inquiry as a follow-up to Hershey and Chase's demonstration that phage nucleic acid "presides over the replication of the infecting particle." His new incorporation experiments would "answer the question of *how long* after infection the parental nucleic acid still continues to preside in this way, or restating the question in another way, of *how soon* after infection the parental nucleic acid has accomplished its mission."[133]

Stent's mixed infections also allowed for recombination between different mutants, enabling him to assess marker rescue from inactivated phage. But this also meant that his experiments had many variables in play at the same time. As Stent wrote Gus Doermann at Oak Ridge (who was working on analogous experiments with ultraviolet radiation there), "I have done one Gargantuan experiment so far, permitting phage growth for 0, 3, 5, and 7 minutes before freezing everything and analyzing the plaque types before and after burst from all the samples from day to day. The results are very interesting, I am sure, but I can't say that I have been able to figure them out."[134] Doermann wrote Stent back that he and his graduate student Franklin Stahl were "working on closely related problems, and our results agree very well."[135]

By the fall of 1953, Stent had obtained results suggesting he could knock out individual genetic loci with radioactive labeling of phage T2 in his "genetic-cum-P^{32} suicide work."[136] He infected bacteria first with

133. Stent, "Mortality Due to Radioactive Phosphorus" (1953), on p. 256.

134. Gunther Stent to A. H. Doermann, 25 Aug 1953, Stent papers, box 4, folder Doermann, A. H.

135. A. H. Doermann to Gunther S. Stent, 28 Aug 1953, Stent papers, box 4, folder Doermann, A. H. Doermann commented also that "our interpretations may perhaps be at variance," but Stent soon abandoned the interpretation Doermann took issue with—that the inactivation of one marker stabilized another marker. See Gunther S. Stent to A. H. Doermann, 10 Sep 1953, Stent papers, box 4, folder Doermann, A. H. Doermann moved in the fall of 1953 from Oak Ridge to Rochester; on the results from his group, see Doermann, Chase, and Stahl, "Genetic Recombination and Replication" (1955).

136. Gunther S. Stent to Cyrus Levinthal, 14 Sep 1953, Stent papers, box 9, folder Levinthal, Cyrus.

a nonradioactive strain of T2 (T2h^+r^+) and second with another strain of ^{32}P-unstable T2 phage (T2*hr*), then he looked at the genetic markers in surviving phage.[137] Focusing on specific loci rather than simply phage viability had made the experiments even more complex to execute. As Stent put it in a letter to Levinthal, "Unfortunately, the significant experiments have to be done in single burst, and at late stages of the decay, perhaps only one in ten bursts is one of interest; the experiments are therefore, frightfully cumbersome, besides lasting a month or so."[138] Stent contended that the decay of phosphorus-32 incorporated in the phage DNA could inactivate specific genetic loci, just as exogenous x-ray exposure could, as shown by Doermann; this drew some skepticism from Hershey.[139] (See figure 7.5.)

These mating experiments exploited the genetic effects of radioactive decay more than the tracing capabilities of the isotope label. But Stent also continued some tracing experiments in the vein of Maaløe and Watson by collaborating with Howard Schachman and Itaru Watanabe on the fate of parental phosphorus during viral multiplication. They used both biophysical and biochemical techniques to follow phage DNA. By ultracentrifuging the contents of the infected bacterial cells they determined when phage-like particles were detectable, and by seeing whether the ^{32}P-containing DNA precipitated in trichloroacetic acid, with or without digesting it with DNase, they assessed when and whether the radiolabeled DNA remained in the form of high-molecular-weight fibers or lower-molecular-weight pieces. They found that much of the parental phosphorus remained associated with high-molecular-weight DNA through the so-called eclipse period, so that direct transfer of phosphorus-32 from parent to progeny was possible.

Watanabe, Stent, and Schachman pointed out that their transfer experiments were compatible with the recent Watson-Crick double-helical model for DNA, already in 1954 an important point of reference.[140] As

137. Stent, "Cross Reactivation" (1953).

138. Gunther S. Stent to Cyrus Levinthal, 14 Sep 1953, Stent papers, box 9, folder Levinthal, Cyrus.

139. Stent, "Cross Reactivation" (1953). On the skepticism Stent's interpretation drew, see Stent to Hershey, 10 Nov 1953, Stent papers, box 7, folder Hershey, Alfred Day #2.

140. Watson and Crick, "Structure for Deoxyribose Nucleic Acid" (1953); idem, "Genetical Implications" (1953). The Watson-Crick structure was viewed as a model, not a certainty. Delbrück raised questions about the topological difficulties of replicating double helical DNA. In response to this and experimental results from ultracentrifuging DNA, Stent's Berkeley colleagues Howard Schachman and Charles Dekker proposed an alternative DNA model

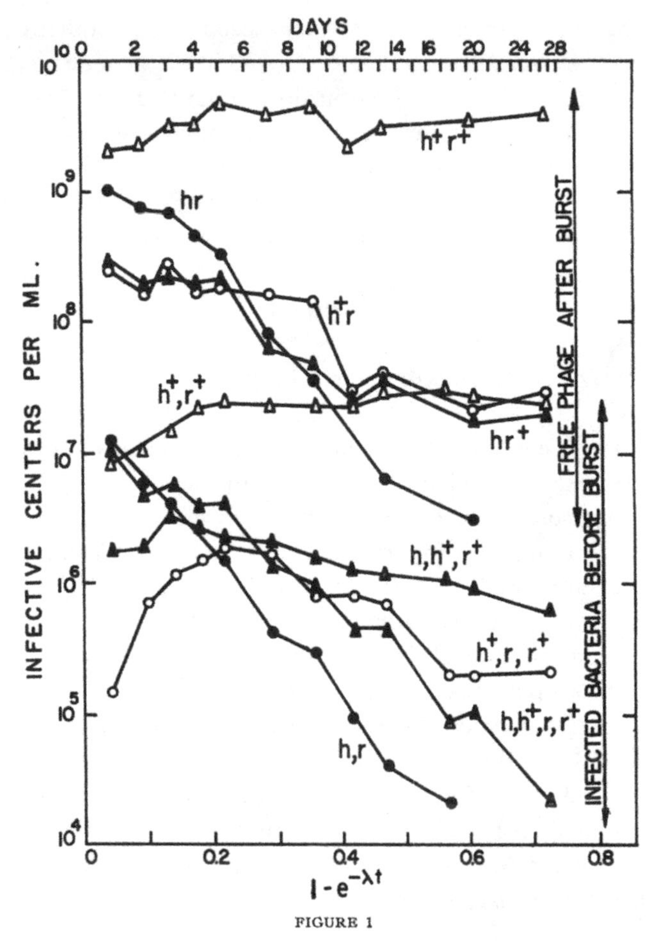

FIGURE 1

DECAY OF HIGH-P^{32}-T2*hr* IN MIXED INFECTION WITH T2h^+r^+

SYMBOL	PLAQUE TYPE	ALLELES LIBERATED BY INFECTED BACTERIUM PLATED BEFORE BURST	PHAGE TYPE PLATED AFTER BURST
●	Large, clear	h,r	hr
▲	Small, clear	h,h^+,r^+	hr^+
○	Large, turbid or mottled, turbid	h^+,r,r^+	h^+r
△	Small, turbid	h^+,r^+	h^+r^+
◬	Mottled, clear	h,h^+,r,r^+	. . .

FIGURE 7.5. Figure depicting the decay of highly labeled ^{32}P-T2*hr* in mixed infection with T2*h*+*r*+. Reproduced from Gunther S. Stent, "Cross Reactivation of Genetic Loci of T2 Bacteriophage after Decay of Incorporated Radioactive Phosphorus," *Proceedings of the National Academy of Sciences, USA* 39 (1953): 1234–41, on p. 1237.

they asserted, "One is at liberty to suppose that the replication of bacteriophage DNA occurs by means of a process, such as that proposed by Watson and Crick in which there is a direct material continuity between parent and daughter structures."[141] But their results also failed to rule out other alternatives. Some DNA from the infecting particles was broken down into low-molecular-weight material, keeping alive the possibility of indirect transfer, in which the progeny were synthesized from existing small molecules in the cell. The question of generational transfer thus remained unresolved. As Stent wrote Latarjet, "I am always chasing after that elusive problem of the fate of the phosphorus of a parental DNA molecule during its replication and still have not given up hope."[142]

From Suicide Experiments to Semiconservative Replication of DNA

The possibility of using radiolabels to trace an atom in a virus's genetic material through the life cycle remained alluring, despite the meager results that had been obtained. The Watson-Crick double helix provided phage researchers with a concrete model for imagining how transfer of material during DNA replication might proceed. Stent's ongoing studies of bacteriophage suicide with graduate student Clarence Fuerst led them to offer a mechanism for the lethality of incorporated phosphorus-32 to DNA. Because, according to the Watson-Crick model, DNA is double-stranded, a single break in the polynucleotide backbone—due to either the replacement of a phosphorus-32 atom by sulfur-32 upon radioactive decay or the energy released by the decay—would not disrupt the DNA molecule. But a second break across from the first would break the double-helical chain.[143] Stent ventured that x-ray ionization inactivated bacteriophage in a similar

involving interrupted chains, which Stent found quite persuasive. Dekker and Schachman, "On the Macromolecular Structure" (1954); Holmes, *Meselson, Stahl, and the Replication of DNA* (2001), ch.1.

141. Watanabe, Stent, and Schachman, "On the State of the Parental Phosphorus" (1954), p. 47.

142. Gunther S. Stent to Raymond Latarjet, 12 Apr 1954, Stent papers, box 9, folder Latarjet, Raymond, p. 2.

143. Stent and Fuerst, "Inactivation of Bacteriophages" (1955), esp. pp. 454–56.

way, by disrupting the DNA double-helix; this was soon substantiated by Stahl.[144]

The AEC found Stent's ongoing project sufficiently promising to offer him research support for radioisotopes beginning in 1955.[145] The AEC had launched an extensive extramural grants program for genetics, through which it supported almost 50% of federally funded research in this field during the 1950s.[146] The agency was especially interested in using results from radiation genetics to help establish an acceptable limit for exposure to low-level radiation, given ongoing atomic weapons testing and emerging public concerns about the safety of radioactive fallout.[147]

Stent extended the principle of phosphorus-32 suicide from bacteriophage to bacteria, showing that incorporation of the label at sufficiently high activities could kill bacterial cells.[148] As in the case of the bacteriophage experiments, this suicide investigation aimed at shedding light on the nature of genetic reproduction, in this case the partitioning of the bacteria's DNA between daughter cells. Stent's questions in this vein were reminiscent of those posed by Watson and Maaløe about phage:

> It has been shown that the DNA of *E. coli* cells retains its phosphorus atoms throughout subsequent bacterial growth and multiplication. How are these phosphorus atoms distributed over the nuclei of daughter cells? Do some descendant nuclei contain only atoms assimilated de novo and are others endowed exclusively with phosphorus atoms of parental origin, or are the atoms of the parental nucleus dispersed among all the nuclei in its line of descendance? The fact that it is the decay of DNA-P^{32} atoms which is mainly responsible for the death of the bacterial cells offers a method of resolving this question.[149]

144. Stahl, "Effects of the Decay" (1956).

145. E. V. McGarry to F. S. Harter, 15 Nov 1955, Stent papers, box 7, folder Isotopes.

146. Beatty, "Genetics in the Atomic Age" (1991); idem, "Genetics and the State" (1999).

147. See chapter 5 as well as Jolly, *Thresholds of Uncertainty* (2003); Rader, *Making Mice* (2004), ch. 6.

148. These experiments were increasingly laborious and Stent sometimes despaired of being able to use them to understand DNA replication. Gunther S. Stent to A. D. Hershey, 8 Jul 1954, Stent papers, box 7, folder Hershey, Alfred Day #2. The complexity of the experiments is apparent in Stent, "Decay of Incorporated Radioactive Phosphorus" (1955). Even so, this type of investigation was not fruitless: Werner Arber's suicide experiments in 1960 and 1961 contributed to his identification, with Daisy Dussoix, of host-controlled modification of DNA via restriction enzymes; Strasser, "Restriction Enzymes" (2005).

149. Fuerst and Stent, "Inactivation of Bacteria" (1956), p. 84.

Stent and Fuerst found that the bacterial DNA was transferred from parent cells to daughter cells in such a way that "*parental and newly assimilated DNA-phosphorus atoms become intermingled within daughter nuclei*."[150] Yet the dispersal of the radioactive label shed no further light on the mechanism of replication.

Stent also undertook experiments to look at *third*-generation transfer of phosphorus-32 in bacteriophage from parent to progeny. He and his coauthors found that radioactive disintegrations in the second generation attenuated the appearance of the label in the "grandchildren" phage, just as occurred between the first and second generations.[151] In an article with Delbrück, Stent schematized their findings as follows:

Parent (label = 100)–> 1st progeny (label = 50)–>
2nd progeny (label = 25)–> 3rd progeny (label = 12)

Delbrück and Stent attributed the incomplete transfer at each generation to "random losses experienced by the entire parental DNA in the course of the infection, replication, and maturation processes."[152] In fact, it was difficult to disentangle the contributions of these various processes to the transfer of the radiolabel. In addition, the original question of parent to progeny transfer in phage was tied up with (and one might be tempted to say, in retrospect, confounded by) research on genetic markers and recombination. In 1956 and 1957, Levinthal and Stent were disputing whether the transfer of genetic material from phage parents to progeny occurred in big pieces (i.e., the phage chromosome) linked to genetic markers (Levinthal's view), or whether the original genetic material was dispersed in the course of phage reproduction (Stent's view).[153] In 1956, Levinthal published a paper laying out three models for how parent DNA might be distributed to progeny: (I) template-type replication, (II) dispersive

150. Italics in original. The paper goes on to state: "It is, unfortunately, not possible to infer from the present experiments the mechanism by which this dispersion occurs, *i.e.* whether it is due to a partition of the parental atoms in the course of the elementary replication act of the DNA itself, such as demanded by some proposals concerning this process or whether it is due to the randomizing effect of some postreplication event, such as 'crossing-over' or assortment of 'chromosomes.'" (Ibid., pp. 86–87)

151. Stent, Sato, and Jerne, "Dispersal of the Parental Nucleic Acid" (1959).

152. Delbrück and Stent, "On the Mechanism of DNA Replication" (1957), p. 716.

153. Gunther S. Stent to Ole Maaløe, 30 Apr 1956, Stent papers, box 10, folder Maaløe, Ole. See also Holmes, *Meselson, Stahl, and the Replication of DNA* (2001), pp. 100–112.

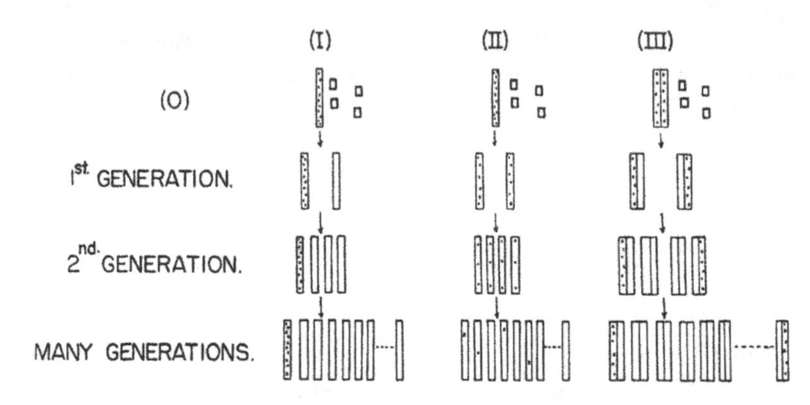

FIGURE 7.6. Levinthal's depiction of three models for DNA replication. The dots represent radioactive label, and the open squares represent the nonradioactive subunits used to build the new structure. (*O*) is the original labeled molecule. I is template-type replication (later called conservative) which leaves the label in one molecule; II is a dispersive type of replication, as proposed by Max Delbrück, and III is a complementary type of replication (later termed semiconservative), as suggested by James D. Watson and Francis H. C. Crick. Figure and legend from Cyrus Levinthal, "The Mechanism of DNA Replication and Genetic Recombination in Phage," *Proceedings of the National Academy of Sciences, USA* 42 (1956): 394–404, on p. 395.

replication, and (III) complementary replication (renamed semiconservative replication).[154] The fact that recombination occurred as well as DNA replication in the case of phage reproduction complicated experimental tests of these models. Levinthal remained convinced that the DNA from T-even phages was bipartite, consisting of one large piece and many (10-20) smaller molecules.[155] (See figure 7.6.)

In the end, the inheritance of genetic material from parent to progeny became clear through a transfer experiment that used neither phage nor a radioactive isotope.[156] In 1957, Matthew Meselson and Franklin Stahl

154. According to Holmes, on the basis of his interviews with Stent, it was Stent who renamed Levinthal's "template-type" replication *conservative*, and "complementary" replication *semiconservative*. Holmes, *Meselson, Stahl, and the Replication of DNA* (2001), p. 109, p. 462n98.

155. Stent, *Molecular Biology* (1963), p. 67.

156. It should be noted that Meselson and Stahl's first attempt to investigate DNA replication using density methods involved labeling phage with 5-bromouracil, which was also being studied for its role in mutagenesis and radiosensitivity, in continuity with Stahl's earlier suicide experiments with ^{32}P-labeled T4 phage. Holmes, *Meselson, Stahl, and the Replication of DNA* (2001), pp. 157–68.

used the heavy isotope nitrogen-15 to label the DNA of *E. coli* that were synchronized (i.e., the cells in the culture would divide at the same time). They then followed the labeled nucleic acid in a cesium chloride gradient in the ultracentrifuge, where the mass difference between nitrogen-15 and nitrogen-14 enabled a differentiation between parental atoms and progeny in the sedimentation pattern. Meselson and Stahl introduced their paper with reference to the radiolabel transfer experiments: "Radioisotopic labels have been employed in experiments bearing on the distribution of parental atoms among progeny molecules in several organisms."[157] In part because *E. coli* reproduced without recombination, unlike the phage in the suicide experiments, Meselson and Stahl could discern a clear pattern of semiconservative DNA replication. Thus "daughter" DNA strands were composed of one parental strand, which served as a template, and one newly synthesized strand. Their 1958 paper was widely cited, not only because of the elegance of their finding, but because biologists viewed it as decisive confirmation of Watson and Crick's double-helical model of DNA structure.

The efforts of phage researchers to use suicide experiments and other radiological techniques to understand viral genetics produced an increasingly arcane scientific literature, much of it now dimly remembered. As Stahl explained by way of a caveat at the beginning of a 1959 review entitled "Radiobiology of Bacteriophage,"

> At times it may appear that the reviewer has forgotten that the primary aim in employing radiation in the study of phage is to elucidate the *normal* state of affairs. However, almost all experiments involving the irradiation of phage have raised far more questions than they have answered. This has resulted in the situation that there now exists, a "radiobiology of bacteriophage," a collection of observations and hypotheses arising *from* irradiation experiments, leading no one knows where, but selfishly demanding an explanation.[158]

Phage researchers, expecting radioisotopes to illuminate the molecular process of gene replication, instead were led on to unanticipated and complex questions about the biological effects of radiation, problems they seemed unwilling to abandon so long as the next experiment beckoned. One cannot help wondering whether the fallout debates of the mid-to-late

157. Meselson and Stahl, "Replication of DNA" (1958), p. 671.

158. Stahl, "Radiobiology of Bacteriophage" (1959), p. 354. Emphases in original.

1950s gave these "suicide experiments" a wry resonance with the cultural anxieties that permeated the nuclear age. At another level, radiolabels enabled researchers to visualize genes in terms of points (wherever the radioactive atom was incorporated), but not wholes. Despite the sensitivity of the technique, one could not necessarily discern whether the transfer of a label was tracing reproduction, recombination, disintegration and reassimilation, or some combination. This line of research reinforced the tendency among biophysicists and phage geneticists to think of reproduction as a molecular—rather than organismal—process.

Conclusions

The availability of radioisotopes in the 1930s, and more abundantly after World War II, facilitated a preoccupation with understanding life at the molecular level, initially the dynamic turnover of small metabolites and subsequently the function and regulation of macromolecules such as enzymes and nucleic acids. As George Whipple commented in 1939 about the usefulness of the radioisotopes E. O. Lawrence was sending him from Berkeley: "As you know, there are so many attractive pathways that turn up as one proceeds along this path of investigation, that it is very difficult to hold a straight course."[159] Whipple was writing at a time when his store of cyclotron-produced radioiron was limited, but once reactor-produced radioisotopes were available for purchase from the AEC, life scientists could chase down all of the promising pathways that this new tool opened up. Yet, the other sense of the metaphor is that certain paths were evident and possible only because of radioisotopes, even if the directions were numerous. In this sense, one might say that radioisotopes channeled experimenters down pathways of molecular knowledge about life, including metabolism and heredity.

The tracer experiments featured here also illustrate how radioisotope work interfaced with other research practices, including autoradiography, chromatography, and genetic survival curves. Radioisotopes made these and other postwar laboratory technologies (such as ultracentrifugation and electrophoresis) easier to use, because the researcher could often

159. George H. Whipple to Ernest O. Lawrence, 26 Jul 1939, EOL papers, series 1, reel 23, folder 15:26A Rochester, University of, 1939. Martin Kamen responded to Whipple's letter on 1 Aug 1939, referring to this passage as "a masterpiece of understatement."

label the molecule of interest and thus follow it through the separation process. The habit of labeling molecular objects of study—hormones, proteins, nucleic acids, or enzyme substrates—persisted for decades, even as motivations changed. By the 1960s and beyond, biochemists shifted from determining the steps of metabolic pathways to studying the control and mechanisms of the enzymes at each step. Molecular biologists moved on from analyzing DNA transfer in gene replication to investigating how genetic information was transmitted into messenger RNA (transcription) and then proteins (translation). Still, their experiments generally started with radiolabeling.

In this sense, the routine tagging of biological objects of study survived beyond the radiotracing impulse of the 1930s through the 1950s, such that a whole battery of laboratory procedures that characterized biochemistry and molecular biology—protein purification, enzyme assay, nucleic acid hybridization, DNA sequencing, to name a few—remained tethered to the supply infrastructure of radioisotopes. Or to put it another way, researchers shifted from tracing radiolabels through biological systems to tracing them through experimental systems.[160] The prevalence of radiolabeling in biochemistry and molecular biology, which persisted until the age of genomics, developed momentum on account of the AEC's distribution program.

This chapter's focus on the uses of radioisotopes in biochemistry and molecular biology also reveals the overlap between these fields.[161] The commonalities are especially strong at the level of scientific practice. Following experimentation with radioisotopes, however, illuminates a much broader set of common practices than those among only biochemists and molecular biologists. Life scientists in a variety of fields, from endocrinology to ecology, emulated the radiotracer work that already characterized biochemical and physiological research on metabolism in the late 1930s. However, as the systems of study included human subjects and natural landscapes, the wider uses of radioisotopes ushered in new kinds of problems.

160. On experimental systems, Rheinberger, *Toward a History* (1997).

161. See, for example, the essays gathered together in de Chadarevian and Gaudillière, "Tools of the Discipline" (1996). On the motivations behind demarcating molecular biology as a separate discipline, see Abir-Am, "Politics of Macromolecules" (1992).

CHAPTER EIGHT

Guinea Pigs

The question of the use of radioactive isotopes and other sources of radiation in human experimentation was discussed and it was the sense of the Committee that Dr. [Shields] Warren should explicitly express their view to Dr. [Robert] Stone to the effect that they would look with disfavor on such experimentation. — Minutes, Ninth Meeting of the AEC's Advisory Committee for Biology and Medicine, 8 May 1948[1]

After World War II, even as radioisotopes failed to live up to the publicity that they would cure cancer, they became valuable tools for investigating disease. One result of the government's reactor-based supply was an escalation of radioisotope research with human subjects in the postwar decade. At the time, researchers using radioisotopes did not always differentiate between therapeutic experiments, in which the subject could conceivably benefit from the treatment, and nontherapeutic experiments, in which no such benefit was expected. For this and other reasons, much of this research was later deemed deeply problematic. Yet the type of human experiments now regarded as ethically suspect is not easily separated out from path-breaking and Nobel Prize–winning work with radioisotopes. This chapter illustrates both the promise and the peril of atomic age research on human subjects—which here include pregnant women, children, soldiers, diabetics, and so-called healthy volunteers—through studies of anemia and diabetes.

One of the largest radioisotope experiments with healthy humans was part of a five-year investigation at Vanderbilt Medical School of maternal nutrition in neonatal and postnatal health. The experiment grew out of long-standing research into iron metabolism, a central aspect of several blood diseases. George Whipple and his coworkers at Rochester were

1. Minutes, 9th ACBM Meeting, 8 May 1948, Washington, DC, OpenNet Acc NV0711621, p. 7.

among the first investigators to use radioactive iron when it became available from E. O. Lawrence's Berkeley cyclotron in the late 1930s. Whipple and his coworkers fed iron-59 to dogs, finding that anemic dogs took up much more iron from their diet than normal dogs.[2] Like the biochemists of the last chapter, Whipple used radioiron as a tracer, but followed it through organ systems rather than along chemical pathways. Studies of the assimilation and metabolism of radioiron proceeded from dogs to humans, first on hospitalized patients at Rochester during the late 1930s and then on a wider range of subjects after the war, as radioactive iron became more available. At Vanderbilt from 1945 to 1947, a team of researchers (including one from Whipple's group) administered oral doses of radioactive iron to over eight hundred pregnant women to track its absorption. The pregnant women derived no benefit from the radioisotope, and many did not seem to understand they were research subjects. The Vanderbilt study has come to symbolize the cavalier treatment of experimental subjects by medical researchers seeking to understand the effects of radiation on humans. While this was not the purpose of administering radioiron to pregnant women in the 1940s, Vanderbilt researchers understood by the 1960s that the investigation also shed light on the sensitivity of developing fetuses to radiation.[3]

In the late 1940s and 1950s, radioisotopes were also being used to study thyroid function, blood circulation, and diabetes in veterans. Nuclear physicist Rosalyn Yalow and physician Solomon Berson conducted research with iodine-131 at the Veterans Administration Hospital in the Bronx. Administering iodine-131-labeled insulin to veterans turned up surprising results about blood-borne antibodies, results that Yalow and Berson utilized to develop a novel binding assay. As this method, dubbed radioimmunoassay, became a standardized diagnostic test, actual guinea pigs came to replace the humans in the experimental set-up. Radioimmunoassays put the tremendous specificity of antibodies, derived from people or guinea pigs, to technological use, enabling users to measure the minute concentrations of specific molecules (e.g., 10^{-10} to 10^{-12} molar) even in the presence of billion-fold higher concentrations of other molecules.[4] This assay became a crucial

2. Hahn et al., "Radioactive Iron" (1938); Hahn et al., "Radioactive Iron" (1939).

3. ACHRE, *Final Report* (1996), pp. 213–16; Hagstrom et al., "Long Term Effects of Radioactive Iron" (1969).

4. On other antibody-based tools and immunodiagnosis, Cambrosio and Keating, *Exquisite Specificity* (1995); Keating and Cambrosio, *Biomedical Platforms* (2003); Silverstein, *History of Immunology* (1989), chapter 12.

tool, widely used not only in medical diagnostics but also in basic laboratory research and in environmental and drug testing. The development of radioimmunoassays is counted as one of the most successful and productive applications of radioisotopes in biomedical research.

In both of these lines of research, human subjects were central to the experimental enterprise. The particular ethical problems of using humans as subjects in atomic energy research has received extensive scrutiny in the past two decades by investigative journalists and by President Clinton's Advisory Committee on Human Radiation Experiments, which published its report in 1995. Humans were unwitting subjects of atomic weapons–related research during and after World War II. The Manhattan Engineer District, and subsequently the AEC, conducted classified experiments in which patients diagnosed with terminal conditions were administered small quantities of radioisotopes such as plutonium, polonium, and uranium to trace their absorption and metabolism. These studies were aimed at assessing how workers and others exposed to these radioactive materials would metabolize them and, if the dose were high enough, what the biological effects might be. In other cases human subjects, usually cancer patients, received heavy doses of radiation in experiments aimed at understanding its effects. Notable (and notorious) examples along these lines were investigations conducted by Robert Stone at the University of California, San Francisco and by Eugene Saenger at the University of Cincinnati. Saenger's program illustrates the convergence of agendas of the military and cancer researchers, as his work on total body irradiation of patients (aimed at reducing painful metastatic growths) was funded in part by the Department of Defense to get information on cognitive and physiological changes that might be experienced by soldiers on a nuclear battlefield.[5] Many similar experiments were made public only in the 1990s.

Yet the majority of human radioisotope experiments, while relying on AEC-produced radioisotopes, were not otherwise government sponsored. Moreover the now-standard label of "human radiation experiments" fails to account for the variety of motives, agendas, and types of radioactive exposure associated with such research. The radioiron experiments analyzed in this chapter were aimed not at understanding how the radioactive decay of iron-59 affected the body, but at elucidating the assimilation

5. On Stone's research, see Jones and Martensen, "Human Radiation Experiments" (2003); on Saenger's research see Kutcher, "Cancer Therapy" (2003) and *Contested Medicine* (2009). Other total body irradiation research is discussed in chapter 9.

and metabolism of ordinary iron, on account of its role as a vital nutrient. The clinical researchers who administered radioiron to human subjects viewed it as a tracer, blithely assuming that such low doses of radioactivity were not harmful.[6] The development of radioimmunoassays is generally seen as a scientific advance rather than a lapse of medical judgment, but it relied no less on the role of human subjects, both patients and "healthy volunteers," in clinical research with radioisotopes.[7] Strikingly, the fact that the bodies of veterans provided crucial experimental material for this discovery fits with a broader pattern of soldiers as "atomic guinea pigs" in the postwar period.[8] But more generally, the use of these institutionally defined populations such as ward patients and veterans for clinical investigation exemplifies David Rothman's depiction of the "gilded age" of medical research. In an era of rapidly growing federal funds for medical research, ethical considerations were entrusted to the investigator, and conflicts between caring for patients and the prerogatives of research were rarely acknowledged.[9] The work at the Bronx Veterans Administration Hospital and Vanderbilt University may represent, in retrospect, the best and worst of civilian human experiments with radioisotopes. Yet both typify in character and scale American medical research of the mid-twentieth century.

In this context of burgeoning clinical experimentation after World War II, the politics of atomic energy shaped how radioisotope studies in humans proceeded in two distinctive ways. The first was the federal government's vigorous promotion of the scientific and medical benefits of radioisotopes, as seen in the foregoing chapters. This enthusiasm, even fanaticism, about the value of radioisotopes extended to research with human subjects, which the AEC generally represented as safe as long as its regulations and guidelines were followed. Consequently, many radioisotope experiments with human subjects were launched due to the availability of the materials and a mindset that the gains in scientific knowledge outweighed minor risks. Most such biomedical investigations had no

6. Many retrospective accounts agree with this general perspective. As J. Newell Stannard puts it, "Once the 'hot' sources were diluted to tracer levels, worries about toxic effects were minimal. On the whole, this sanguine attitude was probably justified for tracer applications." *Radioactivity and Health* (1988), vol. 1, p. 289.

7. The line between "healthy subjects" and patients was itself permeable. Stark, *Behind Closed Doors* (2012), ch. 4.

8. ACHRE, *Final Report* (1996), ch. 10.

9. Rothman, *Strangers at the Bedside* (1991), ch. 3.

connection to weapons development and were published in the open literature. As this chapter illustrates, such work developed in a range of private and public institutions.

The second aspect concerns the military-related underside of radioisotope research, the human experiments on the metabolism of weapons materials and fission products. The leadership of the AEC understood that these investigations were ethically dubious and politically problematic. Such experiments, if publicized, would (justly) give Americans the impression that the federal government was experimenting on "human guinea pigs." For this reason the AEC refused to declassify many research studies involving humans, even if they had no importance for national security.[10] The agency also developed guidelines in 1947, such as obtaining consent for clinical experiments, aimed at protecting patients and other human subjects of medical research with radioactive materials. However, these guidelines do not appear to have been enforced for research beyond the AEC's own laboratories, creating a bizarre situation in which the regulatory framework for experimental subjects seems to have applied only to secret human research with radioisotopes.

The main form of government oversight for "off-Project" clinical research with radioisotopes consisted of a Subcommittee on Human Application in the AEC's Isotopes Division. This group of medical experts reviewed (with veto power) any requests to the AEC for radioactive materials to be used in humans. Most applications survived their scrutiny. By October 1946, just two months after Oak Ridge started shipping isotopes, 94 of the 217 radioisotope orders were for human usage; 90 of them were approved.[11] As for regulating the safe administration and handling of radioisotopes once they reached their destination, the AEC initially looked to local committees at hospitals and research institutions. Over time, a stronger regulatory apparatus for using radioisotopes in laboratories and hospitals developed, both as part of the general codification of federal regulation and in response to the more stringent regulations to safeguard the public that came out of the 1954 Atomic Energy Act. As it turned out, a growing awareness of the health hazards of low-level radiation exposure also emerged over the 1950s and 1960s, by which time extensive human experimentation had already taken place.

10. ACHRE, *Final Report* (1996), p. 49.

11. ACHRE, *Final Report* (1996), p. 175; Isotopes Branch, Research Division, Manhattan District, Reports of Requests Received through 31 Oct 1946, NARA Atlanta, OROO Lab & Univ Div Official Files, Acc 68A1096, box 6, folder Radioisotopes–National Distribution.

Radioiron as a Physiological Tracer

Medical researcher George H. Whipple of the University of Rochester Medical School began physiological research with radioactive iron obtained from E. O. Lawrence; in 1937 he was among the first nonlocal recipients of cyclotron-produced radioisotopes from Berkeley.[12] At the time a cyclotron was being built at Rochester, which during World War II became one of the major centers for research on the biological effects of radiation. In 1943, Stafford Warren, a professor of radiology at Rochester, was appointed medical director of the Manhattan Project. Warren oversaw a wartime contract to Rochester Medical School on uranium toxicology, which subsequently included investigating the acute and chronic toxic effects of plutonium.[13] By the end of the war, this effort had grown to be the second largest program of applied medical and biological research in the Manhattan Engineer District, receiving $1.7 million in 1945–46.[14] The AEC continued this high level of funding to Rochester for research into the biological effects of radiation. The facility for this military-related work was across the street from the medical school and closely guarded. Shortly after the war as part of this project, eleven patients in the "metabolic ward" of Rochester's Strong Memorial Hospital were injected with microgram amounts of plutonium, to follow its metabolism, localization, and excretion from the body.[15] The work at Rochester was analogous to Joseph Hamilton's research at Berkeley on the metabolism of long-lived fission products. At both sites AEC-supported researchers injected hospital patients with weapons-related radiomaterials, experiments kept secret for decades.

These war-related radiation experiments were not directly related to the work of Whipple's group—in fact, the plutonium experiments at Rochester took place later than the radioiron experiments discussed below. Yet

12. See chapter 2.

13. Hacker, *Dragon's Tail* (1987), pp. 49–50, 67.

14. That year only University of Chicago received more, at $2.5 million, though in 1946–47 funding at Chicago was cut to $1.0 million, whereas Rochester received $1.2 million. These funds were out of an overall medical research budget of $4.91 million in 1945–46 and $3.88 million in 1946–47. See Lenoir and Hays, "Manhattan Project for Biomedicine" (2000), p. 38.

15. Nearly all of the unwitting subjects of the government's plutonium studies have been identified in Eileen Welsome's compelling account of medical research associated with atomic energy: *Plutonium Files* (1999).

Whipple's work drew on key resources that were also valued by the Manhattan Project, especially a radiology department up to date with using the newest radiation sources and local nuclear physics expertise, notably physics professor Lee A. DuBridge, who oversaw the cyclotron. Moreover, research with radioiron highlights important continuities between the era of human experiments with cyclotron-produced radioisotopes, applied radiation research under the Manhattan Project, and postwar clinical investigation. Those universities where pioneering nuclear physics and medicine took place were the same institutions tapped by the Manhattan Project and then the AEC to investigate the hazards of radioisotopes related to atomic weapons development.

Whipple had received the 1934 Nobel Prize in Physiology or Medicine with George Minot and William P. Murphy for their work on liver therapy for pernicious anemia. Whipple had first focused on anemia due to blood loss, having developed a "standard anemic dog" whose condition was induced by bleeding and a special diet. Beginning in 1925 and 1930 Whipple and his laboratory technician Frieda Robscheit-Robbins published eighteen papers on "Blood Regeneration in Severe Anemia." They tested the regenerative effects of a variety of food supplements, and found raw or cooked liver to be the most potent. (For their part, Minot and Murphy adapted this dietary approach to their studies of pernicious anemia.) Whipple and Robscheit-Robbins eventually found that the efficacy for treating anemia of the foods they tested, whether derived from animals or plants, correlated with iron content, even though feeding iron salts alone was not as effective as feeding liver.[16] Whipple's research on experimental dogs, some of which were surgically modified to study aspects of the gastrointestinal tract, continued the tradition of such work in physiology exemplified three decades earlier by the contributions of Ivan Pavlov.[17]

The identification of iron in red blood cells, and its role as a key component of hemoglobin in animals from insects to mammals, dated to the nineteenth century.[18] Still the clinical significance of dietary iron for blood

16. Miller, "George Hoyt Whipple" (1995), pp. 382–83. Whipple and Robscheit-Robbins initially concluded that iron was of no value in the treatment of simple anemia, based on the much greater efficacy of liver. However, by 1925 they evaluated iron to be effective for chronic severe anemia. Wintrobe, *Blood, Pure and Eloquent* (1980), p. 174.

17. Todes, "Pavlov's Physiology Factory" (1997); idem, *Pavlov's Physiology Factory* (2002).

18. Edsall, "Blood and Hemoglobin" (1972); Holmes, "Crystals and Carriers" (1995).

diseases was far from clear. The etiological category of "chlorosis" gave way to anemia by the early twentieth century due not to a rationalization of treatment with iron, but to a cultural shift in its meaning for women—and new technologies of diagnosis.[19] Nonetheless, by the mid-1930s, clinicians generally treated "hypochromic anemia" (which could derive from certain nutritional deficiencies or chronic blood loss) with iron supplements, though the medical literature on the efficacy of different forms of iron remained full of contradictions.[20] Interest in the physiological roles of trace elements such as iron, copper, and zinc paralleled the intensive search for vitamins that dominated biochemistry in the 1910s through 1930s.[21] Artificial radioisotopes provided a means to track the assimilation and metabolism of such trace elements, or "micro-nutrients."[22]

The National Academy of Sciences held its fall 1937 meeting at the University of Rochester, where Whipple had an opportunity to present his research on anemia and dietary iron. Ernest O. Lawrence attended the meeting and told Whipple that Martin Kamen at the Berkeley Rad Lab had recently discovered an isotope of iron with a relatively long half-life of 47 days.[23] This half-life was sufficient to trace the fate of the isotope, iron-59, after feeding it to dogs.[24] Moreover, by tagging the dietary iron, researchers could distinguish newly assimilated iron from the iron stores already in the body. Whereas the carbon isotopes used in photosynthesis research were useful because the element is found in all biological molecules, the utility of artificially radioactive iron had to do with its uncommon, but still crucial, presence in the body.

Whipple was among the first researchers to acquire some of this recently discovered isotope of iron. Late in 1937 Lawrence sent a shipment of iron-59 from Berkeley to Rochester. DuBridge told Lawrence that

19. Wailoo, *Drawing Blood* (1997), ch. 1.

20. On treatments for anemia, see Strauss, "Use of Drugs" (1936), esp. 1635. On disagreements in the medical literature, see Hahn, "Metabolism of Iron" (1937).

21. McCance and Widdowson, "Mineral Metabolism" (1944). On vitamin research, see Kamminga, "Vitamins and the Dynamics" (1998).

22. A mid-century survey of the field used the term trace elements "because of its historical associations" while acknowledging the more recent terms "micro-element" and "micronutrient." Underwood, *Trace Elements* (1956), p. 1.

23. Corner, *George Hoyt Whipple* (1963), p. 245.

24. Whipple outlined the investigation he wished to undertake with radioiron from the Berkeley cyclotron in "Memorandum for Dr. Lawrence," 23 Nov 1937, EOL papers, series 1, reel 23, folder 15:26 Rochester, University of, 1935–38.

Whipple was "very much excited over the possibilities" although at first "he was inclined to treat the radio iron as such precious stuff that he was almost afraid to use it at all."[25] Whipple overcame his hesitancy, no doubt stimulated by the arrival of a second allotment of radioiron from Berkeley early in 1938, this one with a higher specific activity.[26] These initial gifts of radioiron from Lawrence and Kamen were acknowledged by the Rochester researchers through coauthorship.[27] The collaborators at Rochester included William F. Bale, a physicist in Stafford Warren's Department of Radiology at Rochester, who counted the radioactivity, and Paul F. Hahn, who worked with Whipple in the Department of Pathology, where he received his PhD after initial training in engineering at MIT.[28]

The group's initial publication in the *Journal of the American Medical Association* presented the background: "Physiologists will admit that our understanding of iron metabolism is in a parlous state. There are diametrically opposed views relating to almost every phase of iron metabolism in the body. It is fair to state that much of this difficulty relates to methods of iron analysis."[29] Radioactive iron afforded Whipple and his coworkers a way to track definitively the element's absorption from feeding bowl to blood cells. As expressed in the second, longer paper, "grateful physiologists have been presented with what may prove to be the 'Rosetta Stone' for the understanding and study of body metabolism."[30] From January 1938 through January 1940, Lawrence and Kamen sent the Rochester group twenty-two samples of radioactive iron, which enabled them to conduct a wide variety of animal experiments on iron uptake and elimina-

25. Lee A. DuBridge to Ernest O. Lawrence, 17 Dec 1937, EOL papers, series 1, reel 23, folder 15:26 Rochester, University of, 1935–38.

26. Kamen had to operate the cyclotron at an elevated voltage to achieve the greater activity, though this generated so much heat that it melted the iron target. Martin D. Kamen to Lee A. DuBridge, 11 Jan 1938, EOL papers, series 1, reel 23, folder 15:26 Rochester, University of, 1935–38. Radioiron samples of higher specific activity followed (Lawrence to Whipple, 5 Apr 1938, same folder).

27. Lawrence initially demurred at being made a coauthor but Whipple insisted, at least on the first paper. Kamen was coauthor on a subsequent paper. Hahn et al., "Radioactive Iron" (1938); Hahn et al., "Radioactive Iron" (1939); Hahn et al., "Radioactive Iron" (1939). See Lawrence to Whipple, 30 Nov 1937; Whipple to Lawrence, 7 Dec 1937; Lawrence to Whipple, 1 Oct 1938; and Whipple to Lawrence, 11 Oct 1938, EOL papers, series 1, reel 23, folder 15:26 Rochester, University of, 1935–38.

28. G. H. Whipple to Warren Weaver, 22 Apr 1946, RAC RF 2–1946, series 200, box 333, folder 2254.

29. Hahn et al., "Radioactive Iron" (1938), p. 2285.

30. Hahn et al., "Radioactive Iron" (1939), p. 739.

tion.[31] Even so, early on the radioiron was in such short supply that these researchers collected the feces of the dogs to recover the excreted radioisotope for reuse.[32]

By feeding experimental dogs iron-59, Whipple and his coworkers demonstrated that absorption depended on whether the animal was already depleted of iron. The six dogs with induced anemia took up the radioactive iron abundantly, whereas three normally nourished dogs absorbed only a trace.[33] The anemic dogs absorbed iron more efficiently (up to 50%) when it was fed to them in several smaller doses of 30 to 40 mg per day.[34] Whipple and his coworkers sacrificed three of the anemic dogs and all of the control dogs to determine the amount and distribution of radioactive iron in the blood, organs, and bone. The remaining dogs had their blood drawn for several days so that the appearance and persistence of the iron-59 could be measured. In anemic dogs, radioactive iron appeared in red blood cells within a few hours of feeding.[35] As the authors commented, "The speed of absorption and transfer of iron to the red cell is spectacular."[36] (See figure 8.1.)

The early experiments included feeding radioiron to a pregnant beagle. Twelve hours after the oral dose, she delivered a stillborn pup. Twenty-four hours later she delivered a live pup, and thirty hours later a third pup, also alive. Both of the live pups were sacrificed and analyzed to see if any of the radioiron had been transferred. Only fractions from the second live pup showed a small amount of radioactivity. As Paul Hahn and William Bale noted in an unpublished report:

> The finding of even a small amount of radio-iron in the second live pup would invite some speculation. Heretofore it has been assumed by most investigators that once entering the blood stream iron is combined with a globulin and as such transported to the usual storage sites. If such were the case, it would be

31. List of shipments with letter from George H. Whipple to E. O. Lawrence, 8 Feb 1940, EOL papers, series 1, reel 23, folder 15:27 Rochester, University of, 1940–59. There is a list of publications from Rochester from experiments using this radioiron in the same folder.

32. It was Joseph Ross's responsibility to extract the radioiron from the dog excrement. Oral history of Joseph F. Ross by Eric Hoffman, 11–12 Jun 1986, Columbia University Oral History Library.

33. Hahn et al., "Radioactive Iron" (1939).

34. Ibid., p. 745. The 50% figure comes from the subsequent paper: Hahn et al., "Radioactive Iron" (1939).

35. Hahn et al., "Radioactive Iron" (1939), p. 747.

36. Ibid., p. 753.

Radioactive Iron Content of Tissues

Tissue or Organ Content = Per Cent of Total Amount Fed

	Anemic								Normal		
Dog No.	**H-9**	**H-8**	**37-116**	**37-227**		**37-204**		**37-202**	**37-77**	**37-144**	**37-214**
Number of feedings	4	2	1	1		1		1	18	5	1
Blood volume	330	350	500	370		440		770	480	630	700
Plasma volume	260	260	390	250		330		620	240	360	400
Iron fed in mg	220	66	130	84		300		115	650	103	60
Counts per minute as fed	770	464	5,730	21,500		6,590		13,000	600	2,120	14,240
Hb. level per cent when fed	39	62	53	68		61		56	178	138	114
Hours after last feeding	20	20	23	4	75	11	26	6	84	23	7
Radio-iron found											
Liver	0.4	0.4	0.5	—	—	—	—	—	0.2	0.03	0.02
Spleen	0.0	0.0	0.1	—	—	—	—	—	0.0	0.02	0.01
Marrow	0.2±	3.0±	2.0±	—	—	—	—	—	0.0	0.03±	—
Plasma	0.0	0.3	0.1	0.7	0.1	0.8	0.2	3.5	0.0	0.01 }	0.05
Red cells	8.7	9.0	1.4	0.9	4.6	0.0	0.4	0.2	0.04	0.06 }	
Total radio-iron	9.3±	12.7±	4.1±	1.6	4.7	0.8	0.6	3.7	0.24	0.15±	0.08

FIGURE 8.1. Table of results from iron-59 feeding experiments with nine dogs, six with induced anemia and three normal. The anemic dogs absorbed the radioactive iron to a much greater degree. "Hb" is an abbreviation for hemoglobin. © Rockefeller University Press. Originally published in P. F. Hahn, W. F. Bale, E. O. Lawrence, and George Whipple, "Radioactive Iron and Its Metabolism in Anemia: Its Absorption, Transportation, and Utilization," *Journal of Experimental Medicine* 69 (1939): 739–53, on p. 744.

necessary to postulate the splitting of iron from the protein in order to explain the passage across the placenta.[37]

Although the researchers expressed an intention to resolve this issue through further pregnancy studies, apparently this was not pursued—nor was this finding published. The researchers' remarks, however, reveal their expectation that radioiron would not be easily transported to the fetus if given to a pregnant dog (or human).

As Whipple observed to Lawrence, the initial experiments opened up so many other possible lines of inquiry, he hardly knew "which way to turn":

37. Paul Hahn and William F. Bale, "Report No. 4, Radioactive Iron Experiments," 9-pg. typescript enclosed with George H. Whipple to Ernest O. Lawrence, 29 Sep 1938, EOL papers, series 1, reel 23, folder 15:26 Rochester, University of, 1935–38, p. 2.

> The questions of distribution in the body, means of transport, storage, utilization by muscles and bone marrow, rapidity of turn-over of radioactive iron, and a dozen other projects, will be calling for study. Also we want to try some material on human cases as soon as may be.[38]

Whipple's team extended his work to human subjects through collaborating with faculty in the Department of Radiology and the Department of Obstetrics and Gynecology. The resulting publication reported on thirty-four cases, including men, women, and children who had been hospitalized for a variety of conditions, such as bleeding ulcers and pernicious anemia, as well as one medical student regarded as a control. Fourteen pregnant women at all stages of pregnancy were included in the study. Half of these were hospitalized for therapeutic abortion (for various medical reasons), and the others for delivery. The study followed the standard ethical practice of commencing medical research on humans only after animal experiments.[39] The authors argued that human studies were necessary "to exclude possible differences between the physiology of the dog and man."[40]

All the subjects were administered radioactive iron, usually as a single oral dose. Their absorption was measured through blood samples, for several days or even months, depending on the length of hospitalization. The researchers listed the amounts of radioactive iron administered in milligrams of iron, rather than specifying the amount of radioactivity in millicuries or microcuries, making it impossible to ascertain how large a dose these patients ingested. By the same token, the Rochester researchers found determining the absolute radioactivity of samples to be difficult; the β-rays given off by iron-59 decay were "soft" and picked up inefficiently by the Geiger-Müller counter.[41] Nowhere in the paper is the radioactivity exposure for human subjects mentioned as a concern—or

38. George H. Whipple to Ernest O. Lawrence, 25 Jan 1939, EOL papers, series 1, reel 23, folder 15:26A Rochester, University of, 1939.

39. Lederer, *Subjected to Science* (1995), pp. 1 and 74.

40. Balfour et al., "Radioactive Iron Absorption" (1942), p. 16.

41. The scientists at Rochester wrote E. O. Lawrence about their difficulties with quantitatively measuring the amount of radioactivity in samples containing radioiron, with only about 2% of the decays registered by the counter. See W. F. Bale and P. F. Hahn, "Radioactive-Iron Experiments," write-up of experiments enclosed with G. H. Whipple to Ernest O. Lawrence, 11 Mar 1938, EOL papers, series 1, reel 23, folder 15:26 Rochester, University of, 1935–38.

an object of study. The point was to understand the role of nutritional iron in these pathological conditions. Moreover, it remains unclear whether the researchers regarded this as therapeutic or nontherapeutic research; information about iron absorption might have been clinically valuable for some of the cases. The results confirmed what had been observed in the experimental dogs. Those patients with anemia due to blood loss, such as through a bleeding ulcer or hemorrhage due to an incomplete abortion, took up two or three times more radioactive iron than healthy subjects. And not all anemic patients took up iron readily, suggesting that some anemic conditions (such as pernicious anemia) were not the result of depleted iron stores.[42] The pregnant women showed the greatest radioiron uptake, two to ten times higher than normal human absorption.[43]

Whipple and his coworkers continued their experiments with dogs through 1944, although the Berkeley Rad Lab was unable to provide radioactive iron past 1942 due to the wartime mobilization.[44] Instead, the MIT cyclotron became the major supplier of radioiron for their continuing research.[45] They extended the investigation of iron absorption to surgically altered anemic dogs, in which various segments of the stomach and small intestine (the jejunum, duodenum, and ileum) were made experimentally accessible through surgically created fistulas and pouches.[46] The results revealed a gradient of absorption, with the highest amounts of radioiron absorbed in the duodenum, less in the jejunum and stomach, and very little in the ileum and colon. Elimination of absorbed iron was a much

42. See Case 7 in Balfour et al., "Radioactive Iron Absorption" (1942), p. 22.

43. Ibid., p. 29. Some pregnant patients absorbed 16% to 27% of the radioactive iron. As Whipple noted to Lawrence when reporting on his preliminary human results, "The pregnant or lactating woman, having used up her reserves of iron in most instances, takes up iron when given by mouth with great alacrity." George H. Whipple to Ernest O. Lawrence, 11 Nov 1941, EOL papers, series 1, reel 23, folder 15:27 Rochester, University of, 1940–59.

44. Martin D. Kamen to George H. Whipple, 9 Feb 1942, EOL papers, series 1, reel 23, folder 15:27 Rochester, University of, 1940–59; "Special Materials," 7-pg. typescript, undated [1946?], EOL papers, series 3, reel 32, folder 21:22 Special Materials. In addition, Whipple's group obtained some radioactive copper from the cyclotron at University of Rochester to study its uptake in normal and anemic dogs. Yoshikawa, Hahn, and Bale, "Red Cell and Plasma Radioactive Copper" (1942).

45. This transition was already underway in 1942. Robley D. Evans to Howland H. Sargeant, 4 May 1942, MIT President's Papers, box 82, folder 1 Robley Evans, 1942. As mentioned above, Rochester's cyclotron did not operate at a high enough voltage to generate radioiron of sufficient specific activity for the physiological experiments.

46. On the development of these surgical techniques, Todes, *Pavlov's Physiology Factory* (2002).

more gradual process; the researchers found the radioiron secreted only slowly, through the bile. Because iron was eliminated so slowly, absorption functioned as the principal regulation of iron levels in the body, through selective uptake by the gastrointestinal mucosal epithelium. Whipple postulated that reserve stores of iron in the body, rather than anemia per se, controlled iron absorption.[47]

Only a handful of researchers besides Whipple's group received iron-59 from the Berkeley cyclotron.[48] This made it hard to challenge the Rochester group's authoritative findings and interpretation. Carl V. Moore at Washington University, for instance, began studies of iron absorption in dogs and humans with a sample of iron-59 from E. O. Lawrence and Martin Kamen in 1940. Over the course of the war, he shifted to using iron-59 produced by the Washington University cyclotron, which had entered the radioisotope supply network.[49] In Moore's lab, many of the human research subjects were laboratory staff. Elmer Brown, a researcher there, recalled that "we all served as subjects in one form or another, and if you look through the papers of Carl, Bill Harrington, etc. the initials are readily identifiable." He explained that this was a something of a tradition in hematology there and claimed that Moore was first among the volunteers. The reason was not that there was "any special mystique about self-experimentation or any derring-do that made risk-taking pleasurable or exhilarating." Rather, researchers were more motivated, reliable, and knowledgeable than patients, and willing to put up with "recurrent discomfort." After "hematology became more sophisticated," he observed, there was a shift to using paid medical students and other hospital volunteers—"and I'm afraid something was lost in the process."[50] While being an experimental subject had its risks, so did aspects of laboratory

47. Hahn et al., "Red Cell and Plasma Volumes" (1942); Hahn et al., "Radioactive Iron Absorption" (1943); Hahn et al., "Peritoneal Absorption" (1944).

48. The others who received shipments of radioiron from the Rad Lab were Erich Baer (Toronto), George Hevesy, I. L. Inman, and Carl V. Moore (Washington University), all in 1940, with a second shipment to Moore in 1941. By comparison, University of Rochester, where Whipple's group was based, received shipments in 1937, 1938, 1939, 1940, 1941, and 1942. Amounts are not specified. "Special Materials."

49. Moore et al., "Absorption of Ferrous and Ferric Radioactive Iron" (1944). For mention of the shipment from Berkeley, see footnote 2 on p. 756. Moore received another shipment from Berkeley in 1941, but it was iron-55: Martin D. Kamen to Carl V. Moore, 21 Jul 1941, E. O. Lawrence papers, Banc Film 2248, series 1, reel 14, folder 10:10 Kamen, Martin D.

50. Elmer B. Brown to Max Wintrobe, 17 Jan 1982, Wintrobe papers, box 70, folder 10, pp. 5 and 6.

work, particularly the in-house preparation of the phosphorus-32 from irradiated cyclotron targets.[51] Yet at centers such as Washington University and Harvard Medical School, researchers seemed to accept the necessity of some radiation exposure in the course of conducting research and providing therapy.[52]

Radioiron in Wartime and Postwar Research

Paul Hahn, who had worked with Whipple on nearly all the radioiron studies in Rochester, moved during the war to a position at Vanderbilt, where he continued to work on iron metabolism. A 1945 paper examined the absorption of iron in its ferrous and ferric forms in nine patients and three dogs, and illustrates how the use of radioisotopes in medical research continued through World War II on a relatively small scale. This publication included several collaborators from Rochester and Louisiana State Medical School and a coauthor (Wendell Peacock) from the Department of Physics at MIT, the source of the iron-59 used in the research.[53] The patients were selected due to their availability through hospitalization or outpatient treatment of anemia; in this respect the study is like that from Rochester in 1942.[54] All were being given iron supplements orally for therapeutic reasons. (In this sense the study was not strictly nontherapeutic research, though the radioactivity itself had no therapeutic value.) Each patient was given three administrations of radioactive iron (specified in milligrams, not curies), alternating ferric and ferrous forms for each subject. Absorption was determined by measuring the radioactivity of the

51. Virginia Minnich to M. M. Wintrobe, 2 May 1983, Wintrobe papers, box 73, folder 12; oral history of Joseph F. Ross conducted by Eric Hoffman, 11–12 Jun 1986, Columbia University Oral History Research Office.

52. See especially Ross oral history, ibid., on his work preparing radium for therapy, and when cyclotron material became available, iron-59 and phosphorus-32. He talks about how the "guys" who had been at Harvard Medical School since the early days of radium and x-ray use had their arms serially amputated up to the shoulder joints due to cancer. At Berkeley, Martin Kamen did most of the radiochemical separation before giving radioisotopes to researchers, and the MIT cyclotron similarly set up facilities for radiochemistry.

53. Hahn et al., "Relative Absorption" (1945). On the supply of radioisotopes from MIT: Robley D. Evans, "Radioactivity Center, 1934–1945," unpublished history, 28 Jun 1945, Evans papers, box 1, folder Radioactivity Center 1934–1945.

54. It is not clear from the text whether the patients were being seen at Vanderbilt, Rochester, or Louisiana State Medical School.

iron from red blood cells obtained through blood samples, four to eight days after the tracer was administered.

The descriptions of the human participants convey the prevailing medical paternalism. Four of the subjects were "colored patients," hospitalized for stomach cancer, uterine tumors, ulcer, and multiple pregnancy iron deficiency. If these patients were informed of the study, there is no mention in the paper. Strikingly, the authors remarked directly on the subjects' race as related to their perceived compliance and intelligence. When results with these four proved equivocal as to which form of iron was better absorbed, the researchers decided to "study several subjects who could be relied upon not to take any iron other than that given for tracer studies." For this purpose they selected "two young white women whose intelligence and dependability were unquestionable."[55] Results with these two, who were presumably informed of their status as subjects so that they would not consume any additional iron, showed uptake of the ferrous form to be more efficient. Similar studies with three outpatients, whose racial status is unremarked, confirmed this finding. Clearly, the ability of researchers to control their subjects' behaviors, including diet, were not without limitations, but neither is it evident that they informed all patients of the study, which could have elicited their deliberate cooperation—or their outright refusal. In these respects, early radioisotope studies were not inherently different from other clinical experiments.[56]

During the war, as part of the scientific mobilization, radioiron was used at MIT in studies of shock, sponsored by the Committee on Medical Research of the Office of Scientific Research and Development. On the battlefield, plasma was widely used to treat battlefield shock, but there was question as to whether this blood fraction, lacking red blood cells, could keep tissues sufficiently oxygenated. Using radioactive iron, researchers found that in shock the red cells are sequestered and not accessible to circulation. Hence treatment of shock with plasma (to increase blood volume) might not be as useful as previously thought, especially for

55. Hahn et al., "Relative Absorption" (1945), p. 196.

56. American physicians and scientists in the first half of the twentieth century recognized the principle of obtaining consent for subjects of nontherapeutic research, yet consent procedures were used erratically and seem to have been sometimes disregarded altogether. Needless to say, it is difficult to document ethics practices that were not managed bureaucratically. ACHRE, *Final Report* (1996), ch. 2 and p. 160; Halpern, *Lesser Harms* (2004), introduction; Rothman, *Strangers at the Bedside* (1991); Stark, *Behind Closed Doors* (2012).

battlefield casualties involving hemorrhage. Most of this research used experimental animals, but there were human experiments as well. In at least one experiment healthy subjects were injected with radioactive iron. The perceived safety of these experiments comes across in a 1948 account of the procedure: "Some of the radioactive iron thus introduced was in turn synthesized by the subject into hemoglobin in his own circulating red cells in a period of two to four weeks and without any reaction of the subject to the small amount of radiation received."[57]

In 1946, once radioactive iron could be purchased from Oak Ridge, researchers at a much wider range of medical schools and hospitals began obtaining it.[58] The AEC's *Fourth Semiannual Report* to Congress featured descriptions of research conducted using Oak Ridge radioisotopes, including fourteen experiments employing radioiron in biology and medicine.[59] Most focused on the absorption and metabolism of iron in various experimental animals. For example, researchers at the Thorndike Memorial Laboratory in Boston sought to investigate how uptake of ingested iron was affected by "environmental conditions" such as accessory food materials, change in pH, oxidizing and reducing agents, and the form of iron given.[60] Such studies generated information of value for improving iron supplementation. A few studies related to animal nutrition or agriculture as well.[61]

Other prominent postwar investigations using radioiron were oriented toward hematology. Clement Finch, nicknamed "Mr. Iron," used radioiso-

57. Burchard, *Q.E.D.* (1948), p. 156.

58. The Rochester group had published a method for determining red cell and plasma volumes with radioiron: Hahn et al., "Red Cell and Plasma Volumes" (1942). This method did not come into general use because one could not label red blood cells in vitro with radioiron, whereas one could label blood serum albumin with radioiodine, and a similar method existed for using this label for ascertaining blood volume. Berlin, "Blood Volume: Methods and Results" (1953).

59. AEC, *Fourth Semiannual Report* (1948), Appendix A. Seventy-three institutions were represented in the section on "Biological and Medical Research" and 39 in "Medical Therapy and Diagnostics," including some duplication.

60. AEC, *Fourth Semiannual Report* (1948), p. 87.

61. For instance, researchers at the University of Hawaii used iron-59 to "determine the physiological mechanism in the chlorosis of pineapple plants when supplied with great amounts of manganese." That mechanism, as they worked out, involved the combination of iron with an enzyme stimulating protein formation in the chloroplasts. AEC, *Fourth Semiannual Report* (1948), p. 106.

topes extensively in his research on blood diseases.[62] As a medical student at the University of Rochester in the late 1930s, Finch had been introduced to problems of iron metabolism and pathologies by doing research with George Whipple (and Paul Hahn). A fellowship with hematologist Joseph Ross at Boston University after World War II brought him into contact with researchers at Harvard and MIT using cyclotron-produced radioiron.[63] He was appointed to the hematology staff at Peter Bent Brigham Hospital (where he had also been an intern) and investigated the uptake of iron given intravenously to normal subjects and to patients with a variety of blood disorders.[64]

In 1949, Finch moved to the University of Washington to start a hematology department at the medical school. There, supported by grants from the National Institutes of Health and the AEC, he investigated iron turnover in the body, attending to the role of iron-containing proteins besides hemoglobin such as ferritin and hemosiderin. In one classic study with healthy and sick human subjects, Finch and his associates used radioiron to measure the rate of hemoglobin synthesis in the bone marrow and the rate of its delivery to red cells in the circulating blood. They showed that patients with pernicious anemia synthesized sufficient amounts of hemoglobin, but most of their red blood cells were destroyed in the bone marrow rather than released into circulation.[65] In another publication on the lifespan of the red blood cell, he and his coauthors represented iron turnover schematically as a circuit, whose flow could be altered by either if from "iron loading" or rerouting iron to the bone marrow. Like biochemical metabolic cycles, their representation emphasized the circular movement of materials as the body regulates biosynthesis and breakdown. (See figure 8.2.)

Carl Moore at Washington University led another prominent hematology group studying iron metabolism and blood disorders. Using radioiron

62. On Finch's nickname, see Altman, "Clement A. Finch" (2010).

63. Ross had worked with Whipple and Hahn at Rochester using radioiron to study blood diseases. Oral history of Joseph F. Ross by Eric Hoffman, 11–12 Jun 1986, Columbia University Oral History Library; oral history of Clement A. Finch by Keith Wailoo, 16 Nov 1990, Columbia University Oral History Library.

64. Finch et al., "Iron Metabolism" (1949). Finch was at Brigham Hospital from 1941 to 1949, except for the one year's fellowship at Boston University with Joseph Ross.

65. Finch et al., "Erythrokinetics in Pernicious Anemia" (1956). On the importance of this experiment and its representation in textbooks, see Wintrobe, *Blood, Pure and Eloquent* (1980), pp. 290–91.

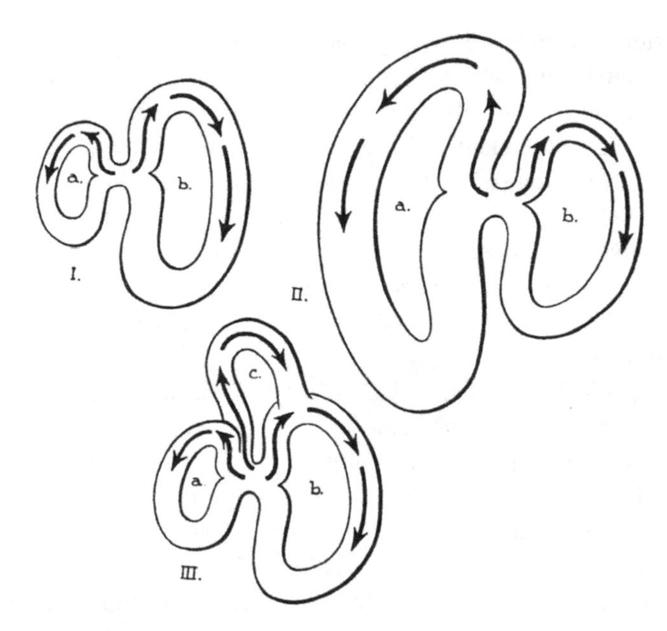

FIGURE 8.2. Iron turnover in the body expressed diagrammatically as circuits. The portions marked *a* represent all body tissue iron with the exception of that going to the red cell mass, including ferritin, hemosiderin, myoglobin and iron-containing cell enzymes; *b* represents that portion synthesized into hemoglobin within the red cell; *c* represents a hypothetical circuit into which iron initially passes, but from which it is rapidly rerouted to the erythron. *I.* Normal iron metabolism in which approximately 80% of the iron goes through the red cell mass (*b*) and a smaller portion goes to other tissues (*a*). *II.* The situation produced by iron loading. Circuit (*a*) is greatly increased so that only a small portion of iron passing through the serum iron compartment goes to the red cell mass (*b*). *III.* An additional circuit (*c*) is shown to explain the difference in results from iron turnover as compared with life span studies. Caption and image from E. Langdon Burwell, Barbara A. Brickley, and Clement A. Finch, "Erythrocyte Life Span in Small Animals: Comparison of Two Methods Employing Radioiron," *American Journal of Physiology* 172 (1953): 718–24, on p. 719.

purchased from Oak Ridge, Moore sought to challenge Whipple's theory that iron uptake was controlled by the intestinal mucosa and regulated by iron stores in the body.[66] Moore used healthy subjects as well as patients in his investigations of how pernicious anemia affected hemoglobin and red blood cell synthesis.[67] Other researchers examined how quickly ra-

66. AEC, *Fourth Semiannual Report* (1948), p. 90. For the evidence Moore assembled to disprove Whipple's theory, see Moore, "Iron Metabolism and Nutrition" (1961).

67. Dubach, Callender, and Moore, "Studies in Iron Transportation" (1948); Moore and Dubach, "Absorption of Radioiron from Foods" (1952).

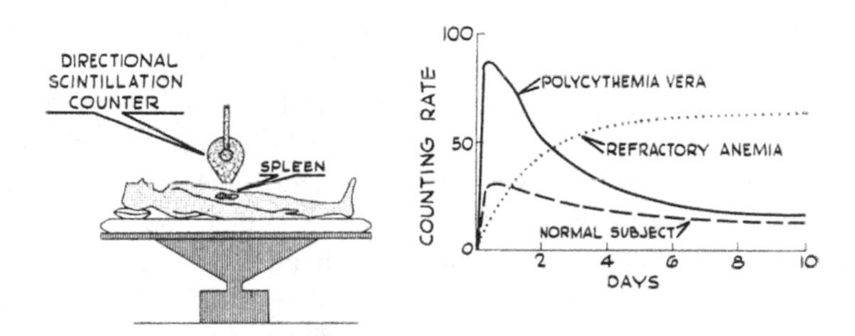

FIGURE 8.3. Diagram showing how the accumulation of radioactivity in the human spleen over time was measured. At time zero, a subject would be given a dose of radioactive iron. The appearance of gamma radiation from the decaying radioiron would be measured by a scintillation counter directed at the spleen. In the chart at right, the rapid appearance of radioiron in the spleen of patients with polycythemia vera, in which red blood cells are overproduced, shows that short-lived cells were being formed (and degraded in the spleen) as well as normal red blood cells of a 120-day life span. In the case of refractory anemia, radioiron accumulates in the spleen as red blood cells are destroyed, but the breakdown products are then retained in the organ. From US Atomic Energy Commission, *Eight-Year Isotope Summary*, vol. 7 of *Selected Reference Material, United States Energy Program* (Washington, DC: US Government Printing Office, 1955), p. 10.

dioiron accumulated in the spleen of patients with other blood disorders, namely, polycythemia vera and refractory anemia.[68] The AEC's *Eight-Year Isotope Summary*, published in 1955, covered recent findings in this area and showed how a directional scintillation counter could be used to determine the amount of radioactivity in the spleens of human subjects given radioiron. Various blood diseases could be distinguished by the markedly different patterns of radioiron accumulation in the spleen. (See figure 8.3.)

68. Whipple's group identified the spleen as a "depot for iron coming in through the gastro-intestinal tract on its way to general body use." Hahn et al., "Radioactive Iron Absorption" (1943), p. 183.

Early research with radioiron was almost by definition nontherapeutic, although in many cases research results may have been informative for treating patients. In part this was because radioisotopes came into the study of blood diseases primarily as a research tool rather than as a therapeutic or diagnostic agent.[69] The uptake of radioisotopes into this area of clinical research introduced novel problems as well as opportunities. Clement Finch stated that early in his career, he tended to perform every experiment on himself before involving other human subjects (like researchers in Carl Moore's lab). As he put it, "This seemed a good way to evaluate the appropriateness for another person."[70] However, he noted that the greater availability of radioisotopes and their routine incorporation into hematology research made this impossible—the radioactive body burden such a researcher would accumulate over time would be too high. Both the postwar influx of federal funding and the widespread uptake of radioisotopes in research studies made the availability of patients as subjects critical.[71]

The Vanderbilt Nutrition Studies

As compared with the clinical studies just discussed, nutritional research with radioiron focused on healthy human subjects. The largest of such efforts was a nutritional study at Vanderbilt Medical School supported by the Tennessee Department of Public Health, the Nutrition Foundation, and the Rockefeller Foundation's International Health Division. The initial impetus for this came from the Rockefeller Foundation's expansion of its public health work, which dated to 1938, to include nutrition.[72] The foundation sought to involve state health departments in the South; the first states to respond were Tennessee and North Carolina.[73] The fact that

69. Oral history of Joseph F. Ross by Eric Hoffman, 11–12 Jun 1986, Columbia University Oral History Library.

70. Oral history of Clement A. Finch by Keith Wailoo, 16 Nov 1990, Columbia University Oral History Library. On self-experimentation as demonstrating "the nobility of the investigator," see Lederer, *Subjected to Science* (1995), p. 127.

71. Rothman, *Strangers at the Bedside* (1991), pp. 51–55.

72. "Nutrition as a Public Health Problem," RAC, RF 1.1, series 200, box 65, folder 789 Vanderbilt University–Nutrition, April–December 1939.

73. John A. Ferrell to W. C. Williams, Commissioner, Tennessee Department of Health, 12 Aug 1938, RAC RF 2–1938, series 100, box 154, folder 1137.

some of the South's disease patterns might be specifically linked to malnutrition had been recognized for some time, pellagra being "the distinctive mark of hunger in the South."[74] In 1939, two populations in Tennessee were selected for dietary surveys and laboratory tests for basic vitamin and mineral deficiencies, conducted by researchers from Vanderbilt University Medical School.[75]

When Hahn moved to Vanderbilt in the fall of 1943 he joined the nutrition project.[76] He was assigned to investigate "the manner by which the body handles nutritional essentials and their mode of action in the human, for the purpose of obtaining information from which tests of the nutritional status can be devised or improved."[77] Though he was an Assistant Professor of Biochemistry, Hahn's position depended on soft money, initially from the Rockefeller Foundation's grant and subsequently from the Nutrition Foundation.[78] A year after Hahn arrived William J. Darby was appointed Assistant Professor of Medicine and director of the reorganized Tennessee-Vanderbilt Nutrition Project. Under Darby's leadership this research effort developed into a permanent Division of Nutrition at Vanderbilt Medical School.[79] Darby's background was in biochemistry as well as vitamin and nutritional deficiencies.

74. Etheridge, "Pellagra" (1988), p. 115. As determined in 1937, lack of the vitamin nicotinic acid (or niacin) causes pellagra.

75. See documents in RAC, RF 1.1, series 200, box 65, folder Vanderbilt University–Nutrition, 1940. A similar program was set up in North Carolina.

76. William D. Robinson to J. A. Ferrell, 14 Oct 1943, RAC, RF 1.1, series 248, box 1, folder 7 Nutrition August–October 1943.

77. "Duties of Personnel," RAC, RF 1.1, series 248, box 1, folder 8 Nutrition November–December 1943.

78. John Ferrell at the International Health Division requested that Hahn's salary, which began in October 1943, be transferred at the end of the year to the non-Rockefeller Foundation portion of the budget (to funds supplied by the Nutrition Foundation). The current director of the Vanderbilt project, William Robinson, had hired two biochemists whereas only one had been budgeted for. Hahn was to be shifted because he had recently been denied a fellowship from the Natural Sciences Division (which had a policy against giving support in someone's first year in a position), and Ferrell felt it would look bad if Hahn's salary were then picked up by the International Health Division. Hahn was also supported by a grant from the TVA to study phosphorus intoxication and malarial parasites (for which Hahn employed radiophosphorus), and by other small grants from the Arthur D. Little Company and the OSRD. See documents in RAC, RF 1.1, series 248, box 1, folder 8 Nutrition November–December 1943 and RG 2–1946, series 200, box 333, folder 2254.

79. William J. Darby, "An Application for a Grant to Assist in the Developing of a Division of Nutrition in Vanderbilt University School of Medicine," with cover note from Ernest

In 1945 Darby launched the Vanderbilt Cooperative Study of Maternal and Infant Nutrition, which involved researchers from five medical school departments as well as the Tennessee-Vanderbilt Nutrition Project.[80] The Rockefeller Foundation had been disappointed with the initial survey-based nutrition studies they had supported in Tennessee, and was delighted with Darby's clinical- and laboratory-based approach.[81] The project aimed at assessing the nutritional status of pregnant women who entered the Vanderbilt Hospital Outpatient Clinic. These women were followed through delivery so that the health of their infants could be evaluated. Darby sought to establish whether there was a correlation between nutritional status and health performance in pregnancy.[82] The number of pregnant women that came through Vanderbilt Hospital would provide a large enough population to identify significant correlations. The women who sought care at the outpatient clinic were generally poor; because of the segregation of health care, they were all white. Data previously collected from the Obstetrical Service of Vanderbilt Hospital revealed that women on the wards showed higher incidence of toxemia, puerperal fever, and fetal mortality than private patients, suggesting that "some factor related to economic status may be involved in determining their appearance."[83] Darby suspected that factor might be nutrition. He secured the continuing support of the Rockefeller Foundation, the Nutrition Foundation, and Vanderbilt University, as well as a grant from the US Public Health Service.[84] (See figure 8.4.)

W. Goodpasture to Hugh H. Smith, 26 Sep 1946, RAC, RF 1.1, series 248, box 1, folder 12 Nutrition 1946.

80. Darby, Densen, et al., "Vanderbilt Cooperative Study" (1953); Darby, McGanity, et al., "Vanderbilt Cooperative Study" (1953). The medical school departments whose research groups became involved were Preventive Medicine, Obstetrics and Gynecology, Pediatrics, Medicine, and Biochemistry.

81. See change in the correspondence in RAC RF 1.1, series 248, box 1, folder 5 Nutrition January–May 1943 and folder 11 Nutrition 1945.

82. Projected Plans for the Activities of Section B of the Tennessee-Vanderbilt Nutrition Project for the Period from July 1, 1946, with letter from William J. Darby to R. H. Hutchison, 11 Aug 1945, RAC, RF 1.1, series 248, box 1, folder 11 Nutrition 1945, p. 3.

83. "Outlines of Plans for a Cooperative Study of Nutritional Status during Pregnancy," enclosed with letter from William J. Darby to Hugh H. Smith, 21 Sep 1945, RAC, RF 1.1, series 248, box 1, folder 11 Nutrition 1945, p. 2.

84. Darby, "An Application," 26 Sep 1946. The Nutrition Foundation was incorporated in Dec 1941 with funding from the food industry to support research on nutrition and its

FIGURE 8.4. Patients register at the obstetrics clinic of Vanderbilt Hospital. As the original caption states, "All white patients attending the prenatal clinic were included in the study," i.e., the Vanderbilt Cooperative Study of Maternal and Infant Nutrition. Rockefeller Foundation Photographs, series 248, box 59, folder 1351. Courtesy of Rockefeller Archive Center.

Anemia in pregnancy was a well-known medical problem, and so it is not surprising that the study included attention to iron deficiency.[85] Darby planned to conduct an iron absorption test on each pregnant subject, to be scheduled alongside a vitamin B excretion test.[86] The study included a

---

dissemination through grants-in-aid and a publication, "Nutrition Reviews." See Nutrition Foundation pamphlet, RAC. RF 2–1943, series 200, box 246, folder 1700.

85. On anemia in pregnancy, Larrabee, "Severe Anemias" (1925); Bland, Goldstein, and First, "Secondary Anemia" (1930); Elsom and Sample, "Macrocytic Anemia" (1937).

86. "Outlines of Plans," 21 Sep 1945, p. 5. C. G. King, Scientific Director of the Nutrition Foundation, had already approached D. F. Milam, who was undertaking the Rockefeller Foundation's nutrition study in North Carolina, about the desirability of studying iron deficiency and supplementation. Milam demurred, after consulting with John Ferrell at the International Health Division. But this provides some evidence that the issue was timely. There is no evidence the motivation for including iron absorption studies in Darby's project came from the Nutrition Foundation. See correspondence from Mar 1943 between Ferrell, Milam, and King in RAC, RF 2–1943, series 200, box 246, folder 1700.

battery of other laboratory tests of blood constituents and vitamin excretion to be performed on each woman, but only the iron test involved use of a radioactive isotope. As Darby wrote Hugh Smith at the Rockefeller Foundation, "The radioactive iron studies will be the first extensive series of this sort and should provide much fundamental physiological information concerning the iron requirements during pregnancy."[87] Smith wrote right back expressing his enthusiasm about the planned investigation; his only suggestion was to add a measure of blood volume for a significant number of the women.[88] (See figure 8.5.)

It was evidently Hahn's presence at Vanderbilt that made the iron uptake part of the research possible. Darby, Hahn, and four other collaborators undertook a pilot study of radioiron uptake in 1945. The experiment, involving 189 children from two Nashville school districts, demonstrated to the satisfaction of its authors "the feasibility of employing isotopes in large-scale surveys of the nutriture of populations and has led to the institution of a more extensive similar investigation of the iron requirements during pregnancy."[89] The children were administered trace doses of iron-59 in lemonade; two weeks later blood samples were drawn to measure the absorption of the iron and other blood markers. The MIT Radioactivity Center supplied Hahn with the radioiron.[90]

This study of schoolchildren showed no statistically significant difference in iron absorption or hemoglobin between students in the two districts. This was despite clear socioeconomic disparities and differing levels of serum ascorbic acid and carotene that suggested (along with weight differences) the children in the more affluent district received better nutrition.[91] The publications make no mention of the level of radiation to which the children were exposed. Neither did another Vanderbilt study that investigated iron absorption in newborns in order to understand whether the "physiological" anemia experienced by most infants at 12 to 14 weeks

87. "Outlines of Plans," p. 7.

88. Hugh H. Smith to William J. Darby, 25 Sep 1945, RAC, RF 1.1, series 248, box 1, folder 11 Nutrition 1945.

89. Darby et al., "Absorption of Radioactive Iron by Children" (1947), p. 108. For the pilot study, see Darby et al., "Absorption of Radioactive Iron by School Children" (1946).

90. Robley D. Evans, "Radioactivity Center, 1934–1945," unpublished history, 28 Jun 1945, Evans papers, box 1, folder Radioactivity Center 1934–1945, p. 32. Appendix IX, dated 24 Mar 1945, gives a list of wartime government projects that relied on the Radioactivity Center and a separate list of individuals who were receiving radioisotopes for research or therapy.

91. Darby et al., "Absorption of Radioactive Iron by Children" (1947), p. 110.

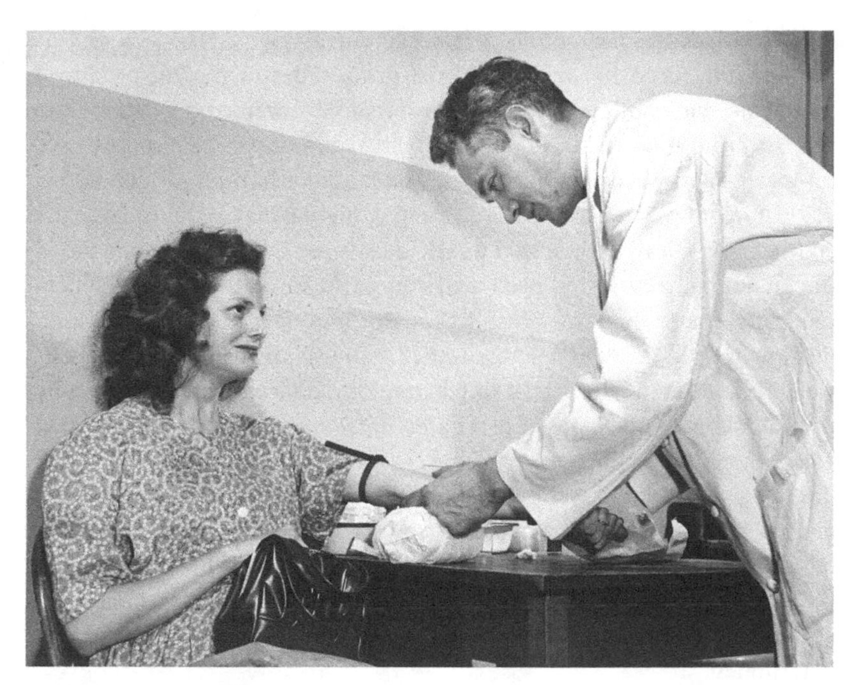

FIGURE 8.5. A patient in the Vanderbilt Cooperative Study of Maternal and Infant Nutrition having her blood drawn for tests. This photograph was taken in 1949, after the radioiron experiments had ceased. Rockefeller Foundation Photographs, series 248, box 59, folder 1351. Courtesy of Rockefeller Archive Center.

is caused by an inability to absorb iron. Fourteen full-term infants and ten premature infants were administered 200,000 counts per minute of iron-59 through a feeding tube and blood was drawn two to six weeks later from each baby to measure the amount of radioactive iron incorporated into red blood cells. All of the infants absorbed iron, between 0.4% and 8.2% of the test dose, the same range absorbed by normal children.[92] Researchers seemed to have regarded the radioiron in this amount as a tracer, with no significant biological radiation effects.

Paul Hahn designed the portion of the Vanderbilt study assessing iron absorption in over 800 pregnant women, which began in September 1945.[93] This was before radioisotopes were available for sale from Oak Ridge. Hahn obtained the radioiron used initially from MIT, but purchased

92. Oettinger, Mills, and Hahn, "Iron Absorption" (1954).

93. ACHRE, *Final Report* (1996), p. 214, note c.

AEC-produced radioiron when it became available. There was an important difference between cyclotron-produced and reactor-produced radioiron: one could generate pure iron-59 in a cyclotron, whereas the reactor-produced version contained about 10% iron-55. The iron-55 gave off lower-energy radiation but possessed a longer half-life, which in retrospect ought to have given those planning human experiments cause for concern.[94] Ironically, whereas the AEC's Isotopes Division recommended the agency make cyclotron-produced iron-59 available, the Commissioners requested that "whenever possible, . . . isotopes distributed by the Commission are produced in reactors rather than cyclotrons."[95] The pregnant women in the study received a single dose of radioiron with a therapeutic iron supplement during their second prenatal visit, which took place anywhere from the tenth to thirty-fifth week of pregnancy. The radioisotope dose was considered low by the standards of the 1940s; each woman was given a solution containing 200,000 to 1,000,000 counts per minute of radioiron.[96] The amounts of nonradioactive therapeutic iron varied among the women, as part of the experimental design. On the next prenatal visit, a blood sample was drawn from which the absorption of radioiron was determined.

When Hahn presented early results at a December 1946 meeting of the American Federation for Clinical Research in New Orleans, his talk was picked up by the Nashville papers. One headline read "Iron Doses with

94. "Availability of Radioactive Isotopes" (1946), esp. Table 4; Welsome, *Plutonium Files* (1999), ch. 22 and esp. p. 223, which cites a 1947 letter from Hahn to a physician in Florida stating that the half-lives of both iron-55 and iron-59 were too long for therapeutic use (though he clearly approved using them in tracer experiments).

95. Extract from Status Report, 1–15 May 1949, "Program for Production and Distribution of Cyclotron-Produced Isotopes – AEC 195," NARA College Park, RG 326, E67A, box 45, folder 3 Distribution of Radioisotopes–Domestic. As subsequently assessed by the Isotopes Division, "Production of radiochemically pure Fe 59 by pile irradiation does not compare favorably with its production by fast neutrons from the cyclotron." "A Review of the Possibility of Reactor Production of Isotopes Currently Produced by Cyclotron Bombardment," report by the Manager, Oak Ridge Operations Office, 20 Feb 1950, NARA College Park, RG 326, E67A, box 45, folder 3 Distribution of Radioisotopes–Domestic, p. 4.

96. By my calculations, this would be in the range of 0.1 to 0.5 microcuries if counting were completely efficient. One contemporary textbook author specifies "tracer quantities" as in the microcurie range. Siri, *Isotopic Tracers* (1949), p. 450. The investigators did not estimate the fetal effective dose as part of their study; a recent estimate is that it was a few hundred millirems (a rem being equivalent to a roentgen for x-ray and gamma radiation). ACHRE, *Final Report* (1996), p. 214, notes e–f.

Radioactive Isotopes Aid to Pregnancy, Experiment Shows."[97] Further preliminary findings were presented by Hahn at a meeting of the American Society for Experimental Pathology in 1947.[98] Hahn and his coworkers determined that the overall average uptake of the iron in the pregnant women was 28.5%, with the percentage uptake increasing as gestation progressed, from 17% during the first quarter to 36% in the last. In 10% of the cases, blood was drawn from the umbilical cord at delivery as well as from the mother. Results from analyzing these blood samples showed that the fetus took up about a tenth as much iron as the mother. This finding confirmed the suggestive, unpublished observation made by Hahn and Bale with an experimental dog in 1938 that radioiron could cross the placenta. Given the widespread assumption among obstetricians that the placenta was generally impermeable, this might well have been a surprising result.[99]

The full publication on this Vanderbilt study did not appear until 1951; Hahn was the lead author.[100] It presented results from 466 of the pregnant patients, including only those who returned for the follow-up visit and whose blood samples met certain criteria for radiochemical purity. Although the authors are listed as members of several departments at Vanderbilt Medical School, Hahn was no longer there. In 1946, the nutrition project was refinanced, leaving a gap in the coverage for Hahn. William Darby was unable to persuade the Department of Biochemistry, where Hahn had been promoted to Associate Professor, to pick up one-half of his salary (the other half came from the Nutrition Foundation grant). Nor could the Rockefeller Foundation be convinced to cover the salary of someone with a faculty appointment.[101]

97. *Nashville Banner*, 13 Dec 1946. The other news story ran as "VU To Report On Isotopes" in the *Nashville Tennessean*, 14 Dec 1946, both cited in ACHRE, *Final Report* (1996), p. 214, note h.

98. Hahn et al., "Iron Uptake" (1947).

99. Dally, "Thalidomide" (1998). On Hahn and Bale's work, see p. 269.

100. Hahn et al., "Iron Metabolism in Human Pregnancy" (1951).

101. Entry for 9 Jan 1946, Hugh H. Smith Diary, RAC RF 12.1, box 130, 1946 volume, p. 5. Ernest Goodpasture, Dean of the School of Medicine, made an effort to secure support for Hahn's program from Warren Weaver at the Rockefeller Foundation, but Weaver claimed his division's budget was already overloaded. For his part, Goodpasture contended that "the situation with reference to Dr. Hahn has no reflection on the work which he has done here for it has been excellent." Goodpasture to Weaver, 9 Apr 1946, RAC RF 2–1946, series 200, box 333, folder 2254; Weaver to Goodpasture, 10 Mar 1947 and Goodpasture to H. Marshall Chadwell, 9 Oct 1947, RAC RF 2–1947, series 200, box 374, folder 2521. This documentation

As a result of this funding impasse, Hahn moved within the city in 1947 to Meharry Medical College, a historically black medical school. From the dating, it appears that when he left Vanderbilt the radioiron tests in the Cooperative Study of Maternal and Infant Nutrition ceased.[102] The following year Hahn reported to the AEC that he was conducting a study of iron absorption in 1,000 hospital admissions at Meharry (again, an institutionally available population), with radioiron purchased from Oak Ridge, though complete results were never published.[103] In fact, Hahn was shifting gears from using radioisotopes in tracer studies to using them as radiation sources, beginning with the treatment of leukemia with radioactive manganese and the use of colloidal gold-198 for radiation therapy.[104] Most of his subsequent publications focused on this topic, on which he also edited a book.[105] Strikingly, during the first year of the AEC's radioisotope program, the Vanderbilt research group of Hahn and his coworker C. W. Sheppard received "the largest number of shipments and the largest total activity of material."[106]

Little about Hahn's personality comes through in the correspondence about his brief appointment at Vanderbilt. Whipple wrote to Warren Weaver, "Occasionally he is a little too enthusiastic but that is a good fault."[107] Officers at the Rockefeller Foundation never criticized Hahn's

---

does not resolve whether personality, performance, or institutional priorities played the largest role in ending Hahn's position at Vanderbilt.

102. It is possible that the group's recent finding that fetuses took up one-tenth as much radioiron at their mothers informed this decision, but it seems more likely to have resulted from the personnel change.

103. Preliminary results are published in the AEC, *Fourth Semiannual Report* (1948), p. 77.

104. Hahn began this work while at Vanderbilt; C. S. Robinson to E. W. Goodpasture, 11 Apr 1946 and "VU Doctors Discover Therapy for Leukemia," clipping from Nashville Tennessean, 23 Apr 1946, RAC RF 2–1946, series 200, box 333, folder 2254.

105. Hahn and Sheppard, "Selective Radiation" (1946); Sheppard, Goodell, and Hahn, "Colloidal Gold" (1947); Hahn et al., "Direct Infiltration" (1947); Hahn, *Therapeutic Use* (1956).

106. "Background Material on Activity in First Year of Distribution of Pile-Produced Radioisotopes," Press Release, 2 Aug 1947, AEC Records, NARA College Park, RG 326, E67A, box 45, folder 3 Distribution of Radioisotopes–Domestic, p. 8. This same document (p. 10) features Hahn's research into radiogold therapy, covered in chapter 9, not the tracer experiments with radioiron (which undoubtedly used much less radioactive material).

107. G. H. Whipple to Warren Weaver, 22 Apr 1946, RAC RF 2–1946, series 200, box 333, folder 2254. A Rockefeller Foundation officer, who visited with Hahn in 1957, wrote that "it was a bit difficult to see him through a blue haze of tobacco smoke which I should judge has

use of radioisotopes in human subjects; indeed, this aspect of the project is barely mentioned in the grant correspondence. Interestingly, Joseph Ross, a hematologist who worked alongside Hahn in Rochester, claims that Whipple discouraged Hahn from pursuing a medical degree, saying that his training in biochemistry was sufficient for the radioisotope work he wished to do. According to Ross, this became a source of professional frustration to Hahn because he "was never really able to step out and become a clinical investigator."[108]

The 1951 publication by the Vanderbilt researchers on "Iron Metabolism in Human Pregnancy" opens by citing the 1942 Rochester investigation of absorption of radioiron in hospital patients including pregnant women. The large-scale study is presented as a more careful examination of Whipple's striking result that pregnant women absorb two to ten times the amount of iron taken up by nonpregnant individuals. The Vanderbilt study differed from that of the Rochester group not only in scale, but also in the use of prenatal outpatients as the experimental subjects. Recall that the pregnant women who had been given radioiron at Rochester were either full-term or were being hospitalized for a therapeutic abortion. Nearly all of the fetuses of women enrolled in the Vanderbilt study, by contrast, were exposed to radiation from ingested radioiron over a significant portion of the gestational term. At the time this did not apparently concern the medical researchers, nor their funding organizations. As Ann Dally has observed, there was "mass denial by the medical profession in the mid-20th century about the permeability of the placenta."[109] Not until the 1950s did evidence of the adverse health effects of low levels of ionizing radiation prompt a wider reevaluation of the safety of human exposures to tracer-level doses of radioisotopes.

By the 1960s, new scientific evidence for the heightened hazards of low-level radiation to developing fetuses prompted Vanderbilt researchers to reassess the health effects on the children whose mothers had taken radioiron. In 1956, Alice Stewart and her coworkers at Oxford published

---

time to dissipate only during periods when Hahn is on extended vacations from work." He did not advise rushing into funding any fellowship programs or research aid at Meharry. O. L. Peterson, diary entry, 17 Oct 1957, RAC RF 2–1957, series 200, box 25, folder 203.

108. Ross attributes Hahn's eventual alcoholism to this frustration. Oral history of Joseph F. Ross by Eric Hoffman, 11–12 Jun 1986, Columbia University Oral History Library.

109. Dally, "Thalidomide" (1998), p. 1197. She argues that this did not change until the 1960s with the very public evidence of the damage of thalidomide on developing fetuses.

a controversial finding: they had found an increased incidence of leukemia and other malignancies in children born to women who had received diagnostic x-rays while pregnant.[110] Critics observed that these children were not a "medically unselected population"; their mothers had conditions that had warranted diagnosis and that might have somehow predisposed their children to cancer. The Vanderbilt Cooperative Study of Maternal and Infant Nutrition offered a medically unselected population against which the correlation between fetal radiation exposure and subsequent occurrence of cancer could be tested. Ruth Hagstrom of the Department of Preventive Medicine at Vanderbilt headed this study, in which most of the women who had received radioiron were located, contacted, and surveyed; their cancer incidence was compared to 705 women in the same clinic population who had not received radioiron. Hagstrom and her coauthors found no higher rate of malignancy among the mothers, but did find a small increase in cancer among children exposed in utero to the radioiron. Three of 634 children whose mothers had taken radioiron had been diagnosed with cancer, as opposed to none in the control group. The researchers reported the discrepancy as "small, but statistically significant."[111]

In 1993, new investigative journalism into government-sponsored research that exposed subjects to radiation brought this experiment to light again.[112] In 1996, women who had been subjects in this radioisotope study brought a class-action lawsuit against Vanderbilt University and the Rockefeller Foundation, for hazards to their unborn children through radiation exposure. They settled for $10 million and a formal apology.[113] In 1945 the heightened sensitivity of developing fetuses to radiation was not appreciated by physicians; ironically, this study ended up providing important confirming evidence of the particular susceptibility to radiation in development.[114]

110. Stewart et al., "Malignant Disease in Childhood" (1956).

111. Hagstrom et al., "Long Term Effects" (1969), p. 1. The *P* value of .03 was indeed small.

112. The most important contributor was Eileen Welsome, who won a Pulitzer Prize for her coverage of the plutonium injection experiments. The various news stories are cited in LeBaron, *America's Nuclear Legacy* (1998), pp. 98–99.

113. Proctor, "Expert Witnesses" (2000); Rothman, "Serving Clio and Client" (2003).

114. The second set of Vanderbilt researchers published a related paper in a volume that attests to the new attention to fetal sensitivity to radiation: Dyer and Brill, "Fetal Radiation Dose" (1969). See also Dyer et al., "Maternal-Fetal Transport" (1969).

The Question of Oversight

What government oversight existed in the early postwar years for use of radioisotopes in human research? Notably, the Vanderbilt study began before the founding of the AEC, and there was no regulation of cyclotron-produced radioisotopes.[115] (Similarly, the Vanderbilt pilot study of schoolchildren relied on radioiron obtained privately from the MIT cyclotron.) Even once Hahn began using radioiron purchased from Oak Ridge, there is no indication that the large-scale nutritional studies at Vanderbilt raised concern. As mentioned, both a local isotope committee and the AEC's Subcommittee on Human Application had to approve all requests for radioisotopes to be used in humans. Hahn's request to use radioiron must have been accepted.[116] At the outset of the isotope distribution program, the main ethical concern of the Subcommittee on Human Application was allocation—since radioisotopes were still regarded as a scarce resource, this group was charged with setting priorities, both among various possible human uses and between human uses and research applications.

115. Hahn's other publications point to the MIT cyclotron as his source of iron-59, as do records at MIT. This is also true of the experiments in which institutionalized children at the Fernald School in Massachusetts were given tracer doses of iron-59 by MIT nutrition researchers, in a project funded by General Mills. This iron tracer study was conducted in 1946 with iron-59 from the MIT cyclotron. By contrast, a related study with radiocalcium, undertaken between 1950 and 1953, used calcium-45 from Oak Ridge. See Massachusetts Task Force, *Report on the Use of Radioactive Materials* (1994); West, "Radiation Experiments on Children" (1998).

116. Hahn filed a request for radioiron (listed as "Fe 55,59") in the summer of 1946 before government distribution began; the purpose was given as "study of iron metabolism in animals & humans." Isotopes Branch, Research Division, Manhattan District, Reports of Requests Received through 31 Jul 1946, NARA Atlanta, OROO Lab & Univ Div Official Files, Acc 68A1096, box 6, folder Radioisotopes–National Distribution Report of Requests Received to July 31, 1946, p. 4. This is also how it is listed by the AEC, *Isotopes: A Five-Year Summary* (1951), p. 227. The "Agreement and Condition for Order and Receipt of Radioactive Materials" signed by Hahn, Charles W. Sheppard, James P. B. Goodell, and Ernest W. Goodpasture (as Dean of Vanderbilt Medical School) and received by the Isotopes Branch on 12 Aug 1946 lists the departments in which radioactive materials will be used, namely Biochemistry, Medicine, Pediatrics, Obstetrics, and Surgery. NARA Atlanta, OROO Files Relating to K-25, X-10, Y-12, Acc 67A1309, box 14, Certificates. I could not find records for the Subcommittee on Human Application, and so could not ascertain whether Hahn's application for their approval specified his intention to use pregnant subjects.

However, as the Oak Ridge facility increased production, supply kept up with demand—and in fact exceeded it. In retrospect, the ethics of allocation were much less significant than issues of safety and informed consent. Here the government's sense of urgency in developing atomic energy, first for new weapons then for civilian application, ran ahead of scientific understanding of the hazards of exposure—and sometimes ahead of protection against known dangers.

Certain human experiments with radioactive materials did prompt the leadership of the AEC to grapple with its responsibility to human subjects. In order to gather information on the hazards of fissionable material and the radioactive by-products of atomic weapons production, the leaders of the Manhattan Project had decided that some human experiments were necessary.[117] The information thus obtained was to be used for occupational safety, since thousands of workers in the agency's facilities were exposed to radioactive materials of unknown danger. These experiments began during the war, and were conducted by researchers contracted by the Manhattan Engineer District. Between 1945 and 1947 a total of eighteen hospital patients in San Francisco, Chicago, and Rochester were injected with small amounts of plutonium to study how the body metabolized and excreted this substance. Similar experiments were conducted with polonium and uranium.[118]

The medical advisors that the AEC inherited from the Manhattan District, particularly Stafford Warren, strongly advised that the civilian agency continue this program of research in the postwar period. The AEC's Interim Medical Committee, chaired by Warren, met in January 1947 and approved the establishment of biomedical research contracts to twelve universities and national laboratories, many of which featured research of this nature. At the University of Rochester, Andrew Dowdy oversaw a "study of the metabolism of plutonium, polonium, radium, etc., in human subjects." Related projects with specific mention of human subjects or clinical investigation were contracted for 1947–48 at the University of California at Berkeley (under Joseph Hamilton and Robert Stone), Monsanto Chemical Company for the agency's Dayton facility, and Los

117. ACHRE, *Final Report* (1996), p. 7.

118. Ibid., ch. 6; Welsome, *Plutonium* Files (1999). Strikingly, the AEC's polonium injection experiments with human subjects were published: Silberstein et al., "Studies of Polonium Metabolism" (1950). The uranium injection experiments are discussed in ch. 9.

Alamos (in collaboration with Rochester).[119] These experiments were referred to in agency documents as "human tracer experiments."

The Commission continued these contracts, even though the plutonium injections "provoked a strong reaction at the highest levels" during the transition to the civilian agency at the beginning of 1947.[120] After all, during these same months the American Medical Association was advising the prosecution of Nazi doctors in the Nuremberg trials.[121] The AEC's leadership immediately grasped the political problems that government research on humans could cause, if publicized. In response, the Commission decided to keep information about these experiments secret, and to set in place guidelines for any future investigations. In April 1947, the General Manager of the AEC, Carroll Wilson, outlined the procedures to be followed by its contracted researchers in "obtaining medical data of interest to the Commission in the course of treatment of patients, which may involve clinical testing." The AEC would require that "prior to treatment, each individual patient, being in an understanding state of mind, was clearly informed of the nature of the treatment and its possible effects, and expressed his willingness to receive the treatment."[122] Furthermore, two doctors had to certify in writing that this had occurred. This was a clear departure from how such experiments had been handled under the Manhattan Engineer District, but it does not seem to have been intended to apply to "off-Project" biomedical experiments with radioisotopes.[123] The Advisory Committee on Human Radiation Experiments did not even find evidence that this policy was communicated to institutions conducting

119. Stafford L. Warren, Report of the 23–24 Jan 1947 Meeting of the Interim Medical Committee, US Atomic Energy Commission, OpenNet Acc NV0727195, quote from p. 8.

120. ACHRE, *Final Report* (1996), p. 152.

121. Dr. Andrew C. Ivy acted as the American Medical Association's official consultant to the prosecutors. ACHRE, *Final Report* (1996), ch. 2; Annas and Grodin, *Nazi Doctors* (1992); Glantz, "Influence of the Nuremberg Code" (1992).

122. Carroll L. Wilson to Stafford L. Warren, 30 Apr 1947, reproduced in ACHRE, *Final Report, Supp. Vol. 1* (1995), pp. 71–72. See also ACHRE, *Final Report* (1996), pp. 47–48.

123. Welsome (*Plutonium Files* [1999], p. 193) takes this letter and one other by Wilson penned that fall as evidence of general AEC policy about human experimentation, but the Advisory Committee on Human Radiation Experiments (ACHRE, *Final Report* [1996], p. 48) contends that such an interpretation is inconsistent with the centrality of nontherapeutic experiments in the AEC's isotope program and contracted medical research. I find the Advisory Committee more careful and persuasive on this matter.

medical research for the AEC, with the possible exception of the University of California, San Francisco.[124]

The issue of tracer research using radioisotopes with human subjects came up again in the fall of 1947, when the AEC's Oak Ridge Operations Office asked Wilson what legal responsibilities the government bore for "human administration of isotopes." A memorandum conveying this and other policy concerns was also sent to the agency's newly formed Advisory Committee for Biology and Medicine. It outlined explicitly the advantages and disadvantages of "tracer studies" conducted "on-Project," that is, by the AEC's contractors:

Pro–

1. Tracer research is fundamental to toxicity studies.
2. The adequacy of the health protection which we afford our present employees may in a large measure depend upon information obtained using tracer techniques.
3. New and improved medical applications can only be developed through careful experimentation and clinical trial.
4. Tracer techniques are inherent in the radioisotope distribution program.

Con–

1. Moral, ethical and medico-legal objections to the administration of radioactive materials without the patient's knowledge or consent.
2. There is perhaps a greater responsibility if a federal agency condones human guinea pig experimentation.
3. Publication of such researches in some instances will compromise the best interests of the Atomic Energy Commission.
4. Publication of experiments done by Atomic Energy Commission contractors' personnel may frequently be the source of litigation and be prejudicial to the proper functioning of the Atomic Energy Commission Insurance Branch.[125]

A ban on nontherapeutic human experiments would necessitate that the agency forego important research, as "tracer techniques are inherent in

124. The evidence on this has to do with the inclusion of documents adhering to some of these requirements in the medical file of Elmer Allen, the last patient to be injected with plutonium, and the only one after this letter. ACHRE, *Final Report* (1996), p. 48. In my own research, I have found no evidence that these guidelines applied to any "off-Project" research.

125. Medical Advisor's Office of Oak Ridge Directed Operations to Advisory Board on Medicine and Biology, 8 Oct 1947, reproduced in ACHRE, *Final Report, Suppl. Vol. 1* (1995), pp. 80–88, quote from pp. 82–83.

the radioisotope distribution program."[126] The author of this memorandum viewed these "human tracer experiments" with radioisotopes as essential to the agency and its mission, particularly in providing knowledge of the metabolism and effect of radioisotopes encountered in the course of atomic weapons production—hence the focus on toxicity studies, insurance matters, and employees at the atomic energy installations.[127] At the same time, the memorandum raised broader issues about whether the Commission had legal responsibilities for how radioisotopes were applied, and whether the government should enforce safe handling.

In the end, the AEC's stated policies—of obtaining informed consent and restricting human experimentation to instances in which a therapeutic benefit might be obtained—were applied only to experiments undertaken "within the Project," and particularly those aimed at establishing radiological safety for continuing atomic weapons development. (Even there, whether the policy was applied uniformly is unclear at best.) Human plutonium injection experiments ceased within a few months of the formulation of this policy, and the agency tried to rein in researchers affiliated with its own program—particularly Joseph Hamilton at Berkeley and Robert Stone in San Francisco—who seemed to disregard both possible hazards and newly articulated government requirements for patient consent in their human research. The Advisory Committee for Biology and Medicine argued that these researchers should follow the AEC's medical research guidelines for all of their research, even projects not contracted to the agency: "It was felt a responsibility devolved upon those holding AEC contracts to be sure that no question could arise with regard to activities in treatment of patients, whether done under AEC auspices or not."[128]

By contrast, the civilian biomedical researchers who purchased radioisotopes from Oak Ridge were not scrutinized at the agency level beyond

126. Ibid., p. 83.

127. I am convinced by the wording of this document that it was meant to refer specifically to the secret experiments in which human subjects deemed terminally ill were injected with fissionable material (such as plutonium) or fission products.

128. Memorandum from Shields Warren to Carroll L. Wilson, 14 Oct 1948, Report of the 12th Meeting of the Advisory Committee for Biology and Medicine, 8–9 Oct 1948, Hanford, Washington, OpenNet Acc NV0711671, p. 2. This issue came up again in the minutes of the following meeting, and here Robert Stone is mentioned by name (Minutes, 13th ACBM Meeting, 10–11 Nov 1948, Los Alamos, NM, OpenNet Acc NV0711689, p. 17). He was the researcher they most worried about. Stone had problems getting his research proposals through the UCSF Cancer Board because of concerns about his use of human subjects; Jones and Martensen, "Human Radiation Experiments" (2003), p. 93.

having their projects vetted by the Subcommittee on Human Application. One of this subcommittee's responsibilities was "to protect, as far as possible, patients from indiscriminate use of radioisotopes."[129] In general, they permitted the administration of radioactive materials in "normal human subjects" provided that the subject was aware of the study and consented to participate, and the experiment was justified by animal studies.[130] Not that the agency stringently enforced these guidelines. When a non-AEC researcher inquired of the agency about "permission forms" and "medical-legal aspects" of his approved nontherapeutic investigation of phosphorus-32, the Division of Biology and Medicine deferred to the Isotopes Division, which said it could be of "little assistance" and advised him to follow the procedures of his local institutional review committee (or "Medical Isotope Committees," as the agency called them).[131] In a shift of policy in 1949, the AEC made clear that its regulations, including required approval by its Subcommittee on Human Applications, would apply even to research projects conducted with radiomaterials produced in the laboratories where they were to be used (i.e., from local cyclotrons rather than purchased from Oak Ridge).[132] However, experienced researchers such as John Lawrence at Berkeley resisted oversight even by local committees, much less the Isotopes Division.[133]

Given the AEC's concerns about public relations, it is striking that its leadership did not worry about bad publicity that would result from irresponsible uses of its radioisotopes by private physicians or university researchers. But in legal terms, the agency saw its responsibilities as limited. Once a purchaser had been "deemed 'qualified' by the Isotopes Branch" and the Subcommittee on Human Application, "the Commission bears little if any responsibility for human administration." The agency judged the relevant precedent to be that of experimental drug tests, for which physicians rather than drug companies were held liable: "In practically

129. Isotopes Division Circular D-4, Radioisotopes for Use in Medicine, 6 Dec 1948, DC-LBL files, box 1, folder 4 Chemistry Program: Isotopes: General Correspondence, p. 1.

130. These criteria were articulated in a document dated 29 Mar 1948; ACHRE, *Final Report* (1996), p. 51.

131. ACHRE, *Final Report* (1996), p. 51 and, on the local committees, pp. 179–83 and 192n61.

132. A. Tammaro and Paul C. Aebersold, "Use of Radioisotopes in Human Subjects," 5 Oct 1949, DC-LBL files, box 1, folder 4 Chemistry Program: Isotopes: General Correspondence. ("Since this procedure has not been uniformly followed in the past, we are writing to acquaint you with the appropriate details.")

133. Jones and Martensen, "Human Radiation Experiments" (2003).

every case the physician involved has taken the sole responsibility and therefore may or may not be guilty of malpractice, but the institution is exonerated."[134] The AEC apparently regarded its safety guidelines and Advisory Field Service for isotope users as sufficient to meet its mandate. The ironic consequence was that weapons-related human research was regulated by the AEC for its adherence to more stringent standards of medical ethics than civilian experimentation with its isotopes.

Radioisotopes at the Veterans Administration Hospital in the Bronx

Another government agency, the Veterans Administration, had its own complex relationship with "human radiation experiments." The atomic bomb tests at Bikini in 1946, which went by the name Operation Crossroads, provided motivation for the Veterans Administration (VA) to develop a new kind of radiological expertise.[135] The tests, which involved over 200,000 servicemen, resulted in unanticipated levels of radioactive contamination on the test ships and service vessels.[136] General Groves and other government officials expressed worry that the men who participated in the Navy's operations at Bikini might bring lawsuits against the government over injuries resulting from their involuntary radiological exposure.[137] In 1947, a newly created "Central Advisory Committee" recommended that the VA establish an Atomic Medicine Division to deal with disability claims and other litigation filed by veterans exposed to radiation from atomic bomb testing.[138] The committee envisioned a Radioisotope Program as part of this Atomic Medicine Division, to facilitate "research

134. Medical Advisor's Office of Oak Ridge Directed Operations to Advisory Board on Medicine and Biology, 8 Oct 1947, reproduced in ACHRE, *Final Report, Suppl. Vol. 1* (1995), pp. 80–88, on p. 82.

135. ACHRE, *Final Report* (1996), pp. 299–300.

136. In fact, the third of the three planned tests was canceled altogether, due in part to the radiological catastrophe caused by the second underwater blast. Weisgall, *Operation Crossroads* (1994).

137. As it turned out, such lawsuits did not materialize until four decades later; see the last section of this chapter as well as Weisgall, *Operation Crossroads* (1994).

138. In their account, the Advisory Committee on Human Radiation Experiments mentions that the "Central Advisory Committee" advised the VA to keep the creation of the Atomic Medicine Division confidential, but publicize the Radioisotope Program. The very name of the committee had been selected so as not to disclose that "there might be problems in connection with alleged service-connected disability claims." George M. Lyon to

aimed at bringing veterans the benefits of medical breakthroughs connected with the use of radioisotopes."[139] Although the division did not, in the end, materialize, the Radioisotope Program did, beginning with six VA Hospitals establishing Radioisotope Units supplied by the AEC.[140] By 1953, the number of these units had grown to thirty-three, employing 202 staff.[141]

This venture was part of the AEC's broader plan to cultivate clinical facilities for medical uses of radioisotopes. Early on, the agency funded research programs and new facilities at a few medical schools, such as UCLA and the University of Rochester (the latter of which, as we have seen, had already been funded to investigate human radiation effects by the Manhattan Project).[142] The AEC also constructed cancer research hospitals at Argonne and Oak Ridge National Laboratories, referring to one such hospital in 1949 as a "clinical proving ground."[143] While many of these facilities were still under construction, the prospect of establishing Radioisotope Units at VA Hospitals opened up a significant—and higher volume—clinical venue for the utilization of radioisotopes.[144]

One of the original six units was the Radioisotope Service at the VA Hospital in the Bronx, New York, which in 1947 hired the young nuclear physicist Rosalyn Yalow to help them set up this venture.[145] (See figure 8.6.)

Committee on Veterans Medical Problems, 8 Dec 1952 (ACHRE document VA-05294-A), as quoted by ACHRE, *Final Report* (1996), p. 300 and p. 314, footnote 177.

139. Lenoir and Hays, "Manhattan Project for Biomedicine" (2000), p. 54. Lenoir and Hays state that the Atomic Medicine Division was founded in 1947, but, according to the Advisory Committee on Human Radiation Experiments, "the feared claims from Crossroads did not materialize and . . . the confidential Atomic Medicine Division was not activated." ACHRE, *Final Report* (1996), p. 300.

140. The VA Hospitals initially involved in the radioisotope program were in Framingham, Massachusetts; the Bronx, New York; Cleveland, Ohio; Hines, Illinois; Minneapolis, Minnesota; and Van Nuys, California. Correspondence between officials at the AEC's Isotope Division and the Veterans Administration about the founding of these Radioisotope Units is in the NARA-SE, Manhattan Engineer District/Clinton Engineering Works General Research Correspondence, Acc 67B0803, box 177, folder AEC 441.2 (R-Veterans Administration).

141. Lenoir and Hays, "Manhattan Project for Biomedicine" (2000), p. 57.

142. Ibid.

143. AEC, *Sixth Semiannual Report* (1949), p. 91.

144. During the same time period that VA Hospitals provided key sites for the development of clinical uses of radioisotopes, they were similarly important to the emergence of clinical trials. Marks, *Progress of Experiment* (1997).

145. Although women physicists were uncommon in the 1940s, since the beginning of the century there had been a good number investigating radioactivity. For excellent analysis of the historiography of women in this field of specialization, see Rentetzi, "Gender, Politics, and

By 1949 Yalow had established a laboratory there and was investigating the usefulness of radioactive phosphorus and sodium in diagnosing tumors.[146] As she shifted to doing research at the hospital full-time, she asked Dr. Bernard Straus, Chief of Medicine at the hospital, if he knew of any physicians with whom she might collaborate. He recommended a young clinician named Solomon Berson, who joined Yalow in 1950 as soon as he completed his residency in internal medicine.[147] The two rapidly established a tight partnership in medical research using radioisotopic tracers, especially iodine-131 to study thyroid physiology in VA hospital patients. They also used red blood cells labeled with potassium-42 or phosphorus-32 to measure blood volume.[148] These investigations relied on being able to inject patients with isotopically labeled material. Yalow and Berson's publications do not generally include information about the relationship of their research to the treatment and care of these patient-subjects. Equally vague in their papers are provisions for informed consent. Their focus on improving the measurement of iodine-131 in patients made sense given the clinical consumption of radioisotopes. The AEC sold more radioiodine (as measured in curies) than any other isotope; it was used to diagnose thyroid function and treat thyroid conditions.[149]

Radioactivity Research" (2004). Yalow's biographer offers thoughtful reflection on her experiences not only as a woman scientist, but also as a child of Jewish immigrants. As he notes, the New York world of medicine was dominated by Columbia's College of Physicians and Surgeons and Cornell Medical School, with few Jewish students and almost no Jewish faculty. The Bronx VA Hospital's Department of Medicine appointed a Jewish physician as its chief in 1946, the first of many such appointments. The concentration of Jews, women, and even African-Americans made the department a target of investigation in 1954, as part of the McCarthy era. This was the political and cultural backdrop to Yalow and Berson's early work together, a collaboration of two ambitious, bright Jewish scientists. Straus, *Rosalyn Yalow* (1998).

146. One submitted paper based on these experiments was reported to the AEC for the bibliography included in the agency's publication *Isotopes* (1949), p. 89: Roswit et al., "Use of Radioactive Phosphorus" (1950).

147. See Yalow, "Radioactivity in the Service of Humanity" (1985) and Straus, *Rosalyn Yalow* (1998), p. 5 ff. From 1947 to 1950 Yalow was working as a consultant at the Bronx VA Hospital while teaching at Hunter College; in 1950 she became Assistant Chief of the hospital's Radioisotope Unit.

148. Berson et al, "Determination of Thyroidal and Renal Plasma" (1952); Berson and Yalow, "Use of K^{42} or P^{32} Labeled Erythrocytes" (1952); Berson et al., "Biological Decay Curve" (1952); Berson et al., "Tracer Experiments" (1953); Berson and Yalow, "Distribution of I^{131} Labeled Human Serum Albumin" (1954).

149. AEC, *Isotopes* (1949), p. 5. For more on the early uses of AEC-produced iodine-131, see Creager, "Nuclear Energy" (2006).

FIGURE 8.6. Rosalyn Yalow preparing an "atomic cocktail," 1948. National Archives, RG 326-G, box 1, folder 4, AEC-48-183.

Within a few years, Yalow and Berson applied their precise methods of measuring blood proteins to see whether radioisotopically labeled insulin disappeared more rapidly from the bloodstream of diabetic than normal subjects. This study put to the test a suggestion by one prominent expert on diabetes that the disease resulted from the abnormally rapid degradation of serum insulin.[150] Yalow and Berson administered radioiodine-labeled insulin to both diabetic and nondiabetic patients as well as to some "healthy volunteer laboratory personnel" of the Bronx VA Hospital.[151] To their surprise, they observed the very opposite of the effect they expected: insulin persisted in the bloodstream of most diabetics far longer than in that of normal subjects. This nullified the rapid degradation theory of diabetes, but it raised another question of why diabetic patients exhibited such a low turnover of insulin in their bloodstream.

Yalow and Berson could differentiate the effects of diabetes from its treatment with insulin by analyzing another patient population at the hospital: psychiatric patients receiving insulin "shock therapy." (From the 1930s through the 1960s, psychiatrists in the United States used insulin to induce hypoglycemic shock, even coma, as a way to treat thousands of schizophrenic patients.[152]) Blood samples from two insulin-receiving schizophrenic patients contained the same long-lasting insulin (bound to antibody) that characterized the insulin-treated diabetic patients; persistence of insulin in the bloodstream correlated with whether the patient had received exogenous insulin.[153] Yalow and Berson concluded it was the immunogenicity of beef or pork insulin that was responsible for the presence of the insulin-binding antibodies.

Yalow and Berson's contention that patients treated with animal-derived insulin developed antibodies against it was highly controversial. Their manuscripts were rejected from the journal *Science* and initially from the *Journal of Clinical Investigation*, because peer reviewers were unpersuaded that a molecule as small as insulin could be immunogenic.[154]

150. The expert was Dr. I. Arthur Mirsky, who put forth this theory in 1952 about adult-onset diabetes in his lectures at the Laurentian Hormone Conference ("Etiology of Diabetes"). His theory of rapid degradation took into account that "the pancreata of adult diabetics contained almost normal amounts of insulin." Yalow, "Radioactivity in the Service of Humanity" (1985), p. 58.

151. Berson et al., "Insulin-I^{131} Metabolism in Human Subjects" (1956), p. 170.

152. James, "Insulin Treatment in Psychiatry" (1992).

153. Yalow, "Development and Proliferation" (1999).

154. Yalow states that the concept that small peptides could elicit plasma antibodies "was not acceptable to immunologists until the 1950s." Yalow: "Radioimmunoassay: A Probe" (1978), p. 1237. In the 1945 edition of his book on immunochemistry, Karl Landsteiner illustrates the uncertainty as to whether such serum proteins should be called antibodies or not. He states that insulin has an antigenic capacity but that it is low ("fortunately for therapeutic use"). The passage continues: "After continued administration of various hormones neutralizing agents can appear in the serum for which the term 'anti-hormones' has been adopted. . . .

As the editor of the *Journal of Clinical Investigation* explained in his rejection letter, the "experts in this field" were adamant that the authors had not demonstrated that the globulin was an antibody raised against the administered insulin.[155] (See figure 8.7.) To get their paper accepted, Yalow and Berson had to replace the term antibody, which appeared in the title, with "insulin-binding globulin."[156]

Yalow and Berson utilized several physical-chemical techniques to support their interpretation that the insulin-binding globulin was a specific antibody, found only in patients who had been treated with animal-derived insulin. They used radioelectrophoresis to show that radioactively labeled insulin migrated with serum gamma globulin (plasma antibodies) in insulin-treated patients. They also employed chromatography and ultracentrifugation to analyze how labeled insulin interacted with serum proteins from patients, both those who had received insulin treatment and those who had not. All the results showed a consistent pattern: in insulin-treated patients most of the radioactivity from labeled insulin was associated with serum gamma globulin protein (i.e., IgG antibodies).[157] If the initial reviewers of Yalow and Berson's paper were skeptical, their experiments won over the field. Berson received the American Diabetes Association's first Lilly Award in 1957. (Yalow won it in 1961.) This was the first of many prizes recognizing their contributions, including the Nobel Prize for Yalow in 1977.[158]

As part of these studies, Yalow and Berson sought to determine the maximum binding capacity of the serum antibodies against insulin. They recognized that the binding of radiolabeled insulin to a fixed amount of antibody is a quantitative function of the amount of insulin present. When a small amount of radioactively labeled insulin was added to insulin antibody, all of it was bound by the antibody. The addition of unlabeled in-

---

The moot question in this line of research—supposing the mechanism is always the same—is whether the neutralizing substances found in the serum are antibodies in the serological sense or, according to Collip's theory, antagonistic hormones present in small amounts also in normal animals. . . . In favor of the antibody theory is the fact that neutralizing sera have not been reliably demonstrated against low-molecular hormones, e.g. adrenalin or estrin." Landsteiner, *Specificity of Serological Reactions* (1945), p. 37.

155. Yalow, "Radioimmunoassay: A Probe" (1978), p. 1238.

156. Straus, *Rosalyn Yalow* (1998), p. 8.

157. Berson et al., "Insulin-I^{131} Metabolism in Human Subjects" (1956).

158. Berson passed away five years before Yalow received the Nobel Prize. Yalow, "Radioimmunoassay: A Probe" (1978).

September 29, 1955

Dr. Solomon A. Berson
Radioisotope Service
Veterans Administration Hospital
130 West Kingsbridge Road
Bronx 68, New York

Dear Dr. Berson:

I regret that the revision of your paper entitled "Insulin-I131 Metabolism in Human Subjects: Demonstration of Insulin Transporting Antibody in the Circulation of Insulin Treated Subjects" is not acceptable for publication in THE JOURNAL OF CLINICAL INVESTIGATION. - - - - - - - - - - - - - -

- - - - - - - - - - - - - - - - - The second major criticism relates to the dogmatic conclusions set forth which are not warranted by the data. The experts in this field have been particularly emphatic in rejecting your positive statement that the "conclusion that the globulin responsible for insulin binding is an acquired antibody appears to be inescapable". They believe that you have not demonstrated an antigen-antibody reaction on the basis of adequate criteria, nor that you have definitely proved that a globulin is responsible for insulin binding, nor that insulin is an antigen. The data you present are indeed suggestive but any more positive cleaim seems unjustifiable at present.

- -

Sincerely,

Stanley E. Bradley

Stanley E. Bradley, M.D.
Editor-in-Chief

FIGURE 8.7. Excerpts of a letter of rejection from the *Journal of Clinical Investigation*, reproduced from Rosalyn S. Yalow, "Radioimmunoassay: A Probe for the Fine Structure of Biologic Systems," *Science* 200 (1978): 1236–45, on p. 1238. © The Nobel Foundation 1977.

sulin prevented the binding of labeled insulin in proportion to the total amount of insulin present. This meant that one could add labeled insulin to a solution containing both insulin antibody and an unknown amount of insulin, and calculate precisely the concentration of insulin based on how much of the labeled insulin was bound by antibody. This is the principle of radioimmunoassay, although it was three more years before Yalow and Berson published a paper showing how the technique could be used to measure insulin levels in human plasma. They claimed that their technique could measure human insulin in the range of 0.25–1.0 μ-units of hormone activity.[159] The sensitivity of this method improved upon existing bioassay techniques by about two orders of magnitude, and enabled users to measure levels of human serum insulin directly using a very small amount of blood.[160]

The Technological Trajectory of Radioimmunoassays

Yalow and Berson were not the only investigators to realize that competitive binding, radioactive labels, and specific antibodies could be used in concert for quantitative assays. In the United Kingdom, Roger Ekins was working in the Department of Physics Applied to Medicine at the Middlesex Hospital Medical School in London. He was collaborating with some hormone biochemists to develop techniques to measure serum thyroxine (thyroid hormone), and realized that the recently isolated specific thyroxine-binding globulin (antibody) and radiolabeled hormone could be used for the purpose of such an assay. As he has recounted since, his idea was greeted with skepticism by his peers and he was refused funding to purchase the radiolabel needed to test it. However, in 1957 a hospital patient provided him with the opportunity to carry out his experimental plan. The patient had a thyroid tumor and was being treated with massive doses of iodine-131. Ekins observed that the resulting radioactivity in the patient's bloodstream was largely bound by blood proteins—specifically, antibodies to the patient's now-radioactive thyroid hormone. Ekins used blood samples from this patient to show how unlabeled (exogenous) thyroid hormone could compete off the radioactive endogenous thyroxine bound to serum antibody.[161] As in the case of Yalow and Berson's work, the postwar use of radioisotopes (especially radioiodine) in clinical medi-

159. The detection in terms of concentration was 1.25–5.0 μ-units insulin per milliliter. Yalow and Berson, "Assay of Plasma Insulin in Human Subjects" (1959); Yalow and Berson, "Immunoassay of Endogenous Plasma Insulin in Man" (1960); Yalow and Berson, "Immunoassay of Plasma Insulin in Man" (1960).

160. The serum levels of insulin measured using radioimmunoassay were significantly lower than levels of insulin-like bioactivity determined using other methods; this discrepancy led to a fifteen-year controversy over the forms and level of insulin activity. See Kahn and Roth, "Berson , Yalow, and the *JCI*" (2004).

161. Ekins, "Estimation of Thyroxine" (1960); idem, "Immunoassay, DNA Analysis" (1999).

cine provided the context in which the radioimmunoassay method originated.[162] Human subjects again provided the experimental material for the discovery. Both the antigens being studied—insulin and thyroxine—were hormones.[163]

Through the late 1960s, endocrinology remained the major arena of application for radioimmunoassays.[164] Yalow and Berson extended their method to develop assays for a variety of other peptide hormones, including growth hormone, ACTH, parathyroid hormone, and gastrin.[165] These assays benefited medical practice as well as research: being able to detect various hormones in human plasma down to picomolar concentrations greatly expanded the diagnostic capabilities of clinical endocrinology. In fact, it brought many blood hormones within range of direct measurement—hormones such as insulin and thyroxine were not concentrated enough for detection by the antibody-based agglutination tests of the 1950s.[166]

This sophisticated detection method relied on a very old-fashioned substance, antisera from laboratory animals, as its source of antibodies. In Yalow and Berson's laboratory, the antisera came from guinea pigs, each of which was inoculated with a distinct antigen, such as gastrin or growth hormone. Yalow's biographer writes that in the early hours of the morning, she would spend time with her guinea pigs, offering them lettuce from home, cradling them in the crook of her arm, and cajoling them to produce the best antisera in the world.[167] (See figure 8.8.) Her laboratory did not make their valuable antisera available for sale, but they did provide precious vials to the scientists who came from all over the world to the Bronx to learn their methods.

162. On the medical uses of radioisotopes associated with Great Britain's atomic energy program, see Kraft, "Between Medicine and Industry" (2006).

163. A third research group developed a radioimmunoassay for another hormone, glucagon, in 1959: Unger et al., "Glucagon Antibodies" (1959).

164. For an account of postwar endocrinology that attends to other significant applications of radioisotopes, see Fragu, "How the Field of Thyroid Endocrinology Developed" (2003).

165. Glick et al., "Immunoassay of Human Growth Hormone" (1963); Roth et al., "Hypoglycemia" (1963); Berson and Yalow, "Radioimmunoassay of ACTH in Plasma" (1968); Yalow et al., "Radioimmunoassay of Human Plasma ACTH" (1964); Berson et al., "Immunoassay of Bovine and Human Parathyroid Hormone" (1963); Yalow and Berson, "Radioimmunoassay of Gastrin" (1970).

166. See Yalow, "Radioimmunoassay: A Probe" (1978), pp. 1239–40.

167. Straus, *Rosalyn Yalow* (1998), p. 13.

FIGURE 8.8. Rosalyn Yalow holding one of the antisera-producing guinea pigs in her laboratory. Reproduced from Eugene Straus, *Rosalyn Yalow, Nobel Laureate: Her Life and Work in Medicine* (New York: Plenum Trade, 1998), p. 15. Copyright © 1998 Straus, Eugene. Reprinted by permission of Plenum Trade, a member of the Perseus Books Group.

In the early 1970s, radioimmunoassays began to reach a much wider range of users—and began to go by the acronym RIA. The emergence of commercial radioimmunoassay reagents and "kits" registered and reinforced this trend, while taking advantage of new technical developments. In particular, iodine-125 overtook iodine-131 as the label of choice, and the longer half-life of this isotope in turn meant a reasonable shelf life for commercial assay kits.[168] New England Nuclear, which had been providing radiolabeled reagents to researchers since 1956, moved into the area of radioimmunoassay products and featured this technique in their 1973 Annual Report.[169] Their report focused on the uses of RIA for endocrine diagnosis (e.g., for angiotensin 1, whose levels were affected by hyperten-

168. Charlton, "Overcoming the Radiological and Legislative Obstacles" (1979).

169. For more on New England Nuclear, see chapter 6.

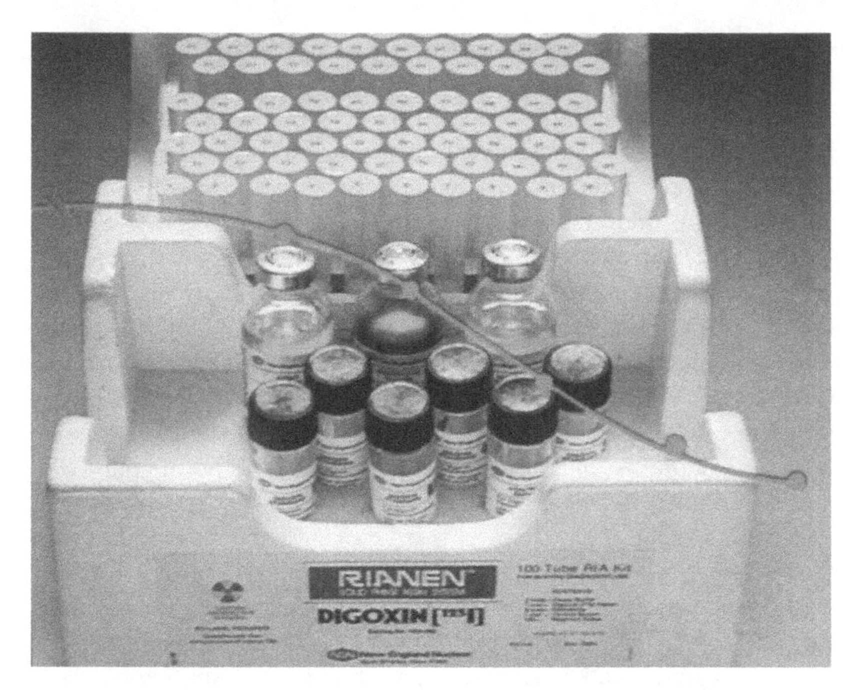

FIGURE 8.9. Picture of a Digoxin ^{125}I radioimmunoassay kit from New England Nuclear. Image and caption from New England Nuclear 1977 Annual Report, courtesy of Paul McNulty.

sion) and drug dosage (e.g., for digoxin in patients with congestive heart failure).[170] The Red Cross used an RIA test for hepatitis-associated antigen to screen donated blood, and RIA diagnostic tests for tumor markers became widely used in oncology.[171] (See figure 8.9.)

RIA attained an impressive scale of use: Around 52 million radioimmunoassays were performed in 1975.[172] Although these assays used radioactivity, they were done in test tubes—the patients were never exposed to the radioisotope. The technique of radioimmunoassay took advantage of radioisotopes as tracers to identify and quantify antigens with unrivaled sensitivity. Radioimmunoassays thus extended the capability of scientists to access previously unseen molecular agents, demonstrating the power of radiolabels in technologies for detection and diagnostics.

170. Copy of New England Nuclear 1973 Annual Report, courtesy of Paul McNulty.

171. Block, "Overview of Radioimmunoassay Testing" (1979); Herberman, "Immunodiagnostics for Cancer Testing" (1979); Yalow, "Radioimmunoassay in Oncology" (1984).

172. Landon, "Look at the Future" (1979).

Conclusions

As these sketches of medical research on anemia and diabetes make clear, the incorporation of radioisotopes into clinical investigation took off in the years following World War II, building on strong patterns of research with cyclotron-produced radiomaterials and accelerated by the US government's promotion of radioisotopes as peacetime dividends of atomic energy. In retrospect, one might say that the uptake of radioisotopes into medical research ran ahead of regulations to protect the human subjects of these studies. Such a statement is clearly correct, but also presumes the necessity of federal government regulation to protect subjects of medical investigations. As many historians have shown, that presumption was itself a product of history, emerging in the 1960s.[173]

What is striking about the correspondence among leaders of the AEC early in 1947 was that although they were concerned that secret Manhattan Project experiments with terminally ill patients, if published, would stain the new civilian agency's reputation, they were not equally worried about the government's perceived responsibilities for radioisotopes used in "off-Project" laboratories, even when the materials were purchased from Oak Ridge. Once projects were approved by the Subcommittee on Human Application, the AEC trusted physicians to follow the prevailing principles of voluntary participation and consent, and looked to hospitals and other local institutions to monitor adherence to these standards of safety. The negative publicity which the AEC feared in 1947 did in fact materialize nearly a half-century later, but the changed public expectations about regulation and responsibility meant that the government was condemned for not protecting the subjects of medical research with AEC-provided radiomaterials just as much as for sponsoring military-related experiments in which patients were unwittingly injected with plutonium. Whether the project related to the hazards of atomic weaponry or was part of mainstream clinical investigation, many Americans viewed subjects of radioisotope research as "human guinea pigs." To give a more historical interpretation, as I have attempted here, does not make the ethical issues around human experiments any less vexing. Even those physicians engaged in the best medical science of their day began targeting vulner-

173. E.g., Rothman, *Strangers at the Bedside* (1991); Stark, *Behind Closed Doors* (2012).

able populations (to use a later term) as research subjects in the 1940s and 1950s.

In addition to the civilians who were involved in biomedical experiments with radioisotopes, veterans occupy a distinct historical position as subjects of investigation. In the United States, "GI guinea pigs" were significant in two kinds of activity related to atomic energy.[174] First and foremost, military personnel were unwitting subjects in the dozens of atomic test explosions that took place between 1946 and 1963.[175] To the degree that the planners of these tests regarded military personnel as human subjects, it was not to investigate the biological effects of radiation but to study the psychological and physiological reactions of servicemen to the atomic blasts. These research subjects (often involved in training maneuvers at test sites) formed a small contingent of the 200,000 people who experienced radiation exposure in conjunction with American atomic weapons testing. Servicemen who experienced radiation as an occupational hazard at test sites were still used as sources of biomedical data.[176] In the 1960s and 1970s, hundreds of veterans who participated in nuclear tests filed claims with the Veterans Administration for service-connected radiation injuries.[177] Although the VA ruled in favor of only a handful of claimants, the veterans' cause became a political issue. In 1988, Congress passed legislation establishing compensation for radiation-exposed veterans irrespective of whether injury could be proven.

The other arena of exposure related to military service was strictly clinical: veterans were at the frontlines of medical experiments with radioisotopes at VA Hospitals in the late 1940s and 1950s. The AEC took advantage of the government-controlled hospital infrastructure for military personnel in order to establish clinical research sites for nuclear medicine. This meant that veterans were in a position to benefit from the most recent advances in nuclear medicine, but they were also part of the AEC's "clinical proving ground" for radioisotopes. The government's motivation in this case was not so much military as political—the agency was eager

174. Uhl and Ensign, *GI Guinea Pigs* (1980).

175. See ACHRE, *Final Report* (1996), ch. 10, "Atomic Veterans." Not all of the radiation exposure was intentional on the part of the military, and there was little attention (or scientific understanding) of the long-term effects of low-level radiation. (Not that this eases the plight of afflicted servicemen.)

176. ACHRE, *Final Report* (1996), pp. 283–84.

177. Weisgall, *Operation Crossroads* (1994), p. 278.

for medical breakthroughs that would show atomic energy's peaceful uses in the midst of an emerging nuclear arms race.

Although radioimmunoassays could have been developed without the use of human subjects, both in the United States and the United Kingdom, the novel binding assays in fact arose as part of clinical research programs where patients were being treated with radioisotopes. In effect, patients were the experimental vessels for observing the competitive binding behavior of antibodies in the presence of marked antigen. Yalow herself emphasized the serendipitous nature of RIA, emerging as it did from her studies of insulin metabolism in a hospital's Radioisotope Service.[178] Yalow and Berson's technique effectively replaced the human body—which was both the site for binding reactions and the source of antibodies—with test tubes on the one hand and animal sera on the other. In the 1980s, antibodies synthesized in vitro (monoclonal antibodies) began replacing animal-derived antibodies. Whereas clinical investigation with radioiron by Whipple and Hahn went on the usual pathway from animal research to human experiments, the trajectory of Yalow's radioiodine work was one of *disembodiment*, a progressive liberation from diabetic veterans and then actual guinea pigs to a method employing purely synthetic constituents. Or to put it another way, RIA externalized clinical observations, rendering what had been an in vivo experiment into an in vitro assay. This examination of the technique's history has put the veterans and guinea pigs back into the picture, to highlight the complex legacies of government atomic energy policy and clinical research for postwar biomedicine.

178. E.g., Yalow, "Radioimmunoassay: A Probe" (1978) and idem, "Radioimmunoassay: Its Relevance to Clinical Medicine" (1981).

CHAPTER NINE

Beams and Emanations

From no other aspect of atomic energy application does the general public in the United States benefit more directly and extensively than from the use of radioisotopes and radiation in medical research, diagnosis, and therapy. To place a dimension on the extent of use, over 4 million persons were diagnosed or treated using radioisotopes in 1970. —E. E. Fowler, 1972[1]

Radioisotopes put nuclear medicine at the technological vanguard of postwar health care. Initially, the same radioisotopes served as both diagnostic tracers and sources of therapy—particularly iodine-131 and phosphorus-32—but over the course of the 1950s there emerged a differentiation of medical isotopes by purpose. High-intensity, longer-lived radioisotopes such as cobalt-60 and cesium-137, which served as external radiation sources, dominated radiotherapy. Novel diagnostic methods began using shorter-lived radioisotopes, from iodine-132 to fluorine-18, in synergy with instruments to detect radioactivity deep within the human body.

Concerns about patient safety shaped the emergence of these new diagnostic tools decisively and explicitly. As the AEC put it in a 1960 report, "The benefits to the patient of a diagnostic or therapeutic procedure using radioactivity must be measured against both the potential and direct biological effects of the irradiation."[2] In part, this attention to minimizing dose reflects simple chronology: many novel radioisotope-based methods were developed in the late 1950s and 1960s, against a background of growing public and medical concern about the hazards of low-level and long-term radiation exposure.[3]

1. Fowler, "Recent Advances" (1972), p. 253.
2. AEC, *Radioisotopes in Science and Industry* (1960), p. 17.
3. See chapter 5.

Histories of nuclear medicine often bypass the late 1940s and early 1950s to highlight the generation of diagnostic tools.[4] In particular, the invention of scintillation-detector scanners and the introduction of technetium-99m vastly increased the sophistication with which radioisotopes were used to visualize organs and evaluate function.[5] Focusing on the earlier applications of radioisotopes in diagnostics, however, highlights the commonality of the tracer method across biology and medicine. Moreover, the AEC's research facilities, notably Brookhaven National Laboratory and Argonne Cancer Research Hospital, contributed crucially to the development of novel instruments and isotopes used in both radiotherapy and diagnosis.

The AEC represented nuclear medicine, more than any other field of application of radioisotopes, as the "humanitarian" face of atomic energy. Yet no area in biology and medicine reveals the overlap between the AEC's civilian and military interests as clearly as does nuclear medicine.[6] This chapter focuses on two case studies that illustrate the synergy of cancer medicine and atomic weapons–related research around the clinical uses of radioisotopes. First, the incorporation of cobalt-60 and cesium-137 sources into teletherapy machines offered improvements on earlier modes of radiotherapy with radium. These new intensive-radiation machines were also employed in experimental therapy with total body irradiation, often with military-related funding. Irradiated cancer patients became sources of classified information on how humans respond biologically and cognitively to sublethal doses of radiation.

Second, investigation into the detection and treatment of brain tumors with various radioelements, particularly by William Sweet and Gordon Brownell, laid the groundwork for the subsequent development of positron-emission tomography. At the same time, Sweet's related research into neutron-capturing isotopes for treating such tumors provided the basis for secret AEC studies of uranium metabolism in dying patients.[7] In part, military and atomic weapons–related specialists paid attention to nuclear medicine because of the availability of cancer patients as research

4. E.g., Miale, "Nuclear Medicine" (1995).

5. The "m" at the end of this isotope designation means "metastable"; it is a nuclear isomer of technetium-99, into which it transitions through a nuclear rearrangement, emitting a gamma ray.

6. Kutcher, *Contested Medicine* (2009); Leopold, *Under the Radar* (2009).

7. Whittemore and Boleyn-Fitzgerald, "Injecting Comatose Patients" (2003).

subjects. But it also reflected the fact that the same government institution was promoting the civilian benefits of atomic energy while building and managing a massive nuclear weapons stockpile. Nuclear medicine reveals how both a common organizational structure and key materials—not least radioisotopes—linked the AEC's civilian and military agendas.

From Magic Bullets to Better Beams

In the early years of the AEC, physicians and government officials alike hoped that radioisotopes, administered therapeutically, would concentrate in specific tissues to provide a source of internal radiation, eradicating tumors in situ. This approach did not live up to anticipation.[8] Radioiodine, which localized efficiently in the thyroid, could be used to treat hyperthyroidism, but proved disappointing for many cases of thyroid cancer.[9] As an alternative tack, some researchers began directing radioisotopes into specific areas of the body by means of their preparation. Colloidal particles, for instance, tended to concentrate in the liver and spleen. Medical researchers tried using colloidal preparations of radioactive zirconium, niobium, and yttrium for therapy, although yttrium particles below a certain size collected in the bone marrow. The most promising radioisotope for this purpose proved to be gold-198. In 1947, Paul Hahn developed a technique for infusing colloidal gold-198 into the chest or abdominal cavities to reduce accumulation of fluid due to malignancies.[10] Colloidal radiogold became a standard palliative treatment for metastatic growths, but did not cure the original cancer.[11] (See figure 9.1.)

8. As John Lawrence and Cornelius Tobias wrote in a 1956 article ("Radioactive Isotopes"): "It is now 20 years since we began to use artificially produced radioactive isotopes in cancer research and in medical research. Actually, our first therapeutic trials were in 1936. In the early days the hope was that at the end of 20 years one could report outstanding examples of selective localization of radioactive compounds in neoplastic tissue, making it possible to give the tumor tissue many times more irradiation than the surrounding, normal tissue. . . . However, one who has worked in the field from the beginning and who is asked to sum up his experiences can't help saying that the therapeutic achievements have been disappointing" (p. 185).

9. Aebersold, "Development of Nuclear Medicine" (1956), p. 1033.

10. Stannard, *Radioactivity and Health* (1988), vol. 3, pp. 1763–64.

11. As J. H. Muller commented, "These procedures are now widely accepted as efficient palliative treatment of malignant effusions. . . . It must be pointed out that all of these patients were desperate cases, who would have died within a few weeks or months if they had not had

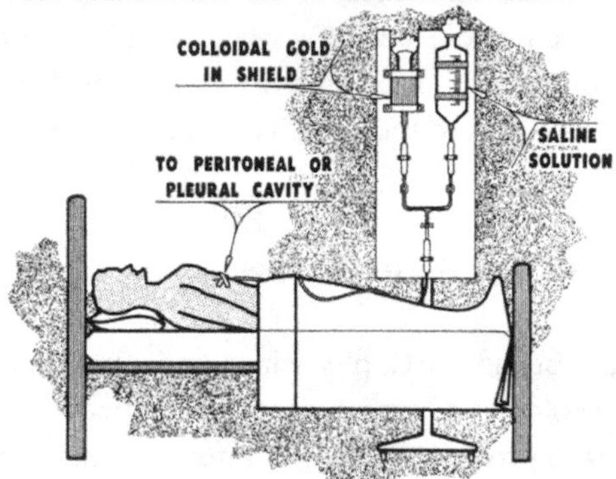

FIGURE 9.1. Radioactive Gold–Au 198 in Therapeutic Use. This technique was being used in around 250 hospitals at the time of this publication. From US Atomic Energy Commission, *Eight-year Isotope Summary*, vol. 7 of *Selected Reference Material, United States Energy Program* (Washington, DC: US Government Printing Office, 1955), p. 26.

By the end of the 1940s, the AEC was shifting away altogether from representing radioisotopes as "medical bullets" against cancer, stressing instead that isotopes were potential substitutes for radium in therapy. Cobalt-60 offered particular promise.[12] Like radium-226, cobalt-60 emits gamma radiation. However, the radioelements are chemically quite different. Radium tends to be metabolized and retained in the human body, like calcium. When localized to the bone, radium can be stored there indefinitely, irradiating the marrow with alpha particles. Cobalt-60, by contrast, does not emit alpha radiation, nor does it concentrate in the body. Scientists had found that when cobalt-60 was ingested, it was rapidly excreted.[13]

---

this newly developed therapy." In fact, most of the patients he described treating with radiogold had stage III, IV, or V ovarian cancer; some also received radium or x-ray therapy and most lived between one and two years. Muller, "Intraperitoneal Application" (1956), pp. 270 and 291. See also Andrews, "Treatment of Pleural Effusion" (1956).

12. See AEC Press Release 158, "AEC to Make All Radioisotopes Available for Cancer Research Without Charge," 27 Feb 1949, RG 326, NARA College Park, E67A, box 64, folder 5 Research in Biological and Medical Science; AEC, *Sixth Semiannual Report* (1949), p. 90.

13. Studies on the metabolism of cobalt-60 were conducted by researchers in the University of California Radiation Laboratory. AEC, *Sixth Semiannual Report* (1949), p. 94.

In addition, the half-life of cobalt-60 was relatively long, 5.3 years. Perhaps most importantly, its price was negligible compared to radium, which cost $15,000–$20,000 per gram in the late 1940s.[14] Commissioner Lilienthal referred to cobalt-60 as "virtually costless" when prepared in the agency's nuclear reactors.[15]

Initial interest in cobalt focused on its properties as a metal in comparison to radium. Typically, radium was embedded in a wax form designed to fit inside the bodily cavity being targeted: inside the mouth, outside of the jaw, or near the cervix.[16] The radium itself was usually dispensed in nonpliable tubes, in which radiologists had to distribute the material carefully so as not to create "hot spots" leading to overexposure. By contrast, cobalt-60 was pliable and could be molded into whatever shape was most suitable for treatment. In fact, it could be prefabricated into its final shape before being irradiated.[17] Cobalt-60 was also magnetic, so that it could be handled with electromagnetic holders, enabling users to keep it away from their bodies.[18] On nearly every front, cobalt-60 seemed superior to radium-226 for radiotherapy. The *New York Times* carried a story on cobalt-60 under the headline "Atom Bomb By-Product Promises to Replace Radium as Cancer Aid."[19] And as Shields Warren put it in 1950, radium had become as "outmoded as a Model T Ford" for treating cancer.[20]

As it turned out, rather than replacing radium-226 in established forms of cancer treatment, cobalt-60 drove the development of a new mode of delivering radiation, teletherapy. One of the limitations with conventional radium therapy was the difficulty of reaching tumors deep within the body. To irradiate growths that were not on or near the surface of the body, radium was usually applied as "seeds," placed within body cavities or implanted surgically. In teletherapy, an approach initially worked out with radium (despite its limitations), the radiation source was housed in a machine and its emitted radioactivity focused into a beam that could

14. AEC, *Atomic Energy and the Life Sciences* (1949), p. 93.

15. Leviero, "Atom Bomb By-Product" (1948), p. 19. Radium has a much longer half-life, at 1,500 years, but the lower cost of cobalt-60 compensated for the difference.

16. AEC, *Atomic Energy and the Life Sciences* (1949), p. 93.

17. Ibid.

18. Ibid., p. 94.

19. Leviero, "Atom Bomb By-Product" (1948), p. 1.

20. "Cobalt Put Above Radium in Cancer" (1950), p. 81.

be directed at a specific point within the body.[21] It was, in effect, a way of combining radium therapy with the delivery method used in x-ray therapy.[22] The intensity of gamma radiation available through cobalt-60, when sufficiently focused for long-range delivery, could reach internal tumors without doing extensive damage to surrounding tissue or skin. The effectively single wavelength of cobalt-60's gamma rays improved upon x-rays, which have varying wavelengths and as a consequence irradiated normal as well as cancerous tissue when applied at high intensity. After World War II, therapeutic interest in harnessing cobalt-60, newly available on account of nuclear reactors, gave the United States an opportunity to overtake Europe in the development of radiation-based cancer treatments. As it turned out, however, the United States lagged behind Canada in bringing telecobalt units to market.

The initial impetus for developing cobalt-60 teletherapy in the United States came from M. D. Anderson Hospital for Cancer Research in Houston, founded in 1944. R. Lee Clark, first Director and Surgeon-in-Chief, and Gilbert H. Fletcher, head of Radiology, were impressed by the advanced state of radium therapy in Europe.[23] While visiting radiology centers abroad, Fletcher met Leonard Grimmett, a physicist at Hammersmith Hospital in London who designed one of the first teleradium units in the 1930s. Soon recruited to M. D. Anderson, Grimmett arrived in Houston in 1949 with the idea of building a teletherapy unit around a 1,000-curie source of radioactive cobalt.[24] Shields Warren, Director of the AEC's Division of Biology and Medicine, did not think that housing a 1,000-curie cobalt-60 source was feasible; the Navy had experienced difficulties building a unit with just 100 curies of cobalt. Consequently, the AEC was not forthcoming with support for the project, though the Damon Runyon Memorial Fund for Cancer Research began providing funding (initially, raised through a special football game).[25]

21. One limitation of radium teletherapy was cost—to sequester 4 grams of radium in a teletherapy machine could cost $200,000. Radium also had a complicated decay scheme and posed problems with shielding. This is why hospitals relied on x-ray machines for irradiating deep tumors, but this alternative had its own technical problems (especially the continuous spectrum of radiation) and was still expensive. Schulz, "Supervoltage Story" (1975).

22. M. D. Anderson Hospital, *First Twenty Years*, pp. 193–94, p. 213.

23. Leopold, *Under the Radar* (2009), p. 65.

24. M.D. Anderson Hospital, *First Twenty Years* (1964), pp. 213–215.

25. Ibid., p. 212.

Seeking another avenue for AEC support of Grimmett's teletherapy idea, Lee approached the Oak Ridge Institute of Nuclear Studies (ORINS), an association of southern universities chartered in 1946 to help make the facilities and resources of Oak Ridge National Laboratory available to university researchers in the region. This was the organization that the AEC contracted to offer its radioisotope training courses in Oak Ridge, and it hosted graduate students and faculty members who came to Oak Ridge to participate in the AEC's research programs.[26] Initially, fourteen universities were members of ORINS; by 1950 that number had reached twenty-four.

ORINS developed its medical program and facilities in response to the AEC's cancer program. In 1948 the Commission contracted with ORINS to develop and operate a 32-bed clinical research hospital and laboratory in Oak Ridge for testing radiomaterials in the treatment and diagnosis of cancer.[27] Marshall Brucer became Director of the Medical Division of ORINS, which was also overseen by a Board of Medical Consultants from six of its affiliated medical schools.[28] The hospital facility and new laboratory building opened in 1950; the research program was in full swing by that summer. The emphasis was on using radioisotopes for cancer therapy. Physicians at ORINS's affiliated medical schools would refer only patients for whom this approach seemed promising.[29] As one description of the research program put it, "There would be few things more damaging to the program than to be forced to tell patients, after they arrive, that there

26. A Chronology of the Clinical Studies Program at the Oak Ridge Institute of Nuclear Studies/ Oak Ridge Associated Universities, 11 Jan 1994, DOE OpenNet Acc NV0707053. This document also lists the members of ORINS in 1948: University of Alabama, University of Arkansas, Auburn University, Catholic University of America, Duke University, Emory University, University of Florida, University of Georgia, Georgia Institute of Technology, University of Kentucky, Louisiana State University, University of Mississippi, University of North Carolina, University of Tennessee, University of Texas at Austin, Tulane University, Vanderbilt University, and University of Virginia. Because ORINS was a nonclassified facility, it was much easier to bring researchers and students there than to Oak Ridge National Laboratory, where only cleared US citizens could visit.

27. AEC, *Sixth Semiannual Report* (1949), p. 90; Statement by W. R. Bibb before the Subcommittee on Investigations and Oversight Committee on Science and Technology, US House of Representatives, 23 Sep 1981, OpenNet Acc NV0707565, p. 5. The ORINS clinical facility was housed in an unused wing of a community hospital that the AEC owned. (It was operated by contract.) A laboratory building was added and the hospital wing renovated.

28. Bruner and Andrews, "Cancer Research Program" (1950).

29. Ibid., p. 579.

is no suitable isotope therapy for their particular neoplasm."[30] The hospital was designed to be a "model medical radioisotope laboratory," with instrumentation and shielding for the safe handling of radioactivity.[31] As a "radioisotope hospital," the facility had some special rules. Visitors were restricted during a period of therapy, to avoid radiation exposure from a "hot" patient.[32] In addition, "nurses, technicians, maids, and orderlies" had to be specially trained for radiological safety.[33]

Brucer was keenly interested in developing teletherapy with cobalt-60, and involved Clark and Grimmett in meetings with ORINS members and AEC officials at Oak Ridge in December 1949. As a result, Grimmett's preliminary design was provisionally accepted, enabling him to request a small cobalt unit to be shipped to Houston for measurement work. Brucer, however, also wished to enlist a wider field of contributors and invited researchers "throughout the country to submit designs for a cobalt-60 unit."[34] In February 1950, twelve universities presented plans for teletherapy units at a conference in Washington, DC. Grimmett's was judged the most feasible design and selected for official development by the AEC.

Grimmett's design called for a 1,000-curie irradiator of cobalt-60 in the form of four slabs, each about the size of a quarter. The slabs were to be positioned inside a tungsten alloy cylinder, eighteen inches in diameter, which was in turn mounted on a rotatable disk, also of tungsten. The gamma-ray beam emerged through an aperture at the base of the cylinder. The disk could be rotated by remote control to either expose or cover the cobalt-60.[35] General Electric X-ray Corporation (GE) was contracted in July 1950 to fabricate the irradiator head and design a cantilevered supporting mechanism for the unit.[36] The company's expertise in x-ray and high voltage equipment made them a logical choice for designing the machine, and GE was already an AEC contractor. As it turned out, Grimmett died on May 27, 1951, just a few days before GE completed the unit.[37] (See figure 9.2.)

30. Ibid., p. 577.
31. Ibid., p. 583.
32. Ibid., p. 580.
33. Ibid., p. 581.
34. M. D. Anderson Hospital, *First Twenty Years* (1964), p. 215.
35. Ibid., pp. 215–16.
36. Ibid., p. 216.
37. Ibid., p. 217.

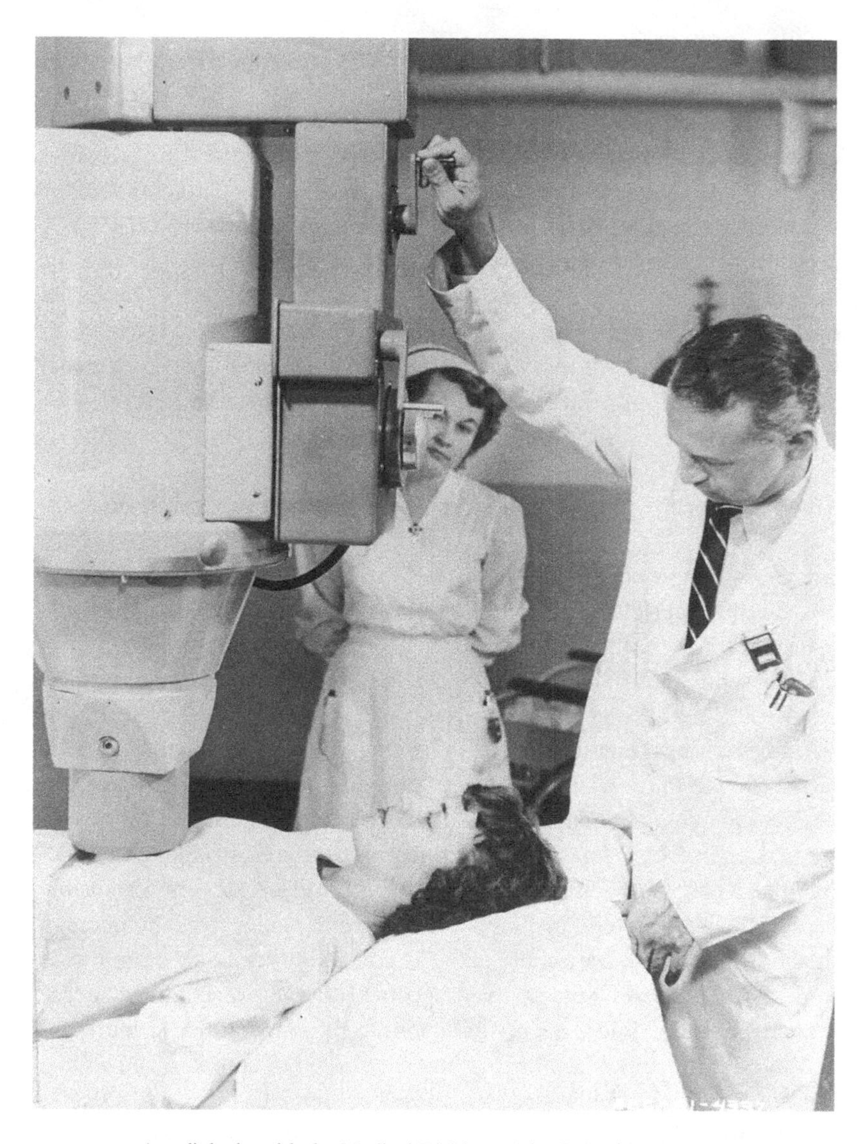

FIGURE 9.2. A radiologist with the Medical Division of the Oak Ridge Institute of Nuclear Studies demonstrates how a cobalt-60 teletherapy machine is adjusted. This appears to be the original machine designed by Leonard Grimmett and built by General Electric. Credit: Houston, ORINS. National Archives, RG 326-G, box 3, folder 3, AEC-51-4337.

The cobalt slabs were irradiated not in Oak Ridge, but in the Canadian nuclear reactor in Chalk River. They were scheduled to be in the reactor for two years, in order to allow a sufficient percentage of the metal to be converted into cobalt-60 so that the sample reached 1,000 curies in activity. The Korean War interfered with these plans, as the Canadian reactor was needed for military use, so the cobalt reached an activity of only 805 curies before being removed in July 1952 and shipped to Oak Ridge for assembly into the irradiator head.[38] The unit was kept at ORINS for initial testing, including animal experiments. Then in September 1953 the unit was shipped to Houston and installed in Texas Medical Center for clinical use. The first patients were treated on February 22, 1954.[39]

The Teletherapy Evaluation Board's Promotion of Cobalt-60

Brucer was impressed by this initiative from M. D. Anderson but wished to solicit broader participation in developing teletherapy. In 1952, he established a Teletherapy Evaluation Board of twenty representatives from medical schools around the country. (Most were from departments of radiology.) The board's stated purpose was "to investigate, develop, and evaluate radioisotopes for teletherapy."[40] In effect, rather than trusting this new technology to the market, ORINS assembled regional experts to assist in its development and commercialization. The machine GE had fabricated for M. D. Anderson was only a starting point—the board planned to develop and test its own prototype.[41] Commercial interests were ostensibly kept in check by forbidding individuals from companies to be members of the board, although they could serve as consultants. Five subcommittees were constituted to assess various aspects of the new technology: source evaluation and shield design, small source design, rotational methods, housing design, and clinical program. The design of standardized source capsules posed particular problems for commercialization, and ORINS collaborated with the AEC's Isotopes Division in working with manufactur-

38. Ibid.

39. Leopold, *Under the Radar* (2009), p. 67.

40. Minutes of August 8 [1952] Meeting of Teletherapy Evaluation Board [the first official meeting], copy from DOE Archives, OpenNet Acc NV0720833, p. 3.

41. Brucer, "Teletherapy Evaluation Board" (1953).

ers to solve them.[42] Cobalt-60 was not the only radioisotope being considered as a radiation source for teletherapy; cesium-137 and europium-152 were other candidate isotopes.[43]

The funding of ORINS's Teletherapy Evaluation Board was somewhat unconventional: institutions represented on the board were expected to contribute $2,500 apiece toward its research effort, though the AEC paid for the radiation source and provided $10,000 annually toward operating costs. The contributing universities would obtain teletherapy machines of their own, fabricated to their own specifications, for clinical trials.[44] One-quarter of the teletherapy machine time at ORINS-affiliated institutions was to be used in research laid out by the Teletherapy Evaluation Board. The financial arrangements of the Teletherapy Evaluation Board drew criticism from members of the AEC's Advisory Committee for Biology and Medicine (ACBM), which reviewed the grant application. At their meeting in June 1954, Gioacchino Failla (dubbed "the dean of American medical physicists") observed that the only real criterion for joining was a monetary contribution.[45] Perhaps as a result, "those that contributed were not the best institutions in the field."[46] In his view, there was not a first-class radiologist on the board, and Brucer ran the program in a way that antagonized "influential individuals."[47] He suspected the National Cancer Institute would do a better job managing such a program.

Despite the ACBM's reservations, using radioisotopes to combat cancer was politically expedient, so the agency continued supporting the ORINS Teletherapy Evaluation Board. Other AEC-supported approaches using radioisotopes in cancer therapy had not panned out. Paul Hahn's

42. Leopold, *Under the Radar* (2009), p. 70.

43. Minutes of August 8 Meeting of Teletherapy Evaluation Board, addendum draft contract for p. 2. Europium-152 proved to be difficult to prepare, but the board considered a unit that would use a radiation source of 7500 curies of cesium-137. Minutes of the Meeting of the Executive Committee of the Teletherapy Evaluation Board, ORINS, New York University College of Medicine Building, 3 Jun 1953, OpenNet Acc NV0718571.

44. See Marshall Brucer, "A Proposal for a Research Contract," draft version, undated, to be returned with suggestions and corrections by 15 May 1953, ORINS Teletherapy Minutes and Related Documents, File 1 of 3, OpenNet Acc NV0718558, pp. 1135470–72.

45. Goldsmith, "Rosalyn S. Yalow" (2012), p. 21N. Failla specialized in high-voltage therapy; see del Regato, *Radiological Oncologists* (1993), ch. 14.

46. Minutes, 45th ACBM Meeting, 25–26 Jun 1954, Washington, DC, OpenNet Acc NV0712007, p. 8.

47. Ibid., p. 9. Failla was at Columbia and one wonders if there was some regional bias underlying his opinion.

continuing project at Meharry Medical College investigating radiogold to treat tumors seemed at a dead end. As Failla expressed at the next ACBM meeting, "Nothing new has come out of Dr. Hahn's project recently and likewise there has been nothing new come out from many other projects elsewhere."[48] The following year, the ACBM again questioned "the advisability of continuing with further support to teletherapy," but funding remained in place.[49] In part, the issue devolved to whether the AEC's responsibilities in biomedicine were to support research or therapy. To this question, the Director of the Division of Biology and Medicine, Charles Dunham, gave a clear answer: "We do not support therapy for therapy's sake."[50] Teletherapy, however, remained at an experimental (thus fundable) stage.

To investigate the therapeutic efficacy of the new cobalt-60 machines, the Clinical Program Subcommittee developed an "approved treatment pattern" to be followed by all radiologists involved.[51] The protocol prescribed the length of treatment and dose for tumors of the brain, esophagus, nasopharynx, posterior tongue, tonsil, intrinsic larynx, throat, bone, and lung. In addition, standard data forms were developed and distributed to ensure that the Teletherapy Evaluation Board obtained complete information for its assessment. The same forms were to be used for treatment of tumors with 250 kilovolt x-ray machines, for comparison with the radioisotope-based teletherapy machines. By 1955 thirteen teletherapy machines were in operation at ORINS's participating medical schools, three of which were kilocurie machines and the others of which were the smaller hecto-

48. Minutes, 46th ACBM Meeting, 17–18 Sep 1954, Washington, DC, OpenNet Acc NV0411742, p. 3.

49. Minutes, 53rd ACBM Meeting, 30 Nov–2 Dec 1955, Washington, DC, OpenNet Acc NV0411747, p. 24.

50. Ibid., p. 25. Dunham was referring to the AEC policy of subsidizing radioisotopes used in all biomedical research (by covering 80% of production costs for domestic purchasers, as of 1 Jul 1955), but not those used in "routine clinical treatment." See AEC 398/14, Decision on AEC 398/12, Proposed Subsidy of Radioisotopes Program, 23 Aug 1955, NARA College Park, RG 326, E67B, box 28, folder 5 Isotopes Program 3: Distribution vol. 1.

51. Minutes of the Teletherapy Evaluation Board Clinical Program Subcommittee Meeting, 6 Nov 1954, ORINS Teletherapy Minutes and Related Documents, File 1 of 3, OpenNet Acc NV0718551, p. 1135200. As decided in the initial meeting of the Executive Committee of the Teletherapy Evaluation Board, the Clinical Program Subcommittee would comprise the entire Executive Committee plus six other members. Minutes of the Meeting of the Executive Committee of the Teletherapy Evaluation Board, ORINS, Medical Division Library, Oak Ridge, Tennessee, 25 Jan 1953, OpenNet Acc NV0718571, p. 1137072.

curie machines. In addition, a 1,500-curie rotating cesium unit was being tested at the ORINS hospital, alongside its two cobalt machines.[52]

The Clinical Program Subcommittee compiled patient treatment data from all of these sites. Early results were encouraging: of twenty-one patients with head, neck, or oral cavity cancers treated in ORINS-affiliated medical schools, thirteen showed no evidence of the disease after treatment (i.e., the tumors had regressed), three died of unrelated conditions, and five were alive with the disease.[53] The Clinical Program Subcommittee planned to analyze teletherapy results over a five- to ten-year period. Not that the profession waited for these long-term results. The Teletherapy Evaluation Board found that the availability of a cobalt-60 machine at its affiliated medical school hospitals led to a 10%–30% increase in patient treatment referrals.[54] (See figure 9.3.)

The limiting factor for the dissemination of teletherapy units proved to be the supply of radioactive sources. In contrast with nearly all of the other isotopes the AEC produced, demand actually outstripped supply for cobalt-60 in the mid-1950s.[55] In part, this was attributable to the concentrated radioactivity that needed to be produced for each source. In 1953, the Teletherapy Evaluation Board instructed the member medical schools to place orders for their radioactive sources (2,000 curies for cobalt-60 or 7,000 curies for cesium-137) with the AEC as early as possible, since they were filled on a "first come, first served" basis. Requests to the Subcommittee on Human Application for human research approval were to be filed at the same time, so that the machines could be used in clinical investigation as soon as the radioactive sources became available.[56] So as

52. Minutes, 51st ACBM Meeting, 5–7 May 1955, Oak Ridge, Tennessee, OpenNet Acc NV0708697, p. 4; Comas and Brucer, "First Impressions" (1957). The cesium-137 unit was also described as a kilocurie teletherapy machine.

53. Minutes of the Fourth Annual Teletherapy Evaluation Board Meeting, 3–4 Mar 1955, ORINS Teletherapy Minutes and Related Documents, File 1 of 3, OpenNet Acc NV0718558, p. 1135438. The time after therapy when the tumors were assessed varied from one month to five months.

54. Ibid., p. 1135433.

55. To make matters worse, a leak in the Canadian reactor set their cobalt-60 production schedule behind 12–18 months. C. H. Hetherington, "Current Production, Schedules, and Problems of Canada," Third Industrial Conference on Teletherapy, 14–15 May 1954, OpenNet Acc NV0718572, p. 1137234.

56. Minutes of the Meeting of the Executive Committee of the Teletherapy Evaluation Board, 3 Jun 1953, ORINS Teletherapy Minutes and Related Documents, File 1 of 3, OpenNet Acc NV0718558, p. 1135549.

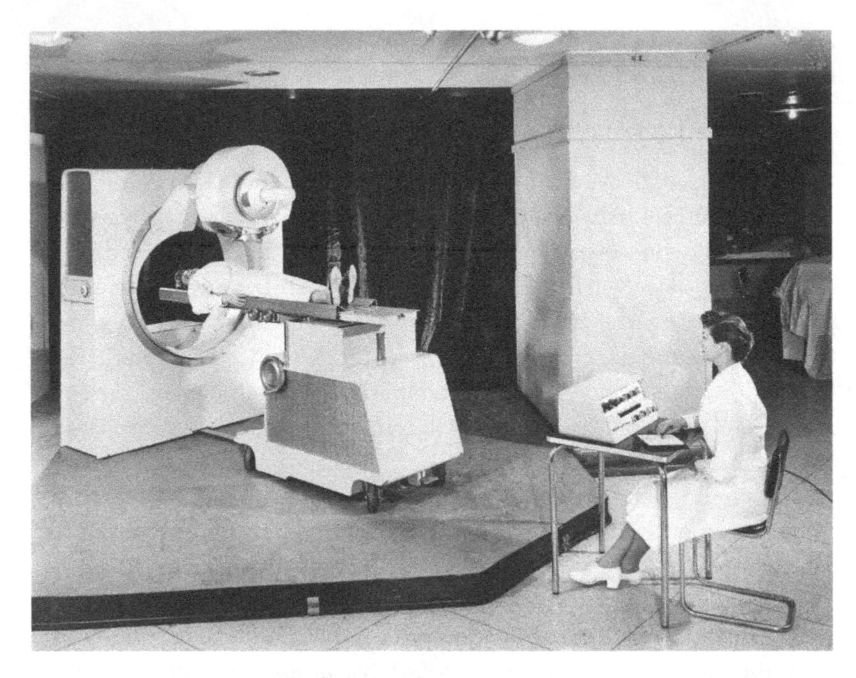

FIGURE 9.3. A 1957 cobalt-60 teletherapy unit, developed and manufactured by Westinghouse Electric Corporation at the request of the Oak Ridge Institute of Nuclear Studies. The drum-like portion of the unit directly above the woman posing as a patient contained seven fixed portals for the cobalt-60 rays (the cobalt being fixed in the drum). Credit: Westinghouse Electric Corporation. National Archives, RG 326-G, box 6, folder 5. AEC-57-5799. Reproduced by permission of Westinghouse Electric Corporation.

not to discourage industrial development of teletherapy equipment, the AEC developed an allocation policy allowing companies to order sources in increments of five.[57]

The main radioisotope production reactor in Oak Ridge did not have the neutron flux necessary to produce cobalt-60 of the desired specific activity. The AEC had to turn to its newer reactors (which had been built for the agency's own research and production programs), particularly the large Materials Testing Reactor in Idaho. The production of cobalt-60 for

57. To receive these sources, companies were required to document the readiness of five domestic users, with authorization dated prior to July 1, 1954, to take delivery of the machines within six months. Distribution Policy for High Specific Cobalt 60 Sources, 1 Mar 1954, Open-Net Acc NV0717416, pp. 1137166–67.

cancer therapy had high enough priority in the AEC to override other work for the Idaho reactor in the short term. It was used to generate 22,000 curies by the end of the year, to meet existing, year-old demand. However, the AEC did not expect to produce any high-activity cobalt in 1956 or the first half of 1957; the demand for cobalt-60 had to be balanced against the AEC's production of fissionable material for weapons. In response to the lapse, the Advisory Committee for Biology and Medicine requested the agency to accelerate cobalt-60 production, ramping up to 50,000 curies a year so that demand would not be met elsewhere: "Apparently, if the AEC falls short of this goal, the Atomic Energy of Canada Limited will capture the lead in supplying cobalt for medical purposes at an international level."[58]

Competition from the Canadians in the field of cobalt teletherapy equipment was not new. The first two cobalt-60 teletherapy units were installed in Saskatchewan and Ontario in October 1951 with kilocurie sources from the Chalk River reactor.[59] Atomic Energy of Canada Ltd., a national company, could manufacture both the cobalt source and the machines, selling integrated units. As a consequence, their cobalt teletherapy unit was on the market before any American model. Six of the first ten cobalt units installed in hospitals in the United States were imported from Canada. Moreover, the Canadians were able to market radiotherapy equipment to the USSR and China, countries to which American companies could not export nuclear technology.[60]

By contrast with the Canadian and British reliance on state-owned companies to bring atomic technologies to market, the US AEC was endeavoring to demonstrate its friendliness to free enterprise. In the development of teletherapy as well as in the commercialization of radiolabeled compounds and radiopharmaceuticals, the agency vigorously encouraged industrial involvement.[61] The main mechanism for this was a series of meetings among ORINS members, AEC Isotopes Division officials, and representatives from fifteen companies. Brucer opened their first joint

58. Minutes, 49th ACBM Meeting, 11–12 Mar 1955, Washington, DC, OpenNet Acc NV0411744, pp. 8–9 (quote on p. 9); Paul C. Aebersold, "Current Production, Schedules, and Problems of Cobalt Production in the United States," Third Industrial Conference on Teletherapy, 14–15 May 1954, OpenNet Acc NV0718572, p. 1137230.

59. Aebersold, "Progress against Cancer" (1955), p. 788.

60. Leopold, *Under the Radar* (2009), pp. 71–72; Funigiello, *American-Soviet Trade* (1988), ch. 5.

61. See chapter 6.

meeting by stating that "the United States Atomic Energy Commission and the Oak Ridge Institute of Nuclear Studies are not in the x-ray business and do not intend to get into it."[62] Through such meetings, as Eileen Leopold astutely notes, the AEC was "essentially providing the services of a trade association" by furnishing venues for firms to work out design and production problems and agree on commercial standards.[63]

Despite the limited US cobalt-60 supplies in the mid-1950s, hospitals began to purchase teletherapy units in significant numbers. These cobalt "bombs," as they were termed, were touted as "the greatest advance in radiotherapy equipment since the 1920s."[64] By 1955 at least thirty teletherapy units were in operation in the United States, all but one of which used a cobalt-60 radiation source. (The exception contained cesium-137.) Fifteen more machines were on order, awaiting the irradiation of the cobalt-60 wafers at the National Reactor Testing Station in Idaho.[65] By 1957, the number of teletherapy units in hospitals had increased to 110. The sales of teletherapy machines continued to grow in the late 1950s, with commercial units from GE and Westinghouse entering the market.

The massive kilocurie teletherapy units were permanently installed, not mobile, requiring significant space (usually a dedicated room) as well as extensive shielding. These were sophisticated machines, in which the radiation source could be rotated around the patient to direct gamma rays at the tumor without unnecessary exposure of skin or normal tissues. Already in the mid-1950s some models incorporated electronic computers into their controls to direct the timing mechanism during treatment.[66] Purchasing a teletherapy unit represented a major investment for hospitals,

62. Minutes of the Joint Meeting of the Oak Ridge Institute of Nuclear Studies–Isotopes Division–and the X-Ray Industry on Teletherapy and Human Radiographic Problems with Isotopes, 3 Oct 1953, OpenNet Acc NV0718572, p. 1. Aebersold made the same point on p. 14.

63. Leopold, *Under the Radar* (2009), p. 70. The third such meeting, which took place on 14–15 May 1954, was referred to as the Third Industrial Conference on Teletherapy. The minutes are in OpenNet Acc NV0718572 with those of the prior meetings.

64. M. D. Anderson Hospital, *First Twenty Years* (1964), p. 212.

65. AEC, *Eight-year Isotope Summary* (1955), p. 30.

66. Ibid., p. 29.

often taking several years to pay for itself.[67] By 1975, there were around 970 cobalt teletherapy machines operating in hospitals.[68]

Total Body Irradiation and the Dangers of Teletherapy

Teletherapy machines had been designed to target radiation to internal tumors, but they could also be used for Total Body Irradiation (TBI), or the delivery of external radiation uniformly to the entire body. This approach had been developed in the 1930s at Memorial Sloan-Kettering Cancer Research Institute, using high-voltage x-ray machines. The development of cobalt and cesium teletherapy machines after the war increased its use. In general, two situations could prompt a radiologist to prescribe TBI: a disseminated cancer, such as leukemia or lymphoma, in which the cells to be eradicated were found throughout the body; or a highly metastasized cancer, in which TBI could be used to shrink, if not eliminate, painful growths, even if the cancer were radioresistant. TBI could thus be used either for primary therapy or palliative care. Before the establishment in the 1960s of chemotherapy to target widely dispersed cancer cells, TBI offered one of the few treatment options available.[69] By the late 1950s, researchers were also experimenting with TBI in conjunction with bone marrow transplantation to treat some cancers.[70] At M. D. Anderson, the presence of the cobalt-60 teletherapy machine became a rationale for administering TBI treatment.[71] After 1957, an increasing number of the patients at ORINS were receiving TBI rather than targeted radiotherapy.[72]

TBI was also of interest to the military, because it could be used to provide information on how the human body responded to sublethal doses of radiation. During the Manhattan Project, TBI experimental therapy

67. Leopold, *Under the Radar* (2009), p. 78. Henry Jaffee, a member of the Teletherapy Evaluation Board, told the group of the cobalt-60 machine installed at Cedars of Lebanon Hospital in Los Angeles. The total cost of installation was $42,000. Twenty patients could be treated a day. Minutes of the Fourth Annual Meeting of the Teletherapy Evaluation Board, 3–4 Mar 1955, Detroit, Michigan, ORINS Teletherapy Minutes and Related Documents, File 1 of 3, OpenNet Acc 0718558, p. 1135423.

68. Leopold, *Under the Radar* (2009), p. 77.

69. ACHRE, *Final Report* (1996), chapter 8.

70. Kraft, "Manhattan Transfer" (2009).

71. Kutcher, *Contested Medicine* (2009), p. 68.

72. ACHRE, *Final Report* (1996), pp. 248–50.

on terminal cancer patients at Sloan-Kettering was conducted to obtain such data for the military, despite its poor record in treating the cancers in question. Similar studies were undertaken during the war at the University of California Hospital in San Francisco (under the direction of Robert Stone) and at the Chicago Tumor Institute.[73] As Stone later admitted, "the fact that Manhattan District was interested in the effects of total body irradiation was kept a secret."[74] In the postwar period, leaders in the military continued to be interested in the physiological and cognitive responses of the human body to intensive, full-body radiation exposure. Preparing troops for atomic warfare was a paramount concern. What dose of radiation would render soldiers ineffective in combat or unable to carry out complex orders? The use of high-dose radiation in cancer therapy provided an experimental opportunity to answer this question.

Even before its pioneering cobalt-60 teletherapy unit was installed in its hospital, M. D. Anderson entered into a contract with the US Air Force to obtain experimental data on the effects of human whole-body radiation from cancer patients undergoing TBI.[75] The Air Force's interest in this data related to their commitment to the Nuclear Energy for the Propulsion of Aircraft (NEPA) project.[76] The prospect of placing a nuclear reactor on board an aircraft raised questions about the potential radiation effects on the pilot. The medical advisory committee for NEPA believed that only human experiments with radiation would provide reliable information. The proposal from NEPA to use prison volunteers for such research drew the following concern from Joseph Hamilton, himself no stranger to human experimentation: "I feel that those concerned in the Atomic Energy Commission would be subject to considerable criticism, as admittedly this would have a little of the Buchenwald touch."[77] So prisoners were taken off the table as experimental subjects.

The best remaining candidates for this research were cancer patients being treated with radiation. Sick civilians would have to be experimental stand-ins for healthy soldiers.[78] In fact, representatives of the Air Force

73. Ibid., p. 252.

74. Robert Stone to Alan Gregg, 4 Nov 1948, as quoted in ibid., p. 231.

75. ACHRE, *Final Report* (1996), p. 235.

76. Bowles, *Science in Flux* (2006), pp. 38–40.

77. Joseph G. Hamilton to Shields Warren, 28 Nov 1950, as quoted in Leopold, *Under the Radar* (2009), p. 85.

78. This was the case not only at M. D. Anderson but also in Eugene Saenger's Department of Defense–funded TBI treatment studies at University of Cincinnati Hospital; see Kutcher, *Contested Medicine* (2009), ch. 4.

had already met with Clark, Fletcher, and Grimmett of M. D. Anderson to discuss a joint research project with the Aviation School of Medicine involving cancer patients.[79] NEPA's requirements set many parameters of the proposed research, including the range of radiation doses that patients would receive. Before and after the treatment, patients would undergo a battery of cognitive and motor skills tests to assess the effects of radiation on abilities needed to fly a plane. The contract for NEPA research at M. D. Anderson began in 1951 (two years before the cobalt teletherapy machine arrived) and lasted until 1956. Thus from the outset of its use of the cobalt-60 machine for cancer patients, M. D. Anderson was also using it for military research. Patients selected for the NEPA study were not necessarily expected to benefit from radiation treatment, and many were indigent. Results were published, albeit in the magazine of the Air Force School of Aviation Medicine, where the NEPA contract was not explicitly acknowledged.[80] Investigative journalists, government officials, and scholars have all criticized the ethics of these experiments.[81]

Postwar TBI experiments of this sort were not restricted to M.D. Anderson. The proliferation of "cobalt bombs" in the late 1950s expanded the opportunities for the Department of Defense (DoD) to support cancer research that would also yield information of value to the military. The Army sponsored a study of patients being treated with TBI for cancers (including radioresistant ones) from 1954 to 1963 at the Baylor University College of Medicine in Houston, and another at Memorial Sloan-Kettering Institute for Cancer Research in New York. For its part, the Navy conducted studies of patients receiving TBI at its own hospital in Bethesda, Maryland, using a cobalt-60 teletherapy unit.[82] In each of these cases, metabolic data were collected on subjects after TBI to look for biological markers of radiation (to find a "biological dosimeter"). Similar research on a longer and larger scale was undertaken under a DoD contract by Eugene Saenger at the University of Cincinnati College of Medicine.[83]

79. Leopold, *Under the Radar* (2009), p. 86.

80. Ibid., p. 95.

81. E.g., US Congress, House, *Oversight: Human Total Body Irradiation* (1981); ACHRE, *Final Report* (1996). For a nuanced analysis of the history of appraisals of TBI: Kutcher, *Contested Medicine* (2009).

82. ACHRE, *Final Report* (1996), pp. 238–39.

83. This study, which lasted from 1960 to 1971, has been the most controversial and scrutinized of the military research programs on cancer patients; see ACHRE, *Final Report* (1996), pp. 239–48. Kutcher provides excellent analysis of both the primary sources and the historiography in *Contested Medicine* (2009).

NASA, in their quest to understand the biological effects from irradiation that humans would encounter in space, was also interested in TBI-treated patients. From 1957 to 1974 the ORINS hospital treated almost two hundred patients with TBI, collecting data from some of these patients for NASA.[84] Unlike the DoD-contracted research, the physicians at ORINS selected patients with radiosensitive cancers, who could reasonably be expected to benefit from the TBI therapy. But like for the DoD, these sick patients were proxies for others, in this case astronauts.

These studies took place in the context of a greater use of TBI generally, made possible by the many cobalt-60 and cesium-137 teletherapy machines installed in hospitals in the late 1950s and 1960s. Thus teletherapy, which had been developed with strong AEC support in the name of advancing targeted cancer treatment, contributed to the greater use of generalized radiation, particularly as a therapy of last resort. Moreover, this application of radioisotopes reveals the strong and often obscure connections between the military and civilian aspects of postwar uses of atomic energy.

Beyond military-related research, it is important to recognize teletherapy was still a new treatment method in the 1950s, and the rapid diffusion of this technology entailed health risks. The AEC would allocate a teletherapy source only to "a licensed physician in good standing with the local medical society and with at least three years experience in radiation therapy" (including x-ray therapy).[85] Even so, this safeguard, and the requirements that manufacturers design the equipment to comply with the National Committee on Radiation Protection's safety guidelines, did not prevent incidents in which radiotherapy patients were overexposed. The teletherapy machines produced beams of very intense radiation, and not every hospital had a "first class physicist" to carry out a proper radiological survey and monitor exposure.[86] For their part, manufacturers of x-ray and teletherapy equipment refused to "guarantee the radiological safety of the machine and its installation."[87]

84. ACHRE, *Final Report* (1996), pp. 248–50.

85. Minutes of the Joint Meeting of the Oak Ridge Institute of Nuclear Studies, p. 15.

86. James E. Lofstrom, "Essential Data A Radiologist Must Get from the Manufacturer of Cobalt Therapy Units," Third Industrial Conference on Teletherapy, 14–15 May 1954, OpenNet Acc NV0718572, p. 1137249.

87. J. A. Reynolds of Picker X-ray Corporation, in discussion following Paul C. Aebersold, "Allocation Policy for Teletherapy Source Units," Third Industrial Conference on Teletherapy, 14–15 May 1954, OpenNet Acc NV0718572, p. 1137244. Another industry repre-

Compelling evidence of how dangerous these machines were comes from a lawsuit filed by Irma Natanson over injuries she sustained from treatment with cobalt teletherapy following mastectomy in 1955.[88] She suffered severe and disabling radiation burns, resulting in the removal of several of her ribs and some chest tissue, and extensive skin grafts. Natanson sued her radiologist, John R. Kline, for failure to warn her of the risks of treatment and for failure to administer the treatment properly. The jury rejected the charges of negligence and malpractice, but upon appeal, the court in 1960 found Kline guilty of negligence in not sufficiently disclosing the nature, consequences, and possible hazards of radiation treatment.[89] Discussions of patient safety in the meetings conducted by the AEC and ORINS on the industrial development of teletherapy revolved around whether the equipment manufacturers or the radiologists should be held liable for accidents. Strikingly, though physicians and industry representatives contended with each other over which party was responsible for safety, no one looked to the government to take on a greater role.[90]

More broadly, the potentially hazardous application of radiation sources to treat cancer is a story much older than the military development of atomic energy. Through the AEC's development of megacurie cobalt-60 and cesium-137, radioisotopes did not supersede external radiotherapy but intensified this older mode of treatment. Radiation injuries were not uncommon in the early twentieth century as x-rays and radium were widely—and sometimes carelessly—used in clinics and industries. The history of cobalt teletherapy thus exposes the connections between the pre–World War II uses of radiation and radioisotopes in medicine and

sentative from the same firm said, "I should begin by pointing out a policy of ours. A policy, which I believe is widely accepted by the x-ray industry. We believe that the responsibility for the administration of a safe dose rests entirely on the radiologist. It is his responsibility to find out what comes out of his machine, and to check it periodically. The manufacturer may run calibration data on output, depth dose, and isodose curves for his own information, for use in advertising or sales promotion; however, it is recognized that such data is merely typical, and is not to be used in connection with the treatment of a patient. We will, therefore, not furnish our data with the machine, because to do so would be to encourage its use without further checking, and this seems very risky." J. B. Stickney, "Essential Data a Manufacturer Must Supply to the Radiologist," p. 1137247.

88. This case is at the heart of Leopold, *Under the Radar* (2009).

89. *Natanson vs. Kline* is usually cited for its importance to the legal status of informed consent. Faden and Beauchamp, *History and Theory of Informed Consent* (1986).

90. See the contributions and discussion in Third Industrial Conference on Teletherapy, 14–15 May 1954, OpenNet Acc NV0718572.

postwar developments, continuities that were conceptual but also technical and industrial (one thinks of GE as a company producing x-ray machines). At the same time, the featuring of cobalt-60 machines as part of the "Atoms for Peace" program points to the distinctive political framing of this technology as a medical—and civilian—dividend of an undeniably military infrastructure.[91]

Radioisotopes in Medical Diagnostics

While the harnessing of atomic energy against cancer focused on teletherapy, radioisotopes also proved to be immensely useful for medical diagnosis. As the AEC noted in their 1948 report to Congress, researchers and physicians took advantage of the rapid absorption of certain radioelements to locate tumors.[92] This was an application of tracer methodology to medicine, analogous to tagging a compound with radiocarbon to follow the steps of a metabolic pathway, or adding a radioelement to a lake to track its cycling through an ecosystem. Diagnostic radioisotopes also built upon the tradition of using x-rays to visualize anatomical structures while moving the source of radiation from without the body to within it.[93] Tumors were not the only target. By 1955, radioisotopes were being used in a wide variety of diagnostic tests. Radioisotope-based diagnostics frequently permitted a more dynamic view of organ function or circulation than previous methods. Moreover, tracer applications of radioisotopes generally required much smaller amounts of radioactivity than the dosage required in therapeutic applications.[94] In fact, ideally the use of a radioisotope in a diagnostic test could be of a low enough radiation dose to entail negligible hazard to the patient.

91. US Delegation to the International Conference on the Peaceful Uses of Atomic Energy, *International Conference* (1955), vol. 1, pp. 298–99; Leopold, *Under the Radar* (2009), pp. 68–69.

92. AEC, *Fourth Semiannual Report* (1948), p. 23.

93. On the use of x-rays in medical diagnosis, see Pasveer, "Knowledge of Shadows" (1989).

94. There could be a thousand-fold difference between these kinds of applications. Human Radiation Studies: Remembering the Early Years, Oral History of Biochemist Waldo E. Cohn, Ph.D. Conducted January 18, 1995 through the Department of Energy by Thomas Fisher, Jr. and Michael Yuffee, and published at http://www.hss.doe.gov/healthsafety/ohre/roadmap/histories/0464/0464toc.html.

Diagnosis of the concentration of elements in the body through dilution of a radioisotope began in the 1930s. For example, Hamilton's early work with sodium-24 laid the groundwork for diagnostic tests using this radioelement to assess total exchangeable body sodium.[95] Using the same general rationale, chromium-51 or iron-59 were used to measure red cell mass, and ^{131}I-labeled serum albumin was used to measure plasma volume. One could also use radioisotopes to measure rate of flow in the circulatory system. Sodium-24 could be employed to assess cardiac output and diagnose peripheral vascular disorders, building on Hamilton's early study of its absorption and movement through the bloodstream to the extremities. Other diagnostic tests followed the metabolism of radiolabeled compounds in the human body. For example, patients with pernicious anemia did not excrete as much vitamin B-12 into urine as normal subjects. Consequently, reliable diagnosis could be achieved by administering cobalt-60–labeled vitamin B-12 followed by a precise urine test.[96]

Perhaps the best-known radioisotope-based diagnostics involved physiological localization, as exemplified by the use of iodine-131 since the late 1930s to study thyroid physiology and dysfunction.[97] In 1956, Aebersold estimated that "over half a million thyroid studies have been done with iodine 131 since that time."[98] Similarly, the localization of phosphorus-32 in tumors had been noted by John Lawrence and other researchers during the 1930s, and many diagnostic procedures for various forms of cancer were developed using this isotope. The challenge of using radioactivity to locate a tumor or measure function in an internal organ, however, was detection—the tissues of the human body both absorbed and interfered with radiation given off by isotopes.

One way of circumventing problems with detection was to administer diagnostic isotopes in conjunction with surgery, when a patient's body was opened up and radioactivity could be measured directly. With grant support from the AEC, William H. Sweet, a neurosurgeon at Massachusetts General Hospital, developed a method for locating brain tumors with phosphorus-32 (procured from Oak Ridge).[99] He and his coworkers

95. Aebersold, "Development of Nuclear Medicine" (1956), p. 1031.

96. Ibid., p. 1032.

97. Iodine-131 also became the treatment of choice for Graves disease, a type of hyperthyroidism. On the early uses of iodine-131 for diagnosis and therapy, see chapter 2.

98. Aebersold, "Development of Nuclear Medicine" (1956), p. 1031.

99. Minutes, 23rd ACBM Meeting, 8–9 Sept 1950, Washington, DC, OpenNet Acc NV0708842, p. 16. Others were trying similar approaches with radiolabeled compounds, e.g., Moore, "Use of Radioactive Diiodofluorescein" (1948).

found that phosphorus-32 localized well in brain tumors of patients, concentrating four- to seventy-fold over its presence in gray and white matter in the brain, although its beta rays were not energetic enough to exit the scalp or skull and be detected.[100] Consequently phosphorus-32 could serve as a diagnostic marker only once the brain was exposed during surgery. Charles V. Robinson of Harvard Medical School developed miniature Geiger counters, which Bertram Selverstone (working with Sweet) adapted to be placed directly into the brain during a craniotomy.[101] In their study, thirty-three patients diagnosed with glioblastomas received one to four millicuries of phosphorus-32 intravenously up to three days before the operation. The surgeon, having accessed the patient's brain by removing the skull, would insert the 1–3 mm-long probe at various depths. An increase in the counting rate from five to one hundred times over that of normal brain tissue indicated the presence of the brain tumor. For twenty-nine of the thirty-three patients first studied, the tumor was pinpointed successfully.[102] This method of locating brain tumors provided a dramatic improvement over the conventional method of putting air into the fluid-containing cavity inside the head and looking for distortions.[103] Just a few years after the method was first published, Sweet asserted that so-called Selverstone-Robinson probes "are now used routinely at operations."[104] (See figures 9.4 and 9.5.)

Sweet also sought to develop an isotope-based detection method that could be used on patients whose heads were still intact. He and his co-workers experimented with several radioisotopes that emitted gamma rays—since these could exit the skull. So, for example, potassium-42 localized to tumors and could be detected; however, the radioisotope was also taken up by the muscle and scalp tissues.[105] Another complication

100. On the level of concentration in brain tumors, Interview with Dr. William Sweet by Gil Whittemore, ACHRE Staff, 8 Apr 1995, OpenNet Acc NV0751118, p. 5; see also Selverstone, Sweet, and Robinson, "Clinical Use of Radioactive Phosphorus" (1949).

101. Early, "Use of Diagnostic Radionuclides" (1995), p. 651; Selverstone, Solomon, and Sweet, "Location of Brain Tumors" (1949); Sweet, "Use of Nuclear Disintegration" (1951).

102. Selverstone, Sweet, and Robinson, "Clinical Use of Radioactive Phosphorus" (1949), p. 649. One of the patients that went through this surgery, in the end, had no tumor, and made a complete recovery.

103. Interview with Dr. William Sweet, p. 8.

104. This assertion was made at a meeting in September 1953, though published two years later. Sweet and Brownell, "Use of Radioactive Isotopes" (1955), p. 211.

105. Selverstone, Sweet, and Ireton, "Radioactive Potassium" (1950).

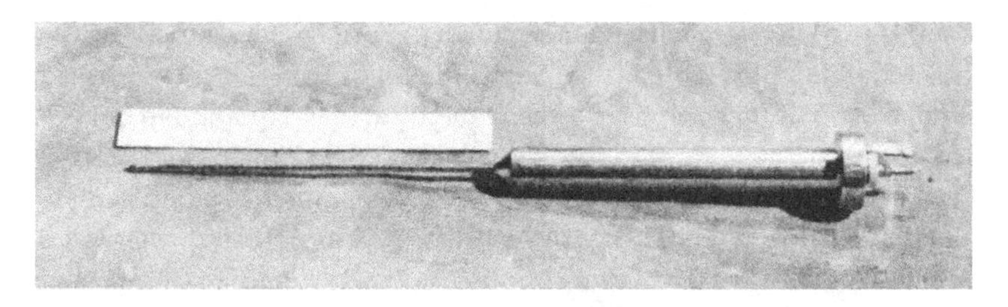

FIGURE 9.4. A Selverstone-Robinson Probe Counter. From William H. Sweet and Gordon L. Brownell, "The Use of Radioactive Isotopes in the Detection and Localization of Brain Tumors," *Radioisotopes in Medicine*, ed. Gould A. Andrews, Marshall Brucer, and Elizabeth B. Anderson (Washington, DC: US Government Printing Office for the Atomic Energy Commission, 1955), pp. 211–218, on p. 211.

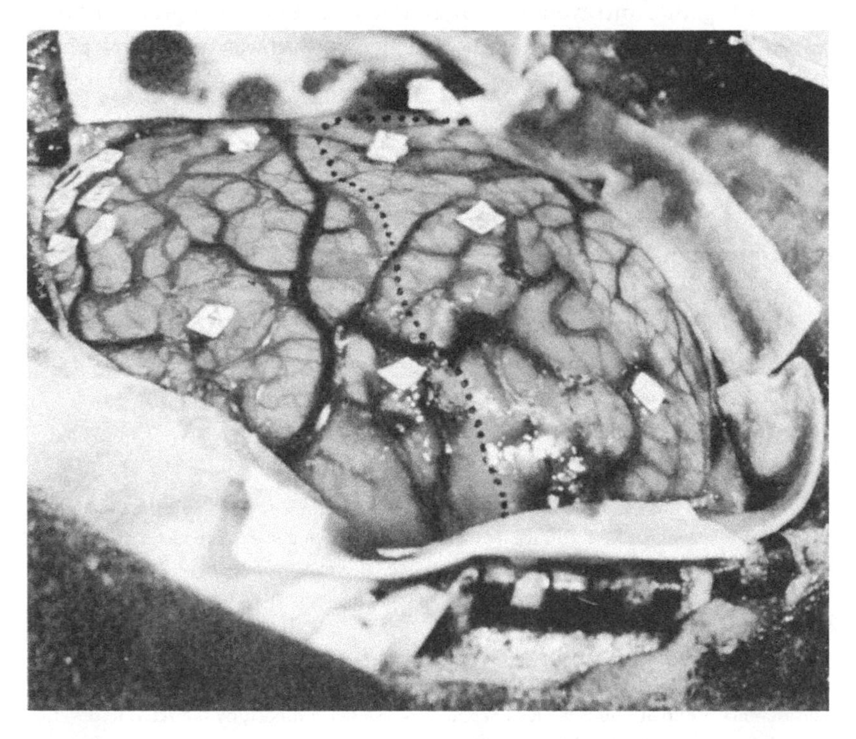

FIGURE 9.5. Operation Field of Patient T.F. Capital letters indicate sites of insertion of probe counter (shown in Figure 9.4). From William H. Sweet and Gordon L. Brownell, "The Use of Radioactive Isotopes in the Detection and Localization of Brain Tumors," *Radioisotopes in Medicine*, ed. Gould A. Andrews, Marshall Brucer, and Elizabeth B. Anderson (Washington, DC: US Government Printing Office for the Atomic Energy Commission, 1955), pp. 211–218, on p. 211.

was that the brain tissue itself tended to scatter gamma rays, obscuring the signal.[106] Gordon Brownell (the hospital physicist) had the idea of trying isotopes that emit positrons, which are identical to beta-radiation but opposite in sign. Once an emitted positron inevitably encounters an electron, the particles are annihilated and give off two oppositely-directed gamma photons.[107] Brownell and Sweet showed that by using two detectors on opposite sides of the head to pick up the coincident photons, one could determine the location of the positron emitter.[108] Using scintillation counters rather than Geiger-Müller counters further improved the sensitivity and efficiency of detection.[109]

In the early 1950s, the most promising positron-emitting radioelement for this kind of diagnostic test was arsenic; arsenic-72 and arsenic-74 were injected together intravenously at 20 microcuries per kilogram of body weight.[110] (The amount of metallic arsenic involved was well below pharmacologic intoxication.) By 1953, a total of 300 patients had undergone brain scans at Massachusetts General Hospital with radioarsenic. In 99 of 133 brain tumor patients tested, radioactivity concentrated detectably in intracranial lesions, whose locations were verified by surgery.[111] By 1956, the technique had been further refined in brain scans with cyclotron-produced arsenic-74.[112] The typical amount of arsenic-74 used for a brain scan was 2.3 millicuries, which exposed patients to a dose equivalent to the thirteen-week tolerance set for workers in atomic energy facilities. The researchers regarded the hazard as small compared with the

106. Sweet, "Use of Nuclear Disintegration" (1951), p. 876.

107. Photons are the same form of radiation given off by x-rays, so detection technologies were well developed. Martin Kamen and Samuel Ruben had explored uses of positron-emitting radioisotopes in the late 1930s, including carbon-11, nitrogen-13, oxygen-15, and fluorine-18, but dropped these once they discovered the long-lived isotope carbon-14. Dumit, "PET Scanner" (1998).

108. Brownell and Sweet, "Localization of Brain Tumors" (1953).

109. Wrenn, Good, and Handler, "Use of Positron-Emitting Radioisotopes" (1951). This neurosurgery group at Duke Medical School was also being funded by the AEC to develop methods for brain tumor localization with radioisotopes.

110. In one publication, the source of the radioarsenic was listed as the MIT cyclotron rather than Oak Ridge. Brownell and Sweet, "Localization of Brain Tumors" (1953). Animal tests were used to look at the localization of various isotopes in tumors and in various tissues: Locksley et al., "Suitability of Tumor-Bearing Mice" (1954).

111. Sweet and Brownell, "Use of Radioactive Isotopes" (1955), p. 214.

112. Brownell and Sweet, "Scanning of Positron-Emitting Isotopes" (1956).

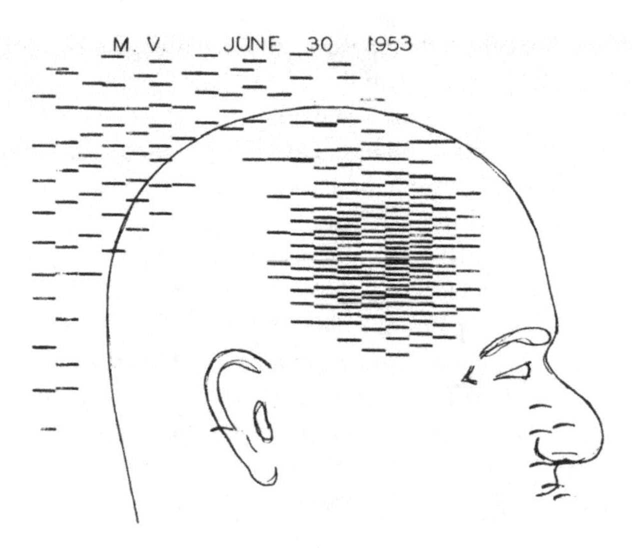

FIGURE 9.6. A brain scan showing the asymmetry in radioactivity between the right to left sides of the head of a patient injected with radioarsenic. The cluster of bars indicates an area of increased counting; surgery revealed a tumor where the radioactivity was concentrated. From William H. Sweet and Gordon L. Brownell, "The Use of Radioactive Isotopes in the Detection and Localization of Brain Tumors," *Radioisotopes in Medicine*, ed. Gould A. Andrews, Marshall Brucer, and Elizabeth B. Anderson (Washington, DC: US Government Printing Office for the Atomic Energy Commission, 1955), pp. 211–218, on p. 216.

danger posed by the brain tumor.[113] (See figure 9.6.) The potential market for users of positron-emission scanning in medical diagnostics was sufficiently large that by 1957 the Atomic Instrument Company of Cambridge, Massachusetts, was developing a commercial scanner.[114]

In the early 1950s, Sweet was also conducting experiments to see if radioisotopes could be employed therapeutically against brain tumors. This work was supported by grants from both the Atomic Energy Commission and the American Cancer Society.[115] As in the development of positron-emission tomography, this project involved taking advantage of novel properties of certain isotopes, in this case a radioelement's ability to "capture" a neutron. Physicians had been attempting to harness neutron

113. Mealy, Brownell, and Sweet, "Radioarsenic in Plasma" (1959), p. 317.

114. Sweet and Brownell, "Localization of Intracranial Lesions" (1955), p. 1188.

115. See acknowledgments for Javid, Brownell, and Sweet, "Possible Use of Neutron-Capturing Isotopes" (1952).

beams as a tool for cancer therapy since the 1930s, beginning with the efforts of John Lawrence, but with limited success. Plant radiation studies inspired an alternative approach combining aspects of internal radioisotope therapy and external radiation therapy.[116] In 1950, Alan Conger and Norman Giles at Oak Ridge National Laboratory showed that many of the chromosomal aberrations resulting from irradiating lily plants with slow neutrons were attributable to trace amounts of boron in the lily bulbs.[117] Boron-10 (a stable isotope) has a strong avidity for slow neutrons; even traces of this isotope would "capture" neutrons. Upon absorbing a neutron, boron-10 disintegrates into lithium-7 and releases 2.4 MeV in a high-energy alpha particle that travels only 5–9 mm. Sweet and his colleagues realized that if one could localize sufficient boron-10 to a tumor, then irradiate the area with a stream of slow neutrons, the tumor might be preferentially destroyed.[118] Or, as Sweet put it in a retrospective account, "the cell closest to the ^{10}B [boron-10] atom bears the brunt of its atomic explosion."[119]

Sweet and Brownell oversaw experimental trials of this therapy in collaboration with scientists at Brookhaven National Laboratory (BNL).[120] Between 1951 and 1953, ten patients diagnosed with glioblastoma multiforme were treated using boron-10 neutron capture therapy; the boron had been irradiated atop the newly completed BNL reactor.[121] Glioblastomas were invariably fatal, justifying (in the eyes of these researchers) "a new approach to control."[122] The longest survival time after treatment in this first set was 186 days, although some patients showed improvement

116. Brownell et al., "Reassessment of Neutron Capture Therapy" (1972), p. 827.

117. Conger and Giles, "Cytogenetic Effect of Slow Neutrons" (1950). Boron was not the only element capturing neutrons—also implicated were hydrogen and nitrogen, both of which were much more abundant. These three elements were responsible for 99% of the ionization dose from neutron capture. The contribution of boron was nonetheless striking given its very low concentration in the plant tissue.

118. The concentration of boron-10 was enabled by the fact that the normal blood-brain barrier was less pronounced in brain tumors. Sweet's group also experimented with lithium-6, another stable isotope that captured neutrons and released high-energy alpha particles upon subsequent disintegration. Luessenhop et al., "Possible Use" (1956).

119. Sweet, "Early History" (1997), p. 19.

120. For an account of these experiments from the Brookhaven side, see Crease, *Making Physics* (1999), pp. 182–92.

121. Some patients received glycerol in conjunction with borax to try to improve uptake by the tumorous tissue. Sweet and Javid, "Possible Use" (1952).

122. Farr et al., "Neutron Capture Therapy" (1954), p. 280.

following treatment that suggested temporary regression of the tumor.[123] A second series of patients was subsequently treated using higher doses of neutrons. One patient survived eighteen months.[124]

Altogether, out of twenty-one patients treated by Sweet and his collaborators with neutron capture therapy, eight showed evidence that tumor growth had been retarded.[125] The Advisory Committee for Biology and Medicine met in Boston in June 1953, and heard about various cancer research programs at Harvard Medical School. As the minutes of this meeting note,

> Much interest was displayed by the Committee in the work of Dr. William Sweet and his group on the external localization of brain tumors employing positron emitting isotopes such as radio arsenic. Dr. Sweet stated that he and his staff are working closely with the medical staff at the BNL on the neutron capture technique in the treatment of brain tumors. It was the opinion of the Committee that the work underway at Harvard Medical School is basic research of top quality.[126]

Optimism that the method could be improved for wider use led to another set of clinical trials at Massachusetts General Hospital in collaboration with MIT (for the reactor irradiation) in 1961–62.[127] However, with hindsight Brownell and Sweet have described all these early studies as "uniformly discouraging."[128] A significant amount of boron remained in the blood; this was activated by the neutron beam and the resulting radiation

123. Ibid., particularly patients 3977 and 4055 on p. 284. For the group's further clinical studies: Farr et al., "Neutron Capture Therapy of Gliomas" (1954); Godwin et al., "Pathological Study of Eight Patients" (1955).

124. Summary Factsheet Human Experimentation, SFS8.001, Neutron Capture Therapy, OpenNet Acc NV0704284.

125. Struxness et al., "Distribution and Excretion of Hexavalent Uranium" (1956), p. 186.

126. Minutes, 38th ACBM Meeting, Cancer Research Institute, Boston, 26–27 Jun 1953, OpenNet Acc NV0711916, p. 3.

127. Sweet claimed that verbal consent had been obtained on behalf of each patient treated with boron neutron capture therapy at Brookhaven and MIT. In the MIT trials the boron was administered to most of the patients as phenylboronic acid in an attempt to improve its localization to the tumors. Sweet, "Early History" (1997), p. 23.

128. Brownell et al., "Reassessment of Neutron Capture Therapy" (1972), p. 827.

was carried to many sites in the body.[129] The lifespan of the patients was not significantly increased, and postmortem pathological analysis indicated radiation damage to normal brain tissue.[130]

The Boston Project

Sweet's research on this novel form of cancer therapy opened the door to undertaking experiments that served the military side of the AEC. Mouse experiments conducted by Cornelius Tobias and medical physicists at Berkeley suggested that uranium-235 might be used instead of boron-10 for neutron capture. Uranium-235 is the most fissile isotope of uranium; it had been used to fuel the original atomic bomb ("Little Boy") detonated over Hiroshima. Mice injected with uranium-235 and then bombarded with a sublethal dose of slow neutrons at the Oak Ridge reactor (the same one used for producing radioisotopes) died within three weeks. Comparison with controls showed that the deaths were not due to the chemical toxicity of uranium but to the induced fission events. The energy given off by the disintegration of uranium was over fifty times that given off by boron-10. Sweet's group hoped to harness the tremendous power of uranium fission to destroy brain tumors in situ.[131]

Needless to say, fissionable uranium isotopes could not simply be purchased from Oak Ridge. Access to uranium was provided, however, through a collaboration established between Sweet's group at Massachusetts General Hospital and scientists in the Health Physics division at Oak Ridge National Laboratory (ORNL). This collaboration, dubbed the "Boston Project," involved injecting various isotopes of uranium into a number of patients with fatal brain cancer (glioblastomas). The aim, so far as cancer research was concerned, was to see if uranium could be used instead of boron for neutron capture therapy. However, these patients were not scheduled for neutron irradiation, and so would not benefit from the potential therapy. In other words, this was clearly nontherapeutic research. The goal was simply to see if the uranium concentrated sufficiently

129. Stannard, *Radioactivity and Health* (1988), vol. 3, p. 1770. This information comes from an interview Stannard conducted with Eugene P. Cronkite of Brookhaven National Laboratory, 15 Sep 1982.

130. Godwin et al., "Pathological Study of Eight Patients" (1955).

131. Tobias et al., "Some Biological Effects" (1948).

in brain tumors for neutron capture therapy to be feasible as a treatment (for future patients, at least).[132] For this purpose, any isotope of uranium would suffice.[133] Between 1953 and 1957, eleven comatose, terminally ill cancer patients received uranium-233, uranium-235, or uranium-238 through the Boston Project.[134] Unlike in the case of earlier plutonium injection experiments, the patients or their nearest relatives agreed to the study and results were published in the open medical literature.[135]

The Boston Project was prompted by another set of research aims, unrelated to neutron capture therapy. For the AEC, determining the metabolism and toxicity of uranium was a pressing issue, and the limited information from early Manhattan District studies was worrisome. Uranium causes kidney damage, and animal data revealed considerable species difference in tolerance—mice tolerated uranium doses one hundred times higher than rabbits, and some mouse strains two-hundred-fold higher.[136] The human studies done at the University of Rochester Medical School on the metabolism of fissionable materials included injection experiments of uranyl nitrate into six patients.[137] The patients were not terminally ill, so the study produced data on excretion of uranium, but not on its distribution through the body's organs. The Rochester data, limited as it was, raised questions about whether the AEC's maximum permissible exposure levels for uranium were set too high for the safety of atomic energy workers, particularly those at the Y-12 uranium isotope separation plant at Oak Ridge.[138]

Sweet was professionally (and geographically) close to Shields Warren, who became a member of the Advisory Committee for Biology and

132. ACHRE, *Final Report* (1996), p. 159.

133. There are no stable isotopes of uranium.

134. Boston–Oak Ridge Uranium Study: Chronology of Significant Events, MMES/X-10, CF Human Studies Project Files, DOE Info Oak Ridge, ACHRE document ES-00298. According to Whittemore and Boleyn-Fitzgerald ("Injecting Comatose Patients" [2003], p. 173), the use of uranium isotopes, which were tightly controlled, meant that this experiment did not go through the usual AEC approval channels, including review by the Subcommittee on Human Applications.

135. Struxness et al., "Distribution and Excretion of Hexavalent Uranium" (1956); Luessenhop et al., "Toxicity in Man of Hexavalent Uranium" (1958); Bernard, "Maximum Permissible Amounts of Natural Uranium" (1958).

136. Luessenhop et al., "Toxicity in Man of Hexavalent Uranium" (1958), p. 84.

137. Stannard, *Radioactivity and Health* (1988), vol. 1, p. 100; ACHRE, *Final Report* (1996), p. 158.

138. ACHRE, *Final Report* (1996), p. 158.

Medicine (ACBM) after retiring as Director of the AEC's Division of Biology and Medicine in 1949. Sweet's recollection was that either Warren or A. Baird Hastings, another close colleague who participated in the ACBM, suggested that he collaborate with ORNL to undertake a study of the human metabolism and excretion of uranium along with his neutron capture therapy studies.[139] The Boston Project included a complex transit of materials between Oak Ridge and Boston. Two uranium nitrate solutions (one of uranium-233, one of uranium-238) were sent on October 31, 1953, via commercial airlines from ORNL to Massachusetts General Hospital. Two more uranium solutions were personally delivered by ORNL scientists who flew to Boston on another commercial airliner on December 8, 1953. Patients were injected from November 1953 through January 1956. Specimens of blood, urine, feces, and bone collected from the injected patients were shipped, also by commercial airline, from the hospital back down to Oak Ridge for chemical analysis.[140] Strikingly, the first patient selected to receive uranium had been given phosphorus-32 previously, and Luessenhop asked Struxness at ORNL whether this would interfere with the uranium analysis planned.[141]

The Boston Project was truly dual-purpose, providing information to the AEC about the occupational hazards of its atomic weapons production facilities while also providing preliminary experimental information on a possible cancer therapy. That said, the use of terminally ill and unconscious patients to obtain such data raised ethical questions, even at the time. The use of patients so close to death made autopsy material available, which was of great value for the ORNL analysis.[142] By the same token, this situ-

139. Interview with Dr. William Sweet, pp. 27–28.

140. These shipments are detailed in Boston–Oak Ridge Uranium Study: Chronology of Significant Events, MMES/X-10, CF Human Studies Project Files, DOE Info Oak Ridge, ACHRE document ES-00298. For more detail on the preparation and transport of the uranium, see Whittemore and Boleyn-Fitzgerald, "Injecting Comatose Patients" (2003).

141. Letter from Alfred J. Luessenhop to Edward G. Struxness, 3 Nov 1953, MMES/X-10, Health Sci. Research Div., 1060 Commerce Park, Rm. 253, DOE Info Oak Ridge, ACHRE document ES-00320. According to Whittemore and Boleyn-Fitzgerald ("Injecting Comatose Patients" [2003], p. 175), the experiment went ahead but the presence of phosphorus-32 did "interfere with tissue analysis." They take this decision as indicative of the haste with which the research was undertaken.

142. That the use of material from autopsies was part of the planned research is clear from the documentation, e.g., Project Boston [handwritten notes], MMES/X-10, Health Sci. Research Div., 1060 Commerce Park, Rm. 253, DOE Info Oak Ridge, ACHRE document ES-00325.

ation created a conflict of interest for the physician-researcher for whom the rapid death of a patient could be scientifically valuable. For this very reason, in 1953 the British Medical Council was campaigning against the use of comatose patients as subjects in medical research.[143] Furthermore, the dose level of uranium Sweet administered to his unconscious patients increased over the duration of the study from four milligrams in the initial patient to over fifty milligrams in the eighth. Most likely, disappointing results on the localization of uranium in the brain tumors led Sweet to experiment with higher doses. However, when Karl Z. Morgan, head of the ORNL Health Physics Division, learned in 1957 that the patients were being given doses of uranium many times higher than the permissible body burden, he canceled the project.[144] As it turns out, the research team's original guidelines for selecting patients were not followed systematically. The Advisory Committee on Human Radiation Experiments found that at least one patient in the experiment did not have a brain tumor, but was hospitalized for a subdural hematoma after a head injury. This patient, who was apparently never identified by the hospital, was injected with sufficient uranium to cause mild kidney failure, and his autopsy also reported evidence of radiation on the "liver, spleen, kidneys and bone marrow."[145]

Two aspects of the Boston Project deserve emphasis here. One is that, as we have seen in research with radioiron, many of the leading institutions in the development of radioisotope studies and nuclear medicine were also involved in human radiation experiments for the AEC—notably the University of California, Berkeley, UCSF Medical School, University of Rochester Medical School, UCLA Medical School, and, in this case, Harvard Medical School and Massachusetts General Hospital. Given the concentration at these sites of both expertise and infrastructure for the clinical use of radioisotopes, their involvement in military-related work is not especially surprising. In most of these cases, clinicians and scientists were being supported by the AEC for conventional biomedical research involving radiomaterials at the same time that they were contracted for

143. Mann, "Radiation: Balancing the Record" (1994), p. 473.

144. This is according to a 1995 interview with Karl Morgan; ACHRE, *Final Committee* (1996), p. 159. In 1958 Morgan was interested in continuing the study but using lower doses of uranium, amounts that "would more closely approximate the dose you get from inhalation." Quote from Whittemore, Interview with Dr. William Sweet, p. 38. See also Whittemore and Boleyn-Fitzgerald, "Injecting Comatose Patients" (2003).

145. Quote from autopsy report, ACHRE, *Final Report* (1996), p. 159.

applied human research on the radiobiology and toxicology of fissile materials. The agency was, after all, manufacturing plutonium bombs even as it sought to cultivate medical applications of atomic energy. In the case of Sweet's program, interest in the medical application of uranium for neutron capture therapy grew out of a pre-established medical research program. In fact, the attempts to get neutron capture therapy working with radioisotopes continued long after the experiments with uranium ceased.

Second, the results, though they confirmed concerns about permissible dose levels set by the NCRP and the AEC for uranium, did not change the government's occupational safety standards. The autopsies of patients in the Boston Project indicated that uranium was retained in the kidneys in humans at higher levels than in experimental animals.[146] This result suggested, according to Karl Morgan, the head of the ORNL Health Physics Division, that the industrial standards for maximum permissible concentration of uranium compounds in air were too high, perhaps even by ten-fold.[147] However, Morgan's concerns about uranium exposure in the AEC's plants were not heeded. As the Advisory Committee on Human Radiation Experiments pointed out, the occupational safety standards for uranium exposure were actually relaxed in the years following the Boston Project.[148]

Scanning the Body

The human body posed particular constraints on the capacities of Geiger counters and radiographic film. Ultimately, novel radioisotope-based techniques for visualizing processes within the body depended on the development of new instruments. A new generation of scanning machines,

146. Bernard, "Maximum Permissible Amounts of Natural Uranium" (1958), p. 289.

147. As Bernard (ibid., p. 289) put it, "These data would indicate that the safe burden in the kidney, dictated by considerations of chemical toxicity, is one-tenth the burden deemed permissible from radiological considerations." An alternative explanation for the autopsy results, as Morgan noted, was that the retention of uranium was a by-product of the high dose injected, but Bernard's carefully argued paper does not support this. Morgan expressed a strong interest in continuing the collaboration with Sweet's group in order to obtain further data on uranium retention in humans, but this never transpired. Karl Z. Morgan to William H. Sweet, 16 Jul 1958, MMES/X-10, Director's Files, 1958–Health Physics, copy in Oak Ridge Information Center, ES-00283.

148. ACHRE, *Final Report* (1996), p. 159.

following upon the early positron-emission scanner developed by Sweet and Brownell, made possible a shift in radioisotope-based diagnostics from tests of physiological functionality to imaging. Most of these devices detected gamma radiation, enabling a different set of radioisotopes to come to the fore in nuclear diagnostics. Few were isotopes of elements used in the body; they were selected instead for their characteristics as radiation emitters. Most were short-lived, some having half-lives as short as a few hours. This dramatically reduced the radiation exposure received by patients undergoing diagnostic tests with isotopes.[149]

In 1949 Benedict Cassen at UCLA designed a scintillation counter for the in vivo localization of iodine-131.[150] Because it was 10 to 20 times more sensitive than Geiger counters for detecting the gamma rays emitted by radioiodine, this detector required less iodine-131 to be administered to the patient. Two years later, Cassen developed a point-by-point counting grid and incorporated more sensitive inorganic calcium tungstate detector crystals.[151] He also mounted the probe on a moving mechanism, to yield what was called the rectilinear scanner. In 1959, Picker X-ray Company began producing this instrument commercially, using even more sensitive high-density sodium iodide crystals in the detectors. This machine enabled organs beyond the thyroid to be visualized.

Hal Anger of the Donner Laboratory at Berkeley subsequently built a device with ten scintillation detectors in a row that could scan the whole body. Anger's innovation was to use radiation emitted from the body to construct an image rapidly, by "focusing" the gamma rays through apertures.[152] The development of this kind of gamma-ray "camera" drove the iconographic orientation of scanning devices for radioisotopes, relying on "ever-larger sodium iodide crystals, the use of larger banks of improved detector phototubes, tomographic applications, and highly sophisticated linkages with computers."[153] The Anger scintillation camera became

149. As one account puts it, "The raison d'être for the developments of this era, let us emphasize again, was the reduction of the radiation dose to the patient. Thus, toxicological considerations were the chief motivation. Increases in speed and convenience were the second major factor." Stannard, *Radioactivity and Health* (1988), vol. 3, p. 1773.

150. Early, "Use of Diagnostic Radionuclides" (1995), p. 652.

151. Myers and Wagner, "How It Began" (1975), p. 11.

152. Early, "Use of Diagnostic Radionuclides" (1995), p. 654; Myers and Wagner, "How It Began" (1975), p. 12.

153. Myers and Wagner, "How It Began" (1975), p. 12.

commercially available, from Nuclear Chicago Corporation, in 1968.[154] Only radioisotopes that emitted gamma radiation worked in conjunction with these scintillation-detector scanners, and those that did not emit alpha or beta radiation were preferred from a safety point of view, since they exposed the patient to less ionizing radiation.

Over the course of the 1950s concerns emerged about the side effects of iodine-131, even at diagnostic doses. This attention arose from the fact that some patients who had received therapeutic doses of iodine-131 when it first became available subsequently developed leukemia, a disease whose association with radiation exposure was widely acknowledged.[155] Researchers at Brookhaven National Laboratory sought to identify an iodine isotope that would expose patients to less radiation after administration. For short-lived isotopes, generators—in which the isotope of interest was harvested on-site from its isotopic precursor—were preferable to both reactors and cyclotrons as a means of production. In 1958, Brookhaven developed and marketed a generator for iodine-132, with a half-life of 2.3 hours, produced from the decay of tellurium-132.[156]

On one occasion the Brookhaven researchers isolated a trace impurity in their iodine-132 preparation; it turned out to be technetium-99m, a radioisotopic isomer possessing a half-life of merely six hours.[157] Technetium-99m, which emits a gamma ray, could be produced from a generator similar to that for iodine-132, but using molybdenum-99 (rather than tellurium-132) as a source. (See figure 9.7.) Scientists at Brookhaven employed technetium-99m to study thyroid physiology, but there appeared to be little interest in medical applications. As one account notes, the AEC decided against filing a patent on the technetium-99m separation process "on the grounds that they could foresee no use for it."[158]

In 1961 Argonne Cancer Research Hospital ordered the first technetium-99m generator from Brookhaven, for use by Paul Harper and Katherine Lathrop. Their successful utilization of technetium-99m in

154. Miale, "Nuclear Medicine" (1995).

155. Furth and Tullis, "Carcinogenesis" (1956), p. 13; Blom et al., "Acute Leukaemia" (1955); Seidlin et al., "Occurrence of Myeloid Leukemia" (1955). Animal experiments also pointed to the carcinogenic effects of iodine-131 administration: Edelmann, "Relation of Thyroidal Activity" (1955).

156. Tellurium-132 itself has a half-life of only 3.2 days. Iodine-132 had been identified in 1954: Stang et al., "Production of Iodine-132" (1954). Iodine-125 was also advocated as a substitute for iodine-131. Stannard, *Radioactivity and Health* (1988), vol. 3, p. 1773.

157. Public Affairs Office, "Celebrating 50 Years" (1997).

158. Ibid., p. 44N.

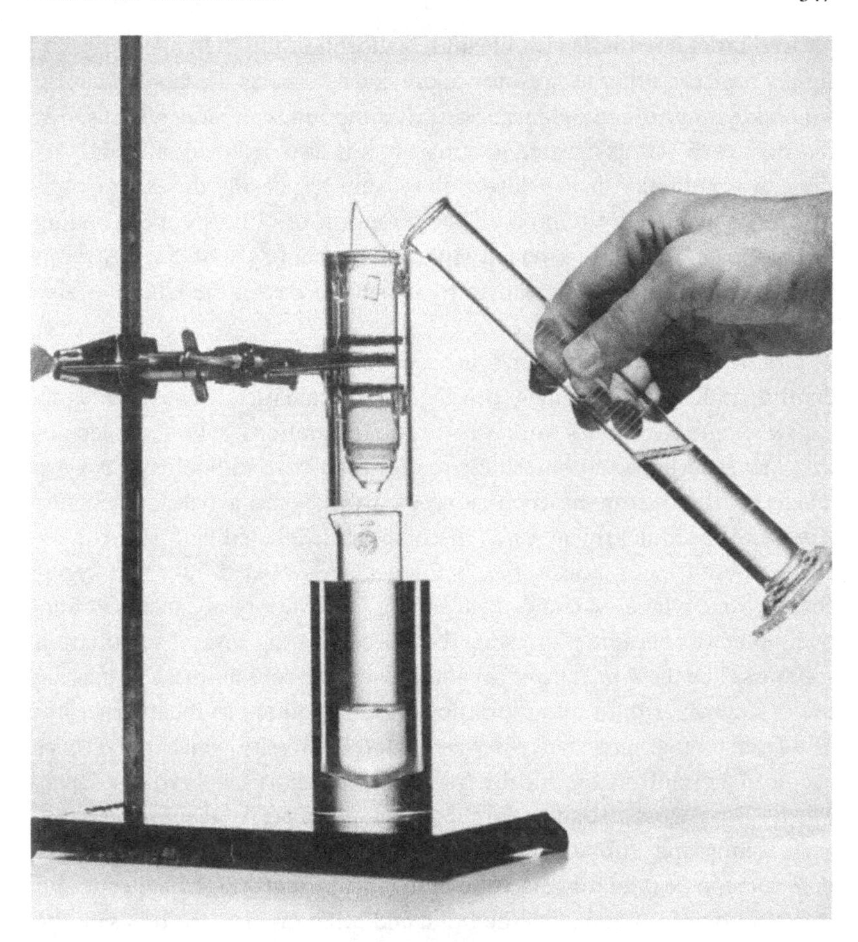

FIGURE 9.7. Picture of a researcher eluting technetium-99m from a generator. From Powell Richards, "Nuclide Generators," in *Radioactive Pharmaceuticals*, ed. Gould A. Andrews, Ralph M. Kniseley, and Henry N. Wagner Jr. (Springfield, VA: Division of Technical Information, 1966), p. 160.

brain scanning stimulated greater interest in this unusual isotope.[159] In 1966, Powell Richards at Brookhaven improved on the design for generators, dubbing them "cows," since they enabled one to "milk" short-lived isotopes from their longer-lived progenitors.[160] By the 1970s technetium-99m had become the most ubiquitous radioisotope in medical diagnostics,

159. Harper, Andros, and Lathrop, "Preliminary Observations" (1962); Miale, "Nuclear Medicine" (1995).

160. Stannard, *Radioactivity and Health* (1988), vol. 3, p. 1708.

on account of its emission profile and its short half-life.[161] In many cases, it simply replaced other radioisotopes previously used as the label of choice for a wide range of radiopharmaceutical compounds already being used in diagnostics.[162] But its emitted gamma ray was also detected efficiently by the new scintillation-based detection systems, especially the Anger camera, so its availability reinforced the adoption of radioisotope scanning technologies. Technetium-99m is still used in about 85% of diagnostic imaging procedures, which amounts to 40,000 per day in the United States alone.[163]

Other imaging technologies proceeded apace. The use of positron-emitting radioisotopes in diagnosis improved dramatically two decades after Sweet and Brownell's work, due to two innovations.[164] First, molecules that localized better to particular tissues than bare radioelements were labeled with positron-emitting isotopes. By 1980, an article in *Scientific American* listed nearly 30 kinds of compounds labeled with short-lived, positron-emitting radioisotopes.[165] Fluorine-18–labeled fluorodeoxyglucose became a label of choice in scanning for a variety of cancers.[166] Second, detectors became more sensitive, incorporating several scintillation counters (like the Anger camera) and converging collimators, and making use of new algorithms and more powerful computers to reconstruct images from tomographic data. At Washington University, which in the 1960s acquired a cyclotron specifically for producing short-lived radioisotopes, Michel Ter-Pogossian and coworkers developed a positron emission transverse tomograph, dubbed PETT III—and eventually shortened to PET.[167] PET came to be (and remains) one of the most widely used diagnostic and research tools for visualizing the brain and other organs, and relies on the availability of radioisotopes and radiolabeled compounds.[168]

161. Technetium-99m, with a half-life of six hours, decays into another isotope, technetium-99, with a half-life of 210,000 years. The long half-life of its decay product means that few of its atoms will decay over a patient's lifetime.

162. Early, "Use of Diagnostic Radionuclides" (1995), p. 657.

163. Wald, "Radioactive Drug for Tests" (2009).

164. On the later uses and meanings of positron emission tomography, see Dumit, *Picturing Personhood* (2004).

165. Ter-Pogossian et al., "Positron-Emission Tomography" (1980).

166. This radiolabeled compound was produced beginning in 1979. Kevles, *Naked to the Bone* (1997), p. 210.

167. Ter-Pogossian, "Origins of Positron Emission Tomography" (1992), p. 145.

168. Phelps et al., "Application of Annihilation Coincidence" (1975); Tilyou, "Evolution of Positron Emission Tomography" (1991).

Conclusions

Nuclear medicine remains one of the most enduring contributions of the AEC's radioisotope distribution program, its diagnostic methods still heavily dependent on radiolabels. Other technologies for visualizing the body, such as magnetic resonance imaging and computed tomography (CT) scanning (the latter using x-rays), provide better anatomical detail, but radioisotope-based methods such as PET scans offer unrivaled functional information. In addition, for certain kinds of therapy, such as ablation of the thyroid gland for hyperthyroidism, radioisotopes still provide the treatment of choice. The chairman of the Nuclear Regulatory Commission estimates that one-third of Americans admitted to the hospital will be diagnosed or treated with radiation or radioactive materials.[169]

Ironically, the resounding success of the AEC's radioisotope program in catalyzing the emergence of nuclear medicine subverted the importance of the government's reactor-based supply system. The short-lived radioisotopes used in conjunction with new scanning devices generally could not be produced in reactors, but required cyclotrons or generators. Hospitals that wished to offer the latest in nuclear medicine, including PET, needed to build or buy an expensive infrastructure for producing radioisotopes on site. This both rejuvenated older technologies, particularly cyclotrons, and led to the development of new ones, including radioisotope "cows" or generators. The situation for radiotherapy displayed the same pattern. After two decades in which reactor-produced cobalt-60 and cesium-137 teletherapy machines dominated radiotherapy, by the 1980s the instrument of choice became the linear accelerator.[170] Thus in both diagnosis and therapy, nuclear medicine grew out of, but then outgrew, the AEC's reactor-based radioisotope supply.

The other irony is that although AEC prized medical advances as the ultimate "humanitarian" deployment of atomic energy, the development of nuclear medicine reveals the agency's deeply embedded military priorities. The same institutions that were central to the development of radioisotope-based technologies for diagnosis and therapy (most notably of cancer) were also contracted by the AEC to perform human experiments aimed at either documenting the metabolism, distribution, and excretion of fissionable materials such as plutonium and uranium, or

169. Jaczko, "Regulator's Perspective on Safety" (2011), p. 1.

170. "Therapeutic Uses of Radiation" (1995), p. 67.

documenting the cognitive and biochemical changes induced by TBI. The AEC often facilitated the merging of military and civilian priorities in radiation research, as seen in the TBI research conducted for the NEPA project at M. D. Anderson. Similarly, the agency provided funding for the development of M. D. Anderson's cobalt-60 machine, which the hospital then used in research for the military.

Some scholars see this incursion of the military into the development of nuclear medicine as having corrupted the world of medical research.[171] This perspective, however, idealizes the ethical situation of ordinary clinical experiments. As Gerald Kutcher has astutely pointed out, the ethical conflicts in Eugene Saenger's TBI research on cancer patients in Cincinnati—which aimed to advance military interests and cancer medicine at the same time—actually reflect the broader, indeed endemic, tensions between the prerogatives of research and therapy in postwar medicine.[172] Patients in experimental therapy necessarily serve as proxies for others, usually those who will be diagnosed with their conditions at a later time in history. It may be repellent to have terminally ill patients serving as stand-ins for soldiers on an irradiated battlefield, but the ethical problem of medical research in which subjects are not the direct beneficiaries of the knowledge they help bring into being characterizes nearly all postwar clinical investigation, particularly in cancer medicine. The development of radioisotope-based cancer therapies, largely funded by the AEC, brings this pervasive but often imperceptible ethical dilemma into clear focus.

171. E.g., Leopold, *Under the Radar* (2009). This is analogous to similar arguments for the military influence on physics after World War II: Forman, "Behind Quantum Electronics" (1987).

172. Kutcher, *Contested Medicine* (2009), e.g., p. 6.

CHAPTER TEN

Ecosystems

Man's opportunity to learn more about environmental processes through the use of radioactive tracers balances the possible troubles he may have with environmental contamination. Radioactive tracers have already been well exploited by the physiologist, but the ecologist is just beginning to develop techniques for studies in "community metabolism" as it becomes clear that with proper precautions radioactive tracers can be used as safely in the field as in the laboratory. —Eugene P. Odum, 1959[1]

The availability of radioisotopes impacted not only biomedicine but also environmental science, particularly through the emergence of ecosystems ecology.[2] In the field, ecologists used radioisotopes to physically trace the movement of materials and energy through ecosystems, emulating how biochemists and physiologists used radioisotopes in the laboratory to elucidate pathways of metabolism. The AEC supported much of this research in order to track the effects of effluents and radioactive waste from its plants. The three most important sites for the development of radioecology were part of the agency—Hanford Works, Oak Ridge National Laboratory, and the University of Georgia research station at the Savannah River plant. The ecosystems approach, fostered by the AEC, was subsequently used to understand the spread of other kinds of contaminants through the environment.

Arthur Tansley introduced the term "ecosystem" in 1935 in an article reflecting on the contributions of Frederic Clements.[3] Clements's

1. Odum, *Fundamentals of Ecology* (1959), p. 469.

2. On this issue, Kwa, *Mimicking Nature* (1989), ch. 1; Kwa, "Radiation Ecology" (1993); Hagen, *Entangled Bank* (1992), ch. 6. I follow Hagen's line of argument but focus specific attention on the material basis for this synergy, namely the radioisotopes that became critical tools for tracing out the patterns of circulation in ecosystems.

3. Tansley, "Use and Abuse" (1935). The article was part of a festschrift for ecologist Henry C. Cowles.

ecological theory relied on two key ideas: first, building on Henry C. Cowles's work, that plant formations follow a predetermined pattern of succession, leading to the climax community; and second, that a plant community functions as a complex organism with its own life cycle and evolutionary development.[4] Tansley was critical of Clements's theory for treating animals and plants as "members" of the same biotic community; this was "to put on an equal footing things which in their whole nature and behaviour are too different."[5] A more sensible overarching unit, in Tansley's view, should encompass nonliving as well as living components:

> The more fundamental conception is, as it seems to me, the whole *system* (in the sense of physics), including not only the organism-complex, but also the whole complex of physical factors forming what we call the environment of the biome—the habitat factors in the widest sense.[6]

In addition, preferring materialist causes to the vitalistic and idealistic overtones of community-as-organism, Tansley attributed the self-regulatory aspects of ecosystems to the stable interaction of their physical, chemical, and biological components.[7] In this sense the ecosystem was, as Frank Golley has put it, "a machine theory applied to nature."[8] Nonetheless, as historians of ecology have pointed out, the neologism tended to be used as simply a new term for biotic community or complex organism, retaining Clements's organicism.[9]

4. Kingsland, "Defining Ecology as a Science" (1991), p. 5. For an earlier articulation of biotic community in the work of Karl Möbius, see Nyhart, "Civic and Economic Zoology" (1998); idem, *Modern Nature* (2009).

5. Tansley, "Use and Abuse" (1935), p. 296.

6. Ibid., p. 299; emphasis in original.

7. Tansley was particularly critical of how Clements's concept had been further developed by South Africans Jan Smuts and John Phillips. Smuts, *Holism and Evolution* (1926); Phillips, "Biotic Community" (1931); Worster, *Nature's Economy* (1977), pp. 239 and 301; Anker, *Imperial Ecology* (2001), ch. 4.

8. Golley, *History of the Ecosystem Concept* (1993), p. 2.

9. Scholars have interpreted the relationship between Clements's and Tansley's contrasting "community" and "ecosystem" in different ways. Ronald Tobey emphasizes the discontinuity signaled in the introduction of ecosystem, whereas Tansley's biographer Harry Godwin views Tansley's concept as "qualifying without disabling" Clements's organismal one. Like Hagen, I am struck by the continuities embedded in Tansley's innovation. Tobey, *Saving the Prairies* (1981); Godwin, "Sir Arthur Tansley" (1977); Hagen, *Entangled Bank* (1992), esp.

The material basis of ecological knowledge suggests another reason why the ecosystem continued to be figured as an organism. With radioisotopes, the ecosystem could be treated like an organism or cell whose chemical pathways could be traced out. Whereas Tansley had suggested "ecosystem" to rid ecology of the term "biotic community," the use of isotopes to follow metabolic pathways in both physiology and ecology kept the organismal conception in play. As G. Evelyn Hutchinson noted in a book review of *Bio-Ecology* by Frederic Clements and Victor Shelford: "If, as is insisted, the community is an organism, it should be possible to study the metabolism of that organism."[10] The following year, Hutchinson made good on this analogy, publishing an article entitled "The Mechanisms of Intermediary Metabolism in Stratified Lakes."[11]

At the same time, Hutchinson shared with Tansley a commitment to the critical role of nonliving environmental components in the functioning of a "community." His pioneering use of isotopes to trace the development and metabolism of aquatic communities in Connecticut showed how elements moved among sediment, microorganisms such as plankton, and larger organisms such as fish. These nutrient cycles rendered Tansley's ecosystem in concrete terms. In addition, the representational practices involved in mapping the patterns of circulation in an ecosystem manifested the epistemological links between metabolic biochemistry and biogeochemical studies of ecosystems.[12] Hutchinson's understanding of ecosystems spread through American ecology after World War II due in part to the work of AEC scientists at Hanford, Oak Ridge, and Savannah, Georgia, who followed radioisotopes through aquatic and terrestrial systems to understand the cycling of nutrients and the movement of radioactive waste.

Bodies of Water

Freshwater ecology, or limnology, proved a more promising arena for the application of Tansley's unit of "ecosystem" than the British scientist's

pp. 79–80; Anker, *Imperial Ecology* (2001), ch. 4; Kingsland, *Evolution of American Ecology* (2005), pp. 184–85.

10. Hutchinson, "Bio-Ecology" (1940), p. 268; Hagen, *Entangled Bank* (1992), ch. 4.

11. Hutchinson, "Mechanisms of Intermediary Metabolism" (1941).

12. Not that the resemblance is exact: the ecological diagrams included nonliving components and stressed the flow of energy as well as materials.

own field of terrestrial plant ecology.[13] Stephen Forbes, in his classic article "The Lake as a Microcosm," had described the lake as "an organic complex" in equilibrium as early as 1887.[14] Along similar lines, August Thienemann had treated the lake as a unit, describing it as a "biosystem" in 1918.[15] The relative boundedness of freshwater bodies made the movement of materials through organisms, and more specifically, through different levels of organisms, tractable for investigation. Raymond Lindeman's doctoral dissertation at the University of Minnesota included analysis of the results from extensively sampling the inhabitants of Cedar Creek Bog, a shallow body of water "lying in the transition between late lake succession and early terrestrial succession."[16] With the collaboration of his wife, Eleanor Hall Lindeman, he surveyed a wide variety of organisms—including aquatic plants, phytoplankton, zooplankton, insects, crustaceans, and fish—enabling what Robert Cook termed "a very intimate understanding of the movement of nutrients from one trophic level to another."[17] The concluding chapter of Lindeman's dissertation related these findings conceptually to the ecological concepts of "community" (as treated by Clements, Thienemann, and, more critically, Tansley) and succession.

Lindeman studied with Hutchinson as a Sterling postdoctoral fellow from late 1941 until June 1942, the month of his untimely death at age twenty-seven. While in New Haven, Lindeman revised this final theoretical chapter of his dissertation and submitted it for publication in *Ecology*.[18] Both men had been trying to develop a framework for understanding how energy, once captured from sunlight, is transferred from plants to other organisms, and then along the food chain. Through their collaboration, Lindeman was able to utilize some of Hutchinson's key concepts by interpreting these nutritional and energy relationships in the aquatic environment he understood so deeply.[19] Their notion of trophic dynamics was a way of recasting Charles Elton's food chain, which represented feeding

13. Golley, *History of the Ecosystem Concept* (1993), p. 36.

14. Forbes, "Lake as a Microcosm" (1887); Schneider, "Local Knowledge" (2000).

15. Thienemann, "Lebengemeinschaft und Lebensraum" (1918); McIntosh, *Background of Ecology* (1985), p. 195.

16. Cook, "Raymond Lindeman" (1977), p. 22.

17. Ibid.

18. Lindeman, "Trophic-Dynamic Aspect of Ecology" (1942).

19. Hagen, *Entangled Bank* (1992), pp. 88–90. Hutchinson's developing theory is reflected in an unpublished 1941 manuscript, *Lecture Notes in Limnology* (that Lindeman refers to as *Recent Advances in Limnology*). Cook, "Raymond Lindeman" (1977), p. 217n52.

relationships as a "pyramid of numbers." Rather than understanding these relationships governing prey and predator as related to animal size, Lindeman focused on the movement of material and energy between the different trophic levels—viewed as producers and consumers.[20]

Lindeman's published paper drew on data from his own work and those of others to calculate the productivity of various food groups and food cycles, but much of the paper was a theoretical contribution.[21] As he noted, earlier studies of trophic levels tended to restrict analysis to the flow of materials and energy between living components, namely the food cycle. He took a more broadly biogeochemical (and Hutchinsonian) approach: "Upon further consideration of the trophic cycle, the discrimination between living organisms as parts of the 'biotic community' and dead organisms and inorganic nutritives as parts of the 'environment' seems arbitrary and unnatural."[22] At the heart of Lindeman's schematic diagram of food-cycle relationships was "non-living nascent ooze," much of which "is rapidly reincorporated through 'dissolved nutrients' back into the living 'biotic community.' "[23] He contended that the data of dynamic ecology are best understood in terms of Tansley's notion of ecosystem, "composed of physical-chemical-biological processes active within a space-time unit of any magnitude, i.e., the biotic community *plus* its abiotic environment."[24] Lindeman differentiated three trophic groups: producers (autotrophic plants which synthesized complex organic substances from simple inorganic compounds, using energy from photosynthesis), primary consumers (herbivores), and secondary consumers (predators). These categories had been employed before, but Lindeman showed how they were related to energy flows. Energy was lost as it flowed through higher trophic levels, so consumers at these higher levels must necessarily be more efficient at retaining energy.

Lindeman's article also aimed at showing how attention to the flow of energy between different trophic levels related to ecological succession. In this sense, his articulation of ecosystems theory retained the underlying

20. Lindeman, "Trophic-Dynamic Aspect of Ecology" (1942), p. 408. Odum continued this reworking of Elton, referring to the "pyramid of biomass." Kwa, "Radiation Ecology" (1993), p. 222.

21. The paper's emphasis on theory almost prevented it from being published; see Cook, "Raymond Lindeman" (1977).

22. Lindeman, "Trophic-Dynamic Aspect of Ecology" (1942), p. 399.

23. Ibid., pp. 399–400.

24. Ibid., p. 400.

metaphor of embryological development that had informed the superorganismic "community" of Clements and John Phillips.[25] However, his attention to how nutrients entered into and cycled through the biotic and abiotic parts of the system illuminated a different temporal dimension: the "metabolism" of the ecosystem. Here his work drew on biogeochemistry as it had been established by Vladimir Vernadsky.[26] Hutchinson had already been drawing on Vernadsky's approach through a decade's work on the movement of elements in aquatic systems.[27] Here the organizing metaphor was one of homeostasis and equilibrium, rather than growth and development.[28] The inspiration was not embryology, but physiology.

Hutchinson had previously appropriated the biochemical notion of metabolism for an aquatic body. The fourth paper in his study of lakes in Connecticut, published in 1941, examined the phosphorus cycle in Linsley Pond "as a specific example of intermediary metabolism."[29] He argued that the phosphorus cycle in the lake was "ideally closed"; it could be effectively "understood without reference to anything but the events in the water and the mud with which it is in contact."[30] In this respect the body

25. As Cook observes, this "organismic approach paralleled the whole conceptual framework being established in developmental biology at this time," citing Haraway, *Crystals, Fabrics and Fields* (1976). Cook, "Raymond Lindeman" (1977), p. 26n30.

26. Golley, *History of the Ecosystem Concept* (1993), p. 56.

27. Slack, "G. Evelyn Hutchinson" (2008). Some of this work was undertaken by Hutchinson's first graduate students, particularly Gordon Riley, who studied the copper cycle in Linsley Pond: Riley, "General Limnological Survey" (1939); idem, "Plankton of Linsley Pond" (1940). Hutchinson's initial foray into limnology was in South Africa where he held a research position in 1926 and 1927 at the University of Witwatersrand in Johannesburg. This research was sponsored by Lancelot Hogben, who was then professor of zoology at the University of Capetown.

28. Golley, *History of the Ecosystem Concept* (1993), p. 59. Sharon Kingsland argues that population ecology may be understood in terms of "a conflict between historical and ahistorical thinking," the latter usually involving mathematics. As she puts it, "The very act of imposing mathematics (or any model) on nature often involved a rejection of history in favor of a harmonious, unifying concept." This fits very closely with the tension between the metaphors of development and physiology in ecosystems ecology, and in fact it was the view of ecosystems as self-regulating that was consonant with the application of mathematics to understand how they maintained equilibrium. Kingsland, *Modeling Nature* (1985), p. 8.

29. Hutchinson, "Mechanisms of Intermediary Metabolism" (1941), p. 56. Hutchinson was also investigating the biogeochemistry of particular metals: Hutchinson, "Biogeochemistry of Aluminum" (1943).

30. Hutchinson, "Mechanisms of Intermediary Metabolism" (1941), p. 56.

of water was like the body of an organism, whose chemical interrelations could be studied in situ:

> A considerable body of information is available as to the total quantity of various important substances present in lakes. Observations on the oxygen deficit and various studies of the photosynthetic and katabolic activity of the plankton have given some information, often, however, of a very relative nature, as to the total metabolism of lakes. The intermediary aspect of metabolism, to continue the analogy with the individual organism, is extremely little known.[31]

Strikingly, this interest in the metabolism of the lake led Hutchinson to consider the cycling not only of elements, but also of vitamins, such as thiamin and niacin.[32] Whereas biochemists had focused on vitamins as essential nutrients for individual organisms (whether animal or microbial), Hutchinson examined their circulation and function in aquatic communities.[33]

In the 1941 paper, Hutchinson offered a picture of how phosphorus moved through both the living and the nonliving components of the lake even as it was maintained at a steady state level. Clearly the redox potential of the mud or water affected the form in which the phosphorus was available for uptake; this in turn depended on the relationship of phosphorus and iron.[34] Phosphate liberated from the mud, as it reached the illuminated layers of the lake, was taken up by the phytoplankton. Later the phosphorus settled back down to the bottom of the lake as particulate matter, from dead plankton and from the feces of the zooplankton that fed on the plant cells.[35] The activity of phytoplankton, which grew in response to the supply of phosphate ions from the mud, effectively maintained the persistently low concentration of phosphorus in the surface water.

Hutchinson sought more direct evidence for this pattern of self-regulation of phosphorus in the lake, and developments in the physical sciences at Yale opened a possibility. In 1939, physicist Ernest Pollard

31. Ibid., p. 23.

32. See Hutchinson, "Thiamin in Lake Waters" (1943); Hutchinson and Setlow, "Limnological Studies in Connecticut, VIII" (1946).

33. On the history of nutritional biochemistry, see Kamminga and Weatherall, "Making of a Biochemist, I" (1996); Weatherall and Kamminga, "Making of a Biochemist, II" (1996).

34. Hutchinson, "Mechanisms of Intermediary Metabolism" (1941), p. 52.

35. Hutchinson and Bowen, "Direct Demonstration" (1947), pp. 148–49.

successfully completed the construction of a cyclotron at Yale with the assistance of E. O. Lawrence. This enabled the production of artificial radioisotopes at Yale (as at Berkeley). Hutchinson requested some phosphorus-32 from Pollard for his continuing study of the pond. In 1941, the two of them, with collaborator W. Thomas Edmondson, planned to use some cyclotron-generated phosphorus-32, with its half-life of two weeks, to directly demonstrate the cycling of phosphorus in Linsley Pond. Unfortunately, the night before the experiment the cyclotron broke down, so the researchers obtained only half as much phosphorus-32 as they had calculated they would need. Hutchinson decided to proceed anyway. As Edmondson described their work,

> The operation was not the most efficient possible. We had a small rowboat with a hand-powered winch and several five-gallon glass carboys. . . . I helped take samples (endless turning of the handle of the winch), but my main function, as the owner of a car, was to drive each carboy to Osborn [Laboratory building] as soon as it was filled. There Ann Wollack, the chemical technician, filtered large volumes through membrane filters to find out how much phosphate had been taken up by algae and other small organisms.[36]

Despite the limitations, he recalled, preliminary results were encouraging. "Pollard was able to detect radioactivity in some of the samples, including material from the deep water. All that this showed was that the study could be made, given enough isotope."[37] The fourteen-day half-life of phosphorus-32 suited an experiment lasting a few weeks.

The Linsley Pond tracer experiment was attempted again on June 21, 1946, this time without cyclotron problems. In addition, instrumentation for detecting radioactivity had improved during the war.[38] Approximately ten millicuries of phosphorus-32, made up as sodium phosphate in a bicarbonate solution, were released in twenty-four equal portions into the surface waters of the pond. The sites at which these aliquots were released

36. Letter, W. T. Edmondson to Joel B. Hagen, 22 Mar 1989, courtesy of Hagen.

37. Ibid. Further information on this early attempt is available from an interview by Joel B. Hagen of G. Evelyn Hutchinson, 31 Mar 1989 as cited in Hagen, *Entangled Bank* (1992), p. 221n45; Hutchinson and Bowen, "Direct Demonstration" (1947), asterisk endnote on p. 153; G. Evelyn Hutchinson to Edward S. Deevey, 26 Sep 1944, Hutchinson papers, box 11, folder 193 Deevey, Edward S. 1944–1949.

38. Letter, W. T. Edmondson to Joel B. Hagen, 22 Mar 1989, courtesy of Hagen.

spanned the pond evenly east to west and were concentrated at several locations on the southern half of the body, to compensate for a northern-blowing wind. A week later, vertical water columns were collected in the deep central part of the lake. Each column was divided into four sections, dried, and then its radioactivity counted. Samples of plants were collected and counted two weeks later. Because the radioactivity was so dilute in the collected liquid samples, a large number of counts had to be taken, and voltage fluctuations in the Geiger counter obscured the signal somewhat. Nonetheless, certain features of distribution were clear. Nearly half of the radioactivity had descended below the three meter level, and 10% below six meters, although there was little mixing of the water due to stable thermal stratification. This was consistent with the role of algae in taking up the available phosphorus and subsequently sedimenting to the bottom of the pond, the model Hutchinson had proposed in 1941.[39] In addition, the plant samples showed a concentration of radioactivity a thousand-fold over that of the water.[40]

Still, further work would be needed to clarify the specifics of the phosphate cycle. Hutchinson was among the earliest applicants to purchase radioisotopes from the AEC; he obtained authorization in May 1947 and received his first shipment of 350 millicuries of phosphorus-32 that summer.[41] On July 25, 1947, Hutchinson and Bowen introduced 70 millicuries into the lake waters. This time the radioisotope was distributed "in twenty-five portions while rowing in an approximately circular course between the central deep part of the lake and the margin; this arrangement is believed to have provided adequate opportunity for the mixing of the radiophosphorus with the superficial layer of the lake."[42] Between August 1 and 22, the researchers took weekly temperature readings at various depths and collected samples of pond water over the same range of depths. Each water sample was filtered, and both the filter paper and remaining water dried down to enable determinations of total phosphorus and radioactivity. Unlike in the previous experiment, the amounts of

39. Hutchinson and Bowen, "Direct Demonstration" (1947), p. 148.

40. Ibid., p. 152.

41. Letter from Paul Aebersold, Isotopes Branch, to G. E. Hutchinson, 8 May 1947, NARA Atlanta, RG 326, MED CEW Gen Res Corr, Acc 67B0803, box 178, folder AEC 441.2 (R–Yale Univ.). Hutchinson's radioisotope application with the Manhattan District and Monsanto (as contractor and hence distributor), signed on 27 Nov 1946, is in NARA Atlanta, RG 326, OROO Files for K-25, X-10, Y-25, Acc 671309, box 14, Certificates.

42. Hutchinson and Bowen, "Quantitative Radiochemical Study" (1950), p. 194.

radioactivity being measured were well above the background count of the detector. In addition, significant amounts of radioactivity were recovered at all levels of the lake. Recovery was so remarkable, in fact, as to suggest that nearly all of the released phosphorus was immediately taken up by phytoplankton. The recovery at lower levels indicated that substantial phosphorus-32 was carried down by the sedimentation of seston; some of this phosphorus then passed into littoral vegetation and back into free water.[43] Again, the results confirmed Hutchinson's portrait of the overall metabolism of phosphorus in the lake as maintained at a steady state by the growth and death of algae.[44]

Biochemistry and physiology were not the only sources of ideas of equilibrium and self-regulation—engineering was another. Or rather, physiological regulation had itself been adopted by Norbert Wiener as inspiration for the articulation of cybernetics, a term he coined in 1947.[45] As others have noted, Hutchinson came into direct contact with this development by attending the Macy Foundation conferences on cybernetics between 1946 and 1953.[46] For the initial meeting on "Teleological Mechanisms," Hutchinson contributed his influential paper "Circular Causal Systems in Ecology" (published in 1948), which emphasized the self-regulating features of an ecosystem as seen in element cycling. Two examples featured in his analysis: the global carbon cycle of the biosphere, and the phosphorus cycle in inland lakes.[47]

The basic carbon cycle, as Hutchinson pointed out, is familiar to students of elementary biology—plants take carbon dioxide out of the atmosphere through photosynthesis; consumption of plants by animals keeps the carbon moving through the terrestrial food chain; and through decay of both plants and animals, carbon is lost to sediment and buried. Some carbon makes its way back to the atmosphere directly through respiration or indirectly through bacterial metabolism. Large-scale natural events,

43. Seston is particulate matter in water, composed of both minute living organisms and nonliving matter, which floats and contributes to turbidity.

44. Undated, hand-written notes on this experiment in Hutchinson papers, box 6, folder 96 Bowen, Vaughan T., 1942–1948, 1956. Unbeknownst to Hutchinson, in July 1948 a Canadian group undertook similar experiments with phosphorus-32 in a lake in Nova Scotia. Slack, *G. Evelyn Hutchinson* (2010), p. 162; Coffin et al., "Exchange of Materials" (1949).

45. Weiner, *Cybernetics* (1948).

46. Taylor, "Technocratic Optimism" (1988), pp. 217–23. On the Macy Foundation conferences, see Heims, *Cybernetics Group* (1991).

47. Hutchinson, "Circular Causal Systems" (1948).

such as volcanic eruptions, forest fires, and human activities, namely combustion of fossil fuels, contribute atmospheric carbon dioxide. Quantitative information was harder to come by than the qualitative picture, though Hutchinson drew on recent work by chemist Walter Noddack and geochemist Victor Goldschmidt, as well as that of his former student Gordon Riley.[48] Hutchinson also considered data suggesting that atmospheric carbon dioxide levels had increased since the nineteenth century. Rather than attribute this trend to industrial development, he argued that it was due to changes in the *biological* parts of the system, particularly the deforestation of land through agricultural development.

Hutchinson's treatment of the carbon cycle shows how ecological understandings of the relationships between living organisms could be extended to include the nonliving world. As he noted, theoretical ecologist V. A. Kostitzin had already observed that "a cycle in which the rate of growth of consuming and decomposing organisms depends on the rate of photosynthetic production, and the latter depends on the rate of return of CO_2 to the atmosphere by decomposing and consuming organisms, would tend to oscillate according to Volterra's prey-predator equations."[49] Hutchinson linked this to "oscillations in the CO_2 content of the atmosphere."[50] Overall, Hutchinson cited two self-correcting systems in the carbon cycle: the circulation of carbon through air, sea, and sediments in CO_2-bicarbonate-carbonate, and the biological cycle involving photosynthesis. However, he stopped short of referring to the carbon cycle as "purposive," in Norbert Wiener and Arturo Rosenblueth's cybernetic sense.[51] Like Tansley, Hutchinson was committed to a materialist rather than vitalist understanding of the larger biological system. Even so, Hutchinson's diagram of the global biogeochemical cycle of carbon shows some visual similarity to metabolic pathways of contemporary biochemists, though few of the specific chemical conversions are depicted. It appears to be the metabolism of a superorganism, if not a living system per se. In this picture, isotopes provided a natural tracer, on account of the variation in

48. Noddack, "Der Kohlenstoff im Haushalt der Natur" (1937); Goldschmidt, "Drei Vorträge über Geochemie" (1934); Riley, "Carbon Metabolism and Photosynthetic Efficiency" (1944).

49. Hutchinson, "Circular Causal Systems" (1948), p. 222.

50. Ibid.

51. Rosenblueth et al., "Behavior, Purpose and Teleology" (1943). See also Rosenblueth and Wiener, "Purposeful and Non-purposeful Behavior" (1950).

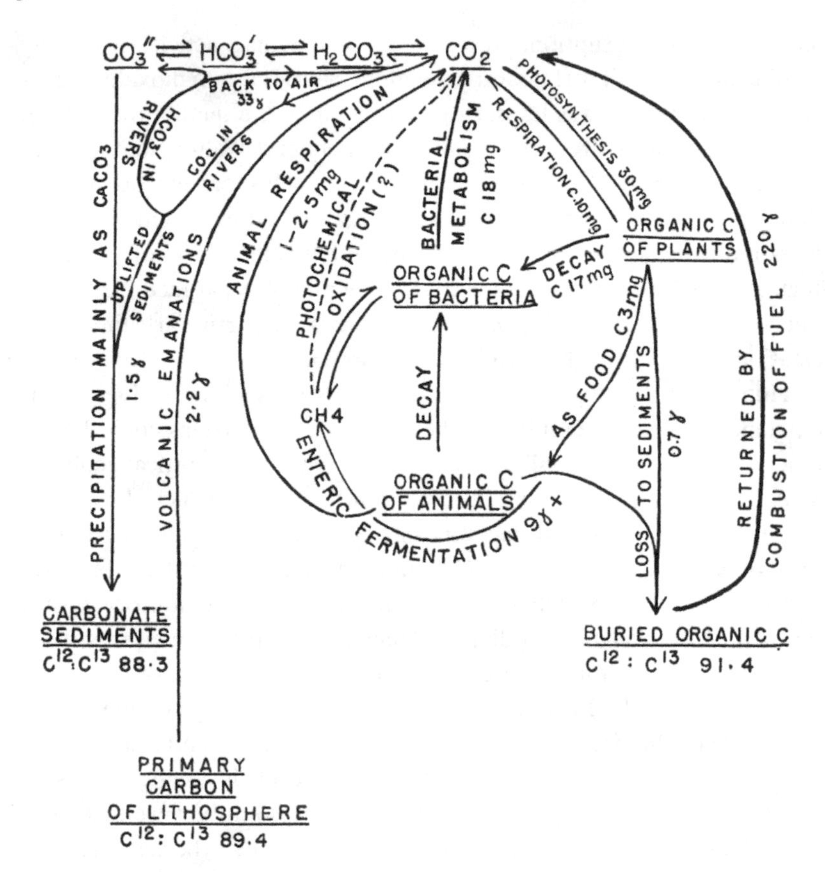

FIGURE 10.1. G. Evelyn Hutchinson's diagram of carbon cycling. G. E. Hutchinson, "Circular Causal Systems in Ecology," *Annals of the New York Academy of Sciences* 50 (1948): 221–46, on p. 223. Reproduced by permission of John Wiley and Sons.

the relative abundance of carbon isotopes that provided additional information.[52] Carbonate precipitated from aqueous solutions was found to be slightly enriched in carbon-13 compared with carbon in plutonic rocks.[53] (See figure 10.1.)

Hutchinson soon requested radioisotopes of other elements besides phosphorus from Oak Ridge, namely iodine, barium, and bromine.[54] Other

52. Nier and Gulbransen, "Variations in the Relative Abundance" (1939).

53. Hutchinson, "Circular Causal Systems" (1948), p. 225.

54. Letter from Paul C. Aebersold to G. Evelyn Hutchinson, 8 May 1947, Subject: Approval of Requests for Radioisotopes, NARA Atlanta, RG 326, MED CEW Gen Res Corr, Acc 67B0803, box 178, folder AEC 441.2 (R–Yale Univ.).

ecologists began to take up this promising method as well. In 1952, E. Steemann-Nielsen published a sensitive technique for using carbon-14 to measure the productivity of aquatic plants.[55] Eugene Odum and coworkers used phosphorus-32 to measure the productivity of marine benthic algae.[56] Isotopes were equally useful in examining the relationships between organisms. Robert Pendleton and A. W. Grundmann used phosphorus-32 to elucidate insect-plant relationships in the thistle.[57] Researchers were also "radio-tagging" specific insects such as mosquitoes and houseflies.[58] By the late 1950s, the intentional release of limited amounts of radioisotopes to trace the uptake and circulation of elements became a vital strategy for work in ecology.[59] In the second edition of his textbook *Fundamentals of Ecology* (1959), Odum added a chapter on "Radiation Ecology" that summarized these trends.[60] In fact, he wrote this new chapter while on a sabbatical leave from the University of Georgia, during which he spent four months with ecologists at Hanford (see below).[61] Ecosystems ecology provided the organizing framework of Odum's textbook, which in turn propagated this theoretical orientation in the 1950s and 1960s and solidified its connection to radioecology—and to the AEC.

Bioconcentration at Hanford

Unlike tracer experiments in biochemistry and nuclear medicine, ecology included "experiments" that researchers did not set in motion.

55. As Eugene Odum describes it: "One of the most sensitive methods for measuring aquatic plant production is done in bottles with radioactive carbon (C^{14}) added as carbonate. After a short period of time, the plankton or other plants are separated from the water, dried and placed in a counting device. With suitable calculations, the amount of carbon dioxide fixed in photosynthesis can be determined from the radioactive counts made." Odum, *Fundamentals of Ecology* (1959), p. 84; Steemann Nielsen, "Use of Radioactive Carbon" (1954); Ryther, "Measurement of Primary Production" (1956).

56. Odum et al., "Uptake of P^{32}" (1958).

57. Pendleton and Grundmann, "Use of P^{32} in Tracing" (1954).

58. Hassett and Jenkins, "Uptake and Effect of Radiophosphorus" (1951).

59. For a retrospective assessment, see Auerbach, "Radionuclide Cycling" (1965). Note that by the mid-1960s, radionuclide had come to replace radioisotope as the preferred term.

60. Odum, *Fundamentals of Ecology* (1959).

61. Odum was seeking greater competence in radiation biology during the year's leave on a National Science Foundation grant. Beside the time at Hanford, he spent another four months at UCLA with the group that studied the effects of radiation on desert ecology at the testing grounds in Nevada. Kwa, "Radiation Ecology" (1993), p. 230.

Radioactivity was entering the environment, often on a large scale, through the AEC's disposal of nuclear waste and through atomic weapons tests, and ecologists were tracking the movement of these radionuclides. In this respect, ecological "tracing" was directly connected to the continuing military uses of atomic energy. Radioactive waste itself provided ecologists with useful tracers.[62]

The US government supported some research on the ecological consequences of nuclear waste during World War II, notably in relation to the building of the major production facilities for plutonium. Late in 1942, leaders of the Manhattan Project sought a remote site for the massive plutonium works which was proximate to supplies of 100,000 kilowatts of electric power and 25,000 gallons per minute of clean, cold water.[63] General Leslie Groves selected an area on the Columbia River in the state of Washington not far from the recently completed Grand Coulee Dam; construction of the Hanford reactors began in April 1943.[64] The decision to construct water-cooled nuclear reactors made the question of water supply even more crucial. The installation was expected to require nearly as much water as a city, raising concerns about the impacts on the Columbia River. Would the volume of water needed to cool the reactors introduce enough heat, toxicity, or radioactivity to detrimentally affect the populations of trout and salmon, which supported a substantial fishing industry? Would it affect the safety of water for human consumption? Hundreds of communities obtained drinking water from the Columbia River and its tributaries. Groves sought the advice of Stafford Warren, Chief of the Medical Section of the Manhattan Engineer District (MED), who concurred on the need for scientific research on the effects of radiation on aquatic life and water quality. Top personnel in the Manhattan Project, including Robert Stone, Eugene Wigner, Arthur Compton, and H. L. Friedell, also discussed the

62. Whicker and Pinder, "Food Chains" (2002).

63. Hewlett and Anderson, *New World* (1962), p. 189; Hines, *Proving Ground* (1962), p. 5. Hines lists the energy requirement as 200,000 kilowatts.

64. Hines, *Proving Ground* (1962), pp. 4–6; Stannard, *Radioactivity and Health* (1988), vol. 2, p. 757. The huge Grand Coulee dam project had been mired in controversy since the 1920s, but its completion led to the building of several plants in Washington and Oregon to produce aluminum during the war. The construction of Hanford continued the federal government's industrial development of the Northwest to aid the war effort, albeit in secrecy. As one historian notes, "Once up and running, the plutonium plants required the entire output of two generators at the Grand Coulee powerhouse." Ficken, "Grand Coulee and Hanford" (1998), p. 27.

possible contamination of the Columbia River at a meeting at the University of Chicago on May 20, 1943, with consensus that the matter warranted investigation.[65]

In August of 1943, Warren approached Lauren R. Donaldson at the University of Washington, an assistant professor of fisheries who had previously worked as a consultant to the Department of the Interior on the construction of the Grand Coulee Dam.[66] He asked Donaldson to begin a study of the effects of radioactivity on aquatic organisms, especially fish. In order to keep the nature of the atomic project secret, funding (to the tune of $65,000) was arranged through the OSRD, not the MED. The contract title—"Investigation of the Use of X-rays in the Treatment of Fungoid Infections in Salmonoid Fishes"—obscured the actual aims of research.[67] So began the Applied Fisheries Laboratory at the University of Washington, Seattle.[68]

There was little knowledge of the biological effects of radiation on aquatic animals for Donaldson's group to build upon.[69] The Applied Fisheries Laboratory conducted a number of fundamental studies assessing the effects of external radiation exposure on eggs, embryos, and adults of various salmon species.[70] These confirmed that the radiation sickness seen in mammals after exposure to high doses also existed in cold-blooded vertebrates, although with a longer latency period.[71] The fish that Donaldson's group investigated were not studied in their native aquatic environment, but rather bred for several generations in the laboratory. A large-scale study of rainbow trout showed high levels of mortality and malformation among embryos whose parents were exposed to high doses (at various levels) of x-rays. The young fish that survived this prenatal exposure grew

65. Hines, *Proving Ground* (1962), p. 8. Stone was director of the new Health Division of the Metallurgical Project at Chicago; Wigner was head of the group in charge of pile design; and Warren's executive officer in the Medical Section was Friedell.

66. Ibid., p. 8; Klingle, "Plying Atomic Waters" (1988).

67. Hines, *Proving Ground* (1962), p. 10.

68. Whicker and Schultz, "Introduction and Historical Perspective" (1982), p. 4; Gerber, *On the Home Front* (2002), p. 116.

69. Hines, *Proving Ground* (1962), p. 12.

70. These experiments made use of a 200 peak kilovoltage Picker-Waite radiation therapy machine, the only radiation source readily available. As Hines notes, "Nowhere was there yet a source of supply of man-made radioisotopes that might have made possible the design of more sophisticated biological studies involving the use of internal emitters even if such studies had been somehow visualized." Ibid., p. 14.

71. Ibid., p. 16.

more slowly than control trout.[72] The researchers also set up long-term studies of the effects of lower-dose radiation, exposing juvenile fish before they returned to sea for two or more years. The Applied Fisheries Laboratory conducted research for a year before the first Hanford reactor went into operation, which resulted in the first significant and ongoing release of radioisotopes into the local environment.[73]

In late 1944, Donaldson persuaded the MED leadership of the importance of beginning research on site at the Columbia River.[74] His graduate student Richard F. Foster was transferred to Hanford in June 1945, to work at a new Aquatic Biological Laboratory.[75] A research plan for aquatic studies at Hanford was formed with input from several MED officials at a June 9, 1945, meeting held at the University of California, Berkeley.[76] Initial laboratory studies by Foster, conducted in the second half of 1945, suggested that the addition of cooling effluent from the Hanford reactors, if diluted sufficiently (at least 1:50) by river water, would not threaten the salmon and trout populations.[77] (See figure 10.2.) More

72. Ibid., pp. 16–17. This project, which involved tabulating the results from 115,454 eggs, was the doctoral dissertation of Richard Foster, subsequently published: Foster et al., "Effect on Embryos and Young of Rainbow Trout" (1949). Seven groups of twenty adult fish each were exposed to whole body doses of 50, 100, 500, 750, 1000, 1500, or 2500 Roentgen, then their offspring followed. Over 500 Roentgen, mortality of offspring was close to 100%.

73. The B reactor was fueled beginning on September 13, 1944, but ceased operating after achieving criticality due to xenon poisoning. This problem was resolved by the end of the year, and Los Alamos received the first shipment of plutonium from Hanford on February 2, 1945. See http://www.cfo.doe.gov/me70/manhattan/hanford_operational.htm.

74. Michele Stenehjem Gerber attributes the impetus for field observations to Stafford Warren, not Donaldson. *On the Home Front* (2002), p. 116.

75. Donaldson was to serve as a consultant for the Aquatic Biological Laboratory. Hines, *Proving Ground* (1962), p. 17.

76. In attendance at the meeting were Stafford L. Warren, H. L. Friedell, A. A. White, Dr. Howland (of the Medical Corps), and Lauren Donaldson. They recommended that Foster determine the effect of the effluent on the fish through a series of dilutions. They estimated the actual dilution in the river to be about 1:500. R. F. Foster, "Some Effects of Pile Area Effluent Water on Young Chinook Salmon and Steelhead Trout," 31 Aug 1946, US AEC Report HW-7-4759, Hanford Engineer Works, Opennet Acc NV0717097, p. 2; L. Donaldson, "Program of Fisheries Experiment for the Hanford Field Laboratory," July 1945, DUH-7287, OpenNet Acc RL-1-336129.

77. Foster, "Some Effects of Pile Area Effluent." It should be noted that if the effluent was not diluted, it was highly toxic to the fish, but the scientists calculated that the dilution factor in the river was at least 1:100. Foster studied the effects on fish of placing them in water with a variety of dilutions with the effluent, from 1:3 to 1:1000.

FIGURE 10.2. Paired intakes of the KE and KW Reactors where water was withdrawn from the Hanford Reach for once-through cooling of the fuel cells. From C. D. Becker, *Aquatic Bioenvironmental Studies: The Hanford Experience 1944–84* (Amsterdam: Elsevier, 1990), p. 43. Copyright Elsevier 1990.

toxic than radioactivity to the fish were the sodium dichromate, calcium chloride, and other process chemicals added to the cooling water to prevent corrosion of the fuel cells. The heating of the river water by the addition of effluent was also a hazard to salmon and trout, which could not survive water warmer than 24°C. Newly hatched salmon ("fry") proved more susceptible to these "adverse factors" in the effluent than adult fish.[78] The earlier the developmental stage, the more sensitive the fish were; eggs at the earliest stages of development exhibited increased mortality when exposed to effluent diluted even 1:500 in river water.[79]

Hanford scientists also studied the levels of radioactivity in the river water. After exiting the reactors, the effluent was held for twenty-four hours in Retention Basins, allowing the decay of short-lived radioisotopes. Most of these radioisotopes were not fission products from the reactor, but normal constituents of river water whose minerals became radioactive

78. Ibid., p. 72.
79. Ibid.

through activation (via neutron capture) when passing through the cooling vessels.[80] Even so, researchers found the concentration of fission products in the river water near Hanford to be higher and more constant than expected based on the discharges from fuel cells. It turned out that this was due to natural "tramp uranium" in the river water being activated by exposure during cooling to the neutron flux of the fuel cells.[81] The residual radioactivity of effluent was measured before the water was returned to the river. In the estimation of Hanford's chief health physicist, Herbert M. Parker, the "water released from the basins has always been considered safe for humans or fish to swim in."[82]

The end of the war lifted the veil of secrecy over the research being done by the University of Washington's Applied Fisheries Laboratory; its connection to the Manhattan Project became public. Donaldson's group expanded their work under the AEC, conducting aquatic surveys after the Crossroads tests at Bikini.[83] In the fall of 1946, General Electric took over from DuPont as contractor of Hanford Works. The Aquatic Biological Laboratory at Hanford was put on more permanent footing as a section within the Radiological Sciences division. Two years later, Foster's group was consolidated with another in the Health Instruments Division

80. Odum, *Fundamentals of Ecology* (1959), p. 467. As Foster observes, an aluminum jacket surrounding the fuel elements kept the cooling water from making direct contact with the uranium rods. Foster, "History of Hanford" (1972), p. 13.

81. Stannard, *Radioactivity and Health* (1988), vol. 2, p. 759.

82. H. M. Parker, "Status of Problem of Measurement of the Activity of Waste Water Returned to the Columbia River," memorandum to S. T. Cantril, 11 Sep 1945, OpenNet Acc NV0719099, p.1. See also Foster, "History of Hanford" (1979), p. 8; Reichle and Auerbach, "U.S. Radioecology Research Programs" (2003), p. 4. One might bear in mind, as Barton Hacker has pointed out, the standard perceptions about radiological safety in World War II America: namely, if radiation exposure did not mean direct harm, it was considered safe enough. Hacker, "No Evidence of Ill Effects" (1991), p. 147.

83. Researchers from the Applied Fisheries Laboratory participated in the surveys of Bikini islands pre- and post-detonation as part of the summer 1946 Nuclear Testing Program in the Pacific. The resurvey in Bikini a year after the detonations revealed the persistence of radioactivity in the flora and fauna of the atoll. See *Life* magazine's feature "What Science Learned at Bikini" (1947). The "Conclusions," authored by Stafford Warren, are entitled "Tests Proved Irresistible Spread of Radioactivity." On the work of Donaldson's group in the Pacific, see Hines, *Proving Ground* (1962); Stannard, *Radioactivity and Health* (1988); Hamblin, *Poison in the Well* (2008), ch. 3.

to comprise a Biology Section, which conducted research on the effects of environmental radioactivity on terrestrial as well as aquatic animals.[84]

In the spring of 1946, Foster presented his work on the ecological effects of the Hanford Engineer Works on the Columbia River at a joint meeting of officials of the Washington State Department of Game and the US Fish and Wildlife Service.[85] That did not mean the work of biologists at Hanford was now unclassified. In the summer of 1946, members of Foster's group were required to sign a statement in which they pledged "not to divulge to any persons, except those authorized by the U.S. government, any information received on the studies . . . conducted at Hanford Engineer Works relative to the effect of plant operations on fish life in the Columbia River."[86] Even so, the work of the Hanford group was remarkably influential within the echelons of the AEC. In their 1947 survey of AEC work on radiation effects inherited from the Manhattan Project, the Interim Medical Committee mentioned the important studies at Hanford on the "transfer of radioactive materials in 'food chains'" from plankton through fish.[87]

Over the next several years, the Hanford group's findings were written up and circulated as classified documents to AEC laboratories and some other government agencies. For example, one hundred fourteen copies of the Biology Group's first annual report (for 1951) were distributed within government channels, including all the major AEC installations and laboratories (e.g., the Berkeley Rad Lab), the US Public Health Service, the Patent Office, and the Naval Medical Research Center.[88] Notably, the recipient list included Eugene Odum's new research station at Savannah. In

84. Becker, *Aquatic Bioenvironmental Studies* (1990), pp. 82–83; Stannard, *Radioactivity and Health* (1988), vol. 1, pp. 434–35. I have tried to reconcile differences between these two accounts in terms of the history of this group. One of the first major studies of the Biology Section concerned the accumulation in sheep of iodine-131, one of the environmental contaminants given off by Hanford Works. This was not a field study, however; experimental sheep were fed iodine-131 and then tracked.

85. Hines, *Proving Ground* (1962), p. 18.

86. "Statements Signed by Hanford Engineer Works Personnel Agreeing Not to Divulge Information Relative to the Effect of Plant Operations on Fish Life in the Columbia River," 18 Jun 1946, OpenNet Acc NV0062060, as quoted in Gerber, *On the Home Front* (2002), p. 116.

87. Stafford L. Warren, Report of the Meeting of the Interim Medical Committee, AEC, 23–24 Jan 1947, OpenNet Acc NV0727195, p. 12.

88. Biology Section, Radiological Sciences Department, "Effect of Hanford Pile Effluent Upon Aquatic Invertebrates in the Columbia River," 19 Jan 1951, OpenNet Acc NV0717092, p. 3.

actuality, the report was declassified soon after its printing, but this did not make it readily available to scientists working outside the channels of the AEC.

Foster and his colleagues in the Aquatic Biology Section continued his research into the effect of the effluent water on fish. Their most significant finding was first reported internally in 1946: fish concentrate radioactivity in their bodies, especially in the liver and kidneys, following exposure.[89] In 1947, Hanford biologist Karl Herde showed that twelve different species of fish caught downstream of the reactor, even twenty miles away, had accumulated radioactivity; his more thorough assessment the next year documented concentrations up to several thousand-fold over that of the river water.[90] This occurred as fish took up essential elements from their environment. Concentrations were especially high for elements, such as phosphorus, that were not naturally abundant. Early on, Hanford researchers inferred that the majority of this radioactivity came from short-lived isotopes manganese-56 and sodium-24 in the effluent. (Sodium-24 has a half-life of just two and a half hours.) Only later did they realize most of the radioactivity in fish was from a longer-lived isotope, notably phosphorus-32.[91] Also, metabolic rate played an important role in the de-

89. John W. Healy, "Accumulation of Radioactive Elements in Fish Immersed in Pile Effluent Water," 27 Feb 1946, Opennet Acc NV0719097. Healy cites an earlier Hanford document: C. Ladd Prossner, Wm. Pervinsek, Jane Arnold, George Svihla and P. C. Tompkins, "Accumulation and Distribution of Radioactive Strontium, Barium-Lanthanum, Fission Mixture, and Sodium in Goldfish," 15 Feb 1945. This paper, however, was part of a laboratory study on the metabolism of mixtures of fission products, as compared with the field observations of bioconcentration at Hanford. One study at Hanford that followed up Healy's was K. E. Herde, "Studies in the Accumulation of Radioactive Elements in *Oncorhynchus tschawytscha* (Chinook salmon) Exposed to a Medium of Pile Effluent Water," 14 Oct 1946, OpenNet Acc NV0719098.

90. Karl E. Herde, "Radioactivity in Various Species of Fish from the Columbia and Yakima Rivers," 14 May 1947, Hanford Health Instruments Section, OpenNet Acc NV0717089; idem, "A One Year Study of Radioactivity in Columbia River Fish," 25 Oct 1948, OpenNet Acc NV0717090. Herde was not in the Aquatic Biology Section, but collaborated with Foster.

91. Gerber, *On the Home Front* (2002), p. 117. As Foster and Rostenbach noted, "In some instances, where the natural supply of an essential element—notably phosphorus—is limited, the radioisotopes may be several hundred thousand times more concentrated in the organism than in the surrounding water." Foster and Rostenbach, "Distribution of Radioisotopes" (1954), p. 638.

FIGURE 10.3. Hanford scientists collecting plankton in nets, to study the movement of radioactivity through Columbia River marine life. A current meter is being used to measure the rate of flow of the river. Credit: Hanford, Washington. National Archives, RG 326-G, box 2, folder 2, AEC-50-3938.

gree of concentration. Young, rapidly growing fish assimilated more radioactivity than older ones.

In 1948, R. W. Coopey, another member of the Hanford group, demonstrated that plankton could concentrate radioactivity in the river water 2,000 to 4,000 times; this constituted the entry point of radioactivity into the aquatic food web.[92] (See figure 10.3.) As Coopey commented, "The concentrating ability of the bottom algae appears to be, in essence, the foundation of the radiobiological problem in the river."[93] As he concluded, the concentrating ability of marine organisms, especially plankton such as algae, made them "a liable source of radioactive contamination in the river

92. R. W. Coopey, "The Accumulation of Radioactivity as Shown by a Limnological Study of the Columbia River in the Vicinity of Hanford Works," 12 Nov 1948, OpenNet Acc NV0717091, p. 2.

93. Ibid., p. 1.

economy."[94] The term economy was not inconsequential; another internal Hanford document circulated in 1947 noted that the "present value of the salmon runs in the Columbia River has been estimated at from $8,000,000 to $10,000,000."[95]

Herde extended this analysis to waterfowl, placing domesticated Pekin ducks in the river for one to fifteen months, feeding them algae from the river, then examining their tissues for radioactivity. The ducks were concentrating substantial amounts of radioactivity in their organs, particularly phosphorus-32. But the highest levels of radioactivity were due to iodine-131 in their thyroids, which they assimilated from plants contaminated by aerial release of radioactive gases from the chemical processing units. In the internal memorandum reporting these results, Herde expressed concern about the Columbia Basin Irrigation Project, which was scheduled to withdraw river water for crop irrigation from twenty miles below where the reactors were distributed.[96]

Parker, the chief health physicist at Hanford, was also concerned about the assimilation of iodine-131 from aerial waste gas release in area livestock. In 1946, he began a program of clandestine monitoring, in which sheep and cattle in the Hanford vicinity were captured and the radioactivity in their thyroids measured with a Geiger-Müller counter.[97] Members of the Environmental Survey Group (including Herde) also visited local farms in the guise of USDA agents in order to obtain thyroid readings on more animals.[98] They found "positive, though low, thyroid readings in nearly every animal examined."[99] The readings were sufficiently low to reassure the Hanford scientists that the radiation level was not damaging to livestock or consumers.

The level of radioactive waste generated by Hanford Works increased steadily in the early postwar period. The beginning of the Cold War arms race meant escalating production of plutonium. In August 1947 Hanford embarked on an ambitious expansion plan for building two new reactors

94. Ibid., p. 11.

95. C. C. Gamertsfelder, "Effects on Surrounding Areas Caused by the Operations of the Hanford Engineer Works," 11 Mar 1947, OpenNet Acc RL-1-374061, p. 5.

96. Gerber, *On the Home Front* (2002), p. 119 and ch. 4. The irrigation project, originally scheduled for 1948, began in the mid-1950s.

97. Ibid., pp. 84–86.

98. Stannard, *Radioactivity and Health* (1988), vol. 2, pp. 760–62.

99. Letter from K. Herde to J. Newell Stannard, 30 Oct 1978, as quoted in Stannard, *Radioactivity and Health* (1988), vol. 2, p. 762.

in addition to the three that were already in operation.[100] In August 1949, when the Soviets detonated their first atomic device, the AEC was already producing more fission fuel yearly than the MED had during World War II. The increased pace of US atomic weapons production that followed the Soviet test only accelerated through the late 1950s under President Eisenhower.[101] By 1955, five new reactors had been added at Hanford (and another five built on the Savannah River in Georgia). The ecological burden on the Columbia River increased accordingly; by 1954 about 8,000 curies of radioactive material were being dumped into the river each day.[102] In 1949 a Columbia River Advisory Group was formed with officials from the US Public Health Service and the health departments of Washington and Oregon.[103] Measures of radioactivity in plankton, in fish, and in the river water itself reached unprecedented levels in 1950. That same year the Public Health Service sent a team to evaluate the safety of the river water for nearby inhabitants.[104]

Hanford's scientists perceived the public health implications of radioactive bioconcentration, even as they continued to insist that there were no current health hazards in the Columbia River. As Parker wrote in a review published in 1948:

> In large bodies of water, the concentration of activity in algae or in colloidal materials with its possible utilization by fish, later used for food, presents a chain of events of great consequence to the public health. Up to the present time, these problems have been by-passed by ultraconservative policy in waste disposal, but future pressure for economical disposal facilities greatly point[s] the need for extensive research on these problems.[105]

100. Hewlett and Duncan, *Atomic Shield* (1969), pp. 141–53.

101. Gerber, *On the Home Front* (2002), pp. 38–42.

102. Herbert M. Parker, "Columbia River Situation–A Semi-Technical Review," 19 Aug 1954, OpenNet Acc RL-1-360700.

103. Gerber, *On the Home Front* (2002), pp. 121–22.

104. Robeck et al., *Water Quality Studies* (1954); Becker, *Aquatic Bioenvironmental Studies* (1990), p. 20. Gerber argues that the first draft of the Public Health Service's report was considered by the AEC to be "highly detrimental to public relations" and was revised to "preserve the present status." She is quoting from Herbert M. Parker, "Columbia River Situation–A Semi-Technical Review," 19 Aug 1954, OpenNet Doc HW-32809, pp. 4–7; Gerber, *On the Home Front* (2002), p. 123.

105. Parker, "Health Physics, Instrumentation, and Radiation Protection" (1948), p. 241. Gerber (*On the Home Front* [2002], p. 296n35) suggests that this review was written for AEC

Yet despite this oblique reference, the Hanford group's finding of the bioconcentration of radioactivity, though regarded by insiders as a major discovery, was not published in the open literature until after the 1954 revision of the Atomic Energy Act—and after a published ecological survey of the watershed near Oak Ridge National Laboratory documented the selective concentration of radioactivity in aquatic life there.[106] Eisenhower's push to develop domestic nuclear power, and the associated declassification and dissemination of information about reactors, gave Hanford's findings about radioactive waste a new relevance. One report anticipated that US production of fission products from nuclear power would reach a ton a day by 2000.[107]

The first International Conference on the Peaceful Uses of Atomic Energy in Geneva in 1955 provided Hanford scientists a world stage on which to announce their findings. In a session on "Ecological Problems Related to Reactor Operation," the group presented two papers, one by Foster and Davis focusing on aquatic life, and one by W. C. Hanson and Harold A. Kornberg on terrestrial animals.[108] The papers appeared in the proceedings in 1956. Davis and Foster published a longer paper entitled "Bioaccumulation of Radioisotopes through Aquatic Food Chains" in *Ecology* in 1958.[109] These papers display a broader and more theoretical orientation to the problems of radioactive contamination than the classified reports written earlier. In the 1956 paper, Foster and Davis asserted:

---

officials and remained classified until it was reprinted in *Health Physics* in 1980, but this is not the case.

106. Krumholz, *Summary of Findings* (1954). In fact, many ecologists, especially those at Oak Ridge, tended to cite the Krumholz finding of bioconcentration in aquatic life. One Hanford publication did appear in 1954, though in the *Journal of the American Water Works Association*, perhaps an obscure location for ecologists: Foster and Rostenbach, "Distribution of Radioisotopes" (1954).

107. This number, provided by E. I. Goodman and R. A. Brightsen in a 1955 report for the American Chemical Society, was calculated on the basis of an estimated US production of 750,000,000 kilowatts of electricity per day from nuclear power. Laurence, "Waste Held Peril" (1955); Hamblin, *Poison in the Well* (2008), p. 62.

108. Foster and Davis, "Accumulation of Radioactive Substances" (1956); Hanson and Kornberg, "Radioactivity in Terrestrial Animals" (1956). There were two other ecological papers at the conference by Hanford scientists, one on absorption of fission products by plants, and the other a review of radiation exposure from environmental sources. Stannard, *Radioactivity and Health* (1988), vol. 2, p. 765.

109. Davis and Foster, "Bioaccumulation of Radioisotopes" (1958).

> The organisms living in the Columbia River which have picked up radioactive substances from the reactor effluent may be utilized as a large-scale experiment in which the isotopes serve as tracers. In this way, studies designed primarily to monitor the level of radioactivity, also provide information on nutrient cycles, metabolic rates, and ecological relationships.[110]

They also pointed out the inadequacy of understanding radioactive contamination through an exclusive reliance on laboratory studies. For example, Foster and Davis found that the fish taken from the Columbia River contained around 100 times as much radioactivity as laboratory fish who were exposed to equivalent mixtures of radioactivity as found in the effluent. This was because the laboratory fish (in contrast to those in the river) were eating uncontaminated food.[111]

In his second edition of *Fundamentals of Ecology*, Odum drew on this group's work at Hanford (familiar to him through the four months he spent there while writing the new chapter on "Radiation Ecology") to illustrate the importance of food webs for the bioconcentration of radioactivity, while interpreting their results explicitly in terms of his ecosystems ecology. As Odum pointed out, there were three routes through which various radioisotopes entered the environment around Hanford, resulting in aquatic, aerial, and terrestrial contamination.[112] The aquatic contamination resulted from the release of radioactivity in the effluent from the reactors. The operation of the plant led to the aerial release of iodine-131 (and other radioisotopes) in waste gases, which contributed to both aerial and terrestrial contamination. Disposal of radioactive waste fluids, in turn, contributed to both terrestrial and aquatic contamination. The food web dispersed and concentrated radioisotopes from these different environmental entry points. For example, phosphorus-32 in the cooling effluent moved from aquatic insects, vegetation, and crustaceans into a variety of waterfowl. River ducks and geese, which principally ate grains, concentrated radiophosphorus less than swallows (which, though not a water fowl, consumed aquatic insects) and diving ducks. Even so, the phosphorus-rich egg yolk of river ducks and geese showed enrichment of phosphorus-32 of 200,000 over its concentration in river water. Iodine-131, released from the Hanford chemical separations facilities into the air

110. Foster and Davis, "Accumulation of Radioactive Substances" (1956), p. 364.
111. Davis and Foster, "Bioaccumulation of Radioisotopes" (1958), p. 531.
112. Odum, *Fundamentals of Ecology* (1959), p. 467.

Table 31. *Concentration of radioisotopes in food chains as shown by the ratio of isotopes in organisms to that in the environment*

1. P^{32}–Columbia River:*					
	Water	*Vege-tation*	*Insects, Crustaceans*	*Vertebrates*	*Eggs*
Swallows, adults	1	–	0.5	75,000	–
Swallows, young	1	–	3.5	500,000	–
Geese and ducks	1	0.1	0.1	7,500	200,000

2. Long-lived mixed fission products in holding pond:*			
	Water	*Vegetation*	*Birds*
Coots	1	3	250 in muscle (mostly Cs^{137}). 500 in bone (mostly Sr^{90}).

3. I^{131} aerial contamination at Hanford:*			
	Vegetation	*Jack Rabbit Thyroids*	*Coyote Thyroids*
Terrestrial food chain	1	500	100

4. P^{32} and other isotopes in Columbia River:†				
	Water	*Phyto-plankton*	*Aquatic insects*	*Bass*
Aquatic food chain	1	1000	500	10

* Data from Hanson and Kornberg, 1956.
† Data from Foster and Rostenbach, 1954.

FIGURE 10.4. Table representing the concentration of radioisotopes in organisms and in food chains, based on data collected by researchers at Hanford Works. From Eugene P. Odum, *Fundamentals of Ecology*, 2nd ed. (Philadelphia: W. B. Saunders, 1959), p. 468.

through waste gases, was taken up by mammals, birds, reptiles, and insects. Rabbits, for instance, showed a concentration of radioiodine in their thyroids as high as 500 times that of the vegetation.[113] (See figure 10.4.)

The publications from Hanford scientists emphasized that radioactive contamination in the Columbia River "never approached hazardous levels."[114] But Odum drew a more cautious conclusion:

113. Hanson and Kornberg's paper, "Radioactivity in Terrestrial Animals" (1956), reported an enrichment of phosphorus-32 in egg yolks of ducks and geese of 1,500,000, but Odum notes that the average was lower, as reflected in his table for *Fundamentals of Ecology* (1959), p. 468.

114. Davis and Foster, "Bioaccumulation of Radioisotopes" (1958), p. 531; see also Foster and Rostenbach, "Distribution of Radioisotopes" (1954), p. 635, in which the authors state, "No effect from the small amounts of radioactivity present has been detected." This same point comes through clearly in "Hanford Science Forum," a television broadcast (sponsored

> Thus, an isotope might be diluted to a relatively harmless level on release into the environment, yet become concentrated by organisms or a series of organisms to a point where it would be critical. In other words we could give "nature" an apparently innocuous amount of radioactivity and have her give it back to us in a lethal package![115]

The reception of the belatedly published Hanford findings was undoubtedly sharpened by the fallout controversy of the mid-1950s, which had focused attention on environmental contamination and the hazards of low-level radiation exposure. Strikingly, in his 1958 paper in *Science* addressing the controversy over strontium-90 from atomic testing fallout, Commissioner Willard Libby cited Foster and Davis's 1956 publication and noted possible environmental concentration mechanisms, even though he denied human health risks.[116]

Radioecology at Oak Ridge and Savannah

Concerns about radioactive contamination in the vicinity of Oak Ridge National Laboratory prompted an ecological research program there as well. Stanley Auerbach arrived in 1954 to join their division, and by 1960 he had built up one of the largest ecology research groups in the country, with twenty-two people on staff.[117] As at Hanford, the revision of the Atomic Energy Act in 1954 and the government's greater emphasis on civilian development of nuclear power gave the ecologists at Oak Ridge

---

by Hanford's contractor, General Electric), which featured an interview with Foster on the work of the Aquatic Biology Operations in a 1957 program. The interviewer introduced the venture as a special kind of "fishing" in the Columbia River. The telecast is available at http://www.archive.org/details/HanfordS1957.

115. Odum, *Fundamentals of Ecology* (1959), p. 467.

116. Kwa, *Mimicking Nature* (1989), p. 83n43: "An early publication by W. F. Libby, member of the Atomic Energy Commission, declared the danger of Strontium-90 unimportant for humans while noting possible concentration mechanisms." That publication was Libby, "Radioactive Fallout and Radioactive Strontium" (1956). According to Gerber, even "AEC Chairman Lewis Strauss expressed worry over the 'Columbia River contamination situation'" (*On the Home Front* [2002], p. 128). Given Strauss's dismissal of concerns about the hazards of radioactive fallout from atomic weapons tests (see chapter 5), this is a rather surprising admission.

117. The ecology group at ORNL continued growing steadily, and had close to 250 employees in 1978. Bocking, *Ecologists and Environmental Politics* (1997), p. 75.

a justification for not only conducting but also publishing their research. Moreover, from the early 1950s "waste disposal" provided the framework under which ecologists were included at Oak Ridge. Along similar lines, Eugene Odum built up a significant ecological research group at the AEC production facility on the Savannah River in Georgia. At both Oak Ridge and Savannah, studies of the movement of radioisotopes, particularly long-lived fission products such as strontium-90 and cesium-137, furthered the importance of ecosystems to ecological research and offered concrete information about how the government and nuclear industry might manage the growing load of radioactive waste.

Beginning in 1951, Oak Ridge National Laboratory disposed of low-level radioactive waste by pumping it into pits in the earth, from which it could seep into surrounding soil. The expectation was that through binding of soil particles, the radioisotopes would become immobilized.[118] In addition, since wartime, nearly all low-level liquid radioactive wastes had been dumped into nearby White Oak Creek or White Oak Lake, a 35-acre impoundment formed in 1943 by a small dam to hold wastes, allowing short-lived radioisotopes to decay before their release into the Clinch River. (See figure 10.5.) Karl Z. Morgan, director of the Health Physics Division, raised concerns as early as 1948 about effects of contaminated waterways on the local populations. This prompted a cooperative project with the Tennessee Valley Authority (TVA) to determine the extent of radioactive contamination, including levels reaching the Clinch River.[119]

Louis A. Krumholz, a fisheries biologist with the TVA, directed the ecological surveys of White Oak Lake and environs from 1950 to 1954. The surveys documented that the plants, phytoplankton, and fish in White Oak Lake accumulated and concentrated radioactivity.[120] In one type of green algae, *Spirogyra*, the concentration factor for radiophosphorus was

118. Ibid., p. 68; Reichle and Auerbach, "U.S. Radioecology Research Programs" (2003), p. 8.

119. The cooperation between ORNL and TVA concerning radioactive waste disposal began earlier. The Health Physics Division and the TVA jointly organized a section of Waste Disposal Research in 1948, involving scientists from the Army Corps of Engineers, the Public Health Service, and the US Geological Survey. The radioecological survey the AEC authorized in 1950 was based in the Health Physics Division and in collaboration with this section. The AEC contracted with the TVA's Fish and Game Branch to conduct the survey. Auerbach, *History of the Environmental Sciences Division* (1993), p. 3.

120. Krumholz, *Summary of Findings* (1954).

FIGURE 10.5. Photograph showing the White Oak Creek drainage basin and White Oak Lake. From Stanley I. Auerbach and Vincent Schultz, eds., *Onsite Ecological Research of the Division of Biology and Medicine at the Oak Ridge National Laboratory*, TID-16890. Washington, DC: AEC Division of Technical Information, 1962, p. 79.

as high as 850,000.[121] There was only one case in which the concentration of radioactivity seemed to have caused damage to an organism: an American elm tree "that selectively concentrated enough radioruthenium to cause the edges of the leaves to curl and die."[122] With this one exception, the surveys did not reveal evidence of deleterious effects on the aquatic or terrestrial populations.[123] However, the lower reaches of White Oak Creek, below the dam, were not as productive as the upper reaches, which supported twice as many genera of benthic organisms. Krumholz noted

121. Ibid., p. 25.

122. Ibid., p. 14.

123. Kwa, "Radiation Ecology" (1993), p. 233; Whicker and Schultz, "Introduction and Historical Perspective" (1982), p. 5; Reichle and Auerbach, "U.S. Radioecology Research Programs" (2003), p. 8. Stannard, *Radioactivity and Health* (1988), vol. 2, p. 762 ff.

that the effects of the heavy silt as well as the waste effluents seemed to inhibit the creek's productivity.[124]

The ORNL administration took the TVA's assessment as confirmation that contamination in the White Oak water system was not at a level sufficient to cause noticeable environmental damage, despite the significant amounts of waste it received.[125] This nearly led to the cessation of any further ecological work at Oak Ridge, in part because, in the eyes of nuclear physicists, "the kind of science they were doing was almost totally alien to what was going on at this laboratory."[126] Health physicist Morgan crusaded for further attention to radioactive waste, arguing that protection of wildlife on the reservation was a legitimate aspect of health physics.[127] He and his assistant Edward Struxness developed a broad funding plan for research on radiation ecology, but the response from the AEC was tepid. In Morgan's memoir, he states that one agency official (unnamed) commented, "Man is the thing we should be interested in protecting; we should protect him and forget about these microorganisms and other forms of life. After all it would be a good thing if radiation destroyed all these microorganisms."[128] However, other factors tipped the balance in Morgan's favor. First, the decision to drain White Oak Lake (for flood-control purposes) created an undeniable opportunity for studying the fate of contamination in the lake bed—a "natural contaminated ecosystem," as one account puts it.[129] Second, President Eisenhower's push for civilian nuclear power development gave the study of environmental radioactivity a new and ur-

124. Krumholz, *Summary of Findings* (1954), p. 26.

125. Krumholz, it should be stressed, did not see his assessment in such a positive light. As he wrote in his conclusions, "The circumstantial evidence against the continued (or increased) contamination of aquatic environments with radiomaterials is very strong. Although the evidence is not unequivocal that the damage to the populations in White Oak Lake was caused by irradiation alone, it can hardly be denied that the constant exposure to radiation may have been a strongly contributing factor." Krumholz, *Summary of Findings* (1954), p. 50.

126. Auerbach goes on to say of the ecological survey work: "The kinds of things we accept in the environment—great variability as a result of multi factors that we've got to deal with—was something that they didn't quite put across well. . . . It created a certain amount of horror in the eyes of the more rigorous physical scientists who ran the ORNL." J. Newell Stannard, transcript of interview with Stanley I. Auerbach, 19 Apr 1979, Stannard papers, box 3, folder 4, quotes from p. 2.

127. Morgan and Peterson, *Angry Genie* (1999), p. 85; Bocking, *Ecologists and Environmental Politics* (1997), pp. 65–68. I picked up the term "crusade" from Stannard, *Radioactivity and Health* (1988), vol. 2, p. 769.

128. Morgan and Peterson, *Angry Genie* (1999), p. 85.

129. Stannard, *Radioactivity and Health* (1988), vol. 2, p. 769.

gent relevance. Former ORNL Director Eugene Wigner, still involved at the upper echelons of the agency, viewed radioactive waste disposal as a key problem for nuclear power. He supported Morgan's proposal.

Ecologist Orlando Park of Northwestern University, with whom Struxness had previously studied, was brought in to consult on Morgan's new program, and Park's former graduate student Stanley Auerbach was hired by ORNL in 1955. Struxness, who had been supervisor of health physics at the Oak Ridge Y-12 plant, was transferred to ORNL to focus on the environmental aspects of radioactive waste management.[130] Auerbach initially pursued laboratory studies, such as investigating the effects of radiation on arthropods in rotting wood and radiostrontium uptake by earthworms.[131] John Wolfe, newly appointed as head of a national ecology program for the AEC Division of Biology and Medicine, visited Oak Ridge in 1956. A field ecologist himself, Wolfe urged Auerbach to take advantage of the newly drained lake as a site for research.[132]

White Oak Lake received, either directly or through seepage from earthen storage pits, lower-activity radioactive wastes, principally containing strontium-90, cesium-137, cobalt-60, and ruthenium-106. As Auerbach has noted, "Even by the standards of that time, the White Oak lake bed was considered to be highly contaminated. . . . At that time this small piece of landscape was considered to be one of the most radioactive sites on Earth."[133] To conduct fieldwork in the lake bed, where radiation dose at waist-high ranged up to 300 millirads per hour in some areas, researchers were required to wear radiological protection gear. Auerbach recalls carrying a "cutie pie" detector and moving around to try to keep himself in areas with less than 25 millirads per hour exposure, to remain below the permissible limit for occupational exposure.[134]

130. Ibid., vol. 2, p. 769.

131. Auerbach, *History of the Environmental Sciences Division* (1993), pp. 5–6. See also Auerbach, "Soil Ecosystem" (1958).

132. Auerbach, *History of the Environmental Sciences Division* (1993), p. 6. In 1958 Wolfe became chief of a new Environmental Sciences Branch of the AEC's Division of Biology and Medicine, a role in which he expanded the agency's funding of ecological research at the national laboratories and at colleges and universities through grants. See Dunham, "Foreword" (1962).

133. Reichle and Auerbach, "U.S. Radioecology Research Programs" (2003), p. 9. Auerbach explains that most of this waste was generated by the intensive work at ORNL on techniques for reprocessing used reactor fuel elements.

134. J. Newell Stannard, transcript of interview with Stanley I. Auerbach, 19 Apr 1979, Stannard papers, box 3, folder 4, p. 10. A "cutie pie" was a diminutive ion chamber detector;

In part due to Odum's role as a consultant for the program, Auerbach adopted an ecosystems approach in studying the succession of vegetation and animal life in the lake bed and the biogeochemical movement of radioisotopes within it.[135] He was also interested in the biological effects of the contaminating radiation on the biota of the lake bed, a concern allied with the health physics orientation of Morgan's ORNL group.[136] The group focused on reactor waste products strontium-90 and cesium-137, investigating their uptake into plants and possible transfer into food chains via insects, birds, and mammals.[137] As Dac Crossley and Henry Howden put it in one of their publications, "The resulting ecosystem on White Oak Lake bed may be envisioned as a gigantic radioactive tracer experiment, which can yield information of interest to both ecologists and health physicists."[138]

Recruits Crossley and Ellis Graham assessed the soil chemistry and surveyed the succession of plants that invaded the lake bed.[139] They found that the contaminating radionuclides were taken up and dispersed by the vegetation, showing that radioactive waste could not be assumed to remain where it had been deposited. However, there was tremendous variability. As Auerbach put it, "We could make no sense out of what we got. There were trees that were hot, there were trees that weren't hot, and we saw quickly that from a scientific point of view—from an inductive point of view, we would get numbers; but we aren't going to get much in the way of useful predictive information."[140]

In order to try to attain more predictive information, they began a more experimental approach, adding either vegetation or radioactivity to the landscape. Following Auerbach's penchant for "controlled field ex-

---

this was among numerous nicknames given radiation detection instruments during World War II.

135. Bocking, *Ecologists and Environmental Politics* (1997), p. 71. Bocking emphasizes that reading Eugene Odum's textbook *Fundamentals of Ecology* (1959) was crucial to Auerbach's awareness of this approach.

136. See, e.g., Dunaway and Kaye, "Effects of Ionizing Radiation" (1963).

137. Auerbach and Crossley, "Strontium-90 and Cesium-137 Uptake" (1958); Crossley and Howden, "Insect-Vegetation Relationships" (1961); Crossley, "Movement and Accumulation" (1963). Paul Dunaway, who joined the group in 1957, worked on mammals.

138. Crossley and Howden, "Insect-Vegetation Relationships" (1961), p. 302.

139. Kwa, "Radiation Ecology" (1993), p. 234; Auerbach, *History of the Environmental Sciences Division* (1993), p. 9; Graham, "Uptake of Waste Sr 90 and Cs 137" (1958).

140. J. Newell Stannard, transcript of interview with Stanley I. Auerbach, 19 Apr 1979, Stannard papers, box 3, folder 4, pp. 8–9.

perimentation," his group planted corn, legumes, and other crops in the contaminated lake bed.[141] Biologists at the University of Tennessee, Knoxville and the University of Georgia collaborated on these studies, their work supported through contracts with the AEC.[142] Other projects reflected the interest at ORNL in entire ecosystems. Here Auerbach followed the precedent of Hutchinson in using artificially produced radioisotopes, readily available at Oak Ridge, to investigate element cycling.[143] In May 1962, ORNL ecologists tagged an entire Oak Ridge forest with cesium-137, with the intent of measuring cesium transfer between the components of the ecosystem. As Auerbach and two coauthors noted, tracer experiments in the United States had not previously been attempted before "on a relatively large scale in field experiments."[144] They applied the radioisotopic material directly to each tree trunk, distributing 467 millicuries across the entire stand. (See figure 10.6.) The group had spent three years developing methods for such a large-scale operation. As the authors of the paper commented, "Utilization of such a large quantity of long-lived radioisotope in the field required careful handling and special procedures in order to avoid unnecessary exposure of personnel or accidental contamination during the tagging operation."[145] The study showed that cesium cycled out of the trees into the litter on the forest floor, but did not rapidly re-enter the system through the roots.[146] (See figure 10.7.)

The ecologists at Oak Ridge were also interested in the biological effects of the radioactive contamination on animals. In one study, ORNL researchers Stephen Kaye and Paul Dunaway captured small wild mammals such as cotton rats to determine their level of exposure from radioactive wastes. They noted that "the radiobiology of standard laboratory animals has been extensively investigated," whereas that of native animals had not. "Are calculations and predictions based upon biological half-lives,

141. The "controlled field experimentation" phrase is in Auerbach, "Soil Ecosystem" (1958), p. 525. Results of the corn planting are given in Auerbach and Crossley, "Strontium-90 and Cesium-137 Uptake" (1958).

142. Willard, "Avian Uptake of Fission Products" (1960); DeSelm and Shanks, "Accumulation and Cycling" (1963); Shanks and DeSelm, "Factors Related to Concentration of Radiocesium" (1963).

143. Johnson and Schaffer, *Oak Ridge National Laboratory* (1994), pp. 99–100.

144. Auerbach, Olson, and Waller, "Landscape Investigations Using Caesium-137" (1964), p. 761.

145. Ibid., p. 762.

146. Stannard, *Radioactivity and Health* (1988), vol. 2, p. 771.

FIGURE 10.6. Oak Ridge National Laboratory ecologists applying cesium-137 to a tree. The entire stand was tagged with 467 millicuries of the radioisotope. From Stanley I. Auerbach and Vincent Schultz, eds., *Onsite Ecological Research of the Division of Biology and Medicine at the Oak Ridge National Laboratory*, TID-16890. Washington, DC: AEC Division of Technical Information, 1962, p. 74.

assimilation factors, critical organs, etc., determined for laboratory rats valid for native rats?"[147] In fact, the country's largest laboratory experiment on radiation effects on rodents was being conducted at ORNL, the "megamouse" experiment, under the direction of William and Lianne Russell in Alexander Hollaender's Biology Division, to determine the mutation rate in mammals.[148] As Auerbach commented on Dunaway and Kaye's research, "While they used techniques that were developed and used by radiobiologists, their study of mammals in a natural environment distinguished their research from that of the Biology Division."[149] One senses

147. Kaye and Dunaway, "Bioaccumulation of Radioactive Isotopes" (1962), p. 205.

148. See Rader, *Making Mice* (2004), ch. 6.

149. Auerbach, *History of the Environmental Sciences Division* (1993), p. 19.

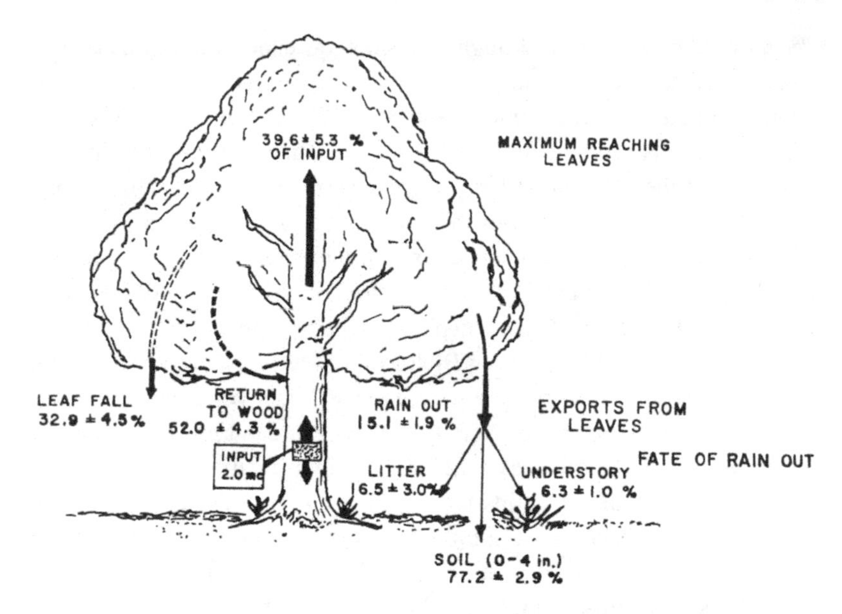

FIGURE 10.7. Schematic diagram of the cycling of cesium-137 out of the tree leaves into the soil and litter of the forest floor, results of the tagging experiment shown in the previous figure. From Stanley I. Auerbach and Vincent Schultz, eds., *Onsite Ecological Research of the Division of Biology and Medicine at the Oak Ridge National Laboratory*, TID-16890. Washington, DC: AEC Division of Technical Information, 1962, p. 79.

the old rivalry between laboratory and field at ORNL, even in an ecology group that was unusually oriented toward the physical sciences.[150]

The field researchers argued that results from their study would also shed light on the exposure of wildlife to contamination from radioactive fallout—which included the same spectrum of radioisotopes that were being released as waste at Oak Ridge. But whereas "fallout contamination in the general environment occurs at relatively low levels which often present radioanalytical difficulties," the contamination at the White Oak Lake bed was "orders of magnitude higher in levels of contamination than fallout, yet . . . low enough so that personnel may work in the area for a practical length of time without being overexposed."[151] Their initial studies of trapped animals showed body burdens to be high enough that pathological

150. On the contentious interplay between lab and field, see Kohler, *Landscapes and Labscapes* (2002).

151. Kaye and Dunaway, "Estimation of Dose Rate" (1963), p. 107.

effects would be expected (though no lesions were identified).[152] But variability was high, and the field experiments were supplemented by setting up fenced-in areas in which uncontaminated animals could be introduced and their assimilation of radioisotopes followed in a controlled fashion.[153] At the same time, the researchers sought to combine population ecology with studies of the movement of radioisotopes and radiation exposure.

Eugene Odum carried out similar radiolabeling experiments on research tracts at the Savannah River nuclear site. When the AEC decided to build new plutonium production plant on this river in Georgia, the agency's newly established Division of Biology and Medicine supported studying this facility's impact on the environment. They invited the University of South Carolina and the University of Georgia to submit proposals for "pre-installation" inventories of the site. Each university received $10,000 a year for three years, and divided the task up between them, with the University of Georgia focusing on animal populations of warm-blooded vertebrates and invertebrates, and the University of South Carolina on botanical work and cold-blooded vertebrates.[154] Odum convinced the AEC that some of the money to Georgia should be used to study secondary succession of vegetation and fauna on farmlands abandoned when the government moved residents off the designated 250,000 acre reservation.[155] After 1954, the Division of Biology and Medicine increased funding for ecological research at Savannah River, enabling Odum eventually to establish a permanent on-site laboratory there, the Savannah River Ecology Laboratory.[156]

Beginning in 1957, the group at Savannah River began field experiments with radioactive tracers, something Odum had expressed interest in doing in an unsuccessful 1951 application to the AEC.[157] As Chunglin

152. In particular, the researchers found and trapped four muskrats living in the settling basin for liquid waste, whose whole-body dose rates were so high that pathologies would be expected. Perhaps due to lack of exposure time, the researchers found no lesions in the two muskrats dissected. Ibid., p. 109.

153. Ibid., p. 111.

154. Odum, "Early University of Georgia Research" (1987), pp. 43–44.

155. Kwa, "Radiation Ecology" (1993), pp. 227–29; Odum, "Organic Production and Turnover" (1960).

156. The on-site laboratory was approved in 1960 and became available in 1961. Kwa, "Radiation Ecology" (1993), p. 229.

157. A copy of Odum's unsuccessful 1951 application is printed as Appendix A (pp. 59–72) to Odum, "Early University of Georgia Research" (1987).

Kwa has noted, Odum was not constrained in the way that Auerbach was by the Oak Ridge focus on radioactive waste and the special concern with fission products such as cesium-137 and strontium-90.[158] He could design his radiolabeling experiment to best measure the movement of material and energy between trophic levels in the terrestrial ecosystem. For this purpose, injecting a specific amount of radioactivity into the system at a particular point allowed the researcher more control over measuring its movement.[159] As an element required for growth in all organisms, phosphorus was a more desirable tracer than strontium or cesium. Odum and Edward Kuenzler devised a method of laying out "hot quadrats" in which all individuals of a kind of plant were labeled with phosphorus-32.[160] They radiolabeled three quadrats, each of a different plant species—heterotheca, rumex, and sorghum—in the spring of 1957.

By following the movement of radiophosphorus into higher trophic levels—namely, animals—the researchers could isolate the food chain: Any animal that became radioactive had to belong to the food chain originating with the tagged plant species. Arthropods and snails closely associated with the vegetation were sampled "by sweeping with a standard sweep net," and crickets, ground beetles, and other insects associated with the substratum were captured with a cryptozoa board (under which these insects congregated). Kuenzler, Odum's collaborator, specialized in wolf spiders, which he caught by " 'shining their eyes' with a flashlight."[161] The researchers also trapped mice to see if the phosphorus reached small mammals. All these animals' bodies were counted for radioactivity. In addition to showing how rapidly phosphorus (and thus energy) was transferred from plants, the primary producers, to grazing herbivores and ants, the primary consumers, the experiment shed light on the eating habits of the snails, whose rapid acquisition of radioactivity indicated that grass was an important food source.[162] The researchers' plots of radioactivity versus time in various species showed "the graphic separation of certain trophic and habitat groups."[163] (See figure 10.8.) Odum's group continued these investigations over longer periods of time, and pointed to the promise

158. Kwa, "Radiation Ecology" (1993), p. 230.
159. See Odum and Golley, "Radioactive Tracers as an Aid" (1963).
160. Odum and Kuenzler, "Experimental Isolation of Food Chains" (1963).
161. Ibid., p. 116.
162. Ibid., p. 118.
163. Ibid., p. 119.

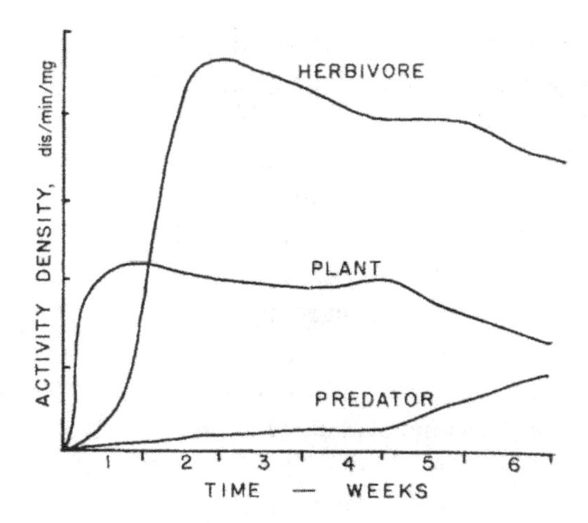

FIGURE 10.8. Typical radioactivity density curves at three trophic levels resulting from the labeling of a single species of herbaceous plant with phosphorus-32 at time zero. The movement of the label from plants to herbivores to predators is clearly visible. All curves are corrected for radioactive decay. From Eugene P. Odum, "Feedback between Radiation Ecology and General Ecology," *Health Physics* 11 (1965): 1257–62, on p. 1260.

of this radiolabeling method for determining "food web diversity for an entire community."[164]

While ORNL and Savannah continued to be major sites for research, the reach of radioecology can be seen in the many symposia volumes published in the late 1950s through the early 1970s as well as in AEC records. (See figure 10.9.) The 1955 and 1958 International Conferences on the Peaceful Uses of Atomic Energy featured many papers by ecologists, mostly oriented toward the problems of radioactive waste from civilian nuclear power development. A Symposium on Radioisotopes in the Biosphere, at the University of Minnesota, October 19–23, 1959, focused principally on the pathways of radioisotopes released into the atmosphere through atomic explosions. Similarly, the International Symposium on Radioecological Concentration Processes, held in Stockholm, April 25–29, 1966, addressed the distribution and movement of radioisotopes in fallout.[165] The importance of radiation studies to general ecology is also evident. The AEC supported three large-scale symposia on radioecology, the first held in Fort Collins, Colorado, September 10–15, 1961, the second in Ann Arbor, Michigan, May 15–17, 1967, and the third at Oak Ridge, Ten-

164. Wiegert and Odum, "Radionuclide Tracer Measurements" (1969), p. 710.

165. Åberg and Hungate, *Radioecological Concentration Processes* (1967).

FIGURE 10.9. A University of Wisconsin researcher preparing to release iodine-131 into the lake at a depth of thirteen feet, in order to determine the physical and biological movements within the deeper layers of the lake. Credit: University of Wisconsin. National Archives, RG 326-G, box 8, folder 2, AEC-62-6582.

nessee, May 10–12, 1971. These three symposia included dozens of papers that feature the use of radioisotopes as tracers through aquatic or terrestrial ecosystems, many of which represented AEC-sponsored work at Oak Ridge and Savannah River. According to Auerbach, the 1961 symposium "brought out the need for training ecologists in the use of radioisotopes as a research tool in ecology."[166] In response, his group organized a special summer course for ecologists, which was first offered in 1962.[167]

166. Auerbach, *History of the Environmental Sciences Division* (1993), p. 21.

167. Ibid. The summer courses for ecologists were short-lived, however, with the last one being offered in 1964. After that, ecologists were expected simply to enroll in the regular course.

Research with isotopic tracers helped put ecosystems ecology on a quantitative footing.[168] Ultimately, computers were used to mathematically simulate ecosystems, a development in which Oak Ridge figured prominently.[169] Jerry Olson, a research ecologist hired at ORNL in 1958, harnessed the computational resources available at a national laboratory to undertake mathematical modeling of nutrient cycling, focusing on the movement of radionuclides through ecosystems.[170] In fact, the 1962 cesium-137 tagging of the Oak Ridge forest was conducted with the hope of inputting the resulting data into Olson's computer model for mineral cycling.[171] Through the 1960s, ORNL ecologists advocated the design of ever-larger computational models to derive information about ecosystems from research data on their components and linkages. The International Biological Program became the vehicle for the development of this kind of large-scale systems biology into the 1970s. This episode, though beyond the scope of this chapter, has been analyzed by others.[172] Here it suffices to underline the important contribution of tracer studies and radiation ecology (not to mention the big-science money and infrastructure available through the AEC) to the emergence of a computational approach.

The AEC's footprint in radiation ecology also included surveys of coral atolls in the Pacific in conjunction with atomic weapons testing, radiation experiments with cesium-137 and cobalt-60 gamma ray sources in forests near Brookhaven National Laboratory and in a Puerto Rican preserve, and ongoing studies of the Trinity site and Nevada Proving Ground by UCLA health physicists interested in the fate of fission products.[173] A

168. Hagen, *Entangled Bank* (1992), pp. 112–15.

169. NASA and the related space industry provided another context in which ecologists became familiar with computer simulations. Anker, "Ecological Colonization of Space" (2005).

170. Olson, "Analog Computer Models" (1963). Olson used the National Laboratory Analog Computer Facility at Oak Ridge. Kwa, "Radiation Ecology" (1993), pp. 235–36, 243; Bocking, *Ecologists and Environmental Politics* (1997), pp. 77–84.

171. Auerbach, Olson, and Waller, "Landscape Investigations Using Caesium-137" (1964).

172. See Kwa, "Radiation Ecology" (1993); Golley, *History of the Ecosystem Concept* (1993), ch. 5; McIntosh, *Background of Ecology* (1985), ch. 6; Coleman, *Big Ecology* (2010), ch. 2.

173. Odum and Odum, "Trophic Structure and Productivity" (1955); Larson, "Continental Close-In Fallout" (1963); Woodwell, "Effects of Ionizing Radiation" (1962); Odum, *Tropical Rain Forest* (1970). For the AEC's impact on oceanography, see Rainger, "Wonderful Oceanographic Tool" (2004); idem, "Going from Blue to Green?" (2004); Hamblin,

wide array of projects by extramural ecologists received support through the AEC's Division of Biology and Medicine, and, after 1958, the Division of Environmental Sciences.[174] The United States was not the only sponsor of radioecological work; the British Atomic Energy Authority had an influential group, and Japanese scientists were also making key contributions.[175] As Odum observed in 1965, a positive feedback loop developed between nuclear power and ecology after World War II.[176]

Conclusions

One might infer from the involvement of the US government in the emergence of radioecology, and the key importance of research at Hanford, that this new field originated as a result of the military development of atomic energy. But as with biochemistry, interests among ecologists in using radioisotopes as tracers preceded the Manhattan Project. For Hutchinson this interest was catalyzed by the success of biochemists and physiologists in elucidating the movement of compounds in the body and metabolic pathways through radiolabeling. Hutchinson, who already viewed aquatic bodies in terms of their "intermediary metabolism," verified the utility of radioisotopes for tracing biogeochemical cycles and visualizing the dynamics of material and energy flow within ecosystems. This disciplinary crossover was consequential in other respects: the physiological understanding that accompanied his use of radiotracers revived the ecosystem as a quasi-organism, not merely a machine.

That said, of course, the interests of the Manhattan Project and subsequently the AEC in managing the disposal of radioactive waste from atomic weapons production nudged ecological research in particular directions, both through patronage relations and through unprecedented experimental opportunities. As we have seen, the amounts of radioactivity being deposited into waterways and landscapes provided ecologists the chance to study the effects of radiation in the field and the movement of

---

"Hallowed Lords of the Sea" (2006); idem, *Poison in the Well* (2008). On the environmental response to Project Chariot in Alaska, see Kirsch, *Proving Grounds* (2005).

174. Kingsland, *Evolution of American Ecology* (2005), ch. 7.

175. Hamblin, *Poison in the Well* (2008), ch. 3.

176. Odum, "Feedback" (1965). See also Rothschild, "Environmental Awareness" (2013); Schloegel, "'Nuclear Revolution' Is Over" (2011).

radioactive materials through the environment. The landscapes around the AEC's atomic weapons plants and national laboratories became experimental test-beds for ecological radiotracers, and problems of radioactive waste there were, in turn, understood in terms of ecosystems.

More broadly, ecological studies of radioactive waste as well as the growing understanding of the dangers of low-level radiation exposure disclosed the environmental and occupational risks associated with using radioisotopes, information the AEC was initially reluctant to accept.[177] Radioactive tracers also made newly visible the problems of containing nuclear contamination; this in turn generated concerns about the disposal of radioactive waste from laboratories and clinics, and even more so from the government's large-scale atomic energy and weapons facilities.[178] The changed political environment prompted new federal regulation that made radioactive materials more challenging to use and dispose of, after the government had done so much to ensure that radioisotopes were widely distributed.[179]

Lastly, the legacy of atomic energy was important to environmentalism in another key respect. Rachel Carson's *Silent Spring*, which launched widespread public concern about environmental contamination with synthetic chemicals, posits the similarity between the hazards of radioactive fallout and the dangers of chemical contaminants: "In this now universal contamination of the environment, chemicals are the sinister and little-recognized partners of radiation in changing the very nature of the world—the very nature of its life."[180] As it turned out, some insecticides were found to exhibit the same trait of bioconcentration as compounds moved up the food chain.[181] DDT became the exemplar of this phenomenon. In fact, it was George Woodwell at Brookhaven National Laboratory, having made his name studying how radiation affects forest ecosystems, who demonstrated that DDT was concentrated up to 1.5-million fold in an aquatic ecosystem on Long Island.[182] As Woodwell has noted, the attention to one part per billion in the environment "in itself was a revolution," and this realization

177. See chapter 6.

178. Mazuzan and Walker, *Controlling the Atom* (1985), ch. 12.

179. Walker, *Containing the Atom* (1992).

180. Carson, *Silent Spring* (1962), p. 6.

181. Woodwell, "Toxic Substances and Ecological Cycles" (1967).

182. The concentration went from a dilution of 0.00005 ppm in water to 75.5 ppm in an immature ring-billed gull. Woodwell, Wurster, and Isaacson, "DDT Residues" (1967). Odum reproduced part of their data in a figure for the third edition of his textbook: Odum, *Funda-*

that biotic studies required measurement in the "range of nanograms and picograms, nanocuries and picocuries" became a defining feature of the study of environmental contamination.[183] Historians have noted how the environmental movement built on the shift from perceiving the invisible danger lurking in radioactivity to that present in synthetic chemicals.[184] This analogy also informed how scientists approached the problem of understanding the ways in which chemical contaminants moved through ecosystems and entered food webs, both dispersing and concentrating in the environment. As two textbook authors noted in 1982, when it came to studying ecological processes involved in the spread of "smog, pesticides, and other chemical substances that may threaten the environment," radioisotopes were a "model pollutant."[185]

mentals of Ecology (1971), p. 74. On Woodwell's work at Brookhaven on forest ecosystems, see Woodwell, "Effects of Ionizing Radiation" (1962).

183. Woodwell, "BRAVO plus 25 Years" (1980), p. 62.

184. Lutts, "Chemical Fallout" (1985). This is not to suggest ecologists saw these hazards as equivalent. See Hagen, "Teaching Ecology" (2008).

185. Whicker and Schultz, "Introduction and Historical Perspective" (1982), p. 2.

CHAPTER ELEVEN

Half-Lives

I will . . . make the prediction that the AEC Isotopes Division does not have a half life of another five or ten years. I am sure that no one is going to agree. — Jerry Luntz, *Nucleonics*, 1954[1]

In 1956 and 1957, the AEC sponsored summer courses for high school teachers at Harvard, Duke, and the University of New Mexico. The courses, which included both lectures and laboratories, trained the teachers to incorporate radiobiology, radioisotopes, and nuclear science into their school curricula. Each participant received a demonstration kit provided by the agency "containing simple radiation detection and measuring equipment, a radioactive source and other basic pieces of equipment to enable him to demonstrate the newly acquired techniques to the science students at his own and neighboring schools."[2] Other high school students learned about the atom on AEC-sponsored trips to the American Museum of Atomic Energy and Oak Ridge National Laboratory.[3] The Com-

1. Jerry Luntz, "The Development of the Atomic Energy Industry," Third Industrial Conference on Teletherapy, 14–15 May 1954, OpenNet Acc NV0718572, p. 1137310.

2. These kits cost $300 each in 1956 (courtesy of the Division of Biology and Medicine), and $476 in 1957 (paid for by the Division of Reactor Development); costs of the courses were also subsidized by the National Science Foundation. Quote from Appendix A to AEC 761/2, 12 Aug 1957, NARA-College Park, RG 326, E67B, box 50, folder 8 Medicine, Health & Safety 21 Education & Training, p. 8; other information from Minutes, 56th ACBM Meeting, 26–27 May 1956, Washington, DC, OpenNet Acc NV0411749, p. 4; "Program Status Report to the Joint Committee on Atomic Energy" 31 Dec 1956, Part VII Biology and Medicine, OpenNet Acc NV0719079, pp. 56–57.

3. Monthly Highlight Report for March 1957, from William G. Pollard, Oak Ridge Institute of Nuclear Studies to Herman M. Roth, AEC Research and Development Division, 1 Apr 1957, OpenNet Acc NV0714090. On the museum at Oak Ridge, Mollela, "Exhibiting Atomic Culture" (2003).

mission invested in these efforts to improve education (and hence public opinion) about the everyday uses of atomic energy, making radioactivity more familiar and less feared. Even as opposition to nuclear weapons testing gathered steam, the AEC assumed that civilian uses of atomic energy would become accepted and established, buoyed by a burgeoning nuclear industry.

Living with Radiation, a booklet that the Commission published in 1959, confidently presented the future atomic bounty. One section, entitled "Benefits Justify Hazard," compared opposition to nuclear power to earlier opposition to electricity with alternating current. As the Commission reassured readers, "Man needed alternating current for the development of his standard of living, and he learned to handle it safely. Man needs the peace time uses of atomic energy, and he will likewise learn to handle them safely."[4] Needless to say, the US government's commitment to building support for nuclear power was not unrelated to its continued development of nuclear weapons. Public acceptance that environmental radioactivity could be safely managed was as important to ongoing peacetime weapons tests as to the construction of dozens of nuclear power plants.[5] Even after the Limited Test Ban Treaty of 1963 (which ended atmospheric atomic weapons tests among signatories, including the United States), the government remained steadfastly committed to a future in which nuclear power contributed a major portion of the country's energy infrastructure. In 1973, the AEC estimated that by the year 2000 nuclear power units would attain a capacity of 1.2 million megawatts, which would have required a total of 1,000 plants.[6]

However, the AEC overestimated its ability to steer public opinion. Critics of nuclear weapons testing had argued that even low-level exposure to radiation could cause cancer and birth defects, and over the 1960s Americans developed increasing misgivings about all sources of radiation.[7] Between 1966 and 1973, as utility companies ordered dozens of nuclear

4. AEC, *Living with Radiation*, vol. 1 (1959), p. 9.

5. On debates within the executive branch about atomic weapons testing, see Greene, *Eisenhower, Science Advice* (2007).

6. American utilities purchased just under 200 nuclear plants from 1955 to 1973. Walker, *Three Mile Island* (2004), p. 7.

7. Walker, *Containing the Atom* (1992), p. 388. The US government's acquiescence to the 1963 ban on atmospheric testing of nuclear weapons appeared to confirm these suspicions.

power plants, public opposition to their construction gathered force.[8] By consistently supporting the nuclear industry rather than shoring up its effectiveness as a regulator, the AEC sabotaged its own credibility. By 1973, the Commissioners themselves agreed that the AEC's statutory conflict of interest could be resolved only by assigning promotion and regulation to separate agencies. The Energy Reorganization Act of 1974 split the AEC into the Nuclear Regulatory Commission (NRC) and the Energy Research and Development Agency (ERDA). In 1977 ERDA assimilated the Federal Energy Administration to become the Department of Energy.[9]

As part of the legacy of the AEC (and the MED that preceded it), radioisotopes followed a strikingly different trajectory than that of nuclear power. Reactor-generated radioisotopes came into circulation a decade before civilian reactors, and continued to be mainstays of science and medicine long after Three Mile Island. Radioisotopes became a routine part of research and therapy prior to the public backlash against nuclear power, and perceptions of the dangers of radioactivity did not diminish their use. (Regulation is another matter, as addressed below.) Their use as molecular tags—particularly in widely used radiolabeled compounds—outlasted the interest in tracing biological systems that prompted their initial application decades earlier. That said, by the early twenty-first century, the presence of radioisotopes in research laboratories had dissipated considerably. This chapter considers the half-life of radioisotopes as scientific and medical tools, offering a brief appraisal of developments in selected fields since the demise of the AEC and some comments on the broader legacy of its radioisotope program.

Radioisotopes and Postwar Knowledge

In biochemistry, molecular biology, and many other areas of biomedical research, radioisotopes continued to be widely used through the end of

8. The trend on the part of utility companies to order nuclear power plants ended after 1973. Other factors besides public opposition, including the inflation (which raised the cost of large capital investments such as nuclear power plants), and frustration with delays in the AEC's licensing process, influenced the preference of utilities to build fossil fuel plants or cancel orders for plants altogether. Ibid., p. 409; Walker, *Three Mile Island* (2004), p. 8. On the political controversies over nuclear power, see also Del Sesto, *Science, Politics, and Controversy* (1979).

9. Walker, *Three Mile Island* (2004), p. 32.

the twentieth century. In a 2001 overview of the history of the experimental life sciences, Daniel Kevles and Gerald Geison refer to radioisotopes as having become "sine qua non in molecular biological research, serving as tags for fragments of DNA employed for purposes ranging from basic gene analysis to forensic genetic fingerprinting."[10] Most of the key techniques associated with genetic engineering, especially DNA sequencing and hybridization methods (e.g., Southern blots, northern blots), used phosphorus-32 or sulfur-35 to label nucleic acids.[11] The "recombinant revolution" of the 1970s and 1980s could not have happened, technologically speaking, without radioisotopes. As mentioned in chapter 8, iodine-125 became a key ingredient of radioimmunoassays, which were widely used in research and in diagnostics. Similarly, enzyme assays frequently depended on ^{14}C- or tritium-labeled compounds to measure activity.[12] For biologists using radiolabeled chemicals, New England Nuclear and Amersham continued to be the leading vendors. In these sorts of applications, radioisotopes were not so much tracers as tags; the label was followed not in vivo, but in vitro.[13]

Several factors have led to the decline since the 1990s in the use of radioisotopes in biochemistry and molecular biology. As scientists developed high-volume automated DNA sequencing machines, they employed fluorescent tags rather than radioisotopes.[14] Many other current methods also favor fluorescent labels, such as assays of gene expression on DNA chips. In the burgeoning field of cell biology, fluorescent proteins are similarly used to visually analyze patterns of gene expression, usually in vivid hues of red and green.[15] In conjunction with fluorescent dyes and probes, real-time polymerase chain reaction and real-time reverse-transcription polymerase chain reaction enable biologists to detect and quantify nucleic acids without radioisotopes.[16] Computer tools have further reordered experimental possibilities. Once DNA sequence databases became available, searches for sequence homologies began to replace nucleic acid hybridization

10. Kevles and Geison, "Experimental Life Sciences" (1995), p. 101.

11. For accounts of the development of some of these techniques: Giacomoni, "Origin of DNA:RNA Hybridization" (1993); García-Sancho, "New Insight" (2010); McElheny, *Drawing the Map of Life* (2010).

12. Rheinberger, "Putting Isotopes to Work" (2001).

13. For a reflection on this shift, see Rheinberger, "In Vitro" (2006).

14. Chow-White and García-Sancho, "Bidirectional Shaping" (2012).

15. Chudakov et al., "Fluorescent Proteins" (2010).

16. VanGuilder, Vrana, and Freeman, "Twenty-Five Years" (2008).

experiments, enabling researchers to identify homologous sequences and compare genomes "in silico" rather than in vitro.[17]

In part, the large-scale abandonment of radioisotopes in laboratories reflects the burden of regulation for licensed radioisotope users and especially the disposal of radioactive waste. But one could also point to epistemological changes. In a postgenomic landscape, the earlier emphasis on tracing single biochemical changes in cells and organisms, and on the role of individual genes in determining biological traits, has given way to a systems approach in biology attuned to networks of molecular interactions and the role of epigenetics.[18] Yet the general picture of life at the molecular level established during the second half of the twentieth century—including the understanding of gene replication, transcription, and translation, and the web of metabolic pathways that an organism regulates in self-maintenance—remains foundational, and is profoundly indebted to experimental methods using radiolabels. In this sense the epistemological footprint of radioisotopes persists, even if they have become a less routine aspect of laboratory life.

In contrast to experimental biology, clinical applications of radioisotopes continue to be numerous and high-volume. Several isotopes still play a part in medical diagnosis and therapy, including iodine-131 and thallium-201, but most prominent is technetium-99m. First used in biology in the early 1960s, technetium-99m has become the dominant medical radioisotope. Because of the high image quality that can be obtained from its emission of gamma radiation, its brief half-life, and the low radiation dose experienced by patients, technetium-99m is employed in a wide variety of diagnostic tests, including cardiac imaging, staging of cancer, thyroid scanning, and detection of bone metastases. By 2010, technetium-99m was used in more than 16 million nuclear imaging procedures yearly in the United States alone.[19]

Technetium-99m is produced on-site in generators supplied with another radioisotope, molybdenum-99. Because molybdenum-99 itself has a short half-life of only sixty-six hours, it cannot be stockpiled. By the early twenty-first century, the global supply of molybdenum-99 came from only five reactors in the world, none of which are located in the United States. The Canadian National Research Universal reactor supplies most of the molybdenum-99 for North America. However, this 51-year-old reactor

17. Auch et al., "Digital DNA-DNA Hybridization" (2010).
18. Kafatos, "Revolutionary Landscape" (2002).
19. Smith, "Looming Isotope Shortage" (2010).

was shut down for an indefinite time due to safety problems on May 14, 2009.[20] This left reactors in Belgium, France, the Netherlands, and South Africa to pick up the slack, and in February 2010 one of these, the Petten reactor in the Netherlands, was shut down for six weeks of maintenance.[21] This precipitated a worldwide shortage of molybdenum-99, prompting clinicians to reduce dosage levels and turn to alternative imaging methods.[22] The Canadian reactor came back on line in August 2010, relieving the immediate crisis, but the question remains whether a handful of half-century-old reactors can be relied upon to supply this critical medical isotope.

Ecological experiments with radioisotopes in aquatic or terrestrial ecosystems gave rise to many important concepts and methods in radioecology, such as determining the pathways of movement of specific radioelements as well as their persistence, accumulation, and possible bioconcentration in food webs.[23] By the 1970s, Oak Ridge National Laboratory and Argonne National Laboratory had expanded the purview of their work in radiation ecology to establish broader programs of environmental research.[24] In part, this more general framework reflected the recognition that nuclear energy introduced ecological problems beyond radioactivity, such as thermal pollution. In addition, a federal court in 1971 ordered a major revision of AEC environmental impact statements for reactor licensing, which put new demands on the Commission's ecologists.[25] For environmental research, the legacy of radioisotopes has to do with the preeminence of an ecosystems approach and the template they provided for understanding environmental contamination generally. As Brookhaven scientist George Woodwell put it in 1970, "The studies of radiation, because of their specificity, provide useful clues for examination of effects of other types of pollution for which evidence is much more fragmentary."[26]

20. Wald, "Radioactive Drug for Tests" (2009).

21. Smith, "Looming Isotope Shortage" (2010).

22. These include replacing technetium-99m with thallium-201, which exposes patients to slightly more radiation and reduces the image quality, and using other imaging methods, such as CAT scans (which still involve radiation exposure), MRI, and ultrasound. Ibid.

23. Whicker and Pinder, "Food Chains" (2002).

24. Westwick, *National Labs* (2003), pp. 286–91; Schloegel and Rader, *Ecology, Environment and 'Big Science'* (2005).

25. Johnson and Schaffer, *Oak Ridge National Laboratory* (1994), pp. 143–45.

26. Woodwell, "Effects of Pollution" (1970), p. 429. It should be noted that radioecology as a specialization continues; see Shaw, "Applying Radioecology" (2005).

Radioisotopes and Regulation

The effects of the AEC's radioisotope supply program can be seen in another context, beyond changes to scientific knowledge and medical practice. Key aspects of the relationship of scientists to the federal government can also be delineated using radioisotopes as historical tracers. Beginning in 1946, the US government (or, strictly speaking, its Oak Ridge contractor) became a vendor of laboratory supplies.[27] But to purchase radioisotopes, researchers and clinicians had to apply for authorization and agree to abide by the AEC's regulations for safe handling and disposal of radioactive waste. (Foreign scientists had additional requirements and obligations, as discussed in chapter 4.) For many American scientists, radioisotope usage may have been the first area in which federal government regulation affected their day-to-day work. Of course, at just this time the imposition of loyalty oaths and security screening also affected the livelihood of many scientists, notably those working for the AEC.[28] The Commission's stipulation of authorization and safety procedures for radioisotope users, in comparison, impinged on the lives of researchers in a more quotidian (and less dramatic) way, from applying for a license to wearing a film badge at work.

The 1946 Atomic Energy Act established a governmental atomic monopoly, in which there would be no private ownership of fissionable materials. Purchasers of radioisotopes from Oak Ridge were expected to abide by the same regulations, including limits on radiation exposure levels, as workers in the AEC's own facilities.[29] But whereas the Commission's contractors could monitor occupational radiation exposure for employees at

27. The AEC explained that it did not sell isotopes directly to purchasers, but through its contractors. ("The Commission makes no direct sales. Sales and distribution service is provided to the public and secondary suppliers through contractor organizations." Isotope Distribution Report by the Director of Research and Chief, Isotopes Division, Oak Ridge Operations, 11 Oct 1951, NARA College Park, RG 326, E67B, box 28, folder 5 Isotope Program Distribution vol. 1.) However, the notion that this was a private transaction is belied by the agency's own summary reports on shipments and their uses.

28. On national security issues as they affected scientists, see Wang, *American Science* (1999).

29. Isotopes Branch Circular No. B-1, General Rules and Procedures Concerning Radioactive Hazards (Excerpts from Clinton Laboratories Regulations), 8 Jan 1947, Evans papers, box 1, folder 1 Isotopes–Clinton Lab. See chapter 6.

its plants and laboratories, regulating radioisotope users all over the country (not to mention the world) was another matter. In general, the AEC trusted universities and hospitals to provide oversight over their own scientists and clinicians. As the agency's first General Manager summarized the Commission's responsibilities, "It cannot police twenty-four hours a day everybody who uses isotopes. It shouldn't. On the other hand, it is necessary to take certain measures."[30]

For radioisotope experiments involving humans, the AEC put in place an extra review committee (the "Subcommittee on Human Application") to approve or deny applications, but delegated oversight to the institutions where research was conducted. In human subjects research, the AEC set important regulatory precedents, including the first use of the term "informed consent" in a document, even as the agency did not impose the guidelines on civilian purchasers of its radioisotopes.[31] These regulations were aimed principally at the AEC's own researchers, particularly those medical researchers contracted to conduct experiments on the biological effects of fission products and radiation. It is unclear how effectively the Commission enforced these guidelines even among its own contractors, but in any case it is ironic that the most conscientious government oversight of human subjects existed for secret research oriented to military issues.

There was some ambiguity in the early years of the isotope program about whether the AEC's regulations applied to radiomaterials that were not obtained from Oak Ridge. According to the Isotopes Division, the AEC's regulations for radioisotope usage did not extend to cyclotron-produced isotopes, unless they were produced in Commission facilities.[32] However, Shields Warren, head of the Division of Biology and Medicine, asserted that the AEC had authority over human uses of radioisotopes,

30. Transcript of the Discussion at the First Meeting of the Medical Review Board, AEC, Washington, DC, 16 Jun 1947, OpenNet Acc NV0709599, pp. 18–19. This is part of a longer statement by Wilson quoted on p. 205.

31. ACHRE, *Final Report* (1996), pp. 49–50. See also Carroll L. Wilson to Stafford L. Warren, 30 Apr 1947, reproduced in ACHRE, *Final Report, Supp. Vol. 1* (1995), pp. 71–72 and, for further discussion of the AEC's regulation of human applications of radioisotopes, chapter 8.

32. "Regulations for the Distribution of Radioisotopes," AEC 398, 22 Jan 1950, Report by the Division on Research and Isotopes Division, AEC Records, National Archives College Park, RG 326, E67A, box 45, folder 1 Regulations for the Distribution of Radioisotopes.

irrespective of their provenance.[33] The issue of the limits of the AEC's authority over radioisotope usage came up in a meeting of the Commissioners in January 1950 when they were discussing expanding the foreign distribution program. The Director of the Division of Research, Kenneth Pitzer, held the opinion that the oversight of radioisotope users should be taken over by a properly regulatory agency, such as the Public Health Service. Commissioner Sumner Pike responded that "the proper role for the Commission appeared to be to furnish advice and aid rather than to regulate or supervise closely the use of isotopes."[34] Thus ambivalence about the AEC's regulatory responsibilities for scientists outside its facilities was voiced at the highest echelons of the organization.

Many of the early purchasers of Oak Ridge radioisotopes had previous experience with cyclotron-produced materials and were not accustomed to any safety-related supervision. In late 1946, Joseph Hamilton, who had been working with radioisotopes from the outset of their availability from cyclotrons, resisted the AEC's requirement that institutions form an isotope review committee.[35] One suspects other researchers simply did not comply with AEC rules and regulations for radioisotopes; enforcement during the early years was lax. Of course, it must also be noted that the same set of scientists who had worked with radioactive materials before and during the war populated the postwar NCRP panels that set the maximum permissible dose levels and the AEC advisory groups that helped formulate government regulation. For example, the NCRP Subcommittee on the Handling of Radioactive Isotopes and Fission Products was chaired by Herbert M. Parker, Hanford's chief health physicist (chapter 10), and included Paul Aebersold, head of the Isotopes Division (chapters 3 and 6),

33. This was in 1949. Jones and Martensen, "Human Radiation Experiments" (2003), p. 99 and p. 108n104.

34. Notes from the 354th AEC Meeting, 18 Jan 1950, NARA College Park, RG 326, box 47, folder 1 Foreign Distribution of Radioisotopes vol. 3.

35. He wrote Paul Aebersold on the grounds that the relevant University of California faculty at Berkeley and San Francisco were so experienced in using isotopes that they were unlikely to request "substances from your organization which are ill-considered, and under the circumstances, which might lead to potentially serious health hazards and problems." He also said that the more experienced senior faculty would "resent the formation of a committee within the institution that would act as a board of review for their research problems." Joseph G. Hamilton to Paul G. Aebersold, 30 Aug 1946, EOL papers, series 1, reel 14, folder 9:26 Isotope Research. For more on this issue, see Jones and Martensen, "Human Radiation Experiments" (2003), p. 99.

Gioacchino Failla, radiotherapist and member of the ACBM (chapter 9), and Hamilton, who conducted early experiments with sodium-24 and worked on the metabolism of fission products during and after the war (chapter 2).[36] Thus there was a recursive quality to radioisotope regulations—scientists were not simply objects of government surveillance, but also the subjects determining what comprised radiological safety in the first place.

The ranks of radioisotope users grew quickly in the postwar years. By 1953, there were estimated to be 7,500 users of radioisotopes in the United States. According to one report, 50%–75% of these individuals received less than 0.05 roentgen of radiation exposure per week, well below the permissible dose of 0.3 rem per week (a rem being equivalent to a roentgen for x-ray and gamma radiation). Only one in three hundred radioisotope users was thought to exceed the maximum permissible dose limit in any given week.[37] Clinical uses of radiation sources involved much more extensive occupational exposure than radioisotope usage; 215,000 medical-technical personnel were potentially exposed to radiation, mostly from x-ray machines. An "appreciable fraction" of radiologists were exposed to more than 0.1 roentgen per day; for patients, an x-ray examination involved an average exposure of 11 roentgens (though usually focused on a small portion of the body).[38] By comparison, Americans were estimated to have a lifetime radiation exposure of 9 roentgens from natural sources.[39] Notably, medical institutions and professionals regulated themselves in the use of radium and x-ray machines, though the NCRP set the recommended standards.[40]

The 1954 Atomic Energy Act expanded the Commission's statutory requirements for radiological protection, in anticipation of the greater number of civilians who would be occupationally exposed to radiation in the nuclear industry. The users of radioisotopes from Oak Ridge would no longer be regulatory outliers, as workers beyond AEC facilities subject to its radiological protection standards. The AEC struggled to manage

36. The other four members were L. F. Curtiss, J. E. Rose, L. Marinelli, and M. M. D. Williams. National Committee on Radiation Protection, *Safe Handling of Radioactive Isotopes* (1949), p. iv.

37. Moeller et al., "Radiation Exposure" (1953), p. 60.

38. Ibid., pp. 57–59.

39. Ibid., p. 57.

40. Walker, *Permissible Dose* (2000), ch. 1.

its expanded promotional and regulatory activities, partitioning the latter into a separate division of the agency. A new Division of Inspection monitored the facilities of civilian licensees for regulatory compliance, though the head of the Isotopes Division fought the transfer into this division of regulatory oversight of radioisotope users.[41] In 1957, the AEC reorganized these responsibilities again, creating a Division of Licensing and Regulation that assimilated radioisotope users as well as civilian licensees of reactors. In 1959, the Advisory Committee for Biology and Medicine considered the AEC's radiological health responsibilities, and expressed the view that, outside of the Commission's own facilities, "surveillance and control" of the use of radioisotopes in research, medicine, and industry should be delegated to another agency.[42] But this did not happen for another fifteen years.

Throughout the 1960s, the AEC increased its resources for handling licenses and regulatory matters, though by the end of that decade its staff could not keep up with the high volume of applications for nuclear reactor construction (which required extensive evaluation on safety and siting grounds).[43] A 1959 amendment to the 1954 Atomic Energy Act allowed states to exercise jurisdiction over radiological protection, though federal radiation standards had to be enforced by participating states. This resulted in a complex system of regulation, in which about half of the US states handled license applications for "byproduct materials." As of December 1966, there were 8,636 licenses for radioactive materials; 4,732 were in agreement states. Physicians and hospitals held 2,994 licenses; industrial firms, 2,992; and federal and state laboratories, 1,840. Colleges and universities held only 672 licenses, though a single institutional license could cover a large number of users.[44]

By the early 1970s, the AEC was politically embattled on account of its regulatory performance. Its perceived lapses included oversight of ra-

41. Mazuzan and Walker, *Controlling the Atom* (1985), pp. 55–56; see chapter 5.

42. Minutes, 74th ACBM Meeting, 26–27 April 1959, Washington, DC, OpenNet Acc NV0710303, p. 3. The ACBM felt that another agency, perhaps the Public Health Service or Food and Drug Administration, should also handle oversight of environmental contamination (including food, air, and water) from nuclear weapons testing and other uses of atomic energy.

43. Walker, *Three Mile Island* (2004), p. 41.

44. Background Information, Commission Meeting with Radioisotopes Licensing Review Panel, 6 Dec 1966, Nuclear Regulatory Commission papers, NARA College Park, RG 431, entry 16, box 12698/12, folder Organization and Management 7 Radioisotopes Licensing Review Panel, 28 Jan 1966 to 24 May 1968.

dioisotope users. Inspections of radioisotope licensees frequently turned up violations. In trying to explain the relatively lax record of compliance to the Comptroller General of the United States, the Commission's Director of Regulations, Manning Muntzing, explained: "About two-thirds of AEC's inspections disclose no violations, and in most of the cases where noncompliance is identified, appropriate corrective action is taken by licensees in response to written notices."[45] He went on to explain that when noncompliance was found, so long as it did not threaten the health or safety of the public or employees, his office sought to obtain "corrective action rather than revoke a licensee's authority to use radioactive materials, which in many cases could deprive the public of an essential service."[46] Nonetheless, Muntzing promised his division would do a better job, especially by using their new ability to fine violators:

> We agree, however, that certain licensees must be provided greater incentive to comply with regulatory requirements than in the past. We intend to accomplish this through a more rigorous enforcement program utilizing all available enforcement sanctions to the extent necessary to achieve this objective. We believe that the recently acquired authority to impose civil monetary penalties, which we have already begun to implement, will provide a necessary incentive.[47]

This letter reflects a changed social climate concerning the trustworthiness of institutions and professionals to regulate themselves. Just as bioethics committees in hospitals were replacing the discretion of physicians, representatives from the AEC (and soon the NRC) were expected to ensure that scientists and medical personnel did not endanger themselves or the public by handling or disposing of radioisotopes recklessly.[48]

The last decades of the twentieth century witnessed a rapid expansion in the regulation of scientists, with the establishment of review boards

45. L. Manning Muntzing, Director of Regulation, to Elmer B. Staats, Comptroller General of the United States, US Government Accounting Office, 23 May 1972, Nuclear Regulatory Commission papers, NARA College Park, RG 431, entry 16, box 12706/34, folder Industrial Development & Regulations 14 Part 20, Vol. 1, 1 Jul 1970 through 30 Jun 1972. Violations could arise not only from radiation exposures to personnel above the permissible limits, but also from evidence of eating in laboratories where radioisotopes were used, insufficient ventilation, or radioactive contamination of equipment or benches above an allowable level.

46. Ibid.

47. Ibid.

48. Rothman, *Strangers at the Bedside* (1991).

for applications of recombinant DNA as well as animal and human subjects, guidelines on scientific misconduct, and control over certain research materials, such as human stem cell lines.[49] The earlier regulation of radioisotope users illuminates the gradual, even hesitant, emergence of governmental oversight of scientific research. Unlike other fields in which new rules or forms of surveillance were introduced in the 1960s through the 1990s, federal regulations existed for radioisotope users since 1946. Yet at that time, neither AEC officials nor influential scientists viewed radiological protection in research institutions as appropriately the agency's responsibility. The debates over fallout from atomic weapons testing undermined public trust of the government to protect citizens from radioactive contamination at just the time that the AEC sought to both promote and regulate a civilian nuclear power industry. These developments not only increased governmental control over radioisotope users, at both the federal and state levels, but also contributed to a new politics of oversight, in which the government was expected to regulate research in order to protect human subjects and the public.

* * *

Through radioisotopes we can see traces of the Manhattan Project across the postwar life sciences and medicine: in the bodies of patients and healthy human subjects, in experimental bacteria, plants, and animals, in the schematic diagrams and theories that represented life in molecular terms, and in pristine and contaminated Cold War landscapes. The radioisotopes and radiation sources present in hospitals and laboratories were intended to represent the peaceful benefits of nuclear knowledge, even as they also provided the AEC and the military with information about their potential to destroy or shorten human life. With a few short decades, the symbolism of atomic humanitarianism had vanished, as production of radioisotopes shifted into the private sector and Americans began to see radioactivity as a threat to their health. Yet even then, radioisotopes remained crucial tools in biology and medicine.

To speak of the half-lives of radioisotopes is to emphasize the complex temporalities of their production, adoption, circulation, and decline. On the one hand, the temporalities were military and political. The conditions for the government distribution of radioisotopes arose from the aftermath

49. Schrag, *Ethical Imperialism* (2010); Stark, *Behind Closed Doors* (2012); Benson, "Difficult Time" (2011).

of the atomic bomb detonations over Hiroshima and Nagasaki, and the establishment of a civilian agency to produce nuclear weapons while promoting nonmilitary uses of atomic energy. Yet the postwar period was not a transition from wartime uses to peacetime uses of atomic energy, but an escalation of both. Through the 1950s, nuclear test detonations put ever-greater amounts of radioactivity into the air, ground, and water even as the total curies of radioisotopes shipped by the federal government increased. At the same time, other trends, not least the growth of a regulatory state, countered the escalating release of radioactivity while also affecting government sales of radioisotopes. Above all, mounting concerns about radiation's dangers made the AEC begin to attend more vigorously to radiological safeguards and environmental remediation. Radioisotope users, in turn, contended with greater government surveillance of their purchases and practices.

On the other hand, the time-scales in play were also epistemological and technological. Radioisotopes offered scientists in many fields the ability to track molecular change through time, bringing temporality itself to the fore as a frontier of biomedical and environmental knowledge. Researchers used isotopes to determine the sequence of reactions in metabolism, the flow of materials and energy through ecosystems, and the activity or turnover of key molecules and markers in the body—from glucose to insulin. But in the late twentieth century, radiotracers began being displaced by other tools and methods, after decades of unrivaled utility. The considerations that have led to the abandonment of radioisotopes are not merely technical. It is hard to imagine an ecologist deliberately releasing a radioactive isotope into a lake anymore, though they continue to use stable isotopes as tracers.[50] Moreover, if radioisotopes can be said to have a half-life of utility, this is not because the technique has inevitably "decayed"—for in fields such as nuclear medicine no such attenuation has taken place. In this respect, the trajectory of radioisotopes as tools in biology and medicine reveals the complex interplay of material possibilities, social and political realities, and public perceptions that all shape patterns of use and consumption. In that sense, to follow radioisotopes is not only to see how their availability as by-products of the atomic age enabled a new sensitivity to temporal processes in the living world, but also to glimpse the evanescent materiality that makes scientific knowledge possible in the first place.

50. Bugalho et al., "Stable Isotopes" (2008).

Bibliography

Åberg, Bertil, and Frank P. Hungate, eds. *Radioecological Concentration Processes: Proceedings of an International Symposium Held in Stockholm, 25–29 April 1966*. Oxford: Pergamon Press, 1967.

Abir-Am, Pnina G. "The Discourse of Physical Power and Biological Knowledge in the 1930s: A Reappraisal of the Rockefeller Foundation's 'Policy' in Molecular Biology." *Social Studies of Science* 12 (1982): 341–82.

———. "The Politics of Macromolecules: Molecular Biologists, Biochemists, and Rhetoric." *Osiris* 7 (1992): 164–91.

Abraham, Itty. *The Making of the Indian Atomic Bomb: Science, Secrecy and the Postcolonial State*. London: Zed Books, 1998.

ACHRE. *See* Advisory Committee on Human Radiation Experiments.

Adamson, Matthew. "Cores of Production: Reactors and Radioisotopes in France." *Dynamis* 29 (2009): 261–84.

Adelstein, S. James. "Robley Evans and What Physics Can Do for Medicine." *Cancer Biotherapy & Radiopharmaceuticals* 16 (2001): 179–85.

Advisory Committee on Human Radiation Experiments (ACHRE). *The Human Radiation Experiments: Final Report of the President's Advisory Committee, Supplemental Volume 1: Ancillary Materials*. Washington, DC: US Government Printing Office, 1995.

———. *The Human Radiation Experiments: Final Report of the President's Advisory Committee, Supplemental Volume 2a: Sources and Documentation—Appendices*. Washington, DC: US Government Printing Office, 1995

———. *The Human Radiation Experiments: Final Report of the President's Advisory Committee*. New York: Oxford University Press, 1996.

Aebersold, Paul C. "The Isotope Distribution Program." *Science* 106 (1947): 175–79.

———. "Isotopes and Their Application to Peacetime Use of Atomic Energy." *Bulletin of the Atomic Scientists* 4, no. 5 (1948): 151–54.

———. "Isotopes for Medicine." *Journal of the American Medical Association* 138 (1948): 1222–25.

———. "Philosophy and Policies of the AEC Control of Radioisotope Distribution." In Andrews, Brucer, and Anderson, *Radioisotopes in Medicine*, 1–11.

———. "Progress against Cancer with Radioisotopes." *Journal of the American Geriatrics Society* 3 (1955): 772–90.

———. "The Development of Nuclear Medicine." *American Journal of Roentgenology, Radium Therapy, and Nuclear Medicine* 75 (1956): 1027–39.

AEC. *See* US Atomic Energy Commission.

"A.E.C. Adds to Funds for Disease Studies." *New York Times*, 11 Dec 1956, p. 41.

"AEC Announces Changes in Prices of 60 Radioisotopes, Irradiation Services," *Isotopes and Radiation Technology* 2, no. 3 (1965): 317.

"AEC Ends Routine ^{85}Sr Production," *Isotopes and Radiation Technology* 2, no. 1 (1965): 93.

"AEC Withdraws from Routine Production of Radioiodine," *Isotopes and Radiation Technology* 1, no. 2 (1963–64): 203–4.

Alonso, Marcelo. "The Impact in Latin America." In *Atoms for Peace: An Analysis after Thirty Years*, ed. Joseph F. Pilat, Robert E. Pendley, and Charles K. Ebinger. Boulder, CO: Westview Press, 1985, 83–90.

Altman, Lawrence K. "Clement A. Finch, a Pioneer in Hematology, Dies at 94." *New York Times*, 6 Jul 2010, p. 24.

American Cancer Society. "Foreword." In *The Research Attack on Cancer, 1946: A Report on the American Cancer Society Research Program by the Committee on Growth of the National Research Council*. Washington, DC: National Research Council, 1946.

Anderson, Thomas F. "Techniques for the Preservation of Three-Dimensional Structure in Preparing Specimens for the Electron Microscope." *Transactions of the New York Academy of Sciences* 13 (1951): 130–34.

Anderson, Warwick. "The Possession of Kuru: Medical Science and Biocolonial Exchange." *Comparative Studies in Society and History* 42 (2000): 713–44.

Andrews, Gould A. "Treatment of Pleural Effusion with Radioactive Colloids." In Hahn, *Therapeutic Use of Artificial Radioisotopes*, 295–317.

Andrews, Gould A.; Marshall Brucer; and Elizabeth B. Anderson, eds. *Radioisotopes in Medicine*. Washington, DC: US Government Printing Office, 1955.

Anker, Peder. *Imperial Ecology: Environmental Order in the British Empire, 1895–1945*. Cambridge, MA: Harvard University Press, 2001.

———. "The Ecological Colonization of Space." *Environmental History* 10 (2005): 239–68.

Annas, George J., and Michael A. Grodin, eds. *The Nazi Doctors and the Nuremberg Code: Human Rights in Human Experimentation*. New York: Oxford University Press, 1992.

Appel, Toby A. *Shaping Biology: The National Science Foundation and American Biological Research, 1945–1975*. Baltimore: Johns Hopkins University Press, 2000.

Arnon, D. I.; P. R. Stout; and F. Sipos. "Radioactive Phosphorus as an Indicator of Phosphorus Absorption of Tomato Fruits at Various Stages of Development." *American Journal of Botany* 27 (1940): 791–98.

Aronoff, S.; H. A. Barker; and M. Calvin. "Distribution of Labeled Carbon in Sugar from Barley." *Journal of Biological Chemistry* 169 (1947): 459–60.

Aronoff, S.; A. Benson; W. Z. Hassid; and M. Calvin. "Distribution of C^{14} in Photosynthesizing Barley Seedlings." *Science* 105 (1947): 664–65.

Artom, C.; G. Sarzana; C. Perrier; M. Santangelo; and E. Segrè. "Rate of 'Organification' of Phosphorus in Animal Tissues." *Nature* (1937): 836–37.

Ashmore, James; Manfred L. Karnovsky; and A. Baird Hastings. "Intermediary Metabolism." In Claus, *Radiation Biology and Medicine*, 738–79.

Asimov, Isaac, and Theodosius Dobzhansky. *The Genetics Effects of Radiation*. Oak Ridge, TN: US AEC Division of Technical Information, 1966.

Aten, A. H. W., and George Hevesy. "Formation of Milk." *Nature* 142 (1938): 111–12.

"The Atom at Work." *Time*, 7 Mar 1955, p. 91.

"The Atomic Energy Act of 1946." *Bulletin of the Atomic Scientists* 2, no. 3–4 (1946): 18–25.

"Atomic Research May End World's Hunger." *Christian Century* 65 (28 Jul 1948): 749–50.

Auch, Alexander F.; Mathias von Jan; Hans-Peter Klenk; and Markus Göker. "Digital DNA-DNA Hybridization for Microbial Species Delineation by Means of Genome-to-Genome Sequence Comparison." *Standards in Genomic Sciences* 2 (2010): 117–34.

Auerbach, S. I. "The Soil Ecosystem and Radioactive Waste Disposal to the Ground." *Ecology* 39 (1958): 522–29.

———"Radionuclide Cycling: Current Status and Future Needs." *Health Physics* 11 (1965): 1355–61.

———. *A History of the Environmental Sciences Division of Oak Ridge National Laboratory*, ORNL/M-2732. Oak Ridge, TN: Oak Ridge National Laboratory Publication No. 4066, 1993.

Auerbach, S. I., and D. A. Crossley. "Strontium-90 and Cesium-137 Uptake by Vegetation under Natural Conditions." *Proceedings of the Second U.N. International Conference on Peaceful Uses of Atomic Energy, Held in Geneva 1 September–13 August 1958*, vol. 18. New York: United Nations, 1958, 494–99.

Auerbach, S. I.; J. S. Olson; and H. D. Waller. "Landscape Investigations Using Caesium-137." *Nature* 201 (1964): 761–64.

"Availability of Radioactive Isotopes." *Science* 103 (1946): 697–705.

Axelrod, Dorothy; Paul C. Aebersold; and John H. Lawrence. "Comparative

Effects of Neutrons and X-Rays on Three Tumors Irradiated *in vitro*." *Proceedings of the Society for Experimental Biology and Medicine* 48 (1941): 251–56.

Badash, Lawrence. *Radioactivity in America: Growth and Decay of a Science*. Baltimore: Johns Hopkins University Press, 1979.

Ball, S. J. "Military Nuclear Relations between the United States and Great Britain under the Terms of the McMahon Act, 1946–1958." *Historical Journal* 38 (1995): 439–54.

Balfour, W. M.; P. F. Hahn; W. F. Bale; W. T. Pommerenke; and G. H. Whipple. "Radioactive Iron Absorption in Clinical Conditions: Normal, Pregnancy, Anemia, and Hemochromatosis." *Journal of Experimental Medicine* 76 (1942): 15–30.

Balogh, Brian. *Chain Reaction: Expert Debate & Public Participation in American Commercial Nuclear Power, 1945–1975*. Cambridge: Cambridge University Press, 1991.

Barker, Crispin R. C. *From Atom Bomb to the "Genetic Time Bomb": Telomeres, Aging, and Cancer in the Era of Molecular Biology*. PhD diss., Yale University, 2008.

Barker, H. A., and M. D. Kamen. "Carbon Dioxide Utilization in the Synthesis of Acetic Acid by *Clostridium thermoaceticum*." *Proceedings of the National Academy of Sciences, USA* 31 (1945): 219–25.

Barker, H. A.; M. D. Kamen; and B. T. Bornstein. "The Synthesis of Butyric and Caproic Acids from Ethanol and Acetic Acid by *Clostridium kluyveri*." *Proceedings of the National Academy of Sciences, USA* 31 (1945): 373–81.

Barker, H. A.; M. D. Kamen; and V. Haas. "Carbon Dioxide Utilization in the Synthesis of Acetic and Butyric Acids by *Butyribacterium rettgeri*." *Proceedings of the National Academy of Sciences, USA* 31 (1945): 355–60.

Bassham, James A. "Mapping the Carbon Reduction Cycle: A Personal Retrospective." *Photosynthesis Research* 76 (2003): 35–52.

Beatty, John. "Weighing the Risks: Stalemate in the Classical/Balance Controversy." *Journal of the History of Biology* 20 (1987): 289–319.

———. "Genetics in the Atomic Age: The Atomic Bomb Casualty Commission, 1947–1956." In *The Expansion of American Biology*, ed. Keith R. Benson, Jane Maienschein, and Ronald Rainger. New Brunswick, NJ: Rutgers University Press, 1991, 284–324.

———. "Genetics and the State." Unpublished paper presented at Penn-Princeton Workshop, 27 Feb 1999.

———. "Masking Disagreement among Experts." *Episteme* 3 (2006): 52–67.

Beck, Ulrich. *Risk Society: Towards a New Modernity*. Translated by Mark Ritter. London: Sage Publications, 1992. Originally published as *Risikogesellschaft: Auf dem Weg in eine andere Moderne*. Frankfurt am Main: Suhrkamp Verlag, 1986.

Becker, C. D. *Aquatic Bioenvironmental Studies: The Hanford Experience 1944–84*. Amsterdam: Elsevier, 1990.

Bennett, Leslie L. "I. L. Chaikoff, Biochemical Physiologist, and His Students." *Perspectives in Biology and Medicine* 30 (1987): 362–83.

Benson, Andrew A. "Identification of Ribulose in $C^{14}O_2$ Photosynthetic Products." *Journal of the American Chemical Society* 73 (1951): 2971–72.

———. "Following the Path of Carbon in Photosynthesis: A Personal Story." *Photosynthesis Research* 73 (2002): 29–49.

———. "Paving the Path." *Annual Review of Plant Biology* 53 (2002): 1–25.

———. "Last Days in the Old Radiation Laboratory (ORL), Berkeley, California, 1954." *Photosynthesis Research* 105 (2010): 209–12.

Benson, A. A.; M. Calvin; V. A. Haas; S. Aronoff; A. G. Hall; J. A. Bassham; and J. W. Weigl. "C^{14} in Photosynthesis." In *Photosynthesis in Plants*, ed. James Franck and Walter E. Loomis. Ames: Iowa State College Press, 1949, 381–401.

Benson, Etienne. "A Difficult Time with the Permit Process." *Journal of the History of Biology* 44 (2011): 103–23.

Benzer, Seymour. "Resistance to Ultraviolet Light as an Index to the Reproduction of Bacteriophage." *Journal of Bacteriology* 63 (1952): 59–72.

Berlin, Nathaniel I. "Blood Volume: Methods and Results." In Andrews, Brucer, and Anderson, *Radioisotopes in Medicine*, 429–36.

Berman, Harold J., and John R. Garson. "United States Export Controls—Past, Present, and Future." *Columbia Law Review* 67 (1967): 791–890.

Bernard, S. R. "Maximum Permissible Amounts of Natural Uranium in the Body, Air and Drinking Water Based on Human Experimental Data." *Health Physics* 1 (1958): 288–305.

Bernstein, Barton J. "Oppenheimer and the Radioactive Poison Plan." *Technology Review* 88, no. 4 (1985): 14–17.

———. "Radiological Warfare: The Path Not Taken." *Bulletin of the Atomic Scientists* 41, no. 7 (1985): 44–49.

Berson, Solomon A., and Rosalyn S. Yalow. "The Use of K^{42} or P^{32} Labeled Erythrocytes and I^{131} Tagged Human Serum Albumin in Simultaneous Blood Volume Determinations." *Journal of Clinical Investigation* 31 (1952): 572–80.

———. "The Distribution of I^{131} Labeled Human Serum Albumin Introduced into Ascitic Fluid: Analysis of the Kinetics of a Three Compartment Catenary Transfer System in Man and Speculations on Possible Sites of Degradation." *Journal of Clinical Investigation* 33 (1954): 377–87.

———. "Radioimmunoassay of ACTH in Plasma." *Journal of Clinical Investigation* 47 (1968): 2725–51.

Berson, Solomon A.; Rosalyn S. Yalow; Abraham Azulay; Sidney Schreiber; and Bernard Roswit. "The Biological Decay Curve of P^{32} Tagged Erythrocytes: Application to the Study of Acute Changes in Blood Volume." *Journal of Clinical Investigation* 31 (1952): 581–91.

Berson, Solomon A.; Rosalyn S. Yalow; Arthur Bauman; Marcus A. Rothschild; and Katharina Newerly. "Insulin-I^{131} Metabolism in Human Subjects: Demonstration of Insulin Binding Globulin in the Circulation of Insulin Treated Subjects." *Journal of Clinical Investigation* 35 (1956): 170–90.

Berson, Solomon A.; Rosalyn S. Yalow; Sidney S. Schreiber; and Joseph Post. "Tracer Experiments with I^{131} Labeled Human Serum Albumin: Distribution and Degradation Studies." *Journal of Clinical Investigation* 32 (1953): 746–68.

Berson, Solomon A.; Roslyn S. Yalow; G. D. Aurbach; and John T. Potts Jr. "Immunoassay of Bovine and Human Parathyroid Hormone." *Proceedings of the National Academy of Sciences, USA* 49 (1963): 613–17.

Berson, Solomon A.; Rosalyn S. Yalow; Joseph Sorrentino; and Bernard Roswit. "The Determination of Thyroidal and Renal Plasma I^{131} Clearance Rates as a Routine Diagnostic Test of Thyroid Dysfunction." *Journal of Clinical Investigation* 31 (1952): 141–58.

Biagioli, Mario. *Galileo Courtier: The Practice of Science in the Culture of Absolutism*. Chicago: University of Chicago Press, 1993.

———. *Galileo's Instruments of Credit: Telescopes, Images, Secrecy*. Chicago: University of Chicago Press, 2006.

Bizzell, Oscar M. "Early History of Radioisotopes from Reactors." *Isotopes and Radiation Technology* 4, no. 1 (1966): 25–32.

Bland, P. Brooke; Leopold Goldstein; and Arthur First. "Secondary Anemia in Pregnancy and in Puerperium." *American Journal of the Medical Sciences* 179 (1930): 48–65.

Block, Elliott. "An Overview of Radioimmunoassay Testing and a Look at the Future." In Schönfeld, *New Developments in Immunoassays*, 1–9.

Blom, P. S.; A. Querido; and C. H. W. Leeksma. "Acute Leukaemia Following X-ray and Radioiodine Treatment of Thyroid Carcinoma." *British Journal of Radiology* 28 (1955): 165–66.

Blumgart, Herrmann L., and Otto C. Yens. "Studies on the Velocity of Blood Flow, I: The Method Utilized." *Journal of Clinical Investigation* 4 (1927): 1–13.

Blumgart, Herrmann L., and Soma Weiss. "Studies on the Velocity of Blood Flow, II: The Velocity of Blood Flow in Normal Resting Individuals, and a Critique of the Method Used." *Journal of Clinical Investigation* 4 (1927): 14–31.

Bocking, Stephen. "Ecosystems, Ecologists, and the Atom: Environmental Research at Oak Ridge National Laboratory." *Journal of the History of Biology* 28 (1995): 1–47.

———. *Ecologists and Environmental Politics: A History of Contemporary Ecology*. New Haven, CT: Yale University Press, 1997.

Boudia, Soraya. "Radioisotopes' 'Economy of Promises': On the Limits of Biomedicine in Public Legitimation of Nuclear Activities." *Dynamis* 29 (2009): 241–59.

———. *Gouverner les risques, gouverner par le risque: Pour une histoire du risque de la société du risque*. Habilitation à diriger des recherches, Université de Strasbourg, 2010.

Bowles, Mark D. *Science in Flux: NASA's Nuclear Program at Plum Brook Station, 1955–2005*. Washington, DC: NASA History Division, 2006.

Boyer, Paul S. *By the Bomb's Early Light: American Thought and Culture at the Dawn of the Atomic Age*. 1985. Reprint, Chapel Hill: University of North Carolina Press, 1994.

Bradley, David. *No Place to Hide*. Boston: Little, Brown, 1948.

Broda, Engelbert. *Radioactive Isotopes in Biochemistry*. Translated by Peter Oesper. Amsterdam: Elsevier Publishing Company, 1960. Originally published as *Radioaktive Isotope in der Biochemie*. Vienna: Verlag Franz Deuticke, 1958.

Broderick, Frank L. "A History of the Australia and New Zealand Association of Physicians in Nuclear Medicine." In *To Follow Knowledge: A History of Examinations, Continuing Education, and Specialist Affiliations of the Royal Australasian College of Physicians*, ed. Josephine C. Wiseman. Sydney, NSW, Australia: The College, 1988, 115–27.

Brownell, Gordon L., and William H. Sweet. "Localization of Brain Tumors with Positron Emitters." *Nucleonics* 11 (1953): 40–45.

———. "Scanning of Positron-Emitting Isotopes in Diagnosis of Intracranial and Other Lesions." *Acta Radiologica* 46 (1956): 425–34.

Brownell, Gordon L.; Brian W. Murray; William H. Sweet; Glyn R. Wellum; and Albert H. Soloway. "A Reassessment of Neutron Capture Therapy in the Treatment of Cerebral Gliomas." *Proceedings of the National Cancer Conference* 7 (1972): 827–37.

Brucer, Marshall. "Teletherapy Evaluation Board." *Radiology* 60 (1953): 738–39.

———. "Radioisotopes in Medicine Immediately After the Great 1946 Deliverance." *Isotopes and Radiation Technology* 4, no. 1 (1966): 59–62.

———. "Nuclear Medicine Begins with a Boa Constrictor." *Journal of Nuclear Medicine* 19 (1978): 581–98.

———. *A Chronology of Nuclear Medicine, 1600–1989*. St. Louis, MO: Heritage Publications, 1990.

Brues, Austin M. "Biological Hazards and Toxicity of Radioactive Isotopes." *Journal of Clinical Investigation* 28 (1949): 1286–96.

———. "Critique of the Linear Theory of Carcinogenesis." *Science* 128 (1958): 693–98.

Bruner, H. D., and Gould A. Andrews. "Cancer Research Program of Oak Ridge Institute of Nuclear Studies." *Southern Surgeon* 16 (Jun 1950): 577–83.

Buchanan, Nicholas. "The Atomic Meal: The Cold War and Irradiated Foods, 1945–1963." *History and Technology* 21 (2005): 221–49.

Bud, Robert F. "Strategy in American Cancer Research After World War II: A Case Study." *Social Studies of Science* 8 (1978): 425–59.

Bugalho, M. N.; P. Barcia; M. C. Caldeira; and J. O. Cerdeira. "Stable Isotopes as Ecological Tracers: An Efficient Method for Assessing The Contribution of Multiple Sources to Mixtures." *Biogeosciences* 5 (2008): 1351–59.

Burchard, John. *Q.E.D.: M.I.T. in World War II.* New York: John Wiley & Sons, 1948.

"Business in Isotopes." *Fortune* (Dec 1947): 121–25, 150, 153–54, 156, 158, 160, 163–64.

Calabrese, Edward J. "Key Studies Used to Support Cancer Risk Assessment Questioned." *Environmental and Molecular Mutagenesis* 52 (2011): 595–626.

Calvin, Melvin. "Intermediates in the Photosynthetic Cycle: A Commentary by Melvin Calvin." *Biochimica et Biophysica Acta* 1000 (1989): 403–7.

———. *Following the Trail of Light: A Scientific Odyssey*. Washington, DC: American Chemical Society, 1992.

Cambrosio, Alberto, and Peter Keating. *Exquisite Specificity: The Monoclonal Antibody Revolution*. New York: Oxford University Press, 1995.

Campos, Luis A. *Radium and the Secret of Life*. PhD diss., Harvard University, 2006.

Cantor, David, ed. *Cancer in the Twentieth Century*. Baltimore: Johns Hopkins University Press, 2008.

Carson, Rachel. *Silent Spring*. Boston: Houghton Mifflin, 1962.

Caufield, Catherine. *Multiple Exposures: Chronicles of the Radiation Age*. Chicago: University of Chicago Press, 1990.

"Chairman Strauss's Statement on Pacific Tests," *Bulletin of the Atomic Scientists* 10, no. 5 (1954): 163–65.

Charlton, John C. "Overcoming the Radiological and Legislative Obstacles in Radioimmunoassay." In Schönfeld, *New Developments in Immunoassays*, 27–37.

Chievitz, O., and G. Hevesy. "Radioactive Indicators in the Study of Phosphorus Metabolism in Rats." *Nature* 136 (1935): 754–55.

———. "Studies on the Metabolism of Phosphorus in Animals." *Kgl. Danske Videnskabernes Selskab. Biologiske Meddelelser* 13, no. 9 (1937): 63–78.

Childs, Herbert. *An American Genius: The Life of Ernest Orlando Lawrence, Father of the Cyclotron*. New York: E. P. Dutton, 1968.

Chodos, Robert B., and Joseph F. Ross. "The Use of Radioactive Phosphorus in the Therapy of Leukemia, Polycythemia Vera, and Lymphomas: A Report of 10 Years' Experience." *Annals of Internal Medicine* 48 (1958): 956–77.

Chow-White, Peter A., and Miguel García-Sancho. "Bidirectional Shaping and Spaces of Convergence: Interactions between Biology and Computing from the First DNA Sequencers to Global Genome Databases." *Science, Technology & Human Values* 37, no. 1 (2012): 124–64.

Chudakov, Dmitriy M.; Mikhail V. Matz; Sergey Lukyanov; and Konstantin A. Lukyanov. "Fluorescent Proteins and Their Applications in Imaging Living Cells and Tissues." *Physiological Reviews* 90 (2010): 1103–63.

Clark, Claudia. *Radium Girls: Women and Industrial Health Reform, 1910–1935*. Chapel Hill: University of North Carolina Press, 1997.

Clarke, Adele E., and Joan H. Fujimura, eds. *The Right Tools for the Job: At Work in Twentieth-Century Life Sciences*. Princeton, NJ: Princeton University Press, 1992.

Clarke, Lee. "The Origins of Nuclear Power: A Case of Institutional Conflict." *Social Problems* 32 (1985): 474–87.

Claus, Walter D., ed. *Radiation Biology and Medicine: Selected Reviews in the Life Sciences*. Reading, MA: Addison-Wesley Publishing, 1958.

"Cobalt Put above Radium in Cancer." *New York Times*, 2 Apr 1950, p. 81.

Coffin, C. C.; F. R. Hayes; L. H. Jodrey; and W. G. Whiteway. "Exchange of Materials in a Lake as Studied by the Addition of Radioactive Phosphorus." *Canadian Journal of Research* 27d (1949): 207–22.

Cohn, Mildred. "Atomic and Nuclear Probes of Enzyme Systems." *Annual Review of Biophysics and Biomolecular Structure* 21 (1992): 1–24.

———. "Some Early Tracer Experiments with Stable Isotopes." *Protein Science* 4 (1995): 2444–47.

Cohn, Waldo E. "Introductory Remarks by Chairman." In *Advances in Tracer Methodology*, vol. 4. New York: Plenum Press, 1968, 1–10.

Cohn, Waldo E., and David M. Greenberg. "Studies in Mineral Metabolism with the Aid of Artificial Radioactive Isotopes, I: Absorption, Distribution, and Excretion of Phosphorus." *Journal of Biological Chemistry* 123 (1938): 185–98.

Coleman, David C. *Big Ecology: The Emergence of Ecosystem Science*. Berkeley: University of California Press, 2010.

Comas, Francisco, and Marshall Brucer. "First Impressions of Therapy with Cesium 137." *Radiology* 69 (1957): 231–35.

Conger, Alan D., and Norman H. Giles. "The Cytogenetic Effect of Slow Neutrons." *Genetics* 35 (1950): 397–419.

"Control of Cancer Instead of Atomic Bombs." *Science News Letter*, 6 April 1946, pp. 213–14.

Cook, Robert Edward. "Raymond Lindeman and the Trophic-Dynamic Concept in Ecology." *Science* 198 (1977): 22–26.

Corner, George W. *George Hoyt Whipple and His Friends: The Life-Story of a Nobel Prize Pathologist*. Philadelphia: J. B. Lippincott, 1963.

Cowan, Ruth Schwartz. "The Consumption Junction: A Proposal for Research Strategies in the Sociology of Technology." In *The Social Construction of Technological Systems: New Directions in the Sociology and History of Technology*, ed. Wiebe E. Bijker, Thomas P. Hughes, and Trevor J. Pinch. Cambridge, MA: MIT Press, 1987, 261–80.

Craig, Campbell, and Sergey S. Radchenko. *The Atomic Bomb and the Origins of the Cold War*. New Haven, CT: Yale University Press, 2008.

Creager, Angela N. H. *The Life of a Virus: Tobacco Mosaic Virus as an Experimental Model, 1930–1965*. Chicago: University of Chicago Press, 2002.

———. "Nuclear Energy in the Service of Biomedicine: The U.S. Atomic Energy Commission's Radioisotope Program, 1946–1950." *Journal of the History of Biology* 39 (2006): 649–84.

———. "Mobilizing Biomedicine: Virus Research between Lay Health Organizations and the U.S. Federal Government, 1935–1955." In *Biomedicine in the Twentieth Century: Practices, Policies, and Politics*. Amsterdam: IOS Press, 2008, 171–201.

Creager, Angela N. H., and Jean-Paul Gaudillière. "Meanings in Search of Experiments and Vice-Versa: The Invention of Allosteric Regulation in Paris and Berkeley, 1959–1968." *Historical Studies in the Physical and Biological Sciences* 27 (1996): 1–89.

Creager, Angela N. H., and Hannah Landecker. "Technical Matters: Method, Knowledge and Infrastructure in Twentieth-Century Life Science." *Nature Methods* 6 (2009): 701–5.

Creager, Angela N. H., and María Jesús Santesmases. "Radiobiology in the Atomic Age: Changing Research Practices and Policies in Comparative Perspective." *Journal of the History of Biology* 39 (2006): 637–47.

Crease, Robert P. *Making Physics: A Biography of Brookhaven National Laboratory, 1946–1972*. Chicago: University of Chicago Press, 1999.

Crossley, D. A., Jr. "Movement and Accumulation of Radiostrontium and Radiocesium in Insects." In Schultz and Klement, *Radioecology*, 103–5.

Crossley, D.A., Jr., and Henry F. Howden. "Insect-Vegetation Relationships in an Area Contaminated by Radioactive Wastes." *Ecology* 42 (1961): 302–17.

Dally, Ann. "Thalidomide: Was the Tragedy Preventable?" *Lancet* 351 (18 Apr 1998): 1197–99.

Darby, William J.; Paul M. Densen; Richard O. Cannon; et al. "The Vanderbilt Cooperative Study of Maternal and Infant Nutrition, I: Background, II: Methods, III: Description of the Sample and Data." *Journal of Nutrition* 51 (1953): 539–64.

Darby, William J.; Paul F. Hahn; Margaret M. Kaser; Ruth C. Steinkamp; Paul M. Densen; and Mary B. Cook. "The Absorption of Radioactive Iron by Children 7–10 Years of Age." *Journal of Nutrition* 33 (1947): 107–19.

Darby, William J.; Paul F. Hahn; Ruth C. Steinkamp; and Margaret M. Kaser. "Absorption of Radioactive Iron by School Children." *Federation Proceedings* 5 (1946): 231–32.

Darby, William J.; William J. McGanity; Margaret P. Martin; et al. "The Vanderbilt Cooperative Study of Maternal and Infant Nutrition, IV: Dietary, Laboratory and Physical Findings in 2,129 Delivered Pregnancies." *Journal of Nutrition* 51 (1953): 565–97.

Daston, Lorraine. "The Moral Economy of Science." *Osiris* 10 (1999): 3–24.

———, ed. *Biographies of Scientific Objects*. Chicago: University of Chicago Press, 2000.

Davis, J. J., and R. F. Foster. "Bioaccumulation of Radioisotopes through Aquatic Food Chains." *Ecology* 39 (1958): 530–35.

Davis, Nuel Pharr. *Lawrence and Oppenheimer*. New York: Simon & Schuster, 1968.

Davis, W. Kenneth; Shields Warren; and Walker L. Cisler. "Some Peaceful Uses of Atomic Energy." *Scientific Monthly* 83 (Dec 1956): 287–97.

de Chadarevian, Soraya. *Designs for Life: Molecular Biology after World War II*. Cambridge: Cambridge University Press, 2002.

———. "Mice and the Reactor: The 'Genetics Experiment' in 1950s Britain." *Journal of the History of Biology* 39 (2006): 707–35.

———. "Mutations in the Nuclear Age." In *Making Mutations: Objects, Practices, Contexts*, ed. Luis Campos and Alexander Schwerin. Berlin: Max Planck Institute for the History of Science Preprint 393, 2010, 179–87.

de Chadarevian, Soraya, and Jean-Paul Gaudillière, eds. "The Tools of the Discipline: Biochemists and Molecular Biologists." *Journal of the History of Biology* 29 (1996): 327–462.

de Chadarevian, Soraya, and Harmke Kamminga, eds. *Molecularizing Biology and Medicine: New Practices and Alliances, 1910s–1970s*, Amsterdam: Harwood Academic Publishers, 1998.

Dekker, Charles A., and Howard K. Schachman. "On the Macromolecular Structure of Deoxyribonucleic Acid: An Interrupted Two-Strand Model." *Proceedings of the National Academy of Sciences, USA* 40 (1954): 894–909.

de la Bruheze, Adri. "Radiological Weapons and Radioactive Waste in the United States: Insiders' and Outsiders' Views, 1941–55." *British Journal for the History of Science* 25 (1992): 207–27.

Delbrück, Max. "Experiments with Bacterial Viruses (Bacteriophages)." *Harvey Lectures* 41 (1946): 161–87.

Delbrück, Max, and Salvador E. Luria. "Interference between Bacterial Viruses, I: Interference between Two Bacterial Viruses Acting upon the Same Host, and the Mechanism of Virus Growth." *Archives of Biochemistry* 1 (1942): 111–41.

Delbrück, Max, and Gunther S. Stent. "On the Mechanism of DNA Replication." In *The Chemical Basis of Heredity*, ed. William D. McElroy and Bentley Glass. Baltimore: Johns Hopkins University Press, 1957, 699–736.

del Regato, Juan A. *Radiological Oncologists: The Unfolding of a Medical Specialty*. Reston, VA: Radiology Centennial, 1993.

Del Sesto, Steven L. *Science, Politics, and Controversy: Civilian Nuclear Power in the United States, 1946–1974*. Boulder, CO: Westview Press, 1979.

de Mendoza, Diego Hurtado. "Autonomy, Even Regional Hegemony: Argentina and the 'Hard Way' toward Its First Research Reactor (1945–1958)." *Science in Context* 18 (2005): 285–308.

DeSelm, H. R., and R. E. Shanks. "Accumulation and Cycling of Organic Matter and Chemical Constituents during Early Vegetational Succession on a Radioactive Waste Disposal Area." In Schultz and Klement, *Radioecology*, 83–96.

Divine, Robert A. *Blowing on the Wind: The Nuclear Test Ban Debate, 1954–1960*. New York: Oxford University Press, 1978.

Doel, Ronald E., and Allan A. Needell. "Science, Scientists, and the CIA: Balancing International Ideals, National Needs, and Professional Opportunities." *Intelligence and National Security* 12 (1997): 59–81.

Doermann, A. H.; Martha Chase; and Franklin W. Stahl. "Genetic Recombination and Replication in Bacteriophage." *Journal of Cellular and Comparative Physiology* 45, suppl. 2 (1955): 51–74.

Dubach, Reubenia; Sheila T. E. Callender; and Carl V. Moore. "Studies in Iron Transportation and Metabolism, VI: Absorption of Radioactive Iron in Patients with Fever and Anemias of Varied Etiology." *Blood* 3 (1948): 526–40.

Dulbecco, Renato. "Experiments on Photoreactivation of Bacteriophages Inactivated with Ultraviolet Radiation." *Journal of Bacteriology* 59 (1950): 329–47.

Dumit, Joseph. "PET Scanner." In *Instruments of Science: An Historical Encyclopedia*, ed. Robert Bud and Deborah Jean Warner. London: Science Museum, 1998, 449–52.

———. *Picturing Personhood: Brain Scans and Biomedical Identity*. Princeton, NJ: Princeton University Press, 2004.

Dunaway, Paul B., and Stephen V. Kaye. "Effects of Ionizing Radiation on Mammal Populations on the White Oak Lake Bed." In Schultz and Klement, *Radioecology*, 333–38.

Dunham, Charles L. "Foreword." In *Onsite Ecological Research of the Division of Biology and Medicine at the Oak Ridge National Laboratory*, ed. Stanley I. Auerbach and Vincent Schultz, TID-16890. Washington, DC: US Atomic Energy Commission Division of Technical Information, 1962, iii–iv.

Dunlavey, Dean C. "Federal Licensing and Atomic Energy." *California Law Review* 46 (1958): 69–83.

Dyer, Norman C., and A. Bertrand Brill. "Fetal Radiation Dose from Maternally Administered ^{59}Fe and ^{131}I." In *Radiation Biology of the Fetal and Juvenile Mammal: Proceedings of the Ninth Annual Hanford Biology Symposium at Richland, Washington, May 5–8, 1969*, ed. Melvin R. Sikov and D. Dennis Mahlum. Springfield, VA: US Atomic Energy Commission Division of Technical Information, 1969, 73–88.

Dyer, N. C.; A. B. Brill; S. R. Glasser; and D. A. Goss. "Maternal-Fetal Transport and Distribution of ^{59}Fe and ^{131}I in Humans." *American Journal of Obstetrics and Gynecology* 103 (1969): 290–96.

Early, Paul J. "Use of Diagnostic Radionuclides in Medicine." *Health Physics* 69 (1995): 649–61.

Echols, Harrison. *Operators and Promoters: The Story of Molecular Biology and Its Creators*, ed. Carol A. Gross. Berkeley: University of California Press, 2001.

Edelmann, Abraham. "The Relation of Thyroidal Activity and Radiation to Growth of Hypophyseal Tumors." In *The Thyroid: Report of Symposium Held June 9 to 11, 1954*. Upton, NY: Brookhaven National Laboratory, 1955, 250–53.

Edgerton, David. *Warfare State: Britain, 1920–1970*. Cambridge: Cambridge University Press, 2006.

Edsall, John T. "Blood and Hemoglobin: The Evolution of Knowledge of Functional Adaptation in a Biochemical System, Part I: The Adaptation of Chemical Structure to Function in Hemoglobin." *Journal of the History of Biology* 5 (1972): 205–57.

Eisenberg, Rebecca S. "Public Research and Private Development: Patents and Technology Transfer in Government-Sponsored Research." *Virginia Law Review* 82 (1996): 1663–727.

Eisenbud, Merril. *Environmental Radioactivity*. New York: McGraw-Hill, 1963.

Eisenbud, Merril, and John H. Harley. "Radioactive Dust from Nuclear Detonations." *Science* 117 (1953): 141–47.

Eisenhower, Dwight D. *Atoms for Peace: Dwight D. Eisenhower's Address to the United Nations*. Washington, DC: National Archives and Records Administration, 1990.

Ekins, Roger P. "The Estimation of Thyroxine in Human Plasma by an Electrophoretic Technique." *Clinica Chimica Acta* 5 (1960): 453–59.

———. "Immunoassay, DNA Analysis, and Other Ligand Binding Assay Techniques: From Electropherograms to Multiplexed, Ultrasensitive Microarrays on a Chip." *Journal of Chemical Education* 76 (1999): 769–80.

Elsom, Katharine O'Shea, and Albert B. Sample. "Macrocytic Anemia in Pregnant Women with Vitamin B Deficiency." *Journal of Clinical Investigation* 16 (1937): 463–74.

Elzen, Boelie. "Two Ultracentrifuges: A Comparative Study of the Social Construction of Artefacts." *Social Studies of Science* 16 (1986): 621–62.

———. *Scientists and Rotors: The Development of Biochemical Ultracentrifuges*, PhD diss., University of Twente, the Netherlands, 1988.

Etheridge, Elizabeth W. "Pellagra: An Unappreciated Reminder of Southern Distinctiveness." In *Disease and Distinctiveness in the American South*, ed. Todd L. Savitt and James Harvey Young. Knoxville: University of Tennessee Press, 1988, 100–119.

Evans, Robley D. "Radium Poisoning: A Review of Present Knowledge." *American Journal of Public Health* 23 (1933): 1017–23.

———. "The Medical Uses of Atomic Energy." *Atlantic Monthly* 177 (1946): 68–73.

———. "Quantitative Inferences Concerning the Genetic Effects of Radiation on Human Beings." *Science* 109 (1949): 299–304.

Faden, Ruth R., and Tom L. Beauchamp. *A History and Theory of Informed Consent*. New York: Oxford University Press, 1986.

Farr, Lee E.; William H. Sweet; James S. Robertson; Charles G. Foster; Herbert B. Locksley; D. Lawrence Sutherland; Mortimer L. Mendelsohn; and E. E. Stickley. "Neutron Capture Therapy with Boron in the Treatment of Glioblastoma Multiforme." *American Journal of Roentgenology, Radium Therapy, and Nuclear Medicine* 71 (1954): 279–93.

Farr, L. E.; W. H. Sweet; H. B. Locksley; and J. S. Robertson. "Neutron Capture Therapy of Gliomas Using Boron^{10}." *Transactions of the American Neurological Association*, 79th meeting, 13 (1954): 110–13.

Feffer, Stuart M. "Atoms, Cancer, and Politics: Supporting Atomic Science at the University of Chicago, 1944–1950." *Historical Studies in the Physical and Biological Sciences* 22 (1992): 233–61.

Fermi, Enrico. "Radioactivity Induced by Neutron Bombardment." *Nature* 133 (1934): 757.

Ficken, Robert E. "Grand Coulee and Hanford: The Atomic Bomb and the Development of the Columbia River." In *The Atomic West*, ed. Bruce Hevly and John M. Findlay. Seattle: University of Washington Press, 1998, 21–38.

Finch, Clement A.; Daniel H. Coleman; Arno G. Motulsky; Dennis M. Donohue; and Robert H. Reiff. "Erythrokinetics in Pernicious Anemia." *Blood* 11 (1956): 807–20.

Finch, Clement A.; John G. Gibson II; Wendell C. Peacock; and Rex G. Fluharty. "Iron Metabolism: Utilization of Intravenous Radioactive Iron." *Blood* 4 (1949): 905–27.

Findlen, Paula. "The Economy of Scientific Exchange in Early Modern Italy." In *Patronage and Institutions: Science, Technology, and Medicine at the European Court, 1500–1750*, ed. Bruce T. Moran. Rochester, NY: Boydell Press, 1991, 5–24.

Fink, Robert M., ed. *Biological Studies with Polonium, Radium, and Plutonium.* National Nuclear Energy Series, Manhattan Project Technical Section. Division 6, University of Rochester Project, vol. 3. New York: McGraw-Hill, 1950.

Finkel, Miriam P. "Mice, Men and Fallout." *Science* 128 (1958): 637–41.

Folley, Jarrett H.; Wayne Borges; and Takuso Yamawaki. "Incidence of Leukemia in Survivors of the Atomic Bomb in Hiroshima and Nagasaki, Japan." *American Journal of Medicine* 13 (1952): 311–21.

Forbes, Stephen A. "The Lake as a Microcosm." *Bulletin of the Peoria Scientific Association* (1887): 77–87.

"Formal Procedure Adopted for Withdrawal from Routine Production, Sale of Radioisotopes." *Isotopes and Radiation Technology* 2, no. 3 (1965): 315.

Forman, Paul. "Behind Quantum Electronics: Natural Security as Basis for Physical Research in the United States, 1940–1960." *Historical Studies in the Physical and Biological Sciences* 18 (1987): 149–229.

Foster, R. F. "The History of Hanford and Its Contribution of Radionuclides to the Columbia River." In *The Columbia River Estuary and Adjacent Ocean Waters:*

Bioenvironmental Studies, ed. A. T. Pruter and D. L. Alverson. Seattle: University of Washington Press, 1972, 3–18.

Foster, R. F., and J. J. Davis. "The Accumulation of Radioactive Substances in Aquatic Forms." *Proceedings of the International Conference on the Peaceful Uses of Atomic Energy, Held in Geneva 8 August–20 August 1955*. Vol. 13. New York: United Nations, 1956: 364–67.

Foster, Richard F., and Royal E. Rostenbach. "Distribution of Radioisotopes in Columbia River." *Journal of the American Water Works Association* 46 (1954): 633–40.

Foster, R. F.; L. R. Donaldson; A. D. Wedlander; K. Bonham; and A. H. Seymour. "The Effect on Embryos and Young of Rainbow Trout from Exposing the Parent Fish to X-Rays." *Growth* 13 (1949): 119–42.

Fowler, E. E. "Recent Advances in Applications of Isotopes and Radiation in the United States." *Isotopes and Radiation Technology* 9, no. 3 (1972): 253–63.

Fragu, Philippe. "How the Field of Thyroid Endocrinology Developed in France after World War II." *Bulletin of the History of Medicine* 77 (2003): 393–414.

Friese, Carrie, and Adele E. Clarke. "Transposing Bodies of Knowledge and Technique: Animal Models at Work in Reproductive Sciences." *Social Studies of Science* 42 (2011): 31–52.

Frohman, I. Phillips. "Role of the General Physician in the Atomic Age." *Journal of the American Medical Association* 162 (1956): 962–66.

Fruton, Joseph S. *Molecules and Life: Historical Essays on the Interplay of Chemistry and Biology*. New York: Wiley-Interscience, 1972.

Fuerst, Clarence R., and Gunther S. Stent. "Inactivation of Bacteria by Decay of Incorporated Radioactive Phosphorus." *Journal of General Physiology* 40 (1956): 73–90.

Fuller, R. Clinton. "Forty Years of Microbial Photosynthesis Research: Where It Came From and What It Led To." *Photosynthesis Research* 62 (1999): 1–29.

Funigiello, Philip J. *American-Soviet Trade in the Cold War*. Chapel Hill: University of North Carolina Press, 1988.

Furth, Jacob, and John L. Tullis. "Carcinogenesis by Radioactive Substances." *Cancer Research* 16 (1956): 5–21.

Galison, Peter. *Image and Logic: A Material Culture of Microphysics*. Chicago: University of Chicago Press, 1997.

Galison, Peter, and Barton Bernstein. "In Any Light: Scientists and the Decision to Build the Superbomb, 1952–1954." *Historical Studies in the Physical and Biological Sciences* 19 (1989): 267–347.

Galison, Peter, and Bruce Hevly, eds. *Big Science: The Growth of Large-Scale Research*. Stanford, CA: Stanford University Press, 1992.

Gallup, George H. *The Gallup Poll: Public Opinion 1935–1971*, 3 vols. New York: Random House, 1972.

García-Sancho, Miguel. "A New Insight into Sanger's Development of Sequencing:

From Proteins to DNA, 1943–1977." *Journal of the History of Biology* 43 (2010): 265–323.

Gaudillière, Jean-Paul. "The Molecularization of Cancer Etiology in the Postwar United States: Instruments, Politics and Management." In de Chadarevian and Kamminga, *Molecularizing Biology and Medicine*, 139–70.

———. "Introduction: Drug Trajectories." *Studies in History and Philosophy of Biological and Biomedical Sciences* 36 (2005): 603–11.

———. "Normal Pathways: Controlling Isotopes and Building Biomedical Research in Postwar France." *Journal of the History of Biology* 39 (2006): 737–64.

Gaudillière, Jean-Paul, and Ilana Löwy, "Introduction." In *The Invisible Industrialist: Manufactures and the Production of Scientific Knowledge*, ed. Jean-Paul Gaudillière and Ilana Löwy. London: Macmillan, 1998, 3–15.

Gerber, Michele Stenehjem. *On the Home Front: The Cold War Legacy of the Hanford Nuclear Site*. 2nd ed. Lincoln: University of Nebraska Press, 2002.

Gest, Howard. "Photosynthesis and Phage: Early Studies on Phosphorus Metabolism in Photosynthetic Microorganisms with ^{32}P, and How They Led to the Serendipic Discovery of ^{32}P-Decay 'Suicide' of Bacteriophage." *Photosynthesis Research* 74 (2002): 331–39.

Giacomoni, Dario. "The Origin of DNA:RNA Hybridization." *Journal of the History of Biology* 26 (1993): 89–107.

Gieryn, Thomas F. *Cultural Boundaries of Science: Credibility on the Line*. Chicago: University of Chicago Press, 1999.

Glantz, Leonard H. "The Influence of the Nuremberg Code on U.S. Statutes and Regulations." In Annas and Grodin, *The Nazi Doctors and the Nuremberg Code*, 183–200.

Glennan, T. Keith. "Radioisotopes: A New Industry." In *Radioisotopes in Industry*, ed. John R. Bradford. New York: Reinhold Publishing, 1953, 3–12.

Glick, S. M.; J. Roth; R. S. Yalow; and S. A. Berson. "Immunoassay of Human Growth Hormone in Plasma." *Nature* 199 (1963): 784–87.

Godwin, Sir Harry. "Sir Arthur Tansley: The Man and the Subject." *Journal of Ecology* 65 (1977): 1–26.

Godwin, H. "Half-Life of Radiocarbon." *Nature* 195 (1962): 984.

Godwin, John T.; Lee E. Farr; William H. Sweet; and James S. Robertson. "Pathological Study of Eight Patients with Glioblastoma Multiforme Treated by Neutron-Capture Therapy Using Boron 10." *Cancer* 8 (1955): 601–15.

Goldschmidt, V. M. "Drei Vorträge über Geochemie." *Geologiska Föreningens Stockholm Förhandlingar* 56 (1934): 385–427.

Goldsmith, Stanley J. "Rosalyn S. Yalow: A Personal and Scientific Memoir." *Journal of Nuclear Medicine* 53, no. 6 (2012): 21N.

Golley, Frank Benjamin. *A History of the Ecosystem Concept in Ecology: More than the Sum of the Parts*. New Haven, CT: Yale University Press, 1993.

Goodman, Jordan; Anthony McElligot; and Lara Marks, eds. *Useful Bodies: Hu-*

mans in the Service of Medical Science in the Twentieth Century. Baltimore: Johns Hopkins University Press, 2003.

Gordin, Michael D. *Five Days in August: How World War II Became a Nuclear War*. Princeton, NJ: Princeton University Press, 2007.

———. *Red Cloud at Dawn: Truman, Stalin, and the End of the Atomic Monopoly*. New York: Farrar, Straus and Giroux, 2009.

Gowing, Margaret. *Independence and Deterrence: Britain and Atomic Energy, 1945–1952*, 2 vols. London: Macmillan, 1974.

Graham, E. R. "Uptake of Waste Sr 90 and Cs 137 by Soil and Vegetation." *Soil Science* 86 (1958): 91–97.

Greenberg, Daniel S. *The Politics of Pure Science*. 2nd ed. 1967; reprinting with new introductory essays and afterword, Chicago: University of Chicago Press, 1999.

Greene, Benjamin P. *Eisenhower, Science Advice, and the Nuclear Test-Ban Debate, 1945–1963*. Stanford, CA: Stanford University Press, 2007.

Griesemer, James. "Tracking Organic Processes: Representations and Research Styles in Classical Embryology and Genetics." In *From Embryology to Evo-Devo: A History of Developmental Evolution*, ed. Manfred D. Laubichler and Jane Maienschein. Cambridge, MA: MIT Press, 2007, 375–433.

Grobman, Arnold B. *Our Atomic Heritage*. Gainesville: University of Florida Press, 1951.

Grover, Will. "All the Easy Experiments: A Berkeley Professor, Dirty Bombs, and the Birth of Informed Consent." *Berkeley Science Review* 5, no. 2 (2005): 41–45.

Groves, Leslie R. *Now It Can Be Told: The Story of the Manhattan Project*. New York: Harper, 1962.

"Growing Demand for Cobalt-60 Stimulates Increased Production by AEC." *Isotopes and Radiation Technology* 3, no. 1 (1965): 76.

"The Growing Market for Nuclear KW." *Fortune* 68 (July 1963): 173–176.

"The H-Bomb and World Opinion." *Bulletin of the Atomic Scientists* 10, no. 5 (1954): 163–65.

Haber, Heinz. *Our Friend the Atom*. New York: Simon and Schuster, 1956.

Hacker, Barton C. *The Dragon's Tail: Radiation Safety in the Manhattan Project, 1942–1946*. Berkeley: University of California Press, 1987.

———. "No Evidence of Ill Effects: Radiation Safety and Weapons Testing in the Manhattan Project, 1945–1946." *Polhem* 9 (1991): 139–49.

———. *Elements of Controversy: The Atomic Energy Commission and Radiation Safety in Nuclear Weapons Testing, 1947–1974*. Berkeley: University of California Press, 1994.

———. "'Hotter than a $2 Pistol': Fallout, Sheep, and the Atomic Energy Commission, 1953–1986." In *The Atomic West*, ed. Bruce Hevly and John M. Findlay. Seattle: University of Washington Press, 1998, 157–75.

Hagen, Joel B. *An Entangled Bank: The Origins of Ecosystem Ecology*. New Brunswick, NJ: Rutgers University Press, 1992.

———. "Teaching Ecology during the Environmental Age, 1965–1980." *Environmental History* 13 (2008): 704–23.

Hagstrom, Ruth M.; S. R. Glasser; A. B. Brill; and R. M. Heyssel. "Long Term Effects of Radioactive Iron Administered during Human Pregnancy." *American Journal of Epidemiology* 90 (1969): 1–10.

Hahn, Paul F. "The Metabolism of Iron." *Medicine* 16 (1937): 249–66.

———, ed. *Therapeutic Use of Artificial Radioisotopes*. New York: John Wiley & Sons, 1956.

Hahn, P. F.; W. F. Bale; R. A. Hettig; M. D. Kamen; and G. H. Whipple. "Radioactive Iron and Its Excretion in Urine, Bile, and Feces." *Journal of Experimental Medicine* 70 (1939): 443–51.

Hahn, P. F.; W. F. Bale; E. O. Lawrence; and G. H. Whipple. "Radioactive Iron and Its Metabolism in Anemia." *Journal of the American Medical Association* 111 (1938): 2285–86.

———. "Radioactive Iron and Its Metabolism in Anemia: Its Absorption, Transportation, and Utilization." *Journal of Experimental Medicine* 69 (1939): 739–53.

Hahn, P. F.; W. F. Bale; J. F. Ross; W. M. Balfour; and G. H. Whipple. "Radioactive Iron Absorption by Gastro-Intestinal Tract: Influence of Anemia, Anoxia, and Antecedent Feeding Distribution in Growing Dogs." *Journal of Experimental Medicine* 78 (1943): 169–88.

Hahn, P. F.; Ella Lea Carothers; R. O. Cannon; et al. "Iron Uptake in 750 Cases of Human Pregnancy Using the Radioactive Isotope Fe^{59}." *Federation Proceedings* 6 (1947): 392–93.

Hahn, P. F.; E. L. Carothers; W. J. Darby; M. Martin; C. W. Sheppard; R. O. Cannon; A. S. Beam; P. M. Densen; J. C. Peterson; and G. S. McClellan. "Iron Metabolism in Human Pregnancy as Studied with the Radioactive Isotope, Fe^{59}." *American Journal of Obstetrics and Gynecology* 61 (1951): 477–86.

Hahn, P. F.; J. P. B. Goodell; C. W. Sheppard; R. O. Cannon; and H. C. Francis. "Direct Infiltration of Radioactive Isotopes as a Means of Delivering Ionizing Radiation to Discrete Tissues." *Journal of Laboratory and Clinical Medicine* 32 (1947): 1442–53.

Hahn, P. F.; Edgar Jones; R. C. Lowe; G. R. Meneely; and Wendell Peacock. "The Relative Absorption and Utilization of Ferrous and Ferric Iron in Anemia as Determined with the Radioactive Isotope." *American Journal of Physiology* 143 (1945): 191–97.

Hahn, P. F.; L. L. Miller; F. S. Robscheit-Robbins; W. F. Bale; and G. H. Whipple. "Peritoneal Absorption: Red Cells Labeled by Radio-Iron Hemoglobin Move Promptly from Peritoneal Cavity into the Circulation." *Journal of Experimental Medicine* 80 (1944): 77–82.

Hahn, P. F.; J. F. Ross; W. F. Bale; W. M. Balfour; and G. H. Whipple. "Red Cell and

Plasma Volumes (Circulating and Total) as Determined by Radio Iron and by Dye." *Journal of Experimental Medicine* 75 (1942): 221–32.

Hahn, P. F., and C. W. Sheppard. "Selective Radiation Obtained by the Intravenous Administration of Colloidal Radioactive Isotopes in Diseases of the Lymphoid System." *Southern Medical Journal* 39 (1946): 558–62.

Hales, Peter Bacon. *Atomic Spaces: Living on the Manhattan Project*. Urbana: University of Illinois Press, 1997.

Hall, B. E., and C. H. Watkins. "Radiophosphorus in the Treatment of Blood Dyscrasias." *Medical Clinics of North America* 31 (1947): 810–40.

Halpern, Sydney A. *Lesser Harms: The Morality of Risk in Medical Research*. Chicago: University of Chicago Press, 2004.

Hamblin, Jacob Darwin. "Hallowed Lords of the Sea: Scientific Authority and Radioactive Waste in the United States, Britain, and France." *Osiris* 21 (2006): 209–28.

———. *Poison in the Well: Radioactive Waste in the Oceans at the Dawn of the Nuclear Age*. New Brunswick, NJ: Rutgers University Press, 2008.

———. " 'A Dispassionate and Objective Effort': Negotiating the First Study on the Biological Effects of Atomic Radiation." *Journal of the History of Biology* 40 (2007): 147–77.

Hamilton, Joseph G. "The Rates of Absorption of Radio-Sodium in Normal Human Subjects." *Proceedings of the National Academy of Sciences, USA* 23 (1937): 521–27.

———. "The Rates of Absorption of the Radioactive Isotopes of Sodium, Potassium, Chlorine, Bromine, and Iodine in Normal Human Subjects." *American Journal of Physiology* 124 (1938): 667–78.

———. "The Use of Radioactive Tracers in Biology and Medicine." *Radiology* 39 (1942): 541–72.

Hamilton, Joseph G., and Gordon A. Alles. "The Physiological Action of Natural and Artificial Radioactivity." *American Journal of Physiology* 125 (1939): 410–13.

Hamilton, Joseph G., and John H. Lawrence. "Recent Clinical Developments in the Therapeutic Application of Radio-Phosphorus and Radio-Iodine." *Journal of Clinical Investigation* 21 (1942): 624.

Hamilton, Joseph G., and Mayo H. Soley. "Studies in Iodine Metabolism by the Use of a New Radioactive Isotope of Iodine." *American Journal of Physiology* 127 (1939): 557–72.

———. "Studies in Iodine Metabolism of the Thyroid Gland in Situ by the Use of Radio-Iodine in Normal Subjects and in Patients with Various Types of Goiter." *American Journal of Physiology* 131 (1940): 135–43.

Hamilton, Joseph G., and Robert S. Stone. "Excretion of Radio-Sodium Following Intravenous Administration in Man." *Proceedings of the Society for Experimental Biology and Medicine* 35 (1937): 595–98.

———. "The Intravenous and Intraduodenal Administration of Radio-Sodium." *Radiology* 28 (1937): 178–88.

Hanson, W. C., and H. A. Kornberg. "Radioactivity in Terrestrial Animals near an Atomic Energy Site." *Proceedings of the International Conference on the Peaceful Uses of Atomic Energy, Held in Geneva 8 August–20 August 1955*. Vol. 13. New York: United Nations, 1956: 385–88.

Haraway, Donna. *Crystals, Fabrics and Fields: Metaphors of Organicism in Twentieth-Century Developmental Biology*. New Haven, CT: Yale University Press, 1976.

Harkewicz, Laura J. *"The Ghost of the Bomb": The Bravo Medical Program, Scientific Uncertainty, and the Legacy of U.S. Cold War Science, 1954–2005*. PhD diss., University of California, San Diego, 2010.

Harper, P. V.; G. Andros; and K. Lathrop. "Preliminary Observations on the Use of Six-Hour Tc^{99m} as a Tracer in Biology and Medicine." *Semiannual Report to the Atomic Energy Commission*. Chicago: Argonne Cancer Research, 1962, 77–88.

Hartmann, Susan M. *Truman and the 80th Congress*. Columbia: University of Missouri Press, 1971.

Hassett, C. C., and D. W. Jenkins. "The Uptake and Effect of Radiophosphorus in Mosquitoes." *Physiological Zoology* 24 (1951): 257–66.

Hecht, Gabrielle. *The Radiance of France: Nuclear Power and National Identity after World War II*. Cambridge, MA: MIT Press, 1998.

Heilbron, John L. "The First European Cyclotrons." *Rivista di Storia della Scienza* 3 (1986): 1–44.

Heilbron, J. L., and Robert W. Seidel. *Lawrence and His Laboratory: A History of the Lawrence Berkeley Laboratory*. Berkeley: University of California Press, 1989.

Heilbron, J. L.; Robert W. Seidel; and Bruce R. Wheaton. *Lawrence and His Laboratory: Nuclear Science at Berkeley*. Berkeley, CA: Lawrence Berkeley Laboratory and Office for History of Science and Technology, 1981.

Heims, Steve Joshua. *The Cybernetics Group*. Cambridge, MA: MIT Press, 1991.

Henderson, Malcolm C.; M. Stanley Livingston; and Ernest O. Lawrence. "Artificial Radioactivity Produced by Deuton [*sic*] Bombardment." *Physical Review* 45 (1934): 428–29.

Henshaw, Paul S. "Atomic Energy: Cancer Cure . . . or Cancer Cause?" *Scientific Illustrated* 2 (Nov 1947): 46–47, 84.

Herberman, Ronald B. "Immunodiagnostics for Cancer Testing." In Schönfeld, *New Developments in Immunoassays*, 38–44.

Herran, Néstor. "Spreading Nucleonics: The Isotope School at the Atomic Energy Research Establishment, 1951–67." *British Journal for the History of Science* 39 (2006): 569–86.

———. "Isotope Networks: Training, Sales and Publications, 1946–1965." *Dynamis* 29 (2009): 285–306.

Herran, Néstor, and Xavier Roqué. "Tracers of Modern Technoscience," introduction to a special collection "Isotopes: Science, Technology and Medicine in the Twentieth Century." *Dynamis* 29 (2009): 123–30.

Herriott, Roger M. "Nucleic-Acid-Free T2 Virus 'Ghosts' with Specific Biological Action." *Journal of Bacteriology* 61 (1951): 752–54.

Hershey, Alfred D. "Reproduction of Bacteriophage." *International Review of Cytology* 1 (1952): 119–34.

———. "Intracellular Phases in the Reproductive Cycle of Bacteriophage T2." *Annales de l'Institut Pasteur* 84 (1953): 99–112.

———. "Conservation of Nucleic Acids during Bacterial Growth." *Journal of General Physiology* 38 (1954): 145–48.

———. "The Injection of DNA into Cells by Phage." In *Phage and the Origins of Molecular Biology*, ed. John Cairns, Gunther S. Stent, and James D. Watson. Cold Spring Harbor, NY: Cold Spring Harbor Laboratory Press, 1966, 100–108.

Hershey, A. D., and Elizabeth Burgi. "Genetic Significance of the Transfer of Nucleic Acid from Parental to Offspring Phage." *Cold Spring Harbor Symposia on Quantitative Biology* 21 (1956): 91–101.

Hershey, A. D., and Martha Chase. "Independent Functions of Viral Protein and Nucleic Acid in Growth of Bacteriophage." *Journal of General Physiology* 36 (1952): 39–56.

Hershey, A. D.; M. D. Kamen; J. W. Kennedy; and H. Gest. "The Mortality of Bacteriophage Containing Assimilated Radioactive Phosphorus." *Journal of General Physiology* 34 (1951): 305–19.

Hershey, A. D.; Catherine Roesel; Martha Chase; and Stanley Forman. "Growth and Inheritance in Bacteriophage." *Carnegie Institution of Washington Year Book* 50 (1951): 195–200. Reprinted in Stahl, *We Can Sleep Later*, 173–78.

Hertz, Saul, and A. Roberts. "Application of Radioactive Iodine in Therapy of Graves' Disease." *Journal of Clinical Investigation* 21 (1942): 624.

Hertz, S.; A. Roberts; and Robley D. Evans. "Radioactive Iodine as an Indicator in the Study of Thyroid Physiology." *Proceedings of the Society of Experimental Biology and Medicine* 38 (1938): 510–13.

Hevesy, G. "The Absorption and Translocation of Lead by Plants: A Contribution to the Application of the Method of Radioactive Indicators in the Investigation of the Change of Substance in Plants." *Biochemical Journal* 17 (1923): 439–45.

———. "Application of Radioactive Indicators in Biology." *Annual Review of Biochemistry* 9 (1940): 641–62.

———. "Historical Progress of the Isotopic Methodology and Its Influences on the Biological Sciences." *Minerva Nucleare* 1, no. 4–5 (1957): 189–200.

Hevesy, G., and E. Hofer. "Diplogen and Fish." *Nature* 133 (1934): 495–96.

Hevesy, G.; H. B. Levi; and O. H. Rebbe. "The Origin of the Phosphorus Compounds in the Embryo of the Chicken." *Biochemical Journal* 32 (1938): 2147–55.

Hewlett, Richard G., and Oscar E. Anderson Jr. *The New World: A History of the United States Atomic Energy Commission, vol. 1, 1939–1946*. University Park: Pennsylvania State University Press, 1962; reprint Berkeley: University of California Press, 1990.

Hewlett, Richard G., and Francis Duncan. *Atomic Shield: A History of the United States Atomic Energy Commission, vol. 2, 1947–1952*. University Park: Pennsylvania State University Press, 1969; reprint Berkeley: University of California Press, 1990.

Hewlett, Richard G., and Jack M. Holl. *Atoms for Peace and War, 1953–1961: Eisenhower and the Atomic Energy Commission*. Berkeley: University of California Press, 1989.

Higuchi, Toshihiro. *Radioactive Fallout, the Politics of Risk, and the Making of a Global Environmental Crisis, 1954–1963*. PhD diss., Georgetown University, 2011.

Hill, Gladwin. "Effect of A-Bomb Is Found Limited." *New York Times*, 30 Mar 1955, p. 16.

Hines, Neal O. *Proving Ground: An Account of the Radiobiological Studies in the Pacific, 1946–1961*. Seattle: University of Washington Press, 1962.

Hoddeson, Lillian; Paul W. Henriksen; Roger A. Meade; and Catherine Westfall. *Critical Assembly: A Technical History of Los Alamos during the Oppenheimer Years, 1943–1945*. Cambridge: Cambridge University Press, 1993.

Holmes, Frederic L. "The Intake-Output Method of Quantification in Physiology." *Historical Studies in the Physical and Biological Sciences* 17 (1987): 235–70.

———. "Manometers, Tissue Slices, and Intermediary Metabolism." In Clarke and Fujimura, *The Right Tools for the Job*, 151–71.

———. *Between Biology and Medicine: The Formation of Intermediary Metabolism*. Berkeley: Office for History of Science and Technology, University of California, 1992.

———. "Crystals and Carriers: The Chemical and Physiological Identification of Hemoglobin." In *No Truth Except in the Details: Essays in Honor of Martin J. Klein*, ed. A. J. Kox and Daniel M. Siegel. Dordrecht, the Netherlands: Kluwer Academic Publishers, 1995, 191–243.

———. *Meselson, Stahl, and the Replication of DNA: A History of "The Most Beautiful Experiment in Biology."* New Haven, CT: Yale University Press, 2001.

———. *Reconceiving the Gene: Seymour Benzer's Adventures in Phage Genetics*, ed. William C. Summers. New Haven, CT: Yale University Press, 2006.

Horwitz, Robert Britt. *The Irony of Regulatory Reform: The Deregulation of American Telecommunications*. New York: Oxford University Press, 1989.

Hughes, Thomas P. *Networks of Power: Electrification in Western Society, 1880–1930*. Baltimore: Johns Hopkins University Press, 1983.

———. "Tennessee Valley and Manhattan Engineering District." In *American Genesis: A Century of Invention and Technological Enthusiasm, 1870–1970*. New York: Viking, 1989, 353–442.

Hutchinson, G. E. "Bio-Ecology." *Ecology* 21 (1940): 267–68.

———. "Limnological Studies in Connecticut, IV: The Mechanisms of Intermediary Metabolism in Stratified Lakes." *Ecological Monographs* 11 (1941): 21–60.

———. "Thiamin in Lake Waters and Aquatic Organisms." *Archives of Biochemistry* 2 (1943): 143–50.

———. "The Biogeochemistry of Aluminum and of Certain Related Elements." *Quarterly Review of Biology* 18 (1943): (I.) 1–29, (II.) 128–153, (III.) 242–262, (IV.) 331–63.

———. "Circular Causal Systems in Ecology." *Annals of the New York Academy of Sciences* 50 (1948): 221–46.

Hutchinson, G. Evelyn, and Vaughan T. Bowen. "A Direct Demonstration of the Phosphorus Cycle in a Small Lake." *Proceedings of the National Academy of Sciences, USA* 33 (1947): 148–53.

———. "Limnological Studies in Connecticut, IX: A Quantitative Radiochemical Study of the Phosphorus Cycle in Linsley Pond." *Ecology* 31 (1950): 194–203.

Hutchinson, G. Evelyn, and Jane K. Setlow. "Limnological Studies in Connecticut, VIII: The Niacin Cycle in a Small Inland Lake." *Ecology* 27 (1946): 13–22.

"Intercomparison of Film Badge Interpretations." *Isotopics* 5, no. 2 (Apr 1955): 8–33.

International Atomic Energy Agency. *International Directory of Radioisotopes, Vol. I: Unprocessed and Processed Radioisotope Preparations and Special Radiation Sources*. Vienna: International Atomic Energy Agency, 1959.

———. *International Directory of Radioisotopes, Vol. II: Compounds of Carbon 14, Hydrogen 3, Iodine 131, Phosphorus 32, and Sulphur 35*. Vienna: International Atomic Energy Agency, 1959.

"Iron Doses with Radioactive Isotopes Aid to Pregnancy, Experiment Shows." *Nashville Banner*, 13 Dec 1946, p. 24.

Jaczko, Gregory B. "A Regulator's Perspective on Safety," remarks presented at American Society for Radiation Oncology Annual Meeting, 2 Oct 2011, http://www.nrc.gov/reading-rm/doc-collections/commission/speeches/2011/.

Jakobson, Max. *Finland in the New Europe*. Westport, CT: Praeger, 1998.

James, F. E. "Insulin Treatment in Psychiatry." *History of Psychiatry* 3 (1992): 221–35.

Javid, Manucher; Gordon L. Brownell; and William H. Sweet. "The Possible Use of Neutron-Capturing Isotopes such as $Boron^{10}$ in the Treatment of Neoplasms, II: Computation of the Radiation Energies and Estimates of Effects in Normal and Neoplastic Brain." *Journal of Clinical Investigation* 31 (1952): 604–10.

JCAE. See US Congress, Joint Committee on Atomic Energy.

Joerges, Bernward, and Terry Shinn, eds. *Instrumentation: Between Science, State and Industry*. Dordrecht, the Netherlands: Kluwer Academic Publishing, 2001.

Johnson, Charles W., and Jackson, Charles O. *City behind a Fence: Oak Ridge, Tennessee 1942–1946*. Knoxville: University of Tennessee Press, 1981.

Johnson, Leland, and Daniel Schaffer. *Oak Ridge National Laboratory: The First Fifty Years*. Knoxville: University of Tennessee Press, 1994.

Jolly, J. Christopher. "Linus Pauling and the Scientific Debate over Fallout Hazards." *Endeavour* 26 (2002): 149–53.

———. *Thresholds of Uncertainty: Radiation and Responsibility in the Fallout Controversy*, PhD diss., Oregon State University, 2003.

Jones, David S., and Robert L. Martensen, "Human Radiation Experiments and the Formation of Medical Physics at the University of California, San Francisco and Berkeley, 1937–1962." In Goodman et al., *Useful Bodies*, 81–108.

Jones, H. B.; I. L. Chaikoff; and John H. Lawrence. "Radioactive Phosphorus as an Indicator of Phospholipid Metabolism, VI: The Phospholipid Metabolism of Neoplastic Tissues (Mammary Carcinoma, Lymphoma, Lymphosarcoma, Sarcoma 180)." *Journal of Biological Chemistry* 128 (1939): 631–44.

Jones, Vincent C. *Manhattan, the Army and the Atomic Bomb*. Washington, DC: Center of Military History, United States Army, 1985.

Judson, Horace Freeland. *The Eighth Day of Creation: Makers of the Revolution in Biology*. New York: Simon & Schuster, 1979.

Kafatos, Fotis C. "A Revolutionary Landscape: The Restructuring of Biology and its Convergence with Medicine." *Journal of Molecular Biology* 319 (2002): 861–67.

Kahn, C. Ronald, and Jesse Roth. "Berson, Yalow, and the *JCI*: The Agony and the Ecstasy." *Journal of Clinical Investigation* 114 (2004): 1051–54.

Kahn, Herman. *On Thermonuclear War*. Princeton, NJ: Princeton University Press, 1960.

Kaiser, David. "Cold War Requisitions, Scientific Manpower, and the Production of American Physicists after World War II." *Historical Studies in the Physical and Biological Sciences* 33 (2002): 131–59.

Kaiser, David I. "The Atomic Secret in Red Hands? American Suspicions of Theoretical Physicists during the Early Cold War." *Representations* 90 (2005): 28–60.

Kamen, Martin D. *Radioactive Tracers in Biology: An Introduction to Tracer Methodology*. 2nd ed. New York: Academic Press, 1951.

———. *Radiant Science, Dark Politics: A Memoir of the Nuclear Age*. Berkeley: University of California Press, 1985.

———. "A Cupful of Luck, a Pinch of Sagacity." *Annual Review of Biochemistry* 55 (1986): 1–34.

———. "Early History of Carbon-14." *Science* 140 (1963): 584–90.

Kamen, Martin D., and Samuel Ruben. "Studies in Photosynthesis with Radio-Carbon." *Journal of Applied Physics* 12 (1941): 326.

Kamminga, Harmke. "Vitamins and the Dynamics of Molecularization: Biochemistry, Policy and Industry in Britain, 1914–1939." In de Chadarevian and Kamminga, *Molecularizing Biology and Medicine*, 83–105.

Kamminga, Harmke, and Mark Weatherall. "The Making of a Biochemist, I: Frederick Gowland Hopkins' Construction of Dynamic Biochemistry." *Medical History* 40 (1996): 269–92.

Kathren, Ronald L. "Pathway to a Paradigm: The Linear Nonthreshold Dose-Response Model in Historical Context: The American Academy of Health Physics 1995 Radiology Centennial Hartman Oration." *Health Physics* 70 (1996): 621–35.

Kay, Lily E. "Laboratory Technology and Biological Knowledge: The Tiselius Electrophoresis Apparatus, 1930–1945." *History and Philosophy of the Life Sciences* 10 (1988): 51–72.

———. *The Molecular Vision of Life: Caltech, the Rockefeller Foundation, and the Rise of the New Biology*. New York: Oxford University Press, 1993.

———. *Who Wrote the Book of Life? A History of the Genetic Code*. Stanford, CA: Stanford University Press, 2000.

Kaye, Stephen V., and Paul B. Dunaway. "Bioaccumulation of Radioactive Isotopes by Herbivorous Small Mammals." *Health Physics* 7 (1962): 205–17.

———. "Estimation of Dose Rate and Equilibrium State from Bioaccumulation of Radionuclides by Mammals." In Schultz and Klement, *Radioecology*, 107–11.

Keating, Peter, and Cambrosio, Alberto. *Biomedical Platforms: Realigning the Normal and the Pathological in Late-Twentieth-Century Medicine*. Cambridge, MA: MIT Press, 2003.

Keller, Evelyn Fox. "From Secrets of Life to Secrets of Death." In *Secrets of Life, Secrets of Death: Essays on Language, Gender and Science*. New York: Routledge, 1992, 39–55.

———. *Refiguring Life: Metaphors of Twentieth-Century Biology*. New York: Columbia University Press, 1995.

Keston, Albert S.; Robert P. Ball; V. Kneeland Frantz; and Walter W. Palmer. "Storage of Radioactive Iodine in a Metastasis from Thyroid Carcinoma." *Science* 95 (1942): 362–63.

Kevles, Bettyann. *Naked to the Bone: Medical Imaging in the Twentieth Century*. New Brunswick, NJ: Rutgers University Press, 1997.

Kevles, Daniel J. "The National Science Foundation and the Debate over Postwar Research Policy, 1942–1945: A Political Interpretation of *Science—The Endless Frontier*." *Isis* 68 (1977): 5–26.

———. *The Physicists: The History of a Scientific Community in America*. 1978; reissued Cambridge, MA: Harvard University Press, 1987.

———. "Cold War and Hot Physics: Science, Security, and the American State,

1945–1956." *Historical Studies in the Physical and Biological Sciences* 20 (1990): 239–64.

Kevles, Daniel J., and Gerald L. Geison. "The Experimental Life Sciences in the Twentieth Century." *Osiris* 10 (1995): 97–121.

Kingsland, Sharon E. *Modeling Nature: Episodes in the History of Population Ecology*. Chicago: University of Chicago Press, 1985.

———. "Defining Ecology as a Science." In *Foundations of Ecology: Classic Papers with Commentaries*, ed. Leslie A. Real and James H. Brown. Chicago: University of Chicago Press, 1991, 1–13.

———. *The Evolution of American Ecology, 1890–2000*. Baltimore: Johns Hopkins University Press, 2005.

Kirsch, Scott. *Proving Grounds: Project Plowshare and the Unrealized Dream of Nuclear Earthmoving*. New Brunswick, NJ: Rutgers University Press, 2005.

Klingle, Matthew W. "Plying Atomic Waters: Lauren Donaldson and the 'Fern Lake Concept' of Fisheries Management." *Journal of the History of Biology* 31 (1988): 1–32.

Kluger, Richard. *The Paper: The Life and Death of the* New York Herald Tribune. New York: Alfred A. Knopf, 1986.

Kohler, Robert E. "The Enzyme Theory and the Origin of Biochemistry." *Isis* 64 (1973): 181–96.

———. "The Management of Science: The Experience of Warren Weaver and the Rockefeller Foundation Programme in Molecular Biology." *Minerva* 14 (1976): 279–306.

———. "Rudolf Schoenheimer, Isotopic Tracers, and Biochemistry in the 1930's." *Historical Studies in the Physical Sciences* 8 (1977): 257–98.

———. *Partners in Science: Foundations and Natural Scientists, 1900–1945*. Chicago: University of Chicago Press, 1991.

———. *Lords of the Fly:* Drosophila *Genetics and the Experimental Life*. Chicago: University of Chicago Press, 1994.

———. "Moral Economy, Material Culture, and Community in *Drosophila* Genetics." In *The Science Studies Reader*, ed. Mario Biagioli. New York: Routledge, 1999: 243–57.

———. *Landscapes and Labscapes: Exploring the Lab-Field Border in Biology*. Chicago: University of Chicago Press, 2002.

Kohman, Truman P. "Proposed New Word: Nuclide." *American Journal of Physics* 15 (1947): 356–57.

Kopp, Carolyn. "The Origins of the American Scientific Debate over Fallout Hazards." *Social Studies of Science* 9 (1979): 403–22.

Korszniak, N. "A Review of the Use of Radio-Isotopes in Medicine and Medical Research in Australia (1947–73)." *Australasian Radiology* 41 (1997): 211–19.

Kozloff, Lloyd M., and Frank W. Putnam. "Biochemical Studies of Virus Reproduction, III: The Origin of Virus Phosphorus in the *Escherichia coli* T_6 Bacteriophage System." *Journal of Biological Chemistry* 182 (1950): 229–42.

Kraft, Alison. "Between Medicine and Industry: Medical Physics and the Rise of the Radioisotope 1945–65." *Contemporary British History* 20 (2006): 1–35.

———. "Manhattan Transfer: Lethal Radiation, Bone Marrow Transplantation, and the Birth of Stem Cell Biology, ca. 1942–1961." *Historical Studies in the Natural Sciences* 39 (2009): 171–218.

Krebs, H. A. "Cyclic Processes in Living Matter." *Enzymologia* 12 (1947): 88–100.

Krige, John. "The Politics of Phosphorus-32: A Cold War Fable Based on Fact." *Historical Studies in the Physical and Biological Sciences* 36 (2005): 71–91.

———. "Atoms for Peace, Scientific Internationalism, and Scientific Intelligence." *Osiris* 21 (2006): 161–81.

———. "Technology, Foreign Policy and International Cooperation in Space." In *Critical Issues in the History of Spaceflight*, ed. Steven J. Dick and Roger D. Launius. Washington, DC: NASA, 2006, 239–62.

———. *American Hegemony and the Postwar Reconstruction of Science in Europe*. Cambridge, MA: MIT Press, 2006.

———. "Techno-Utopian Dreams, Techno-Political Realities: The Education of Desire for the Peaceful Atom." In *Utopia-Dystopia: Conditions of Historical Possibility*, ed. Michael D. Gordin, Helen Tilley, and Gyan Prakash. Princeton, NJ: Princeton University Press, 2010, 151–75.

Krumholz, Louis A. *A Summary of Findings of the Ecological Survey of White Oak Creek, Roane County, Tennessee, 1950–1953*, ORO-132. Oak Ridge, TN: Technical Information Service, 1954.

Kutcher, Gerald J. "Cancer Therapy and Military Cold-War Research: Crossing Epistemological and Ethical Boundaries." *History Workshop Journal* 56 (2003): 105–30.

———. *Contested Medicine: Cancer Research and the Military*. Chicago: University of Chicago Press, 2009.

———. "Fast Neutrons for Cancer Therapy: A Case Study of Failure." Unpublished paper presented to the Program in History of Science at Princeton University, 25 Mar 2010.

Kwa, Chunglin. *Mimicking Nature: The Development of Systems Ecology in the United States*, PhD diss., University of Amsterdam, 1989.

———. "Radiation Ecology, Systems Ecology and the Management of the Environment." In *Science and Nature: Essays in the History of the Environmental Sciences*, ed. Michael Shortland. Oxford: British Society for the History of Science, 1993, 213–49.

Landa, Edward R. "Buried Treasure to Buried Waste: The Rise and Fall of the Radium Industry." *Colorado School of Mines Quarterly* 82, no. 2 (1987): i–viii, 1–77.

———. "The First Nuclear Industry." *Scientific American* 247 (Nov 1982): 180–93.

Landecker, Hannah. *Culturing Life: How Cells Became Technologies*. Cambridge, MA: Harvard University Press, 2007.

———. "Living Differently in Time: Plasticity, Temporality and Cellular Biotechnologies." In *Technologized Images, Technologized Bodies: Anthropological Approaches to a New Politics of Vision*, ed. Jeanette Edwards, Penny Harvey, and Peter Wade. New York: Berghahn Books, 2010, 211–36.

———. "Hormones and Metabolic Regulation, circa 1969." Unpublished paper presented at a meeting of the International Society for the History, Philosophy, and Social Studies of Biology, University of Utah, Salt Lake City, 10–15 Jul 2011.

Landon, John. "A Look at the Future with Regard to Immunoassay." In Schönfeld, *New Developments in Immunoassays*, 118–28.

Landsteiner, Karl. *The Specificity of Serological Reactions*. Rev. ed. Cambridge, MA: Harvard University Press, 1945.

Lapp, Ralph E. " 'Fall-Out': Another Dimension in Atomic Killing Power." *New Republic* 132 (14 Feb 1955): 8–12.

———. *The Voyage of the Lucky Dragon*. New York: Harper, 1958.

Lapp, Ralph E., J. T. Kulp, W. R. Eckelmann, and A. R. Schulert. "Strontium-90 in Man." *Science* 125 (1957): 933–34.

Larrabee, Ralph C. "The Severe Anemias of Pregnancy and the Puerperium." *American Journal of the Medical Sciences* 170 (1925): 371–89.

Larsen, Carl. "Midwest Center for Research on Radiation Effect." *Chicago Daily Sun-Times,* 17 Jan 1955, pp. 1, 4.

Larson, Kermit H. "Continental Close-In Fallout: Its History, Measurement and Characteristics." In Schultz and Klement, *Radioecology*, 19–25.

Laurence, William L. "Atomic Key to Life Is Feasible Now." *New York Times*, 9 Oct 1945, p. 6.

———. "Is Atomic Energy the Key to Our Dreams?" *Saturday Evening Post*, 13 Apr 1946, pp. 9–10, 36–37, 39, 41.

———. "Waste Held Peril in Atomic Power." *New York Times*, 17 Dec 1955, p. 12.

Lavine, Matthew. *A Cultural History of Radiation and Radioactivity in the United States, 1895–1945*. PhD diss., University of Wisconsin, Madison, 2008.

Lawrence, Ernest O. "The Biological Action of Neutron Rays." *Radiology* 29 (1937): 313–22.

Lawrence, Ernest O., and Donald P. Cooksey. "On the Apparatus for Multiple Acceleration of Light Ions to High Speeds." *Physical Review* 50 (1936): 1131–40.

Lawrence, John H. "Some Biological Applications of Neutrons and Artificial Radioactivity." *Nature* 145 (1940): 125–27.

———. "Early Experiences in Nuclear Medicine [1956]." *Journal of Nuclear Medicine* 20 (1979): 561–64.

Lawrence, John H.; Paul C. Aebersold; and Ernest O. Lawrence. "Comparative Effects of X-Rays and Neutrons on Normal and Tumor Tissue." *Proceedings of the National Academy of Sciences, USA* 22 (1936): 543–57.

Lawrence, John H., and William U. Gardner. "A Transmissible Leukemia in the 'A' Strain of Mice." *American Journal of Cancer* 33 (1938): 112–19.

Lawrence, John H., and Ernest O. Lawrence. "The Biological Action of Neutron Rays." *Proceedings of the National Academy of Sciences, USA* 22 (1936): 124–33.

Lawrence, John H., and K. G. Scott. "Comparative Metabolism of Phosphorus in Normal and Lymphomatous Animals." *Proceedings of the Society for Experimental Biology and Medicine* 40 (1939): 694–96.

Lawrence, John H.; K. G. Scott; and L. W. Tuttle. "Studies on Leukemia with the Aid of Radioactive Phosphorus." *New International Clinics* 3 (1939): 33–58.

Lawrence, John H., and Cornelius A. Tobias. "Radioactive Isotopes and Nuclear Radiations in the Treatment of Cancer." *Cancer Research* 16 (1956): 185–93.

Lawrence, J. H.; L. W. Tuttle; K. G. Scott; and C. L. Connor. "Studies on Neoplasms with the Aid of Radioactive Phosphorus, I: The Total Phosphorus Metabolism of Normal and Leukemic Mice." *Journal of Clinical Investigation* 19 (1940): 267–71.

LeBaron, Wayne. *America's Nuclear Legacy*. Commack, NY: Nova Science Publishers, 1998.

Lederer, Susan E. *Subjected to Science: Human Experimentation in America before the Second World War*. Baltimore: Johns Hopkins University Press, 1995.

Lenoir, Timothy, and Hays, Marguerite. "The Manhattan Project for Biomedicine." In *Controlling Our Destinies: Historical, Philosophical, Ethical, and Theological Perspectives on the Human Genome Project*, ed. Philip R. Sloan. Notre Dame, IN: University of Notre Dame Press, 2000, 29–62.

Lenoir, Timothy, and Christophe Lécuyer. "Instrument Makers and Discipline Builders: The Case of Nuclear Magnetic Resonance." *Perspectives on Science* 3 (1995): 276–345.

Leopold, Ellen. *Under the Radar: Cancer and the Cold War*. New Brunswick, NJ: Rutgers University Press, 2009.

Lerner, A. P. "Does Control of Atomic Energy Involve a Controlled Economy?" *Bulletin of the Atomic Scientists* 5, no. 1 (1949): 15–16.

Leviero, Anthony. "Atom Bomb By-Product Promises To Replace Radium as Cancer Aid." *New York Times*, 22 Apr 1948, pp. 1, 19.

Lewis, E. B. "Leukemia and Ionizing Radiation." *Science* 125 (1957): 965–72.

Libby, W. F. "The Radiocarbon Story." *Bulletin of the Atomic Scientists* 4, no. 9 (1948): 263–66.

———. "Radioactive Fallout and Radioactive Strontium." *Science* 123 (1956): 657–60.

Lilienthal, David E. "The Atomic Adventure." *Collier's* 119 (3 May 1947): 12, 82, 84–85.

———. "Private Industry and the Public Atom." *Bulletin of the Atomic Scientists* 5, no. 1 (1949): 6–8.

———. "Free the Atom." *Collier's* 125 (17 Jun 1950), pp. 13–15, 54–58.

———. *The Journals of David E. Lilienthal.* 7 vols. New York: Harper, 1964.

Lindee, M. Susan. "What Is a Mutation? Identifying Heritable Change in the Offspring of Survivors at Hiroshima and Nagasaki." *Journal of the History of Biology* 25 (1992): 231–55.

———. *Suffering Made Real: American Science and the Survivors at Hiroshima.* Chicago: University of Chicago Press, 1994.

Lindeman, Raymond L. "The Trophic-Dynamic Aspect of Ecology." *Ecology* 23 (1942): 399–417.

Livingood, J. J., and G. T. Seaborg. "Radioactive Isotopes of Iodine." *Physical Review* 54 (1938): 775–82.

Livingston, M. Stanley. "Early History of Particle Accelerators." *Advances in Electronics and Electron Physics* 50 (1980): 1–88.

Ljungdahl, L. G., and H. G. Wood. "Total Synthesis of Acetate from CO_2 by Heterotrophic Bacteria." *Annual Reviews of Microbiology* 23 (1969): 515–38.

Locksley, Herbert B.; William H. Sweet; Henry J. Powsner; and Elias Dow. "Suitability of Tumor-Bearing Mice for Predicting Relative Usefulness of Isotopes in Brain Tumors." *Archives of Neurology and Psychiatry* 71 (1954): 684–98.

Lowen, Rebecca S. "Entering the Atomic Power Race: Science, Industry, and Government." *Political Science Quarterly* 102 (1987): 459–79.

Luessenhop, Alfred J.; William H. Sweet; and Janette Robinson. "Possible Use of the Neutron Capturing Isotope Lithium^6 in the Radiation Therapy of Brain Tumors." *American Journal of Roentgenology, Radium Therapy, and Nuclear Medicine* 76 (1956): 376–92.

Luessenhop, A. J.; J. C. Gallimore; W. H. Sweet; E. G. Struxness; and J. Robinson. "The Toxicity in Man of Hexavalent Uranium Following Intravenous Administration." *American Journal of Roentgenology, Radium Therapy, and Nuclear Medicine* 79 (1958): 83–100.

Luria, Salvador. "Reactivation of Irradiated Bacteriophage by Transfer of Self-Reproducing Units." *Proceedings of the National Academy of Sciences, USA* 33 (1947): 253–64.

———. "Reactivation of Ultraviolet-Irradiated Bacteriophage by Multiple Infection." *Journal of Cellular and Comparative Physiology* 39, Supp. 1 (1952): 119–23.

———. "Radiation and Viruses." In *Radiation Biology*, ed. Alexander Hollaender, vol. 2: Ultraviolet and Related Radiations. New York: McGraw-Hill, 1955, 333–64.

Luria, Salvador E., and Raymond Latarjet. "Ultraviolet Irradiation of Bacteriophage during Intracellular Growth." *Journal of Bacteriology* 53 (1947): 149–63.

Lutts, Ralph H. "Chemical Fallout: Rachel Carson's *Silent Spring*, Radioactive Fallout, and the Environmental Movement." *Environmental Review* 9 (1985): 210–25.

M. D. Anderson Hospital and Tumor Institute. *The First Twenty Years of the University of Texas M. D. Anderson Hospital and Tumor Institute.* Houston: University of Texas M. D. Anderson Hospital and Tumor Institute, 1964.

Maaløe, Ole, and Gunther S. Stent. "Radioactive Phosphorus Tracer Studies on the Reproduction of T4 Bacteriophage, I: Intracellular Appearance of Phage-Like Material." *Acta Pathologica et Microbiologica Scandinavica* 30 (1952): 149–57.

Maaløe, Ole, and James D. Watson. "The Transfer of Radioactive Phosphorus from Parental to Progeny Phage." *Proceedings of the National Academy of Sciences, USA* 37 (1951): 507–13.

Maisel, Albert Q. "Medical Dividend." *Collier's* 119 (3 May 1947): 14, 43–44.

Mallard, Grégoire. "Quand l'expertise se heurte au pouvoir souverain: La nation américaine face à la prolifération nucléaire, 1945–1953." *Sociologie du Travail* 48 (2006): 367–89.

———. *The Atomic Confederacy: Europe's Quest for Nuclear Weapons and the Making of a New World Order*. PhD diss., Princeton University, 2008.

Malloy, Sean L. "'A Very Pleasant Way to Die': Radiation Effects and the Decision to Use the Atomic Bomb against Japan." *Diplomatic History* 36 (2012): 515–45.

Mann, Charles C. "Radiation: Balancing the Record." *Science* 263 (1994): 470–73.

March, Herman C. "Leukemia in Radiologists in a 20 Year Period." *American Journal of the Medical Sciences* 220 (1950): 282–86.

Marinelli, L. D.; F. W. Foote; R. F. Hill; and A. F. Hocker. "Retention of Radioactive Iodine in Thyroid Carcinomas: Histopathologic and Radio-Autographic Studies." *American Journal of Roentgenology and Radium Therapy* 58 (1947): 17–32.

Marks, Harry M. *The Progress of Experiment: Science and Therapeutic Reform in the United States, 1900–1990.* Cambridge: Cambridge University Press, 1997.

Martin, A. J. P., and R. L. M. Synge. "Analytical Chemistry of the Proteins." *Advances in Protein Chemistry* 2 (1945): 1–83.

Massachusetts Task Force on Human Subject Research. *A Report on the Use of Radioactive Materials in Human Subject Research That Involved Residents of State-Operated Facilities within the Commonwealth of Massachusetts from 1943 through 1973*. Boston: Commonwealth of Massachusetts, Executive Office of Health & Human Services, Dept. of Mental Retardation, 1994.

Mauss, Marcel. *The Gift: Forms and Functions of Exchange in Archaic Societies*. Translated by Ian Cunnison. London: Cohen & West, 1954.

Mazuzan, George T. "Conflict of Interest: Promoting and Regulating the Infant Nuclear Power Industry, 1954–1956." *Historian* 44, no. 1 (1981): 1–14.

Mazuzan, George T., and J. Samuel Walker. *Controlling the Atom: The Beginnings of Nuclear Regulation 1946–1962*. Berkeley: University of California Press, 1985.

McCance, R. A., and E. M. Widdowson. "Mineral Metabolism." *Annual Review of Biochemistry* 13 (1944): 315–46.

McElheny, Victor K. *Drawing the Map of Life: Inside the Human Genome Project*. New York: Basic Books, 2010.

McEnaney, Laura. *Civil Defense Begins at Home: Militarization Meets Everyday Life in the Fifties*. Princeton, NJ: Princeton University Press, 2000.

McIntosh, Robert P. *The Background of Ecology: Concept and Theory*. Cambridge: Cambridge University Press, 1985.

Mealy, John, Jr.; Gordon L. Brownell; and William H. Sweet. "Radioarsenic in Plasma, Urine, Normal Tissues, and Intracranial Neoplasms." *Archives of Neurology and Psychiatry* 81 (1959): 310–20.

Means, J. H., and G. W. Holmes. "Further Observations on the Roentgen-Ray Treatment of Toxic Goiter." *Archives of Internal Medicine* 31 (1923): 303–41.

Medhurst, Martin J. "Atoms for Peace and Nuclear Hegemony: The Rhetorical Structure of a Cold War Campaign." *Armed Forces & Society* 23 (1997): 571–93.

Medvedev, Zhores A. *Soviet Science*. New York: Norton, 1978.

Meselson, Matthew, and Franklin W. Stahl. "The Replication of DNA in *Escherichia coli*." *Proceedings of the National Academy of Science, USA* 44 (1958): 671–82.

Miale, August, Jr. "Nuclear Medicine: Reflections in Time." *Journal of the Florida Medical Association* 82 (1995): 749–50.

Miller, Byron S. "A Law Is Passed: The Atomic Energy Act of 1946." *University of Chicago Law Review* 15 (1948): 799–821.

Miller, Leon L. "George Hoyt Whipple." *Biographical Memoirs of the National Academy of Sciences*. Washington, DC: National Academies Press, 1995, 371–93.

Mirsky, I. Arthur. "The Etiology of Diabetes Mellitus in Man." *Recent Progress in Hormone Research* 7 (1952): 437–67.

Moeller, Dade W.; James G. Terrill Jr.; and Samuel C. Ingraham II. "Radiation Exposure in the United States." *Public Health Reports* 68 (1953): 57–65.

Molella, Arthur. "Exhibiting Atomic Culture: The View from Oak Ridge." *History and Technology* 19 (2003): 211–26.

Moloney, William C., and Marvin A. Kastenbaum. "Leukemogenic Effects of Ionizing Radiation on Atomic Bomb Survivors in Hiroshima City." *Science* 121 (1955): 308–9.

Moon, John Ellis van Courtland. "Project SPHINX: The Question of the Use of Gas in the Planned Invasion of Japan." *Journal of Strategic Studies* 12 (1989): 303–23.

Moore, Carl V. "Iron Metabolism and Nutrition." *Harvey Lectures* 55 (1961): 67–101.

Moore, Carl V., and Reubenia Dubach. "Absorption of Radioiron from Foods." *Science* 116 (1952): 527.

Moore, Carl V.; Reubenia Dubach; Virginia Minnich; and Harold K. Roberts. "Absorption of Ferrous and Ferric Radioactive Iron by Human Subjects and by Dogs." *Journal of Clinical Investigation* 23 (1944): 755–67.

Moore, George E. "Use of Radioactive Diiodofluorescein in the Diagnosis and Localization of Brain Tumors." *Science* 107 (1948): 569–71.

Morgan, Karl Z., and Ken M. Peterson. *The Angry Genie: One Man's Walk through the Nuclear Age*. Norman: University of Oklahoma Press, 1999.

Morris, John D. "Two Uranium Bars Taken from Plant in a Security Test." *New York Times*, 25 May 1949, pp. 1, 15.

———. "Isotopes Shipment to Norse Stirs Row." *New York Times*, 9 Jun 1949, p. 1.

Mukherjee, Siddhartha. *The Emperor of All Maladies: A Biography of Cancer*. New York: Scribner, 2010.

Muller, H. J. "Artificial Transmutation of the Gene." *Science* 66 (1927): 84–87.

———. "The Menace of Radiation." *Science News Letter* 55 (1949): 374, 379–80.

———. "Some Present Problems in the Genetic Effects of Radiation." *Symposium on Radiation Genetics, Journal of Cellular and Comparative Physiology* 35, suppl. 1 (1950): 9–70.

Muller, J. H. "Intraperitoneal Application of Radioactive Colloids." In Hahn, *Therapeutic Use of Artificial Radioisotopes*, 269–94.

Myers, Jack. "Conceptual Developments in Photosynthesis, 1924–1974." *Plant Physiology* 54 (1974): 420–26.

Myers, William G., and Henry N. Wagner Jr. "How It Began." In *Nuclear Medicine*, ed. Henry N. Wagner Jr. New York: HP Publishers, 1975, 3–14.

Nagai, Takashi. *We of Nagasaki: The Story of Survivors in an Atomic Wasteland*. Translated by Ichiro Shirato and Herbert B. L. Silverman. New York: Duell, Sloan, and Pearce, 1951.

National Academy of Sciences. *The Biological Effects of Atomic Radiation: A Report to the Public*. Washington, DC: National Academy of Sciences–National Research Council, 1956.

National Committee on Radiation Protection (US). *Safe Handling of Radioactive Isotopes, Bureau of Standards Handbook 42*. Washington, DC: US Department of Commerce, 1949.

———. *Permissible Dose from External Sources of Ionizing Radiation, Bureau of Standards Handbook 59*. Washington, DC: US Department of Commerce, 1954.

National Research Council. *The Research Attack on Cancer, 1946: A Report on the American Cancer Society Research Program by the Committee on Growth of the National Research Council*. Washington, DC: National Research Council, 1946.

Neel, James V., and William J. Schull. *The Effect of Exposure to the Atomic Bombs on Pregnancy Termination in Hiroshima and Nagasaki*. Washington, DC:

National Academy of Sciences–National Research Council Publication 461, 1956.

Newman, James R., and Byron R. Miller. *The Control of Atomic Energy*. New York: Whittlesey House, 1948.

———. "The Socialist Island." *Bulletin of the Atomic Scientists* 5, no. 1 (1949): 13–15.

Nickelsen, Kärin. *Of Light and Darkness: Modelling Photosynthesis 1840–1960*. Habilitationsschrift eingereicht der Phil.-nat. Fakultät der Universität Bern, 2009.

———. "The Construction of a Scientific Model: Otto Warburg and the Building Block Strategy." *Studies in History and Philosophy of Biological and Biomedical Sciences* 40 (2009): 73–86.

———. "The Path of Carbon in Photosynthesis: How to Discover a Biochemical Pathway." *Ambix* 59, no. 3 (2012): 266–93.

Nickelsen, Kärin, and Govindjee. *The Maximum Quantum Yield Controversy: Otto Warburg and the Midwest-Gang*. Bern: Bern Studies in the History and Philosophy of Science, 2011.

Nickelsen, Kärin, and Graßhoff, Gerd. "Concepts from the Bench: Hans Krebs, Kurt Henseleit and the Urea Cycle." In *Going Amiss in Experimental Research*, Boston Studies in the Philosophy of Science vol. 267, ed. Giora Hon, Jutta Schickore, and Friedrich Steinle. Dordrecht, the Netherlands: Springer Verlag, 2009, 91–117.

Nichols, K. D. *The Road to Trinity*. New York: William Morrow, 1987.

Nier, Alfred O., and Earl A. Gulbransen. "Variations in the Relative Abundance of the Carbon Isotopes." *Journal of the American Chemical Society* 61 (1939): 697–98.

Noddack, W. "Der Kohlenstoff im Haushalt der Natur." *Angewandte Chemie* 50 (1937): 505–10.

Norris, Robert S. *Racing for the Bomb: General Leslie R. Groves, the Manhattan Project's Indispensable Man*. South Royalton, VT: Steerforth Press, 2002.

"Nuclear Enterprise." *Washington Post and Times Herald*, 21 Oct 1958, p. A14.

"The Nuclear Revolution." *Time*, 6 Feb 1956, pp. 83–84.

Nyhart, Lynn K. "Civic and Economic Zoology in Nineteenth-Century Germany: The 'Living Communities' of Karl Möbius." *Isis* 89 (1998): 605–30.

———. *Modern Nature: The Rise of the Biological Perspective in Germany*. Chicago: University of Chicago Press, 2009.

Oak Ridge National Laboratory. *Swords to Plowshares: A Short History of the Oak Ridge National Laboratory*. Oak Ridge, TN: Oak Ridge National Laboratory Office of Public Affairs, 1993. http://www.ornl.gov/info/swords/swords.shtml.

Odum, Eugene P. *Fundamentals of Ecology*. 2nd ed. Philadelphia: W. B. Saunders, 1959.

———. "Organic Production and Turnover in Old Field Succession." *Ecology* 41 (1960): 34–49.

———. "Feedback between Radiation Ecology and General Ecology." *Health Physics* 11 (1965): 1257–62.

———. *Fundamentals of Ecology*. 3rd ed. Philadelphia: W. B. Saunders, 1971.

———. "Early University of Georgia Research, 1952–1962." In *The Savannah River and Its Environs: Proceedings of a Symposium in Honor of Dr. Ruth Patrick for 35 Years of Studies on the Savannah River*. Aiken, SC: E. I. du Pont de Nemours & Co. Savannah River Laboratory; Springfield, VA: National Technical Information Service, 1987, 43–57.

Odum, Eugene P., and Frank B. Golley. "Radioactive Tracers as an Aid to the Measurement of Energy Flow at the Population Level in Nature." In Schultz and Klement, *Radioecology*, 403–10.

Odum, Eugene P., and Edward J. Kuenzler. "Experimental Isolation of Food Chains in an Old-Field Ecosystem with the Use of Phosphorus-32." In Schultz and Klement, *Radioecology*, 113–20.

Odum, Eugene P.; Edward J. Kuenzler; and Sister Marion Xavier Blunt. "Uptake of P^{32} and Primary Productivity in Marine Benthic Algae." *Limnology and Oceanography* 3 (1958): 340–45.

Odum, Howard T., and Eugene P. Odum. "Trophic Structure and Productivity of a Windward Coral Reef Community on Eniwetok Atoll." *Ecological Monographs* 25 (1955): 291–320.

Odum, Howard T., and Robert F. Pigeon, eds. *A Tropical Rain Forest: A Study of Irradiation and Ecology at El Verde, Puerto Rico*. Oak Ridge, TN: US Atomic Energy Commission Division of Technical Information, 1970.

Oettinger, Leon, Jr.; Willard B. Mills; and Paul F. Hahn. "Iron Absorption in Premature and Full-Term Infants." *Journal of Pediatrics* 45 (1954): 302–6.

Olby, Robert C. *The Path to the Double Helix: The Discovery of DNA*. 1974; reprinted New York: Dover, 1994.

Olson, Jerry S. "Analog Computer Models for Movement of Nuclides through Ecosystems." In Schultz and Klement, *Radioecology*, 121–25.

Osgood, Kenneth. *Total Cold War: Eisenhower's Secret Propaganda Battle at Home and Abroad*. Lawrence: University of Kansas Press, 2006.

"Our Defective Race." *Newsweek* 29 (14 Apr 1947): 56.

Pais, Abraham. *Niels Bohr's Times, in Physics, Philosophy, and Polity*. Oxford: Clarendon Press, 1991.

Palfrey, John Gorham. "Atomic Energy: A New Experiment in Government-Industry Relations." *Columbia Law Review* 56 (1956): 367–92.

Pallo, Gabor. "Scientific Recency: George de Hevesy's Nobel Prize." In *Historical Studies in the Nobel Archives: The Prizes in Science and Medicine*, ed. Elisabeth Crawford. Tokyo: University Academy Press, 2002, 65–78.

Parker, H. M. "Health Physics, Instrumentation, and Radiation Protection." *Advances in Biological and Medical Physics* 1 (1948): 223–85.

Pasveer, Bernike. "Knowledge of Shadows: The Introduction of X-Ray Images in Medicine." *Sociology of Health & Illness* 11 (1989): 360–81.

Patterson, James T. *The Dread Disease: Cancer and Modern American Culture.* Cambridge, MA: Harvard University Press, 1987.

Paul, Septimus H. *Nuclear Rivals: Anglo-American Atomic Relations, 1941–1952.* Columbus: Ohio State University Press, 2000.

Pauling, Linus. "Effect of Strontium-90." *New York Times*, 28 Apr 1959, p. 34.

Pendleton, Robert C., and A. W. Grundmann. "Use of P^{32} in Tracing Some Insect-Plant Relationships of the Thistle, *Cirsium undulatum*." *Ecology* 35 (1954): 187–91.

Perlman, I.; S. Ruben; and I. L. Chaikoff. "Radioactive Phosphorus as an Indicator of Phospholipid Metabolism." *Journal of Biological Chemistry* 122 (1937): 169–82.

Phelps, Michael E.; E. J. Hoffman; N. A. Mullani; and M. M. Ter-Pogossian. "Application of Annihilation Coincidence Detection to Transaxial Reconstruction Tomography." *Journal of Nuclear Medicine* 16 (1975): 210–24.

Phillips, John. "The Biotic Community." *Journal of Ecology* 19 (1931): 1–24.

Pickering, Andrew. *The Mangle of Practice: Time, Agency, and Science.* Chicago: University of Chicago Press, 1995.

Pickstone, John V. "Contested Cumulations: Configurations of Cancer Treatments through the Twentieth Century." *Bulletin of the History of Medicine* 81 (2007): 164–96.

Pontecorvo, Guido. "Genetic Formulation of Gene Structure and Gene Action." *Advances in Enzymology* 13 (1952): 121–49.

Price, Matt. "Roots of Dissent: The Chicago Met Lab and the Origins of the Franck Report." *Isis* 86 (1995): 222–44.

Price, Richard M. *The Chemical Weapons Taboo.* Ithaca, NY: Cornell University Press, 1997.

Proctor, Robert N. "Expert Witnesses Take the Stand." *Nature* 407 (2000): 15–16.

———. *Golden Holocaust: Origins of the Cigarette Catastrophe and the Case for Abolition.* Berkeley: University of California Press, 2011.

Public Affairs Office, Brookhaven National Laboratory. "Celebrating 50 Years of Nuclear Medicine Research." *Journal of Nuclear Medicine* 38, no. 9 (1997): 21N, 44N.

Putnam, Frank W., and Kozloff, Lloyd M. "On the Origin of Virus Phosphorus." *Science* 108 (1948): 386–87.

———. "Biochemical Studies of Virus Reproduction, IV: The Fate of the Infecting Virus Particle." *Journal of Biological Chemistry* 182 (1950): 243–50.

Quist, Arvin S. "A History of Classified Activities at Oak Ridge National Laboratory," 29 Sep 2000, ORCA-7, report from Oak Ridge Classification Associates, http://www.ornl.gov/~webworks/cppr/y2001/rpt/109903.pdf.

Rader, Karen A. *Making Mice: Standardizing Animals for American Biomedical Research, 1900–1955*. Princeton, NJ: Princeton University Press, 2004.

———. "Alexander Hollaender's Postwar Vision for Biology: Oak Ridge and Beyond." *Journal of the History of Biology* 39 (2006): 685–706.

"Radioactive Rays Held Peril to Race: Dr. H. J. Muller, Nobel Prize Winner, Warns of Exposure Changing Germ Cells." *New York Times*, 2 Apr 1947, p. 38.

"The Radioisotope Business . . . Is Booming." *Business Week*, 19 Jan 1963, pp. 50–52.

Rainger, Ronald. "'A Wonderful Oceanographic Tool': The Atomic Bomb, Radioactivity and the Development of American Oceanography." In *The Machine in Neptune's Garden: Historical Perspectives on Technology and the Marine Environment*, ed. Helen W. Rozwadowski and David K. van Keuren. Sagamore Beach, MA: Science History Publications, 2004, 93–131.

———. "Going from Blue to Green? American Oceanographers and the Environment." Unpublished paper presented at the History of Science Society Meeting, 20 Nov 2004.

Rasmussen, Nicolas. "The Mid-Century Biophysics Bubble: Hiroshima and the Biological Revolution in America, Revisited." *History of Science* 35 (1997): 245–93.

———. *Picture Control: The Electron Microscope and the Transformation of Biology in America, 1940–1960*. Stanford, CA: Stanford University Press, 1997.

Rego, Brianna. "The Polonium Brief: A Hidden History of Cancer, Radiation, and the Tobacco Industry." *Isis* 100 (2009): 453–84.

Reichle, D. E., and S. I. Auerbach. "U. S. Radioecology Research Programs of the Atomic Energy Commission in the 1950s." ORNL/TM-2003/280, Dec 2003. http://www.ornl.gov/~webworks/cppr/y2001/rpt/119234.pdf.

Reingold, Nathan. "Vannevar Bush's New Deal for Research: or The Triumph of the Old Order." *Historical Studies in the Physical and Biological Sciences* 17 (1987): 299–344.

Reinhard, Edward H.; Charles L. Neely; and Don M. Samples. "Radioactive Phosphorus in the Treatment of Chronic Leukemias: Long-Term Results over a Period of 15 Years." *Annals of Internal Medicine* 50 (1959): 942–58.

Rentetzi, Maria. "Gender, Politics, and Radioactivity Research in Interwar Vienna: The Case of the Institute for Radium Research." *Isis* 95 (2004): 359–93.

———. *Trafficking Materials and Gendered Experimental Practices: Radium Research in Early Twentieth-Century Vienna*. New York: Columbia University Press, 2008. http://www.gutenberg-e.org/rentetzi/.

"Report of the AEC Industrial Advisory Group." *Bulletin of the Atomic Scientists* 5, no.2 (1949): 51–56.

"Report on Hiroshima: Thousands of Babies, No A-Bomb Effects." *US News & World Report*, 8 Apr 1955, pp. 46–48.

"The Revised McMahon Bill." *Bulletin of the Atomic Scientists* 1, no. 9 (1946): 2–5.

Rheinberger, Hans-Jörg. *Toward a History of Epistemic Things: Synthesizing Proteins in the Test Tube*. Stanford, CA: Stanford University Press, 1997.

———. "Putting Isotopes to Work: Liquid Scintillation Counters, 1950–1970." In Joerges and Shinn, *Instrumentation: Between Science, State and Industry*, 143–74.

———. "Physics and Chemistry of Life: Commentary." In *The Science–Industry Nexus: History, Policy, Implications*, ed. Karl Grandin, Nina Wormbs, and Sven Widmalm, 221–25. Sagamore Beach, MA: Science History Publications, 2004.

———. *An Epistemology of the Concrete: Twentieth-Century Histories of Life*. Durham, NC: Duke University Press, 2010.

———. "In vitro." Unpublished manuscript.

Rhodes, Richard. *The Making of the Atomic Bomb*. New York: Simon & Schuster, 1986.

Ridenour, Louis N. "How Effective Are Radioactive Poisons in Warfare?" *Bulletin of the Atomic Scientists* 6, no. 7 (1950): 199–202, 224.

Riley, Gordon A. "Limnological Studies in Connecticut, I: General Limnological Survey, II: The Copper Cycle." *Ecological Monographs* 9 (1939): 53–94.

———. "Limnological Studies in Connecticut, III: The Plankton of Linsley Pond." *Ecological Monographs* 10 (1940): 279–306.

Riley, Gordon A., with an introduction by G. Evelyn Hutchinson. "The Carbon Metabolism and Photosynthetic Efficiency of the Earth as a Whole." *American Scientist* 32 (1944): 129–34.

Robeck, Gordon G.; Croswell Henderson; and Ralph C. Palange. *Water Quality Studies on the Columbia River*. Special Report, US Public Health Service. Cincinnati, OH: Taft Sanitary Engineering Center, 1954.

Rohrmann, C. A. "Hanford Isotopes Plant." *Isotopes and Radiation Technology* 2, no. 2 (1964–65), pp. 99–123.

Rolph, Elizabeth S. *Nuclear Power and the Public Safety: A Study in Regulation*. Lexington, MA: Lexington Books, 1979.

Rosenblueth, Arturo, and Norbert Wiener. "Purposeful and Non-purposeful Behavior." *Philosophy of Science* 17 (1950): 318–26.

Rosenblueth, Arturo; Norbert Wiener; and Julian Bigelow. "Behavior, Purpose and Teleology." *Philosophy of Science* 10 (1943): 18–24.

Ross, Joseph F. "Radioisotope Division." *Boston Medical Quarterly* 2 (Jun 1951): 38–41.

Roswit, Bernard; J. Sorrentino; and Rosalyn Yalow. "The Use of Radio-active Phosphorus (P^{32}) in the Diagnosis of Testicular Tumors: A Preliminary Report." *The Journal of Urology* 63 (1950): 724–28.

Roth, J.; S. M. Glick; R. S. Yalow; and S. A. Berson. "Hypoglycemia: A Potent Stimulus to Secretion of Growth Hormone." *Science* 140 (1963): 987–88.

Rothman, David J. *Strangers at the Bedside: A History of How Law and Bioethics Transformed Medical Decision Making*. New York: Basic Books, 1991.

———. "Serving Clio and Client: The Historian as Expert Witness." *Bulletin of the History of Medicine* 77 (2003): 25–44.

Rothschild, Rachel. "Environmental Awareness in the Atomic Age: Radioecologists and Nuclear Technology." *Historical Studies in the Natural Sciences*, 43 forthcoming.

Rotter, Andrew J. *Hiroshima: The World's Bomb*. Oxford: Oxford University Press, 2008.

Ruben, S. "Photosynthesis and Phosphorylation." *Journal of the American Chemical Society* 65 (1943): 279–82.

Ruben, S.; W. Z. Hassid; and M. D. Kamen. "Radioactive Carbon in the Study of Photosynthesis." *Journal of the American Chemical Society* 61 (1939): 661–63.

Ruben, S., and M. D. Kamen. "Photosynthesis with Radioactive Carbon, IV: Molecular Weight of the Intermediate Products and a Tentative Theory of Photosynthesis." *Journal of the American Chemical Society* 62 (1940): 3451–55.

———. "Radioactive Carbon in the Study of Respiration in Heterotrophic Systems." *Proceedings of the National Academy of Sciences, USA* 26 (1940): 418–22.

Ruben, S.; M. D. Kamen; and W. Z. Hassid. "Photosynthesis with Radioactive Carbon, II: Chemical Properties of the Intermediates." *Journal of the American Chemical Society* 62 (1940): 3443–50.

Ruben, S.; M. D. Kamen; and L. Perry. "Photosynthesis with Radioactive Carbon, III: Ultracentrifugation of Intermediate Products." *Journal of the American Chemical Society* 62 (1940): 3450–51.

Ruben, Samuel; Merle Randall; Martin Kamen; and James Logan Hyde. "Heavy Oxygen (O^{18}) as a Tracer in the Study of Photosynthesis." *Journal of the American Chemical Society* 63 (1941): 877–89.

Rupp, A. F., and E. E. Beauchamp. "The Early Days of the Radioisotope Production Program." *Isotopes and Radiation Technology* 4, no. 1 (1966): 33–40.

Ryther, John H. "The Measurement of Primary Production." *Limnology & Oceanography* 1 (1956): 72–84.

Santesmases, María Jesús. "Peace Propaganda and Biomedical Experimentation: Influential Uses of Radioisotopes in Endocrinology and Molecular Genetics in Spain (1947–1971)." *Journal of the History of Biology* 39 (2006): 765–94.

———. "From Prophylaxis to Atomic Cocktail: Circulation of Radioiodine." *Dynamis* 29 (2009): 337–63.

———. "Life and Death in the Atomic Era." *Historical Studies in the Natural Sciences* 40 (2010): 409–18.

Schloegel, Judith Johns. "The 'Nuclear Revolution' Is Over: Nuclear Energy and the Origins of Environmental Science at Argonne National Laboratory." Unpublished manuscript, 2011.

Schloegel, Judith Johns, and Karen A. Rader. *Ecology, Environment and "Big Science": An Annotated Bibliography of Sources on Environmental Research at Argonne National Laboratory, 1955–1985*. ANL/HIST-4. Chicago: Argonne National Laboratory, 2005. http://ipd.anl.gov/anlpubs/2005/12/54867.pdf.

Schneider, Daniel W. "Local Knowledge, Environmental Politics, and the Founding of Ecology in the United States: Stephen Forbes and 'The Lake as a Microcosm' (1887)." *Isis* 91 (2000): 681–705.

Schoenheimer, Rudolph. *The Dynamic State of Body Constituents*. Cambridge, MA: Harvard University Press, 1942.

Schoenheimer, Rudolph, and D. Rittenberg. "The Application of Isotopes to the Study of Intermediary Metabolism." *Science* 87 (1938): 221–26.

Schönfeld, H., ed. *New Developments in Immunoassays*. Vol. 26 of *Antibiotics & Chemotherapy*. Basel: S. Karger, 1979.

Schoolman, Harold M., and Steven O. Schwartz. "Aplastic Anemia Secondary to Intravenous Therapy with Radiogold." *Journal of the American Medical Association* 160 (1956): 461–63.

Schrag, Zachary M. *Ethical Imperialism: Institutional Review Boards and the Social Sciences, 1965–2009*. Baltimore: Johns Hopkins University Press, 2010.

Schulz, Milford D. "The Supervoltage Story." *American Journal of Roentgenology, Radium Therapy, and Nuclear Medicine* 124 (1975): 541–59.

Schultz, Vincent, and Alfred W. Klement Jr., eds. *Radioecology: Proceedings of the First National Symposium on Radioecology Held at Colorado State University, Fort Collins, Colorado, September 10–15, 1961*. New York: Reinhold and American Institute of Biological Sciences, 1963.

Schwartz, Rebecca Press. *The Making of the History of the Atomic Bomb: Henry DeWolf Smyth and the Historiography of the Manhattan Project*. PhD diss., Princeton University, 2008.

Schwerin, Alexander von. "Prekäre Stoffe. Radiumökonomie, Risikoepisteme und die Etablierung der Radioindikatortechnik in der Zeit des Nationalsozialismus." *NTM Zeitschrift für Geschichte der Wissenschaften, Technik und Medizin* 17 (2009): 5–33.

———. "Österreichs im Atomzeitalter: Anschluss an die Ökonomie der Radioisotope." In *Kernforschung in Österreich. Wandlungen eines interdisziplinären Forschungsfeldes, 1900–1978*, ed. Silke Fengler and Carola Sachse, 367–94. Vienna: Böhlau, 2012.

"Scientific Monopoly." *New York Herald Tribune*, 21 Jul 1947, section V, p. 18.

Scott, K. G., and S. F. Cook. "The Effect of Radioactive Phosphorus upon the Blood of Growing Chicks." *Proceedings of the National Academy of Sciences, USA* 23 (1937): 265–72.

Scott, K. G., and J. H. Lawrence. "Effect of Radio-Phosphorus on Blood of Monkeys." *Proceedings of the Society for Experimental Biology and Medicine* 48 (1941): 155–58.

Seaborg, Glenn T. "Artificial Radioactivity." *Chemical Reviews* 27 (1940): 199–285.

———. "Artificial Radioactive Tracers." *Science* 105 (1947): 349–54.

Seaborg, Glenn T., and Andrew A. Benson. "Melvin Calvin." *Biographical Memoirs of the National Academy of Sciences* (1998): 1–21.

Seidel, Robert. "Accelerating Science: The Postwar Transformation of the Lawrence Radiation Laboratory." *Historical Studies in the Physical Sciences* 13 (1983): 375–400.

———. "The Origins of the Lawrence Berkeley Laboratory." In Galison and Hevly, *Big Science*, 21–45.

Seidlin, S. M.; L. D. Marinelli; and Eleanor Oshry. "Radioactive Iodine Therapy: Effects on Functioning Metastases of Adenocarcinoma of the Thyroid." *Journal of the American Medical Association* 132 (1946): 838–47.

Seidlin, S. M.; E. Siegel; S. Melamed; and A. A. Yalow. "Occurrence of Myeloid Leukemia in Patients with Metastatic Thyroid Carcinoma Following Prolonged Massive Radioiodine Therapy." *Bulletin of the New York Academy of Medicine* 31 (1955): 410.

Seil, Harvey A.; Charles H. Viol; and M. A. Gordon. "The Elimination of Soluble Radium Salts Taken Intravenously and Per Os." *New York Medical Journal* 101 (1915): 896–98.

Selverstone, B.; A. K. Solomon; and W. H. Sweet. "Location of Brain Tumors by Means of Radioactive Phosphorus." *Journal of the American Medical Association* 140 (1949): 277–78.

Selverstone, Bertram; William H. Sweet; and Richard J. Ireton. "Radioactive Potassium, a New Isotope for Brain Tumor Localization." *Surgical Forum* (1950): 371–75.

Selverstone, Bertram; William H. Sweet; and Charles V. Robinson. "The Clinical Use of Radioactive Phosphorus in the Surgery of Brain Tumors." *Annals of Surgery* 130 (1949): 643–50.

Semendeferi, Ioanna. "Legitimating a Nuclear Critic: John Gofman, Radiation Safety, and Cancer Risks." *Historical Studies in the Natural Sciences* 38 (2008): 259–301.

Serber, Robert, with Robert P. Crease. *Peace and War: Reminiscences of a Life on the Frontiers of Science*. New York: Columbia University Press, 1998.

Serwer, Daniel Paul. *The Rise of Radiation Protection: Science, Medicine and Technology in Society, 1896–1935*. PhD diss., Princeton University, 1976.

Shanks, R. E., and H. R. DeSelm. "Factors Related to Concentration of Radiocesium in Plants Growing on a Radioactive Waste Disposal Area." In Schultz and Klement, *Radioecology*, 97–101.

Shapin, Steven. "The Invisible Technician." *American Scientist* 77 (1989): 554–63.

———. *A Social History of Truth: Civility and Science in Seventeenth-Century England*. Chicago: University of Chicago Press, 1994.

Shapiro, Martin. "APA: Past, Present, Future." *Virginia Law Review* 72 (1986): 447–92.

Shaughnessy, Donald F. *The Story of the American Cancer Society*. PhD diss., Columbia University, 1957.

Shaw, George. "Applying Radioecology in a World of Multiple Contaminants." *Journal of Environmental Radioactivity* 81 (2005): 117–30.

Sheppard, C. W.; J. P. Goodell; and P. F. Hahn. "Colloidal Gold Containing the Radioactive Isotope Au198 in the Selective Internal Radiation Therapy of Diseases of the Lymphoid System." *Journal of Laboratory and Clinical Medicine* 32 (1947): 1437–41.

Silberstein, H. E.; W. N. Valentine; W. L. Minto; J. S. Lawrence; R. M. Fink; and A. T. Gorham. "Studies of Polonium Metabolism in Human Subjects." In Fink, *Biological Studies with Polonium,* 122–53.

Silverstein, Arthur M. *A History of Immunology*. San Diego, CA: Academic Press, 1989.

Simpson, C. A. "X-Ray Treatment of Hyperthyroidism and Toxic Goiter." *Radiology* 3 (1924): 427–31.

Simpson, C. L.; L. H. Hempelmann; and L. M. Fuller. "Neoplasia in Children Treated with X-Rays in Infancy for Thymic Enlargement." *Radiology* 64 (1955): 840–45.

Siri, William E. *Isotopic Tracers and Nuclear Radiations with Applications to Biology and Medicine*. New York: McGraw-Hill, 1949.

Slack, Nancy G. "G. Evelyn Hutchinson." In *New Dictionary of Scientific Biography*, ed. Noretta Koertge. Detroit, MI: Charles Scribner's Sons/Thomson Gale, 2008, vol. 21, 410–18.

———. *G. Evelyn Hutchinson and the Invention of Modern Ecology*. New Haven, CT: Yale University Press, 2010.

Slaney, Patrick David. "Eugene Rabinowitch, the *Bulletin of the Atomic Scientists*, and the Nature of Scientific Internationalism in the Early Cold War." *Historical Studies in the Natural Sciences* 42 (2012): 114–42.

Slater, Leo B. "Instruments and Rules: R. B. Woodward and the Tools of Twentieth-Century Organic Chemistry." *Studies in History and Philosophy of Science* 33 (2002): 1–33.

Sloan, Phillip R., and Brandon Fogel, eds. *Creating a Physical Biology: The Three-Man Paper and Early Molecular Biology*. Chicago: University of Chicago Press, 2011.

Smith, Alice Kimball. *A Peril and a Hope: The Scientists' Movement in America, 1945–47*. Chicago: University of Chicago Press, 1965.

Smith, Michael. "Looming Isotope Shortage Has Clinicians Worried." MedPage-

Today.com. Published 16 Feb 2010. http://www.medpagetoday.com/Radiology/NuclearMedicine/18495.

Smith-Howard, Kendra D. *Perfecting Nature's Food: A Cultural and Environmental History of Milk in the United States, 1900–1970*. PhD diss., University of Wisconsin, Madison, 2007.

Smuts, Jan Christiaan. *Holism and Evolution*. London: Macmillan, 1926.

Smyth, Henry DeWolf. *Atomic Energy for Military Purposes: The Official Report on the Development of the Atomic Bomb under the Auspices of the United States Government, 1940–1945*. Princeton, NJ: Princeton University Press, 1945. Reprint, Stanford, CA: Stanford University Press, 1989.

Soapes, Thomas F. "A Cold Warrior Seeks Peace: Eisenhower's Strategy for Nuclear Disarmament." *Diplomatic History* 4 (1980): 57–71.

Stahl, Franklin W. "Radiobiology of Bacteriophage." In *The Viruses: Biochemical, Biological and Biophysical Properties*. Vol. 2 of *Plant and Bacterial Viruses*, ed. F. M. Burnet and Wendell M. Stanley. New York: Academic Press, 1959, 353–85.

———. "The Effects of the Decay of Incorporated Radioactive Phosphorus on the Genome of Bacteriophage T4." *Virology* 2 (1956): 206–34.

———, ed. *We Can Sleep Later: Alfred D. Hershey and the Origins of Molecular Biology*. Cold Spring Harbor, NY: Cold Spring Harbor Laboratory Press, 2000.

Stang, L. G., Jr.; W. D. Tucker; H. O. Banks Jr.; R. F. Doering; and T. H. Mills. "Production of Iodine-132." *Nucleonics* 12, no. 8 (1954): 22–24.

Stannard, J. Newell. *Radioactivity and Health: A History*, 3 vols. Springfield, VA: National Technical Information Service, 1988.

Stark, Laura. *Behind Closed Doors: IRBs and the Making of Ethical Research*. Chicago: University of Chicago Press, 2012.

Steemann Nielsen, E. "The Use of Radio-active Carbon (C^{14}) for Measuring Organic Production in the Sea." *Journal du Conseil International pour l'Exploration de la Mer* 18 (1954): 117–40.

Stent, Gunther S. "Cross Reactivation of Genetic Loci of T2 Bacteriophage after Decay of Incorporated Radioactive Phosphorus." *Proceedings of the National Academy of Sciences, USA* 39 (1953): 1234–41.

———. "Mortality Due to Radioactive Phosphorus as an Index to Bacteriophage Development." *Cold Spring Harbor Symposia on Quantitative Biology* 18 (1953): 255–59.

———. "Decay of Incorporated Radioactive Phosphorus during Reproduction of Bacteriophage T2." *Journal of General Physiology* 38 (1955): 853–65.

———. *Molecular Biology of Bacterial Viruses*. San Francisco, CA: W. H. Freeman, 1963.

———. *Molecular Genetics: An Introductory Narrative*. San Francisco, CA: W. H. Freeman, 1971.

Stent, Gunther S., and Clarence R. Fuerst. "Inactivation of Bacteriophages by Decay of Incorporated Radioactive Phosphorus." *Journal of General Physiology* 38 (1955): 441–58.

Stent, Gunther S.; Gordon H. Sato; and Niels K. Jerne. "Dispersal of the Parental Nucleic Acid of Bacteriophage T4 among Its Progeny." *Journal of Molecular Biology* 1 (1959): 134–46.

Stepka, W.; A. A. Benson; and M. Calvin. "The Path of Carbon in Photosynthesis, II: Amino Acids." *Science* 108 (1948): 304.

Stevenson, Adlai E. "Why I Raised the H-Bomb Question." *Look* 21, no. 3 (5 Feb 1957): 23–25.

Stewart, Alice; Josefine Webb; Dawn Giles; and David Hewitt. "Malignant Disease in Childhood and Diagnostic Irradiation in Utero." *Lancet* 268 (1956): 447.

Stone, Robert S., ed. *Industrial Medicine on the Plutonium Project: Survey and Collected Papers*. Vol. 20 of National Nuclear Energy Series, Manhattan Project Technical Section. Division 4, Plutonium Project. New York: McGraw Hill, 1951.

Stone, Robert S. "The Concept of a Maximum Permissible Exposure." *Radiology* 58 (1952): 639–61.

Sturdy, Steve. "Looking for Trouble: Medical Science and Clinical Practice in the Historiography of Modern Medicine." *Social History of Medicine* 24 (2011): 739–57.

Straight, Michael. "The Ten-Month Silence." *New Republic* 132, no. 10 (7 Mar 1955): 8–11.

Strasser, Bruno. "Restriction Enzymes in the Atomic Age." Unpublished manuscript, 2005.

———. *La fabrique d'une nouvelle science: La biologie moléculaire à l'âge atomique (1945–1964)*. Florence: Leo S. Olschki Editore, 2006.

Straus, Eugene. *Rosalyn Yalow, Nobel Laureate: Her Life and Work in Medicine: A Biographical Memoir*. New York: Plenum, 1998.

Strauss, Lewis L. *Men and Decisions*. Garden City, NY: Doubleday, 1962.

Strauss, Maurice B. "The Use of Drugs in the Treatment of Anemia." *Journal of the American Medical Association* 107 (1936): 1633–36.

Strickland, Stephen P. *Politics, Science, & Dread Disease: A Short History of United States Medical Research Policy*. Cambridge, MA: Harvard University Press, 1972.

Strong, Leonell C. "The Establishment of the 'A' Strain of Inbred Mice." *Journal of Heredity* 27 (1936): 21–24.

"The Strontium-90 Debate." *America* 97 (15 Jun 1957): 318.

Struxness, E. G.; A. J. Luessenhop; S. R. Bernard; and J. C. Gallimore. "The Distribution and Excretion of Hexavalent Uranium in Man." *Proceedings of the International Conference on the Peaceful Uses of Atomic Energy, Held in Geneva 8 August–20 August 1955*, vol. 10. New York: United Nations, 1956, 186–96.

Sturtevant, A. H. "Social Implications of the Genetics of Man." *Science* 120 (1954): 405–7.

Summers, William C. "Concept Migration: The Case of Target Theories in Physics and Biology." Unpublished paper prepared for the History of Science Society meeting, 26–29 Oct 1995.

Svensson, Hans, and Torsten Landberg. "Neutron Therapy–The Historical Background." *Acta Oncologica* 33 (1994): 227–31.

Sweet, William H. "The Uses of Nuclear Disintegration in the Diagnosis and Treatment of Brain Tumor." *New England Journal of Medicine* 245 (1951): 875–78.

———. "Early History of Development of Boron Neutron Capture Therapy of Tumors." *Journal of Neuro-Oncology* 33 (1997): 19–26.

Sweet, William H., and Gordon L. Brownell. "The Use of Radioactive Isotopes in the Detection and Localization of Brain Tumors." In Andrews, Brucer, and Anderson, *Radioisotopes in Medicine*, 211–18.

———. "Localization of Intracranial Lesions by Scanning with Positron-Emitting Arsenic." *Journal of the American Medical Association* 157 (1955): 1183–88.

Sweet, William H., and Manucher Javid. "The Possible Use of Neutron-Capturing Isotopes such as $Boron^{10}$ in the Treatment of Neoplasms, I: Intracranial Tumors." *Journal of Neurosurgery* 9 (1952): 200–209.

Sylves, Richard T. *Nuclear Oracles: A Political History of the General Advisory Committee of the Atomic Energy Commission, 1947–1977*. Ames: Iowa State University Press, 1987.

Szybalski, Waclaw. "In Memoriam: Alfred D. Hershey (1908–1997)." In Stahl, *We Can Sleep Later*, 19–22.

Tansley, A. G. "The Use and Abuse of Vegetational Concepts and Terms." *Ecology* 16 (1935): 284–307.

Tape, Gerald F., and J. M. Cork. "Induced Radioactivity in Tellurium." *Physical Review* 53 (1938): 676–77.

Taylor, Lauriston S. *Organization for Radiation Protection: The Operations of the ICRP and NCRP, 1928–1974*. Springfield, VA: National Technical Information Service, 1979.

Taylor, Peter J. "Technocratic Optimism, H. T. Odum, and the Partial Transformation of Ecological Metaphor after World War II." *Journal of the History of Biology* 21 (1988): 213–44.

Ter-Pogossian, Michel M. "The Origins of Positron Emission Tomography." *Seminars in Nuclear Medicine* 22 (1992): 140–49.

Ter-Pogossian, Michel M.; Marcus E. Raichle; and Burton E. Sobel. "Positron-Emission Tomography." *Scientific American* 243, no. 4 (1980): 170–81.

"Therapeutic Uses of Radiation." *Radiologic Technology* 67 (1995): 65–68.

Thienemann, A. "Lebengemeinschaft und Lebensraum." *Naturwissenschaftlich Wochenschrift*, N.F. 17 (1918): 281–90, 297–303.

Thomas, J. P. "Russia Grabs Our Inventions." *American* 143 (Jun 1947): 16–19.

Thomas, J. P., and S. V. Jones. "Reds in Our Atom-Bomb Plants." *Liberty: A Magazine of Religious Freedom* (21 Jun 1947): 15, 90–93.

Thompson, E. P. *Customs in Common*. New York: New Press, 1991.

Tilyou, Sarah M. "The Evolution of Positron Emission Tomography." *Journal of Nuclear Medicine* 32, no. 4 (1991): 15N–19N, 23N–26N.

Tivnan, Frank, Jr. "Firm's Annual Report—Output: Half a Pound; Sales: $1 Million." *Boston Sunday Herald*, 6 May 1962, p. 76A.

"To Live—or Die—with It." *Newsweek* 45 (28 Feb 1955): 19–21.

Tobey, Ronald C. *Saving the Prairies: The Life Cycle of the Founding School of American Plant Ecology, 1895–1955*. Berkeley: University of California Press, 1981.

Tobias, C. A.; P. P. Weymouth; L. R. Wasserman; and G. E. Stapleton. "Some Biological Effects Due to Nuclear Fission." *Science* 107 (1948): 115–18.

Todes, Daniel P. "Pavlov's Physiology Factory." *Isis* 88 (1997): 205–46.

———. *Pavlov's Physiology Factory: Experiment, Interpretation, Laboratory Enterprise*. Baltimore: Johns Hopkins University Press, 2002.

Turchetti, Simone. "'For Slow Neutrons, Slow Pay': Enrico Fermi's Patent and the U. S. Atomic Energy Program, 1938–1953." *Isis* 97 (2006): 1–27.

———. "The Invisible Businessman: Nuclear Physics, Patenting Practices, and Trading Activities in the 1930s." *Historical Studies in the Physical and Biological Sciences* 37 (2006): 153–72.

———. "A Contentious Business: Industrial Patents and the Production of Isotopes, 1930–1960." *Dynamis* 29 (2009): 191–218.

———. *The Pontecorvo Affair: A Cold War Defection and Nuclear Physics*. Chicago: University of Chicago Press, 2012.

Tybout, Richard A. *Government Contracting in Atomic Energy*. Ann Arbor: University of Michigan Press, 1956.

Uhl, Michael, and Tod Ensign. *GI Guinea Pigs: How the Pentagon Exposed Our Troops to Dangers More Deadly than War: Agent Orange and Atomic Radiation*. New York: Playboy Press, 1980.

Underwood, E. J. *Trace Elements in Human and Animal Nutrition*. New York: Academic Press, 1956.

Unger, R. H.; A. M. Eisentraut; M. S. McCall; S. Keller; H. C. Lanz; and L. L. Madison. "Glucagon Antibodies and Their Use for Immunoassay for Glucagon." *Proceedings of the Society for Experimental Biology and Medicine* 102 (1959): 621–23.

US Army Corps of Engineers. *Manhattan Project: Official History and Documents*, 14 microfilm reels. Washington, DC: National Archives and Records Service, 1976.

US Atomic Energy Commission (AEC). *Second Semiannual Report*. Washington, DC: US Government Printing Office, 1947.

———. *Fourth Semiannual Report*. Washington, DC: US Government Printing Office, 1948.

———. *Recent Scientific and Technical Developments in the Atomic Energy Program of the United States*. Washington, DC: US Government Printing Office, 1948.
———. *Sixth Semiannual Report*. Washington, DC: US Government Printing Office, 1949.
———. *Isotopes: A Three-Year Summary of Distribution with Extensive Bibliography*. Washington, DC: US Government Printing Office, 1949.
———. *Atomic Energy and the Life Sciences*. Washington, DC: US Government Printing Office, 1949.
———. *Eighth Semiannual Report*. Washington, DC: US Government Printing Office, 1950.
———. *Ninth Semiannual Report*. Washington, DC: US Government Printing Office, 1951.
———. *Tenth Semiannual Report*. Washington, DC: US Government Printing Office, 1951.
———. *Isotopes: A Five-Year Summary of Distribution with Bibliography*. Washington, DC: US Government Printing Office, 1951.
———. *Some Applications of Atomic Energy in Plant Science*. Washington, DC: US Government Printing Office, 1952.
———. *Twelfth Semiannual Report*. Washington, DC: US Government Printing Office, 1952.
———. *Thirteenth Semiannual Report*. Washington, DC: US Government Printing Office, 1953.
———. *Fifteenth Semiannual Report*. Washington, DC: US Government Publishing Office, 1954.
———. *Seventeenth Semiannual Report*. Washington, DC: US Government Printing Office, 1955.
———. *Eighteenth Semiannual Report*. Washington, DC: US Government Printing Office, 1955.
———. *Eight-Year Isotope Summary*, vol. 7 of *Selected Reference Material, United States Energy Program*. Washington, DC: US Government Printing Office, 1955.
———. *Twentieth Semiannual Report*. Washington, DC: US Government Printing Office, 1956.
———. *Twenty-Second Semiannual Report*. Washington, DC: US Government Printing Office, 1957.
———. *Living with Radiation: The Problems of the Nuclear Age for the Layman*, 2 vols. Washington, DC: US Government Printing Office, 1959.
———. *Radioisotopes in Science and Industry*. Washington, DC: US Government Printing Office, 1960.
US Congress, House, *Independent Offices Appropriation Bill for 1948, Hearings*, 80th Congress, 1st session. Washington, DC: US Government Printing Office, 1947.

———. *Second Independent Offices Appropriations for 1954, Hearings*, 83rd Congress, 1st session. Washington, DC: US Government Printing Office, 1953.

———. *Oversight: Human Total Body Irradiation (TBI) Program at Oak Ridge,* Hearing before the Subcommittee on Investigations and Oversight of the Committee on Science and Technology, 97th Congress, 1st session. Washington, DC: US Government Printing Office, 1982.

US Congress, Joint Committee on Atomic Energy (JCAE). *Investigation into the United States Atomic Energy Project*, Hearings before the Joint Committee on Atomic Energy, 81st Congress, 1st session, 23 parts. Washington, DC: US Government Printing Office, 1949.

———. *Report of the Panel on the Impact of the Peaceful Uses of Atomic Energy*, vol. 1. Washington, DC: US Government Printing Office, 1956.

———. *The Nature of Radioactive Fallout and Its Effect on Man*, Hearings before the Special Subcommittee on Radiation, 85th Congress, 1st session, 2 parts. Washington, DC: US Government Printing Office, 1957.

US Congress, Senate. *Independent Offices Appropriation Bill for 1950*, Hearings before the Subcommittee of the Committee on Appropriations on H.R. 4177, 81st Congress, 1st session. Washington, DC: US Government Printing Office, 1949.

US Delegation to the International Conference on the Peaceful Uses of Atomic Energy. *The International Conference on the Peaceful Uses of Atomic Energy, Geneva, Switzerland, August 8–20, 1955: Report, with Appendices and Selected Documents*. 2 vols. 1955.

"U.S. Isotope Export Held Dangerous." *Los Angeles Times*, 9 Jun 1949, p. 14.

"The U.S. Radioisotope Industry–1966." *Isotopes and Radiation Technology* 4, no. 3 (1967): 207–14.

Vaiserman, Alexander M. "Radiation Hormesis: Historical Perspective and Implications for Low-Dose Cancer Risk Assessment." *Dose-Response* 8 (2010): 172–91.

VanGuilder, H. D.; K. E. Vrana; and W. M. Freeman, "Twenty-Five Years of Quantitative PCR for Gene Expression Analysis." *BioTechniques* 44 (2008): 619–26.

Virgona, Angelo. "Radiopharmaceutical Production at Squibb." *Isotopes and Radiation Technology* 4, no. 3 (1967): 222–26.

"VU to Report on Isotopes." *Nashville Tennessean*, 14 Dec 1946, p. 6.

Wailoo, Keith A. *Drawing Blood: Technology and Disease Identity in Twentieth-Century America*. Baltimore: Johns Hopkins University Press, 1997.

Wald, Matthew L. "Radioactive Drug for Tests Is in Short Supply." *New York Times*, 24 Jul 2009, p. A11.

Walker, J. Samuel. *Containing the Atom: Nuclear Regulation in a Changing Environment, 1963–1971*. Berkeley: University of California Press, 1992.

———. *Permissible Dose: A History of Radiation Protection in the Twentieth Century*. Berkeley: University of California Press, 2000.

———. *Three Mile Island: A Nuclear Crisis in Historical Perspective*. Berkeley: University of California Press, 2004.

Wang, Jessica. *American Science in an Age of Anxiety: Scientists, Anticommunism, and the Cold War*. Chapel Hill: University of North Carolina Press, 1999.

Ward, Donald R. "Design of Laboratories for Safe Use of Radioisotopes." *Isotopics* 1, no. 2 (1951): 10–48.

Warren, Shields. "The Therapeutic Use of Radioactive Phosphorus." *American Journal of the Medical Sciences* 209 (1945): 701–11.

———. "The Medical Program of the Atomic Energy Commission." *Bulletin of the Atomic Scientists* 4, no. 8 (1948): 233–34.

———. "You, Your Patients and Radioactive Fallout." *New England Journal of Medicine* 266 (1962): 1123–25.

Watanabe, Itaru; Gunther S. Stent; and Howard K. Schachman. "On the State of the Parental Phosphorus during Reproduction of Bacteriophage T2." *Biochimica et Biophysica Acta* 15 (1954): 38–49.

Watson, James D., and Francis H. C. Crick. "A Structure for Deoxyribose Nucleic Acid." *Nature* 171 (1953): 737–38.

———. "Genetical Implications of the Structure of Deoxyribonucleic Acid." *Nature* 171 (1953): 964–67.

Weart, Spencer R. *Nuclear Fear: A History of Images*. Cambridge, MA: Harvard University Press, 1988.

Weatherall, Mark W., and Harmke Kamminga. "The Making of a Biochemist, II: The Construction of Frederick Gowland Hopkins' Reputation." *Medical History* 40 (1996): 415–36.

Weinberg, Alvin M. "Impact of Large-Scale Science on the United States." *Science* 134 (1961): 161–64.

Weiner, Norbert. *Cybernetics: or Control and Communication in the Animal and the Machine*. Cambridge, MA: MIT Press, 1948.

Weisgall, Jonathan M. *Operation Crossroads: The Atomic Tests at Bikini Atoll*. Annapolis, MD: Naval Institute Press, 1994.

Wellerstein, Alex. "Patenting the Bomb: Nuclear Weapons, Intellectual Property, and Technological Control." *Isis* 99 (2008): 57–87.

———. *Knowledge and the Bomb: Nuclear Secrecy in the United States, 1939–2008*. PhD diss., Harvard University, 2010.

Welsome, Eileen. *The Plutonium Files: America's Secret Medical Experiments in the Cold War*. New York: Delta, 1999.

West, Doe. "Radiation Experiments on Children at the Fernald and Wrentham Schools: Lessons for Protocols in Human Subject Research." *Accountability in Research* 6 (1998): 103–25.

Westwick, Peter J. "Abraded from Several Corners: Medical Physics and Biophysics at Berkeley." *Historical Studies in the Physical and Biological Sciences* 27 (1996): 131–62.

———. *The National Labs: Science in an American System, 1947–1974*. Cambridge, MA: Harvard University Press, 2003.

"What Science Learned at Bikini: Latest Report on the Results." *Life* 23 (11 Aug 1947): 74–87.

Whicker, F. W., and J. E. Pinder. "Food Chains and Biogeochemical Pathways: Contributions of Fallout and Other Radiotracers." *Health Physics* 82 (2002): 680–89.

Whicker, F. W., and Vincent Schultz. "Introduction and Historical Perspective." In *Radioecology: Nuclear Energy and the Environment*. Boca Raton, FL: CRC Press, 1982.

Whittemore, Gilbert F., Jr. *The National Committee on Radiation Protection, 1928–1960: From Professional Guidelines to Government Regulation*. PhD diss., Harvard University, 1986.

Whittemore, Gilbert, and Miriam Boleyn-Fitzgerald, "Injecting Comatose Patients with Uranium: America's Overlapping Wars against Communism and Cancer in the 1950s." In Goodman et al., *Useful Bodies*, 165–89.

Wiegert, Richard G., and Eugene P. Odum. "Radionuclide Tracer Measurement of Food Web Diversity in Nature." In *Symposium on Radioecology: Proceedings of the Second National Symposium, Ann Arbor, Michigan, May 15–17, 1967*, ed. Daniel J. Nelson and Francis C. Evans. Springfield, VA: Clearinghouse for Federal Scientific and Technical Information, 1969, 709–10.

Willard, William K. "Avian Uptake of Fission Products from an Area Contaminated by Low-Level Atomic Wastes." *Science* 132 (1960): 148–50.

Williams, Jeffrey E. "Donner Laboratory: The Birthplace of Nuclear Medicine." *Journal of Nuclear Medicine* 40, no. 1 (1999): 16N, 18N, 20N.

Willstätter, Richard, and Arthur Stoll. *Untersuchungen über die Assimilation der Kohlensäure*. Berlin: J. Springer, 1918.

Wintrobe, Maxwell M. *Blood, Pure and Eloquent: A Story of Discovery, of People, and of Ideas*. New York: McGraw-Hill, 1980.

Wolfe, Audra. *Competing with the Soviets: Science, Technology, and the State in Cold War America*. Baltimore: Johns Hopkins University Press, 2013.

Woodbury, David O. *Atoms for Peace*. New York: Dodd, Mead, 1955.

Woodwell, George M. "Effects of Ionizing Radiation on Terrestrial Ecosystems." *Science* 138 (1962): 572–77.

———. "Toxic Substances and Ecological Cycles." *Scientific American* 216, no. 3 (1967): 24–31.

———. "Effects of Pollution on the Structure and Physiology of Ecosystems." *Science* 168 (1970): 429–33.

———. "BRAVO plus 25 Years." In *Environmental Sciences Laboratory Dedication: Daniel J. Nelson Auditorium, Feb. 26–27, 1979*, ed. S. I. Auerbach and N. T. Milleman. Oak Ridge, TN: Oak Ridge National Laboratory, 1980, 61–64.

Woodwell, George M.; Charles F. Wurster Jr.; and Peter A. Isaacson. "DDT Residues in an East Coast Estuary: A Case of Biological Concentration of a Persistent Insecticide." *Science* 156 (1967): 821–24.

Worster, Donald. *Nature's Economy: A History of Ecological Ideas*. Cambridge: Cambridge University Press, 1977.

Wrenn, F. R.; M. L. Good; and P. Handler. "Use of Positron-Emitting Radioisotopes for Localization of Brain Tumors." *Science* 113 (1951): 525–27.

Wright, Sewall. "Discussion on Population Genetics and Radiation." *Symposium on Radiation Genetics, Journal of Cellular and Comparative Physiology* 35, suppl. 1 (1950): 187–210.

Wyatt, H. V. "How History Has Blended." *Nature* 249 (1974): 803–4.

Yalow, Rosalyn S. "Radioimmunoassay: A Probe for the Fine Structure of Biologic Systems." *Science* 200 (1978): 1236–45.

———. "Radioimmunoassay: Its Relevance to Clinical Medicine." In *Basic Research and Clinical Medicine*, ed. S. Philip Bralow and Rosalyn S. Yalow. Washington, DC: Hemisphere Publishing, 1981, 3–22.

———. "Radioimmunoassay in Oncology." *Cancer* 53 (1984): 1426–31.

———. "Radioactivity in the Service of Humanity." *Interdisciplinary Science Reviews* 10 (1985): 56–64.

———. "Development and Proliferation of Radioimmunoassay Technology." *Journal of Chemical Education* 76 (1999): 767–68.

Yalow, Rosalyn S., and Solomon A. Berson. "Assay of Plasma Insulin in Human Subjects by Immunological Methods." *Nature* 184 (1959): 1648–49.

———. "Immunoassay of Endogenous Plasma Insulin in Man." *Journal of Clinical Investigation* 39 (1960): 1157–75.

———. "Radioimmunoassay of Gastrin." *Gastroenterology* 58 (1970): 1–14.

———. "Immunoassay of Plasma Insulin in Man." *Diabetes* 10 (1960): 339–44.

Yalow, R. S.; S. M. Glick; J. Roth; and S. A. Berson. "Radioimmunoassay of Human Plasma ACTH." *Journal of Clinical Endocrinology and Metabolism* 24 (1964): 1219–25.

Yoshikawa, H.; P. F. Hahn; and W. F. Bale. "Red Cell and Plasma Radioactive Copper in Normal and Anemic Dogs." *Journal of Experimental Medicine* 75 (1942): 489–94.

Zachmann, Karin. "Atoms for Peace and Radiation for Safety: How to Build Trust in Irradiated Foods in Cold War Europe and Beyond." *History and Technology* 27 (2011): 65–90.

———. "Atoms for Food to Achieve 'Freedom from Hunger'? Transnational Food Irradiation Research as an Ingredient of the Cold War." *Deutsches*

Museum Preprint Series 7 (2013). http://www.deutsches-museum.de/verlag/aus-der-forschung/preprint/.

Zallen, Doris T. "The Rockefeller Foundation and Spectroscopy Research: The Programs at Chicago and Utrecht." *Journal of the History of Biology* 25 (1992): 67–89.

———. "The 'Light' Organism for the Job: Green Algae and Photosynthesis Research." *Journal of the History of Biology* 26 (1993): 269–79.

Index

CPSIA information can be obtained
at www.ICGtesting.com
Printed in the USA
LVHW040147290423
745599LV00001B/22

9 780226 323961